Operations Research Proceedings 2002

Selected Papers
of the International Conference
on Operations Research (SOR 2002)

Klagenfurt, September 2–5, 2002

Springer

Berlin
Heidelberg
New York
Hong Kong
London
Milan
Paris
Tokyo

U. Leopold-Wildburger
F. Rendl · G. Wäscher
Editors

Operations Research Proceedings 2002

Selected Papers
of the International Conference
on Operations Research (SOR 2002)

Klagenfurt, September 2–5, 2002

With 120 Figures
and 51 Tables

Springer

Professor Dr. Ulrike Leopold-Wildburger

Universität Graz
Institut für Statistik und Operations Research
Universitätsstraße 15/E3
8010 Graz, Austria

Professor Dr. Franz Rendl

Universität Klagenfurt
Institut für Mathematik
9020 Klagenfurt, Austria

Professor Dr. Gerhard Wäscher

Otto-von-Guericke-Universität Magdeburg
Fakultät für Wirtschaftswissenschaften
BWL VIII: Management Science
Postfach 4120
39016 Magdeburg, Germany

ISBN 3-540-00387-8 Springer-Verlag Berlin Heidelberg New York

Cataloging-in-Publication Data applied for

A catalog record for this book is available from the Library of Congress.

Bibliographic information published by Die Deutsche Bibliothek
Die Deutsche Bibliothek lists this publication in the Deutsche Nationalbibliografie; detailed bibliographic data is available in the Internet at *http://dnb.ddb.de*.

Springer-Verlag Berlin Heidelberg New York
a member of BertelsmannSpringer Science + Business Media GmbH

http://www.springer.de

Printed in Germany

Cover design: Erich Kirchner, Heidelberg

SPIN 10909454 42/3130 – 5 4 3 2 1 0 – Printed on acid-free paper

Preface

This volume contains selected papers presented at the International Conference on Operations Research SOR 2002 held at the University of Klagenfurt from September 2 to September 5, 2002.

The conference was organized under the auspices of the German, the Swiss and the Austrian Operations Research societies

- Gesellschaft für Operations Research e.V. (GOR)

- Schweizerische Vereinigung für Operations Research (SVOR)

- Österreichische Gesellschaft für Operations Research (ÖGOR).

After Vienna (1990), Berlin (1994) and Zürich (1998) this has been the fourth time that the three societies organized a joint conference.

The conference was attended by more than 400 participants from countries all over the world which demonstrates the broad interest in all aspects of Operations Research.

The scientific program of the conference consisted of 4 plenary lectures, 5 semi plenary lectures, and about 320 contributed papers which have been presented in 16 sections. Due to the limited number of pages available for the proceedings volume, the length of each article as well as the total number of contributions had to be restricted.

The decision on the acceptance of papers for the proceedings has been made in close co operation with the section chairmen and was based on their suggestions.

We wish to express our sincere thanks to the chairmen for supporting our editorial work by refereeing the manuscripts and letting us have their advice. We also would like to thank Dr. Werner Müller from Springer-Verlag for his support in publishing this proceedings volume so quickly.

Klagenfurt, November 2002

Ulrike LEOPOLD-WILDBURGER   Franz RENDL   Gerhard WÄSCHER

ORGANIZING COMMITTEE:

F. Rendl (Klagenfurt)

I. Fischer (Klagenfurt)

G. Gruber (Klagenfurt)

A. Wiegele (Klagenfurt)

B. Klinz (Graz)

U. Leopold-Wildburger (Graz)

G. Feichtinger (Wien)

PROGRAM COMMITTEE:

R.E. Burkard (Graz)

E. Fragnière (Lausanne)

K. Frauendorfer (St. Gallen)

R. Hartl (Wien)

W. Kürsten (Jena)

H.-J. Luethi (Zürich)

F. Rendl (Klagenfurt)

T. Spengler (Braunschweig)

G. Wäscher (Magdeburg)

Sections and Chairs

Section 1: Production, Logistics and Supply Chain Management

S. Helber (Clausthal)
W. Jammernegg (WU Wien)

Section 2: Marketing and Data Analysis

L. Hildebrandt (HU Berlin)
U. Wagner (Uni Wien)

Section 3: Transportation and Traffic

S. Voss (TU Braunschweig)
U. Zimmermann (TU Braunschweig)

Section 4: Scheduling and Project Management

P. Brucker (Osnabrück)
E. Pesch (Siegen)

Section 5: Telecommunication und Information Technology

A. Taudes (WU Wien)
R. Gismondi (Telekom)

Section 6: Energy and Environment

A. Haurie (Genf)
A. Tuma (Augsburg)

Section 7: Public Economy, Health, Agriculture, Education

A. Stepan (TU Wien)
L. Zadnik-Stirn (Ljubljana)

Section 8: Banking, Finance, Insurances, Risk Management

A. Oehler (Bamberg)
G. Pflug (Uni Wien)

Section 9: Continuous Optimization

F. Jarre (Düsseldorf)
J.-P. Vial (Genf)

Section 10: Discrete and Combinatorial Optimization

M. Jünger (Köln)
R. Möhring (TU Berlin)

Section 11: Stochastic and Dynamic Programming

W. Römisch (HU Berlin)
R. Schultz (Duisburg)

Section 12: Simulation

P. Chamoni (Duisburg)

Section 13: Control Theory, Systems Dynamics, Dynamic Games

M. Schwaninger (St. Gallen)
G. Tragler (TU Wien)

Section 14: Game Theory, Auctioning and Bidding, Experimental Economics

F. Bolle (Frankfurt Oder)
H.W. Brachinger (Fribourg)

Section 15: Econometrics, Statistics and Mathematical Economics

W. Eichhorn (Karlsruhe)
M. Luptacik (WU Wien)

Section 16: Fuzzy Logic, Multicriteria Decision Making, Decision Theory

A. Scholl (Jena)
B. Werners (Bochum)

Contents

Section 4: Scheduling and Project Management

Section 5: Telecommunication and Information Technology

Section 6: Energy and Environment

The Role of Operations Research in Public Policy

Jonathan P. Caulkins
Carnegie Mellon University
H. John Heinz III School of Public Policy and Management
5000 Forbes Ave.
Pittsburgh, PA 15213-3890
caulkins@cmu.edu

Abstract

This paper reflects on the role Operations Research (OR) has and has not played in public policy. It is based on a plenary talk delivered on September 4, 2002 to the International Conference on OR held in Klagenfurt, Austria. It concludes that OR clearly makes many important contributions to public policy making, but relatively speaking OR makes the smallest contribution to strategic issues for which authority is distributed and for which the "physics" of the underlying system are not central. One reason is that there are so many layers of people between the typical Operations Researcher and the key policy makers. OR models and model results are not effective from such a distance because they are distorted and diluted each time they are translated. An alternative path of influence would be to modify OR curricula so that people with OR training can take jobs "closer" to the key policy makers.

1. Introduction: The Glass if Half-Full and Half-Empty

The goal of this paper is to reflect on and draw lessons from the role Operations Research (OR) has played in public policy. I will not review a series of models. Perusing Pollack et al.'s (1994) very fine volume is a more efficient way to do that. Rather, I will try to characterize what OR has and has not contributed, and why.

My principle qualification for undertaking this is that I am Professor of Operations Research and Public Policy at Carnegie Mellon University, one of the very few academics with both phrases in my job title. My principle limitation is that my experience base and perspectives primarily concern policy in the US. I will tell a story that I believe is true with respect to the US, and allow the reader to determine whether it applies in Austria or elsewhere outside the US.

In organizing these thoughts, it is useful to refer to the metaphor of a glass that can be viewed as half-full or half-empty, depending on what the observer focuses on. In that vein, I begin by pointing out that the glass of OR in the furtherance of public policy is clearly half-full and half-empty.

Given that this paper is addressed to an OR audience, there is limited value in belaboring the many successful applications of OR to public policy. That would be preaching to the choir. The profession can be justifiably proud of its contributions during the Second World War and the Cold War. Public sector applications enjoy a significant "market share" among articles in *Interfaces*, the premier journal for OR applications, and among Franz Edelman Award recipients and finalists, an award described as the "Super Bowl" of OR (Horner, 2002).

Linear programming is in some sense the methodological heart of OR, and two of the classic applications used in textbooks to motivate linear programming are public policy applications: the "diet problem" that played a central role in defining the poverty level and the "school busing problem" that emerged from court orders to desegregate schools in the US.

Rather, for this audience, it is more important to make the case that there is a portion of the glass that is half empty. That case can be made by looking at our conferences, journals, professional society sections, and courses.

The leading North American research-oriented OR conference is INFORMS. Examining the program for this fall's conference is instructive. There are certainly some talks applying OR to public policy problems. Indeed, I will be giving one in a perennially popular session organized by Arnold Barnett entitled "Threats to Life and Limb". More importantly, two prominent keynotes address "Modeling Bioterror Response Logistics" and "Restructuring Electricity Markets". However, the proportion of research talks that intersect public policy is modest, and the vast majority are concentrated in select domains. Of 21 invited session "tracks", three address topics that intersect public policy in important ways (Aviation, Healthcare, and Electricity Markets & Energy Modeling). Similarly for four of the 25 sponsored session tracks (Energy/Natural Resources/Environment, Health Applications, Railroad Applications, and Location Analysis). And clearly not all talks within tracks that intersect public policy actually address public policy. For example, typically most INFORMS presentations on railroad applications strive to improve the operational efficiency and profitability of railroad companies, rather than addressing how the government should regulate railroads.

One might expect the premier *research* conference to focus on methods development, and methods are generally not specific to either public policy or business applications. Perhaps the practitioner conferences are more focused on public policy? In fact, however, that is not the case. In North America the premier OR practitioners' conference is held each May. At the May 2001 Conference, none of the 5 tracks or 41 speakers identified on the conference web page were specific to public policy and management.

Another possibility is that public policy applications are addressed predominantly in special purpose OR conferences. However, of the 83 upcoming conferences listed on the INFORMS web site on August 28, 2002, only two were clearly about public policy (the 23rd Army Science conference and the International Conference on OR for Development) and two were on industries that are heavily regulated (2nd International Conference on Freight Transportation Systems and the International Conference on Telecommunications).

In many respects, the present conference stands out. Three of four plenary talks pertain to public policy, and five of 15 sections have significant intersection with public policy. They are Energy and Environment; Public Economy, Health, Agriculture, and Education; Telecommunications and IT; Transportation and Traffic; and Control Theory, Systems Dynamics, and Dynamic Games. (The last is defined by its methodology, but many of the applications pertain to public policy.)

The story with OR journals is similar. *Operations Research* has four departments that address specific domains that intersect public policy (Environment, Energy, and Natural Resources; Military; Telecommunications; and Transportation), and Interfaces has two (Healthcare and Military). All three major North American journals have departments with names like "Public Sector Applications" (*Management Science*), "Policy Modeling and Public Sector OR" (*Operations Research*), and "Public Sector Department" (*Interfaces*). However, as an area or associate editor for the first two, I can attest that the volume of submissions is not high. The *European Journal of Operational Research* does not have any such dedicated department, although it has had special issues, e.g., one on "Optimizing Public Policy" in 1992.

One could argue that most of the publishing occurs in field or specialty journals, but none of them is focused on public policy generally. Conversely most journals that focus on public policy (e.g., the *Journal of Policy Analysis and Management*, *Policy Sciences*, etc.) are not dominated by articles using OR methods. *Socio-Economic Planning Sciences* is one of the few journals with a broad interest in public policy for which a substantial share of the articles "look like" OR articles.

The same story plays out in the INFORMS subdivisions which are called "sections" and "societies". The two sections that focus explicitly on public policy (Public Programs and Processes and Social Science Applications) are among INFORMS' smallest. Seven other sections and a "society" intersect specific public policy domains (Aviation; Energy, Natural Resources, and the Environment; Health; Location Analysis; Railroad Applications; Telecommunications; Transportation; and the Military Applications society).

I have not made a thorough review of courses at the intersection of OR and public policy, but in general few public policy schools offer, let alone require, OR/MS courses and few OR programs offer, let alone require, courses in public policy. Indeed, as far as I know there are no introductory textbooks on OR and public policy in print. I do have the privilege of teaching a required core course in Management Science at Carnegie Mellon's policy school (Caulkins, 1999), but it is almost one of a kind and I am forced to use a standard Management Science textbook geared toward MBA students.

2. Which Half of the OR and Public Policy Glass is Full?

This review of conferences, journals, sections, and courses not only underscores that the glass of OR and public policy is half-empty as well as half-full, but it also indicates what part is full and what part is empty. Specifically, it suggests distinguishing five categories of roles OR plays in public policy.

Roles OR Plays in Public Policy
(1) OR for organizations and management generally that applies to public problems
(2) OR for "regulated industries"
(3) OR for public resource management
(4) OR applied to government-provided services
(5) OR applied to government policy making

The first category addresses those aspects of government management that strongly parallel operations in the for profit or non-profit sector, such as using supply chain management for blood banks or vehicle routing problem algorithms to manage dial-a-ride services. Roughly speaking, OR is about as central to the management of government activities as it is to the management of typical non-governmental activities. One can write an interesting parallel paper on that topic, but I think the conclusion is again, half-full/half-empty. OR is clearly important to many businesses, but it is sobering to note that MBA programs in the US no longer need to teach OR/Management Science as a core topic, and most exercise their right not to do so.

The second category is probably the largest in terms of journal articles, conference presentations, and the like. These are analyses of industries that are so highly regulated that government has a vested interest in the operation and performance of those industries that goes beyond its interest in the vitality of the economy generally. Principal among these industries are aviation, energy/electricity, healthcare, telecommunications, and transportation. Agriculture is another such example at this conference.

Within these regulated industries, some OR work primarily helps companies, e.g., yield management for airlines. Other work is very directly related to policy: e.g., modeling how far apart planes should be spaced to yield an acceptable risk of collision. Some is intermediate, relevant to both the firms and the regulating bodies (e.g., modeling of ground holds).

The third category pertains to the management of publicly owned environmental and natural resources such as forests, fisheries, and water. These include forestry planting and logging policies, setting harvest quotas for fish, managing irrigation projects, allocating water rights, and multi-objective planning of entire river basins as at the Tennessee Valley Authority.

The fourth category pertains to services provided by the government. Larson and Odoni's (1981) text on Urban Operations Research is a paradigmatic example. It describes, among other things, the "hypercube queuing model" for priority queuing that underpins 911 emergency dispatch systems throughout the US.

Healthcare applications belong in this category as well as among the regulated industries even in the US, not just in countries with nationalized healthcare. The military and the Veteran's Administration directly operate many healthcare facilities that generate important and interesting operational problems.

At least in the US, military applications more generally are an extremely important special case of a government "service" whose provision is guided by OR analysis. Indeed, the origins of OR as a profession are in military applications, and the US military may be the world's largest consumer of OR analysis.

The fifth category is the use of OR to guide government policy in the narrow sense of the word "policy." At some level it is hard to distinguish what is policy and what is management or implementation. If an OR model is used to determine that blood inventories in large hospitals should use a first-in-first-out inventory policy, whereas small hospitals should use last-in-first-out (Pierskalla, 2002), then in some sense OR is directly affecting policy. But the "narrow" definition of policy I have in mind for this fifth category is analysis that guides the writing of laws, regulations, or court decisions that affect the behavior of agents who are not themselves government employees or direct contractors (such as logging firms hired to cut on Forest Service land).

For example, David Paltiel's doctoral dissertation addressed the Food and Drug Administration's drug approval process as a sequential decision analysis (Paltiel and Kaplan, 1993). Ed Kaplan (1994, 1995) modeled the effect of syringe exchange programs (SEPs) on the spread of HIV/AIDS with the goal of informing policy decisions concerning the legality and funding of SEPs. I along with colleagues at RAND, Carnegie Mellon, and the Technical University of Vienna have developed OR models of the spread and control of drugs, crime, and violence that can guide decisions concerning sentencing of repeat criminal offenders and the allocation of resources to various types of drug control operations (cf., Caulkins, 2000). Julia Almeder's award winning thesis is another example (Balta, 2002).

Although there are examples of OR analyses of this sort, to me what is striking is how few there are relative to the number of important and interesting policy problems. Furthermore, at least in the areas with which I am most familiar, these models do not always have a decisive influence. I can cite instances in which the analysis made a concrete difference, but it is even easier to cite examples in which politics interfered, or the concerns of some special interest group prevailed over what was in the common good, or the analysis was simply deemed to be too complicated or too "academic" to weigh heavily in the policy decision process.

This pessimism about the magnitude of OR's impact on public policy making is enhanced if I approach the question in the opposite direction. If I make a list of major US policy actions – other than those involving the military – there are few for which OR analysis played a key role, or at least a role that was sufficiently visible that a casual observer such as myself is aware of that role. (See Table below.) In contrast the President's Council of Economic Advisors weighs in on most important policy issues. The exclusion of military policy and decisions is important. OR analysis does figure prominently in strategic decisions concerning defense, ranging from the cancellation of major weapons system to decisions about preparation and strategy for war.

Recent Major US Domestic Policy Actions

Welfare reform
Bush tax cut (e.g., of estate tax)
Creation of the Department of Homeland Security
Accountability movement in K-12 education
Deregulation of electricity markets
Laissez-faire antitrust policy, especially in telecommunications
Aid to major air carriers after September 11th
"Competitive" approach to attracting firms for local economic development
Stem cell research compromise
Extending public funding of Amtrak

Recent Major US International, Non-Military Policy Actions

IMF intervention in Brazil and Uruguay but not Argentina
NAFTA and other "fast track" trade negotiations
Plan Colombia
Not following the Kyoto agreement

One concise summary of these observations concerning where OR does and does not play a major role in public policy is the following. OR plays an important role if any one of three conditions pertain: (1) The issues pertain to tactical management or are at the implementation level. (2) The "physics" of the system are complex and central. Or, (3) Decision making is centralized in a command and control hierarchy. Where OR plays a secondary role is in *strategic* issues with *distributed authority* in *general domains*.

The distinction between domains in which the "physics" of the system are complex and central vs. other domains is meant to recognize and explain OR's prominent role in aviation, transportation, energy, telecommunications, etc. Few people believe one can think intelligently about appropriate separation distances for aircraft or the best way to manage ground hold delays without some mathematical model. The use of OR or some other quantitative paradigm is central and unavoidable in a way that is not true in welfare policy, education policy, or equal employment law.

The distinction between centralized decision-making and distributed authority is meant to recognize and explain why OR plays a more prominent role in defense policy than in policy toward homelessness. Defense policy is centralized in the Department of Defense. In contrast, homelessness is a concern of multiple federal agencies and importantly of the 50 states, roughly 3,000 counties, and tens of thousands of municipalities in the US. Likewise, the Department of Defense is hierarchical whereas Congress and other legislative bodies are not. It is not literally true, as is sometimes asserted, that all one has to do is convince the top general of the merit of an idea and implementation will take care of itself, but that is closer to being true in military policy than in some other domains.

3. Why is the Other Half Empty?

A natural question to ask is why doesn't OR play a more central role in strategic issues with distributed authority in general domains? Many of the relevant factors are not specific to OR, but pertain to all model-based sciences. For example, it is conventional wisdom that one can lie with statistics. (Recall the axiom that there are three kinds of lies: "Lies, damned lies, and statistics.") The uninitiated may not (and in this context perhaps should not) distinguish numbers produced by OR models and numbers produced by statistical analysis.

In many policy debates, there are diverse stakeholders with competing interests. Ideally in the scientific world, debates emerge when people disagree because they have different understandings and different interpretations but the same objective (to seek the "truth"). In the political world, two parties with completely concordant understanding of the implications of the policy may debate viciously because they have different objectives. One Senator may represent a small urban state; another may represent a rural state in the Plains. They may disagree about what federal gun control policy should be even if they have exactly the same understanding of how that policy would affect crime and access to guns for sport and hunting in the nation as a whole.

Stakeholders typically are not interested in finding the objective truth. They are advocates of a particular position or policy, and they seek to "spin" scientific evidence to their benefit. (Consider Otto Rentz' observation at this conference (2002) that advocacy in the context of environmental policy is "The art of making your neighbor believe you have done more than you did so he will do more of what you didn't do.") In particular, advocates may cite selectively or otherwise distort the results of OR studies and other scientific analyses in order to serve their interests. Recognizing this, scientific evidence cited by stakeholders and advocates is viewed skeptically by other parties. Since the democratic process is fundamentally about stakeholders advocating their positions in public forums, far more time, money, and resources are invested in advocacy-oriented research and reporting of research than is invested by people without a stake in the issue. Hence, most citing of scientific evidence relevant to policy discussions is not objective. Knowing that, people greet with skepticism even the minority of studies that do strive for objectivity.

Exacerbating this problem is the fact that models do not stand up well to cross-examination. It is easy to criticize a model and to do so in a way that undermines its credibility with people who are not familiar with the concept and merits of mathematical modeling. By definition models are simplification of realities, and one cannot scientifically prove the absence of something, so the analyst has a hard time responding to questioning along the following lines. "Professor Smith: Does your model simplify the issue at hand?" The only honest answer is yes, models simplify. Then, "Professor Smith: Can you prove to the committee that this simplification absolutely has *no effect* on your results?" The only honest answer to this is, "No I cannot prove that because the simplification does in fact affect the results"— even if in the analyst's judgment the effects are second-order and do not compromise the fundamental conclusions. The next witness questioned may be an individual who can describe in poignant detail how his or her life was devastated by the policy in question, and then the non-scientists are left to decide whether to trust the colorless judgment of an egg-headed analyst or the heart-wrenching story of one individual's eyewitness account. In such circumstances, anecdotes often trump models.

As an aside, I do not think that this is always bad. Anecdotes and individual accounts can highlight particular issues, and some issues are not fundamentally amenable to mathematical analysis.

None of these observations are particularly novel, but I can use a bit of OR, specifically a network diagram, to capture their essence in a picture that I think may be novel. It makes reference to the 1993 movie "Six Degrees of Separation," based on the John Guare play. The movie's premise is that everyone in the world is connected by no more than six links of the sort, "I know someone who knows someone who knows someone ... who knows you." A common version involves showing that every movie actor is connected to Kevin Bacon by a chain of shared movies "Person in question was in a movie with A who was in a movie with B ... who was in a movie with Kevin Bacon" with no more than six links in the chain. Hence, the idiom that no one – even OR methodologists – is more than six steps removed from Kevin Bacon.

Only somewhat facetiously, I assert that Operations Researchers and strategic public policy makers are a counter-example, at least if we define a link as indicating sufficient personal contact to pass along the sort of sophisticated thinking and ideas that emanate from OR models. The figure below illustrates the links between typical OR methodologists and the policy makers who make strategic decisions. You will note it has seven links. Hence, I claim that senior policy makers who "bring home the pork" are more distant from OR methodologists than is Kevin Bacon. This diagram is influenced by what I have observed in the area of drug and crime policy, but I submit that the gist of the diagram probably holds for welfare policy, education policy, homelessness policy, and many other important public policy domains.

Figure: The Senior Policy Makers Who "Bring Home the Pork" are More Distant from OR Methodologists than Is Kevin Bacon

I do not want to subject this diagram to too literal an interpretation. I personally straddle levels two and three in this diagram but have myself briefed senior policy makers and given testimony before Congress. However, most people, myself included, spend most of their time interacting with people who are at most one level away in this chain, and that creates severe communications problems between one end of the chain and the other.

Recall the children's game of "telephone". A group of children sit in a circle. One whispers something into the ear of the child next to him or her. That child in turn whispers what he or she hears to the next child, and so on around the circle. By the time the message gets back to the original speaker, it has usually been distorted beyond recognition. Sometimes distortion of that magnitude seems to happen as ideas move up or down this multi-layered chain of people separating the OR community from senior policy makers.

When I step back and recognize how many layers there are in this chain, it does not surprise me that OR models are not driving strategic policy making outside the select domains where the "physics" of the system are complex and central and/or there is centralized decision making. (The latter allows analysts to brief directly key decision makers who can then act more or less unilaterally, as in some aspects of military policy.)

4. How Can We Fill the Other Half?

The final question is what might we do to enhance the role OR plays in strategic public policy. Perhaps one should first ask whether that would be desirable. I believe the answer is yes. I think the OR perspective offers a valuable way to integrate complex and competing considerations. Since I'm addressing an OR audience, I will simply assume that at least some of you agree.

So how might one enhance the role of OR in strategic public policy making? As suggested, I do not think it is reasonable to expect models to move all the way down this chain by themselves. Rather, I think we have to think of ways of moving people up or down this chain to reduce the number of links between OR methods and senior policy makers. Conceptually I can think of at least four ways of doing that.

In principle the most direct approach would be to give executive education in OR to senior executives and lawmakers. Practically speaking, that will not work. Most people who have ascended to such pinnacles of power have long since forgotten their mathematics. (Those that remember math are probably working in areas like environmental policy where OR is already affecting decisions at the strategic level.) Furthermore, such senior people are very busy. They would rather attend executive education tailored to their policy domain, not study a methodology.

A second approach would recognize that most leaders have an advanced professional degree. If every such curriculum required one course in OR, that might stimulate understanding of and demand for OR analysis by those people after they move up the career ladder. However, we have already tried this, and it did not work. Business schools in the US used to require a core course in OR, but once accreditation requirements were relaxed and they no longer had to do so, most schools moved decisively away from it because the OR community could not deliver a course that was sufficiently valuable. Among MPA/MPP programs in the US, I think only CMU requires Management Science, and only a few other schools (e.g., the University of Michigan) even offer such classes in house. If this approach did not work for MBA's vis a vis careers in business, it is unlikely to work for MBA's vis a vis government policy, let alone lawyers and government policy.

The third possibility is similar, but one step further back in the education process. One could require an OR course as part of the general education requirements for a first university degree. Every discipline could probably make a parallel case, but the case for OR should not be dismissed. In principle, a general or liberal arts education's main value comes from teaching critical thinking, but the capacity to deliver on that promise has been eroded in many universities, where class sizes are very large and students pick courses from a cafeteria menu of electives (vs. a more regimented and cumulative classic liberal arts education). Learning OR and mathematical modeling may be a more effective and more appreciated way of teaching critical quantitative thinking than the current default, which is a course in calculus. One could even make a case for teaching mathematical modeling earlier, in secondary school. However, making OR a core subject in liberal studies is not likely to happen and, at any rate, is not something the OR community can decide to do unilaterally.

In contrast the fourth possibility, and the one that I want to elaborate on, is something that OR faculty could implement on their own because it pertains to graduate education in OR. I have in mind education at the masters level because a Ph.D. is preparation for academic life, not high-level leadership in the public or private sector. OR masters programs vary across universities, influenced by factors such as whether they sit in a business school or an engineering school, but they have a lot in common. In particular, I do not know of any that were designed specifically with the goal of maximizing how far along this ladder students would get in their career. So it is interesting to ask: how might we change some of our OR masters programs if the stated goal were to push graduates of these programs as far as possible along the trajectory toward strategic leadership, rather than to maximize success in a technical position in the first few years post-graduation?

I do not know what the answer is, but to be provocative I'll suggest the following five possibilities.

*Provide a broader technical training, even at the expense of some depth. E.g., students might be better off taking an extra course each in statistics, negotiations, and GIS, and only three not six courses in optimization.

*Emphasize applications and problem structuring not algorithms, where the applications are real, not textbook problems. E.g., require project courses for real (paying) clients.

*Stress communication skills, including how to communicate intuition and insight, not just technical descriptions, and how to communicate to people who are not themselves trained in OR (in contrast, say, to learning how to write papers for OR journals).

*Recruit a different kind of applicant. Nerds with perfect standardized test scores may not be successful in moving down this chain, even if they get a 4.0 GPA in the program. Admissions criteria might need to weight more than they now do prior experiences such as being captain of a sports team, president of one's high school graduating class, or having started a company or non-profit organization.

*Consider changing the degree title to something like a "Masters in Strategic Analysis" rather than a Masters in OR. The title "Operations Research" has very limited name recognition, not only relative to Coca-Cola but also relative to competing disciplines.

Again, I offer these ideas to be provocative and stimulate discussion, not with any pretense that this is the "right" answer. Other possibilities are worth considering, such as joint degrees in OR and a "soft" discipline or OR and a substantive domain such as Chemical Engineering for environmental policy or an M.D. for health policy.

5. Summary

OR has made important contributions to public policy, but a clear-eyed review suggests that its role has been greater in some areas than in others. Relatively speaking, OR's role has been smaller in the analysis of *strategic* issues with *distributed authority* in *general domains*. A major limitation is that many layers of people now separate the typical Operations Researcher from senior policy makers and lawmakers, and for various reasons it is hard to push models far down that chain. If one wished to expand OR's role in strategic policy making beyond domains (such as defense policy) where it is already prominent, one strategy would be to create OR masters programs that differ in important ways from the typical program of today.

References

1. Balta, Julia, 2002: Optimal control of methadone treatment in preventing blood-borne disease. Masters Thesis, Institute for Econometrics, Operations Research, and Systems Theory. Vienna University of Technology
2. Caulkins, Jonathan P., 1999: The revolution in management science instruction: Implications for teaching public affairs students. The Journal of Public Affairs Education. 5:107-117
3. Caulkins, Jonathan P., 2000: Measurement and analysis of drug problems and drug control efforts. In: Duffee D, McDowall D, Green Mazerolle K, Mastrofski S (eds.) Criminal Justice 2000: Volume 4, Measurement and Analysis of Crime and Justice, USGPO, Washington DC pp 391-449
4. Horner, Peter, 2002: And the winner Is ... OR/MS Today, 29
5. Kaplan, Edward H., 1994: A method for evaluating needle exchange programs, Statistics in Medicine, 13:2179-2187
6. Kaplan, Edward H., 1995: Probability models of needle exchange, Operations Research. 43:558-569
7. Larson, Richard C, Amadeo, Odoni, 1981: Urban Operations Research. Prentice-Hall, Englewood, New Jersey
8. Paltiel, A.D., Kaplan, E., 1993: The epidemiological and economic consequences of AIDS clinical trials. Journal of Acquired Immune Deficiency Syndromes. 6:179-190
9. Pierskalla, William, 2002: Blood bank inventory control – a supply chain approach. Plenary talk at the International Conference on Operations Research, Klagenfurt, Austria, September 3, 2002
10. Pollock, S.M., Rothkopf, M.H., Barnett, A.I.(eds.) 1994. Operations Research and the Public Sector, Handbooks in Operations Research and Management Science, Volume 6, North-Holland, New York
11. Rentz, Otto, 2002: Contributions of Operations Research to Environmental Policy Plenary talk at the International Conference on Operations Research, Klagenfurt, Austria, September 2, 2002

Risk-Return Optimization of the Bank Portfolio

Ursula Theiler

Risk Training, Carl-Zeiss-Str. 11, D-83052 Bruckmuehl, Germany,
mailto:theiler@risk-training.org.

Abstract

In an intensifying competition banks are forced to develop and implement enterprise wide integrated risk-return management systems. Financial risks have to be limited and managed from a bank wide portfolio perspective. Risk management rules must be accomplished from internal and regulatory points of view. Expected returns need to be maximized subject to these constraints, leading to a generalized portfolio optimization problem under different capital limits.

We give a survey on a risk-return optimization model for the bank portfolio that maximizes the expected returns to the planning horizon with respect to internal and regulatory loss risk constraints. We derive consistent planning information that ensures efficient return targets and maximal capital use of the economic and the regulatory capital. The impact of the optimization is shown by an application example.

1 Introduction

In an intensifying competition banks are forced to develop and implement enterprise wide integrated risk-return management systems. Financial risks have to be limited and managed from a bank wide portfolio perspective. Risk management rules must be accomplished from internal and regulatory points of view. Expected returns need to be maximized subject to these constraints, leading to a generalized portfolio optimization problem under different capital limits.

In chapter 2 we give a survey on a risk-return optimization model that maximizes the expected returns of the bank portfolio to the planning horizon subject to internal and regulatory loss risk ceilings. The internal risk constraint is based on the new risk measure of Conditional Value at Risk (CVaR), that has been proved to be appropriate for measuring bank wide loss risk [4,5]. We solve the optimization problem by a CVaR-optimization algorithm by Rockafellar/Uryasev [4,5]. The regulatory capital restrictions represent the 'Basle Rules' of risk limitation [1,2]. We derive consistent planning information from the optimum solution that ensures efficient return targets and a maximal use of the economic and regulatory capital. The impact of the optimization is shown by an application example in chapter 3. We close by a brief summary in chapter 4.

2 Risk-Return Optimization Model for the Bank Portfolio

2.1 Survey

For its planning processes the bank needs to identify risk-return efficient target portfolios, that maximize expected returns to the planning horizon and meet risk constraints from different points of view. From an internal perspective, the bank limits its loss risks by the economic capital available. At the same time, the bank must observe legal loss risk boundaries of the 'Basle rules', that comprise constraints on the risks of the banking book and the trading book, combined with limits on the capital components that cover the different kinds of risks [1,2,7]. We achieve the following general model structure of the bank wide risk-return portfolio optimization problem (P):

(P)		(1)
Objective function: Maximize expected returns		
subject to constraints		
Constraint 1:	Internal risk ≤ Economic capital,	
Constraint 2:	Regulatory risk ≤ Regulatory capital,	
	Constraints on the regulatory capital components,	
Constraint 3:	Position bounds, definition of the feasible solutions.	

We develop the optimization model in three steps. First we define a risk measure that is appropriate to measure the loss risk of the bank portfolio (chapter 2.2). Next we introduce an optimization algorithm for the solution of the basic risk-return optimization problem with respect to the constraints 1 and 3 (chapter 2.3). We then extend the problem to an optimization model for the bank planning process by integrating the regulatory risk constraints (chapter 2.4).

2.2 Definition of the CVaR Risk Measure

While the risk measure of Value at Risk, commonly applied in finance for market risk measurement, lacks the elementary property of sub-additivity, if the loss distributions are not normal, the *Conditional Value at Risk* (CVaR), defined as the conditional expectation beyond the Value at Risk, has been proved to be appropriate for risk measurement of any loss distributions [4,5].

Let $\mathbf{x}$ be the vector of the positions of the bank portfolio assets and $\mathbf{y}$ the vector of the corresponding market prices. We define the Conditional Value at Risk deviation of the portfolio loss $\mathrm{CVaR}_\alpha(L(\mathbf{x},\mathbf{y}))$ as

$$\mathrm{CVaR}_\alpha(L(\mathbf{x},\mathbf{y})) = E[L(\mathbf{x},\mathbf{y}) \mid L(\mathbf{x},\mathbf{y}) \geq \mathrm{VaR}_\alpha(L(\mathbf{x},\mathbf{y}))], \qquad (2)$$

where $L(\mathbf{x},\mathbf{y})$ is defined as the difference of the uncertain from the expected portfolio value at the horizon, $L(\mathbf{x},\mathbf{y})=E[\mathbf{y}]'\mathbf{x}-\mathbf{y}'\mathbf{x}$, and $\mathrm{VaR}(L(\mathbf{x},\mathbf{y}))$ is the α-quantile of the loss function $L(\mathbf{x},\mathbf{y})$.[1] In the case of discontinuities at the α-quantile, CVaR can

[1] We apply the term *CVaR deviation*, as the loss distribution measures the deviation of the uncertain from the expected portfolio value.

be defined as a weighted average of the VaR and the conditional expectation beyond the VaR [5].

In the above definition CVaR deviation is a convex risk measure that ensures the existence of a risk minimum portfolio on a convex set and the solvability of the optimization problem (P). It is appropriate to measure loss risk from any asymmetric and discontinuous loss distribution with discrete probabilities [5]. The following figure shows the loss function and the risk measure of CVaR deviation at the confidence level α:

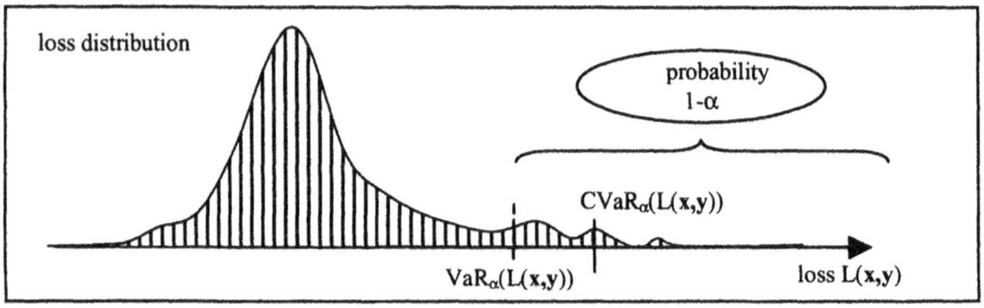

Fig. 1. Loss distribution L(**x**,**y**) and risk measures VaR and CVaR

2.3 Basic Risk-Return Optimization Model

We introduce an optimization algorithm that maximizes expected returns with respect to a CVaR risk constraint that can be applied to portfolios with any loss distribution. We make use of this approach to solve the basic problem of the optimization model (P) to maximize the expected portfolio return subject to the constraints 1 and 3.

Let $\mathbf{x}=(x_1,\ldots,x_n)'$ be the decision variable, i.e. the positions of the single assets. We define a linear objective function for the expected portfolio return $\mu(\mathbf{x})=\boldsymbol{\mu}'\mathbf{x}$, with $\boldsymbol{\mu}=(\mu_1,\ldots,\mu_n)'$ the vector of the expected returns of single assets. The internal loss risk is measured by CVaR deviation of the loss function L(**x**,**y**) and constrained by the maximum amount of economic capital, denoted as ec_cap_max. The area of the feasible solutions is defined by upper and lower position bounds, the vectors **low_bound** and **up_bound** respectively. We solve this generalized risk-return portfolio optimization problem (P) by an algorithm of Rockafellar/Uryasev [4,5]. Based on a scenario generation $\mathbf{y}_1, \ldots, \mathbf{y}_K$ of the market prices of the portfolio assets, the CVaR-constraint is approximated by a set of linear constraints, leading to a linear optimization problem. We achieve the following basic optimization model (P_{CVaR}) that maximizes the expected portfolio returns with respect to the constraints 1 and 3 of the optimization problem (P):

(P_{CVaR}) (3)

Objective Function $\mu(\mathbf{x})=\boldsymbol{\mu}'\mathbf{x}=\sum_{j=1}^{n}\mu_j x_j$,

Constraint 1: Internal Risk Constraint

$$\left.\begin{array}{ll} \text{(i)} & q + \frac{1}{(1-\alpha)} \cdot \frac{1}{K} \sum_{k=1}^{K} z_k \le \text{ec\_cap\_max}, \\ \text{(ii)} & L(\mathbf{x}, \mathbf{y}_k) - q \le z_k, k = 1, \ldots, K, \\ \text{(iii)} & -z_k \le 0, k = 1, \ldots, K, \\ \text{(iv)} & q \in \Re \end{array}\right\}$$

"Internal loss risk (CVaR deviation estimate) ≤ Economic capital"

Constraint 3: Boundaries of the Feasible Solutions

(vi) $\textbf{low\_bound} \le \mathbf{x} \le \textbf{up\_bound}.$

2.4 Extension to an Optimization Model for the Bank Portfolio

We extend the basic optimization model (P_{CVaR}) to an optimization model (P) for the bank planning processes. Beneath the internal loss risk boundaries the bank has to comply with regulatory rules of risk limitation passed by the Basle Committee on Banking Supervision [1,2,7].[2] We give a survey how these constraints are integrated into the optimization model [6].

The *bank book positions*[3] are restricted with respect to their credit risk by a linear constraint. The sum of the risk weighted assets is limited by the regulatory capital resources of the core ('tier 1') capital plus the supplementary ('tier 2') capital. Capital charges of the *trading book* are based on the general market risk and the specific risk, as well as the counterparty risk of the trading book positions. We model linear constraints for the specific and the counterparty risk. With respect to the general market risk of the trading book we assume that the bank applies an internal market risk model to estimate the Value at Risk with the 'Basle parameters' [2,7]. As CVaR represents an upper bound of the corresponding VaR of the loss distribution, we apply a CVaR constraint on the general market risk of the trading book to achieve an upper bound of the regulatory VaR. We limit the sum of the general market risk, measured by the right hand side of the trading book CVaR constraint, the specific and the counterparty risk by the applicable regulatory capital components, that consist of the tier 1 and tier 2 capital elements that are not used to cover bank book risks and the tier 3 capital[4] available. Further we model constraints that limit the regulatory capital components [1,2,6].

3 Application Example

We illustrate the effects of the risk-return optimization by a simplified application example [6]. An XY-Bank consists of four typical bank assets: asset 1 represents high quality bank bonds (rating AA), asset 2 corporate bonds (rating A), as-

[2] We consider the prevailing regulatory risk limitation rules. The model allows a transition to the new 'Basle II' rules [3], that will require different input data, but will not influence the basic structure of the risk constraints.

[3] In brief, the bank book comprises all 'non trading' assets, while the trading book comprises all positions the bank is holding for trading purposes (precise definition see [2]).

[4] The tier 3 capital mainly comprises subordinate short term debt [2,7].

set 3 industrial loans (rating B) and asset 4 a trading portfolio that is dependent on an equity index. In the actual situation, the regulatory capital of the bank is used at 93.80% and cannot be increased in the next business year. The initial portfolio uses 76.20 units of economic capital.

In order to gain additional profits, the managing board considers to raise the economic capital, i.e. the level of internal risk by additional undisclosed reserves of 17.10 units. The managing board wants to know, if the additional economic capital will lead to higher returns in the next business year and will be suitable to improve the portfolio risk-return relations and to meet the internal hurdle-rate of an overall *return on risk adjusted capital* (RORAC), that is defined as the expected portfolio return divided by the portfolio CVaR deviation, of 14.00%.

We apply the optimization model (P) with increasing CVaR levels to generate the efficient line as described in the Fig. 2 below. The initial portfolio is denoted by PF 0, the optimal portfolio at the given level of risk by PF 1 and the optimal portfolio with the increase of the economic capital by PF 2.

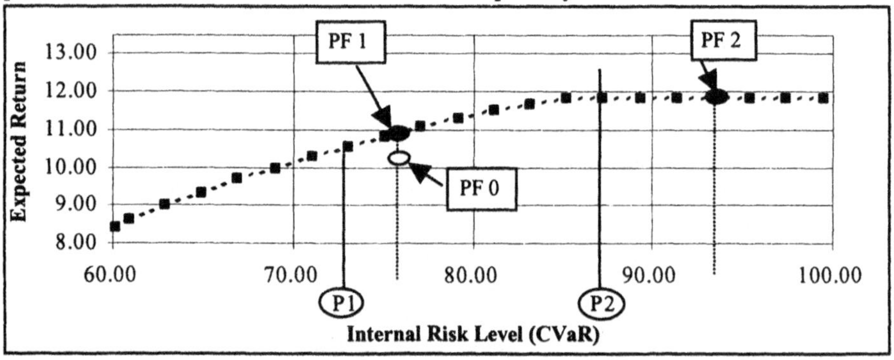

Fig. 2. Efficient Line of the XY-Bank Portfolio Resulting of Applications of (P)

Analyzing the efficient line, we observe that to the left of the CVaR level P1 only the internal risk constraint is active and the regulatory capital is not completely used. To the right of the CVaR level P2 only the regulatory constraint is active, as in this interval we achieve a stationary solution due to the regulatory capital constraint and the economic capital is not totally devoted. In the interval [P1,P2] both capital constraints are active, i.e. both capital resources, the economic and the regulatory capital, are maximal utilized.

A comparison of the optimal portfolios PF1 and PF 2 shows that the increase of the economic capital leads to higher expected returns, however the RORAC of PF 2 of 13.57% does not meet the hurdle rate of 14.00% and is even lower than the RORAC of the initial portfolio PF 0 of 13.98%. Also, the economic capital constraint is not active and the economic capital cannot be maximal used in the portfolio PF 2.

We deduce that an increase of the economic capital is not advisable and that the actual level of risk should be maintained. The expected return of the portfolio PF 1 improves the expected return of the initial portfolio PF 0 by 0.31 units at the given level of internal risk. Its portfolio RORAC of 14.39% complies with the internal hurdle rate. Both capital resources, the economic and the regulatory capital, can maximal be used. Analyzing the risk-return relations along the efficient line we find that a maximal RORAC can be achieved in the interval [68.9,71.0]. However, the implementation of a RORAC optimizing strategy would require to reduce absolute volumes and expected returns. This might be conflicting with other corporate goals and may not be supported by the shareholders.

4 Conclusion

We have introduced a risk-return optimization model for the bank portfolio, that can be applied in the planning processes of the bank in order to identify risk-return efficient target portfolios. It maximizes the expected returns subject to internal and regulatory risk constraints. It is based on the new risk measure of CVaR, which has is appropriate for bank wide portfolio risk measurement. The optimization problem can be solved by linear programming techniques. The optimization model generates consistent planning information. It spots risk-return efficient portfolios and finds intervals of maximal use of both capital resources, the available economic and regulatory capital, and of highest portfolio RORACs, thus providing basic information for a bank wide risk-return management process and contributing to an enhancement of the competitive position of the bank.

References

[1] Basle committee on Banking Supervision (1988): International convergence of capital measurement and capital standards, Basle, July 1988.
[2] Basle committee on Banking Supervision committee (1996): Amendment to the capital accord to incorporate market risks, Basle, January 1996.
[3] Basle Committee on Banking Supervision (2001), Consultative Document: The New Basel Capital Accord, January 2001, Basel, January 2001.
[4] Rockafellar, R. T. and Uryasev, S. (2000): Optimization of Conditional Value-At-Risk, *The Journal of Risk*, Vol. 2, No. 4, pp. 21-51.
[5] Rockafellar, R. T. and Uryasev, S. (2002): Conditional Value-at-Risk for General Loss Distributions, *Journal of Banking and Finance*, 27/7.
[6] Theiler, U. (2002): Optimization Approach for the Risk-Return-Management of the Bank Portfolio, Wiesbaden (in German).
[7] United States Accounting Office (1998): Risk-Based Capital - Regulatory and Industry Approaches to Capital and Risk, Washington, July 1998.

Resource-orientated Purchase Planning and Supplier Selection – Models and Algorithms for Supply Chain Optimization and E-Commerce

Gabriele Reith-Ahlemeier

POM Prof. Tempelmeier GmbH, Am Kapellenbusch 13, 50374 Erftstadt, Germany, http://www.pom-consult.de, gabriele.reith-ahlemeier@pom-consult.de

Abstract. With the internet being available everywhere companies are able to automate parts of their global business processes. In this paper we consider the problem of simultaneous supplier selection and purchase order sizing. Significant cost-savings are possible by integrating these problems in a software-based Business-to-Business environment.

Today a purchasing agent often has to act without any algorithmic assistance. He has to order several items under dynamic demand conditions and with restricted material handling and storage capacities. Moreover, suppliers often offer quantity discounts which can vary over time. They furthermore may only be able to deliver on some special days and may insist on a minimum order volume.

We develop and evaluate three new heuristic procedures – based on a new model formulation – to support the described decision process.

1 Decision Problem

In a typical industrial purchasing environment the purchasing agent has the task of supplier selection and order sizing – given several items with deterministic, dynamic demand for a finite planning horizon. Regarding to the potential suppliers, there might be a preselection process based on qualitative criterias. The remaining suppliers differentiate by fixed ordering costs and all-units or incremental quantity discounts. The discount structures may vary over time, e.g. due to marketing actions. Furthermore, the purchasing agent has to pay attention to product- and supplier-specific delivery times and limited capacities, such as material handling and storage capacities or limited transportation volumes. Finally, some suppliers insist on minimum order volumes and a given delivery schedule.

The purchasing agent's aim is to procure all items in time and for a reasonable price. Relevant cost components are beside the product prices price-dependend holding costs and product-specific fixed purchasing costs as well as fixed purchasing costs that arise independent from the number of items ordered from a certain supplier in a given period.

Only very few literature adresses the described operational purchasing problems including supplier selection. Benton and Rubin [1], [5] present heuristics to solve such problems, but they consider stationary demand and only all-units discount structures. Generally, the literature on order sizing under consideration of quantity discounts comprises only a small number of publications. Recent overviews are provided by Benton and Park [2] and by Munson and Rosenblatt [3]. Up to our knowledge, currently there is no solution approach available that fully solves the above described decision problem.

2 Models

The problem can be modeled as a single level, capacitated dynamic lot sizing problem. Unfortunately, with this formulation it is very complicated to model price-dependend holding costs, as the holding costs are evaluated depending on the stock and as, on account of quantity discounts, we can not evaluate the complete stock with the same price. Therefore, we base our model formulation on the well-known analogy between the plant location problem and the dynamic lot sizing problem. We interprete the demand of an item in a given period as customer location and every possibility to order that demand from a supplier in a certain period and on a certain discount level as a potential plant location. By that, the transportation distance is defined by the number of periods between demand and delivery period – that is exactly the interval an item has to be kept in stock. In this way we are able to trace back the prices that were paid.

The model is structured as follows: The objective function minimizes the above described relevant cost components. Several restrictions are required to ensure the given capacity constraints. Moreover, there are some specials in contrast to standard plant location problems. At first, in every period no more than one discount class of a certain supplier can be taken up. Therefore, it must be ensured that the other classes of that supplier (interpreted as alternative plant locations) are locked. Secondly, backorders are not allowed, what means, that demand must be satisfied in time. So not every supplier's offer (once again interpreted as plant location) can be used to satisfy the demand (as customer location), but only that one from earlier periods. See Reith-Ahlemeier [4] for the exact model formulation.

Note that the model captures planning situations where some suppliers offer all-units discounts and other use an incremental discount scheme for the same item. It is even possible that a supplier switches from all-units to incremental discounts within the planning horizon of the model.

3 Heuristic Procedures

In view of its complexity (belonging to the class of NP-complete problems) the developed model can be solved exactly only for rather small problem instances. Therefore, to solve the model in a routine planning environment, a heuristic solution procedure is required. Such a solution procedure must be fast and – with regard to varying constraints in industrial practice – it must be easily extendible.

We develop a simple local search heuristic as well as two heuristics based on Lagrangean relaxation – one of these additionally using the concept of branching & bounding. Reith-Ahlemeier [4] gives a detailed description of all three heuristics.

Local Search Heuristic. The local search heuristic consists of two phases. The first phase starts with neglecting any capacity constraints. The remaining single-item order quantity problems are solved by the heuristic ISSOS-procedure proposed by Tempelmeier [6]. An initial feasible solution is created by considering each type of capacity constraint in sequence. In case of violation, several options to attain feasibility by splitting, combining or shifting orders are considered, realizing that one, which causes the lowest total costs. The second phase separates the total costs into fixed ordering, variable ordering and holding costs and tries to reduce them by focussing on each component in turn – only considering options that ensure feasibility of the order schedule with respect to all capacity constraints. Fig. 1 summarizes the overall structure of the heuristic.

Fig. 1. Structure of the local search heuristic

Phase I: Initial solution	
	For each item, solve a single-item order size problem, neglecting the capacity constraints.
	If the order schedule violates any capacity constraints, construct a feasible solution by shifting, combining and splitting orders.
Phase II: Improvement steps	
	Step 1: Take measures to reduce the fixed ordering costs.
	Step 2: Take measures to reduce the variable ordering costs.
	Step 3: Take measures to reduce the holding costs.

Lagrange Heuristic. The Lagrange heuristic is based on the relaxation of all capacity constraints and the constraint to ensure demand fullfillment. The Lagrangean multipliers are adjusted via standard subgradient optimization techniques. To provide a lower bound in each iteration the arising relaxed problem is decomposed into several continuous and binary knapsack problems. For that, surrogate constraints – enabling stronger bounds without restricting the solution space – are set up to ensure sufficient order volume. Feasible solutions are developed based on the current Lagrangean multipliers, using the above described local search heuristic. An overview over the steps performed to find the heuristic solution is given in Fig. 2.

Fig. 2. Structure of the Lagrange heuristic

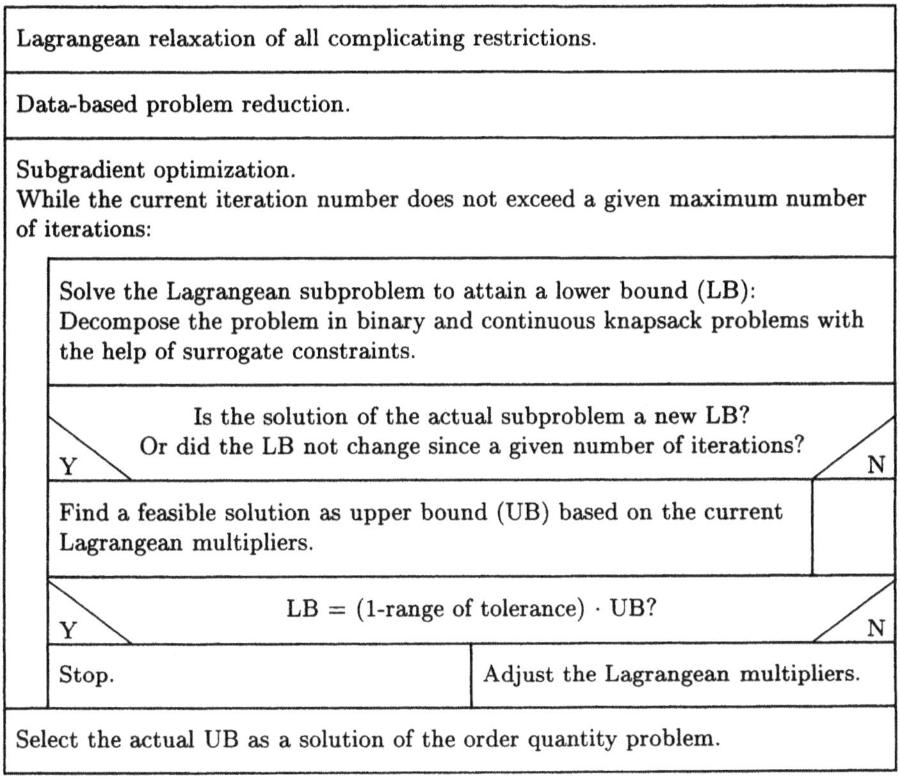

Branch & Bound Heuristic. To develop a branch & bound heuristic the model is relaxed in the same way as for the Lagrangean one. Once again the subproblems are decomposed in continuous and binary knapsack problem. Differing to the Lagrange heuristic the surrogate constraints now limit the number of alternative offers (interpreted as locations) that can be used. The resulting bounds are not as strong as before, but as the subproblems can be solved much faster more iterations are possible. In each iteration, the problem can be reduced by further reducing the number of alternative offers. Furthermore a certain amount of variables can be fixed heuristically, depending on the modification of the total costs. Fig. 3 shows the special elements of the branch & bound heuristic.

Fig. 3. Special structure of the branch & bound heuristic

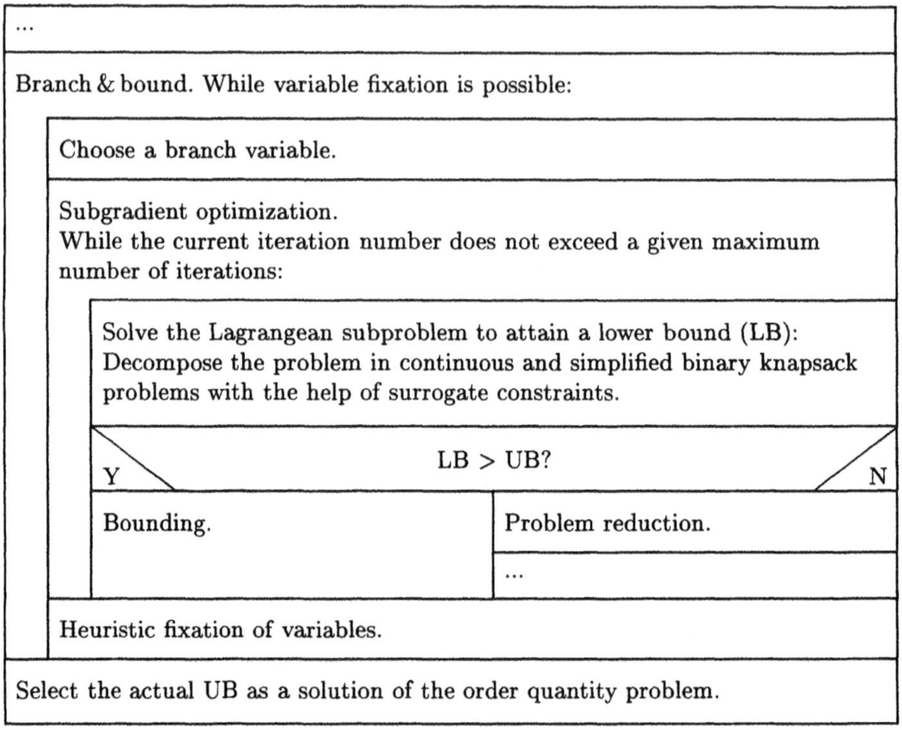

4 Numerical Results

To compute an exact solution for at least small problems we used CPLEX 6.6 for parallel computers running on a UNIX-based host using eight parallel processors. Depending on the problem data, solution times required for finding the exact solutions for problem instances with 3 products, 15 periods, 5 discount levels and 3 suppliers ranged between a few seconds and several hours. The heuristics were implemented in Visual Basic 6.0 on a Pentium-based Windows 2000 personal computer with 400Mhz clock speed. The computation of heuristic solutions only took a few seconds. Not surprisingly, the local search heuristic was the fastest one. But also the computation times of the other heuristics were very small and did not vary as much as for the exact solution procedure (see the coefficient of variation (CV) in Table 1). Although the branch & bound heuristic on average was faster than the Lagrange heuristic, its coefficient of variation is much bigger.

Varying the demand patterns, the order cycles and the capacity restrictions as well as the scheme for delivery periods and for the time-varying price structures we created 1440 instances. Table 1 summarizes the average results.

Table 1. Average solution quality and computation time

Heuristic	Deviation from optimum (%)	Computation time (sec)	CV
Local search	2.47	<1	–
Lagrange	2.05	30.58	0.20
B & B	1.98	16.22	0.90

Even the local search heuristic was able to compute solutions with an average deviation of only 2.5 percent from the exact ones. The other heuristics improved this value to only two percent.

Unfortunately it is not possible to test the solution quality for larger problems. Only the time to compute a heuristic solution can be measured. The extension of the planning horizon to thirty periods by ordering ten products led for the local search heuristic to an average effort of 8.33 seconds. The Lagrange heuristic required an average computation time of 213 seconds, whereas the branch & bound heuristic demanded for 1613.31 seconds. This confirms the observation for the smaller problems, that the latter heuristic is much more sensitive to problem data.

Summarizing the results, the developed heuristics are suitable for application in an interactive environment. Especially the local search heuristic takes very short time to generate a solution with high quality. The advantages of the other heuristics depends on problem data, none of them outperforms the other one.

References

1. Benton, W. C. (1991) Quantity discount decisions under conditions of multiple items, multiple suppliers and resource limitations. International Journal of Production Research 29(10): 1953–1961
2. Benton, W. C., Park, S. (1996) A classification of literature on determining the lot size under quantity discounts. European Journal of Operational Research 92: 219–238
3. Munson, C. L., Rosenblatt, M. J. (1998) Theories and realities of quantity discounts: an exploratory study. Production and Operations Management 7(4): 352–369
4. Reith-Ahlemeier, G. (2002) Ressourcenorientierte Bestellmengenplanung und Lieferantenauswahl – Modelle und Algorithmen für Supply Chain Optimierung und E-Commerce. Books on Demand, Norderstedt
5. Rubin, P. A., Benton, W. C. (1993) Jointly constrained order quantities with all-units discounts. Naval Research Logistics 40: 255–278
6. Tempelmeier, H. (2003) A simple heuristic for dynamic order sizing and suplier selection with time-varying data. Production and Operations Management, to appear

A Combinatorial Approach to Orthogonal Placement Problems

Gunnar W. Klau

Konrad–Zuse–Zentrum für Informationstechnik Berlin (ZIB), klau@zib.de

Abstract. This article presents the main results of a PhD thesis that deals with two families of *NP*-hard orthogonal placement problems. We develop a common combinatorial framework for compaction problems in graph drawing and for labeling problems in computational cartography. Compaction problems are concerned with performing the conversion from a dimensionless description of the orthogonal shape of a graph to an area-efficient drawing in the grid. Map labeling is the task of attaching labels to point-features so that the resulting placement is legible. On the basis of new combinatorial formulations for these problems we develop exact algorithms. Extensive computational studies on real-world benchmarks show that our linear programming-based algorithms solve large instances of the placement problems to provable optimality within short computation time. Often, our algorithms are the first exact algorithms for the respective problem variant.

We analyze two families of orthogonal placement problems that arise in the area of information visualization. The first family, compaction of orthogonal grid drawings, is concerned with performing the conversion from a dimensionless description of the orthogonal shape of a graph to an area-efficient drawing in the orthogonal grid with short edges. This two-dimensional compaction problem emerges in the last phase of a powerful method for high-quality orthogonal graph drawing, the topology-shape-metrics scheme. The second family of problems plays an important role in the area of computational cartography and deals with the task of attaching rectangular labels to point-features such as cities or mountain peaks on a map.

It is common to both drawings of graphs and cartographic maps that they convey complex information about relations of objects as a geometric representation. Moreover, the utility of this representation depends on the quality of the layout process. The overall aim is to generate a drawing or a map of maximum readability that is intuitive to understand and use and which effectively communicates the underlying information. As an example, Fig. 1 shows "good" and "poor" solutions of orthogonal placement problems.

Why are the orthogonal placements in Fig. 1(a) and (c) better than those of Fig. 1(b) and (d)? The edges of the right orthogonal drawing are unnecessarily long, confuse the reader, and increase the amount of drawing space needed. The compact drawing on the left has been enlarged by 25% and is still more area-efficient. Due to the better resolution and shorter edges it is superior to the right drawing. Even more obvious reasons make the left labeling a better one than the right one: In Fig. 1(d), information is lost since not

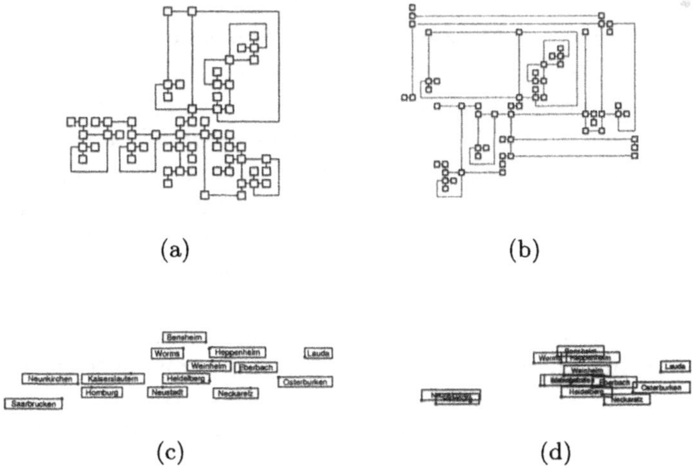

Fig. 1. Good and poor solutions of orthogonal placement problems

all labels are placed. Furthermore, many of the labels overlap which makes it difficult to extract the necessary information.

A further common characteristic to both areas is that most of the respective problems are computationally hard to solve. Everybody who has tried to draw a graph with 20 vertices by hand knows about the difficulties of finding an aesthetically pleasing drawing. Even if this task can be accomplished, it remains a tedious, complicated, and time-consuming process. The same applies to the placement of labels on a map.

Due to space limitations, this article can only give an introduction to the techniques developed in the thesis. We refer to [9] for details.

Compaction in Orthogonal Graph Drawing

The area of *automatic graph drawing* is devoted to the development of algorithms that produce geometrical representations of graphs and to the problems that arise within this context, see, *e.g.*, [5] and [8]. Here, we focus on *orthogonal drawings*, *i.e.*, drawings of graphs in which the edges are represented by sequences of alternating vertical and horizontal line segments. For numerous applications, orthogonality is a convention (*e.g.*, UML-diagrams, circuit layouts, entity-relationship diagrams), and for many other applications orthogonal drawing algorithms produce the best layouts.

Due to the oftentimes conflicting aesthetic criteria, tradeoffs cannot be avoided. For orthogonal drawings, the following ranking is widely accepted: The primary goal is to minimize the number of edge crossings; ideally a graph should be drawn without any crossings all. Secondly, the number of bends should be as low as possible. Finally, besides few crossings and bends, the edges should be short in the drawing, which also leads to good area bounds.

According to the above ranking of criteria, the *topology-shape-metrics scheme*, first mentioned in [2], leads to the best results. It divides the drawing task into three phases. The first phase (*planarization*) aims at minimizing the number of crossings. Initially, it identifies a small set of edges whose removal results in a planar subgraph and then computes a *planar embedding* for this subgraph. Then, the temporarily deleted edges are reinserted at the combinatorial level so that the number of crossings is low. In order to maintain a planar graph, every crossing is replaced by an artificial vertex. The *orthogonalization* phase deals with determining the *orthogonal shape* of the resulting drawing. Here, the optimization goal is to minimize the number of bends that occur along the edges of the drawing. While it is *NP*-hard to minimize this number over all embeddings of a planar graph, the problem can be elegantly solved for a fixed embedding by solving a minimum-cost flow problem [11]. The output of the orthogonalization phase is a so-called *orthogonal representation* that contains the necessary information about the topology and the orthogonal shape of the drawing. Yet, the description is dimensionless and coordinates still have to be assigned to the vertices and bends of the drawing.

The *compaction phase* must transform an orthogonal representation into a drawing with small total edge length or little area. Again, this is an *NP*-hard problem [10]. Previous algorithmic research for this problem can be divided into *constructive* and *improvement heuristics*: Basic constructive heuristics perform a rectangular dissection of the given orthogonal representation [11,6]. In [3] these techniques are extended by introducing the concept of turn-regularity. However, the results still admit room for considerable improvement. Improvement heuristics such as the compression-ridge method [1,4] and graph-based compaction techniques, *e.g.*, [7], originate in the area of VLSI-design and consider the one-dimensional subproblems of reducing the horizontal or vertical edge lengths. In many cases, iterative usage of these heuristics with alternating direction yields considerable improvement.

The key idea of our approach to the two-dimensional compaction problem is to translate it into an equivalent combinatorial problem involving a pair of constraint graphs. Investigating combinatorial properties of these graphs leads to new algorithms that can solve large instances of the compaction problems to optimality in short computation time. Based on the observation that we can treat the horizontal and vertical direction to a great extent separately, each directed graph corresponds to one such direction.

We investigate how the constraint graphs must interact in order to develop a combinatorial characterization of the compaction problem. Thereby, we exploit the fact that, due to the given shape, many relative positions of vertices, edges, and bends are already determined. We introduce the *shape graphs*, that reflect the orthogonal shape of the input. Moreover, we identify a central path- and cycle-based property of constraint graphs, *completeness*, that forms the link between the otherwise unconnected horizontal and vertical graphs. Substantially, a complete pair of constraint graphs consists of two acyclic graphs in which each pair of nodes is separated by one of four paths. Each of these paths corresponds to one of the four possible relative placements of a pair of objects in two dimensions. We show that, for complete placement graphs, the compaction problem reduces to two separate one-dimensional problems for which optimal solutions lead to an optimal solution of the problem in two dimensions.

The shape graphs are uniquely determined by the given orthogonal representation. In case of complete shape graphs, we can solve the two-dimensional compaction problem in polynomial time. Furthermore, we investigate the one-dimensional compaction scheme and demonstrate that instances exist for which a linear number of alternating compaction steps is necessary. Moreover, we show that algorithms within the scheme do not approximate the compaction problem by a constant factor.

In general, the shape graphs are not complete and we identify a set of *potential additional arcs*. We show that the set of *complete extensions* that results from adding certain subsets of potential arcs to the shape graphs is in one-to-one correspondence to the feasible solutions of the original problem. It is the choice of potential arcs that makes the compaction problem difficult at the combinatorial level. However, we can characterize those shape graphs that admit a unique extension and solve the compaction problem in polynomial time for these instances. Otherwise, we translate the combinatorial problem into an integer linear program whose feasible solutions correspond to feasible solutions of the original compaction problem and vice versa. This enables us to optimize over the set of feasible orthogonal drawings for a given instance, and we present both a branch-and-bound and a branch-and-cut algorithm to solve the two-dimensional compaction problem to optimality. We test our implementations on a large set of widely used benchmark-graphs from different test-suites, including a set of 11,582 graphs arising from real-world applications. Our extensive computational study shows that we can solve all real-world problem instances in short computation time.

Map Labeling

Map labeling problems attract many researchers in computer science. On the one hand, this is due to its numerous applications, *e.g.*, in cartography, geographic information systems, point pattern analysis, spatial statistics, and graphical interfaces. On the other hand, many combinatorial optimization problems with beautiful mathematical properties appear in this area. For an overview on this subject see the bibliography [12]. In this article, we focus on *point-feature label placement* in which the task is to place labels adjacent to point features such as cities, mountain peaks, or points in a statistical plot so that no labels overlap. We concentrate on the six different labeling models

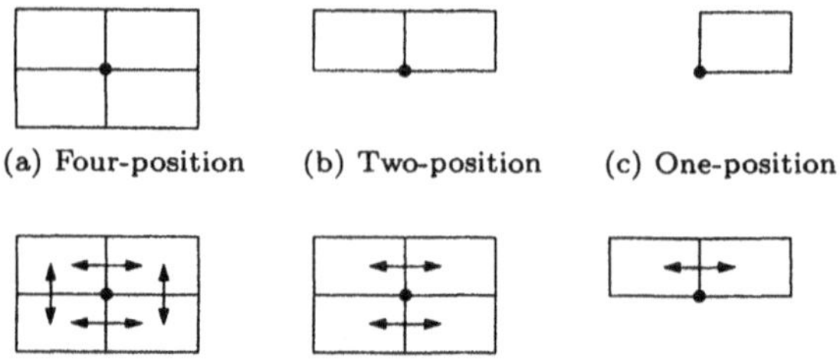

(a) Four-position (b) Two-position (c) One-position

(d) Four-slider (e) Two-slider (f) One-slider

Fig. 2. Axis-parallel rectangular labeling models. A label can be placed in any of the positions indicated by the rectangles and can slide in the directions of the arcs

in Fig. 2. The *discrete* or *fixed-position models* (Fig. 2(a)-(c)) allow only a finite number of positions per label. More natural are *slider models* in which a label may move continuously around its point-feature, see Fig. 2(d)-(f).

In general it is not possible to place all the given labels in their original size without any overlap. We focus on the *label number maximization problem* where the task is to place the maximum number of non-overlapping labels without changing their sizes. Again, we associate a pair of constraint graphs with problem instances; the key idea is the same as for the two-dimensional compaction problems: If these graphs satisfy certain path- and cycle-based properties, we can produce a solution for the original problem by separately assigning values to the nodes of the constraint graphs. These values correspond to the x- and y-coordinates of the labels.

For a given instance of a labeling problem we construct a special pair of *labeling graphs*. We introduce different kinds of arcs whose presence satisfy necessary properties of feasible label placements: The *fixed distance arcs* ensure that the relative position between the point-features remains fix. *Label size arcs* guarantee that every label is represented by a rectangle of width and height as described in the input. By introducing the *proximity arcs* we determine the rectangular region around the appropriate point-feature in which a label can be placed. In order to exclude that a label covers the point-feature it belongs to, we define the *boundary arcs* that are inverse to the proximity arcs. Unlike the previously introduced types of arcs, the boundary arcs belong to the class of *potential arcs* and influence the labeling model. Each discrete or slider model corresponds to requirements on subsets of boundary arcs that have to be present in the labeling graphs. We define a second type of potential arcs in order to control the overlaps between labels. The *label separation arcs* make sure that pairs of labels do not overlap in a placement.

We can now restate the *pure labeling problem* in which all labels have to be placed without scaling and overlaps as the identification of a subset of potential arcs that satisfies the following two properties. First, the set of chosen boundary arcs must comply with the appropriate labeling model. The second property extends the notion of *completeness* as defined for the compaction problems: At least one label separation arc has to be chosen for each label pair, and adding the chosen potential arcs to the labeling graphs must not induce directed cycles of positive weight. We show that the combinatorial reformulation is equivalent to the pure labeling problem by establishing a one-to-one correspondence between feasible solutions. Furthermore, we demonstrate how to adapt the new combinatorial problem to result in equivalent formulations of the label number maximization problem. We find it remarkable that our new approach is independent of the labeling model and results in discrete formulations even if the problems are of continuous nature as in the slider models. The combinatorial formulation for the pure labeling problem admits a straightforward characterization as a zero-one polytope through an incidence vector for the set of potential arcs. We provide an integer linear programming formulation for this polytope by describing feasible solutions of the combinatorial version of the pure labeling problem with classes of inequalities and integrality constraints. For one class of inequalities, the *positive cycle inequalities*, we investigate the corresponding separation problem and show that it is *NP*-complete. We present an extended formulation that evades the class of positive cycle inequalities. Our integer linear programs for the label number maximization problem are not as straightforward. We develop a first formulation with an additional binary variable vector that represents the decision to place or not to place a label. We integrate the new variables in the existing inequalities and show that feasible solutions of the resulting formulations correspond to an overlap-free labeling for a subset of labels. In a second formulation we manage to eliminate the newly introduced decision variables by a substitution step. However, we have to add additional inequalities to adjust the objective function.

We present branch-and-bound and branch-and-cut algorithms and an iterative branch-and-bound scheme for the zero-one and extended formulations. The algorithms work in all labeling models and are the first exact algorithms for the continuous slider models. We provide extensive computational experiments in which we test our new algorithm on a large set of benchmark data. The results show that the exact algorithms produce provably optimal solutions for large instances in reasonable computation time.

We show how to combine our approaches to compaction and labeling problems in order to devise first algorithms for the interesting class of *graph labeling* problems. In particular, we consider a problem that occurs in the area of automation engineering: Simultaneous drawing and labeling of state diagrams. Concluding, we believe that our combinatorial characterizations are expansible and suitable to apply them to many related problems like, *e.g.*, packing or location problems.

References

1. S. Akers, M. Geyer, and D. Roberts. IC mask layout with a single conductor layer. In *Proc. 7th Des. Autom. Workshop*, pages 7–16. ACM/IEEE, 1970.

2. C. Batini, E. Nardelli, and R. Tamassia. A layout algorithm for data-flow diagrams. *IEEE Transactions on Software Engineering*, SE-12(4):538–546, 1986.
3. S. Bridgeman, G. Di Battista, W. Didimo, G. Liotta, R. Tamassia, and L. Vismara. Turn-regularity and optimal area drawings of orthogonal representations. *Computational Geometry: Theory and Applications*, 16(1):53–93, 2000.
4. W. W.-M. Dai and E. S. Kuh. Global spacing of building-block layout. In C. H. Sequin, editor, VLSI *'87*, pages 193–205. Elsevier Science, 1987.
5. G. Di Battista, P. Eades, R. Tamassia, and I. G. Tollis. *Graph Drawing. Algorithms for the Visualization of Graphs.* Prentice Hall, 1999.
6. F. Hoffmann and K. Kriegel. Embedding rectilinear graphs in linear time. *Information Processing Letters*, 29(2):75–79, 1988.
7. M.-Y. Hsueh. Symbolic layout and compaction of integrated circuits. Technical Report UCB/ERL M79/80, Univ. of California, Berkeley, CA, U.S.A., 1979.
8. M. Kaufmann and D. Wagner, editors. *Drawing Graphs: Methods and Models*, volume 2025 of *Lecture Notes in Computer Science.* Springer, 2001.
9. G. W. Klau. *A Combinatorial Approach to Orthogonal Placement Problems.* PhD thesis, Univ. d. Saarlandes, Saarbrücken, Germany, September 2001.
10. M. Patrignani. On the complexity of orthogonal compaction. *Computational Geometry: Theory and Applications*, 19(1):47–67, 2001.
11. R. Tamassia. On embedding a graph in the grid with the minimum number of bends. *SIAM J. Comput.*, 16(3):421–444, 1987.
12. A. Wolff and T. Strijk. The map labeling bibliography. `http://www.math-inf.uni-greifswald.de/map-labeling/bibliography`.

Assigning Frequencies in GSM Networks*

Andreas Eisenblätter

Konrad-Zuse-Zentrum für Informationstechnik Berlin (ZIB)
Takustr. 7, D-14195 Berlin, Germany

Abstract. Mobile communication is a key technology in today's information age. Despite the ongoing improvements in equipment design, interference remains a limiting factor for the use of radio communication. The author investigates in his PhD thesis how to largely prevent interference in GSM networks by carefully assigning the available frequencies to the installed base stations. The topic is addressed from two directions: first, new algorithms are presented to compute "good" frequency assignments fast; second, a novel approach, based on semidefinite programming, is employed to provide lower bounds for the amount of unavoidable interference.

The proposed new methods for automatic frequency planning are compared in terms of running times and effectiveness in computational experiments using instances from practice. For most of the heuristics the running time behavior is suited for interactive planning, and they provide good assignments from a practical point of view. Several of these methods are successfully employed by the German GSM operator E-Plus Mobilfunk GmbH & Co. KG.

The best lower bounds on the amount of unavoidable (co-channel) interference are presently obtained from solving semidefinite programs. These programs arise as nonpolyhedral relaxation of a minimum k-partition problem on complete graphs. The success of this approach is underpinned by revealing structural relations between the solution set of the semidefinite program and a polytope associated with an integer linear programming formulation of the minimum k-partition problem. Comparable relations are not known to hold for any polynomial time solvable polyhedral relaxation of the minimum k-partition problem. The application described is among the first of semidefinite programming to large industrial problems in combinatorial optimization.

1 Introduction

The General System for Mobile communication or, for short, GSM is nowadays the predominant technology for mobile communication. More than half a billion people in over 150 countries use GSM for mobile telephony and for exchanging short text messages (SMS). Within a decade, GSM has grown from a costly service used by few professionals to a mass market with penetration rates higher than 70 % in Finland and Iceland. In some countries, the mobile phone subscribers already outnumber the fixed-line telephone subscriptions.

The mobile communication relies on a radio link between the user's mobile phone and some stationary base station, which is part of a GSM operator's infrastructure, see Fig. 1. Currently, a base station typically serves three

* This presentation is based on the Ph.D. thesis of the author [5].

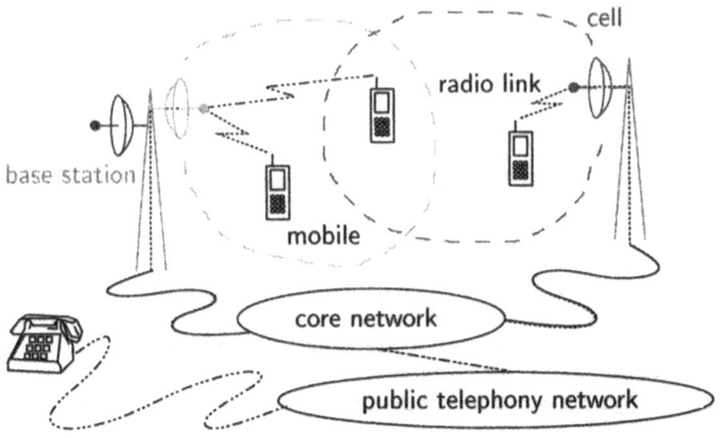

Fig. 1. GSM bits and pieces

different areas (cells) with up to 6 transmitters. Each transmitter uses a frequency slot of 200 kHz, called channel, to handle at most 6–8 users in parallel via time multiplexing (time division multiple access, TDMA). Nearby transmitters have to use different channels (frequency division multiple access, FDMA). As with all forms of radio communication, the limited radio spectrum is a bottleneck. National regulation authorities usually license between 60–120 channels of radio bandwidth to GSM operators.

An operator has to reuse his channels multiple times to operate the several tens of thousands transmitters, which are typically installed in a network. Each radio link, however, requires a signal of sufficient strength which, at the same time, is not suffering too severely from interference by other signals, see Fig. 2. Significant interference may be caused by transmitters using the same channel (co-channel) or an adjacent channel.

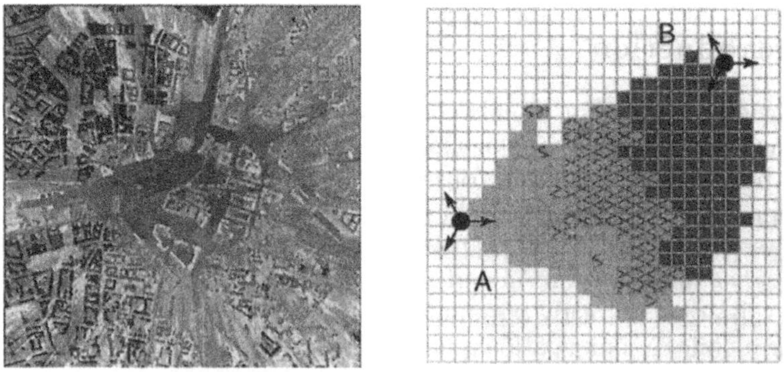

Fig. 2. Field strength and interference: **(a)** inhomogeneous decay of a signal's field strength (path loss) in an urban environment (by courtesy of E-Plus); **(b)** estimation of interference in terms of affected cell area

The reuse of channels is therefore limited, and frequency planning turns into a key issue in fully exploiting the available radio spectrum. Notice that it is customary to use the word frequency as a synonym for channel in this context. By avoiding interference, frequency planning has a significant impact on the quantity as well as on the quality of the radio communication services.

Frequency assignment is usually performed at the end of a chain of planning activities. The placement of the base station as well as the selection and configuration of their antennas are the basis for delivering the desired network coverage. The subsequent decisions on how many transmitters to operate in each cell build the foundation for the desired network capacity. The final step of assigning the frequencies "merely" has to ensure that the coverage and capacity goals can be met: namely, by providing each transmitter with a frequency that is (locally) at most moderately interfered.

The Ph.D. thesis [5] addresses several topics, ranging from the technical background of the GSM frequency planning problem (Chap. 2) over alternative mathematical models (Chap. 3) and heuristic planning methods (Chaps. 4, 5) to quality assessments for the generated frequency plans (Chaps. 6–8). An overview is given in the following. The theory (Chaps. 7, 8) underlying the computation of unavoidable interference for the quality evaluation, however, is not addressed here.

Much of this work is related to a cooperation between the ZIB and the German GSM 1800 network operator E-Plus Mobilfunk GmbH & Co. KG. The focus of the cooperation was primarily on fast frequency planning heuristics for the use in the regular radio planning process at E-Plus. New planning methods were developed at ZIB and integrated into E-Plus' software environment. In 1997, the new software was first used successfully in practice. A series of extensions have meanwhile been implemented.

2 Optimization Model for GSM Frequency Assignment

The frequency planning problem sketched above can be formalized as a combinatorial minimization problem. An undirected graph $G = (V, E)$ is defined together with vertex- and edge-labelings. A vertex is introduced for each transmitter (demand for one frequency). An edge is introduced whenever there is an interdependency between the corresponding transmitters.

The edge-labelings record three types of interdependencies. The *separation* label $d(vw)$ is the minimum required difference of the channels assigned to v and w. Typical values are (0,) 1, 2, 3. The *co- and adjacent channel interference* labels $c^{co}(vw)$, $c^{ad}(vw)$ record how much interference is incurred in case the same channel, respectively adjacent channels are assigned to v and w. Interference is normalized to values between 0 and 1.

Each vertex label A_v specifies the set of *available channels* for the transmitter v. These sets are often genuine subsets of the frequency spectrum C licensed to an operator. Such restrictions arise, for example, along national borders, where a cross-border coordination of the channel use is necessary to prevent (strong) interference between bordering networks. The licensed spectrum itself is usually contiguous.

A *frequency assignment* or simply an *assignment* is a function $y\colon V \to C$. An assignment is *feasible* if every carrier $v \in V$ is assigned an available channel and all separation requirements are met, that is, if

$$y(v) \in A_v \qquad \forall v \in V\,, \tag{1}$$

$$|y(v) - y(w)| \geq d(vw) \qquad \forall vw \in E\,. \tag{2}$$

Finding a feasible assignment is closely related to coloring a graph. Frequency assignment is a generalization of *list colorings* and related to *T-colorings* and *list T-colorings* of graphs [5, Chap. 3]. Drawing on this connection, it is easily shown that finding any feasible frequency assignment is $\mathcal{NP}$-complete in general.

In practice, not just some feasible assignment is of interest, but assignments that minimize the sum of co- and adjacent channel interferences are in demand. The corresponding optimization problem

$$\min_{y \text{ feasible}} \sum_{\substack{vw\in E:\\ y(v)=y(w)}} c^{co}(vw) \quad + \sum_{\substack{vw\in E:\\ |y(v)-y(w)|=1}} c^{ad}(vw) \tag{FAP}$$

is called the *frequency assignment problem.*

This model has proven useful and is largely accepted among researchers and practitioners. From a computational complexity point of view, however, optimal solutions are even very hard to approximate, see [5, Chap. 3] for details. Further models of practical relevance or of theoretical interest are discussed in [1, 4–7, 15].

3 Heuristic Planning Methods

The focus is on planning heuristics, capable of dealing with carrier networks of around 2000 transmitters in a few minutes on a modern PC or workstation. Such methods are well-suited for practical applications, with a particular emphasis on intermediate iterations in the planning cycle.

Seven heuristic planning methods are described [5, Chap. 4]: three greedy-type construction methods, T-COLORING, DSATUR WITH COSTS, and DUAL GREEDY, as well as four improvement methods, ITERATED 1-OPT, K-OPT, VDS, and MCF. The performance of each heuristic (sometimes with augmentations) and its parameter interdependence are extensively analyzed [5, Chap. 5]. Most of the above methods are suited (in combination) for automatic frequency planning in practice. The DUAL GREEDY drops out, because it is slow and produces by far the poorest results.

The concerted acting of various combinations of them is studied on the basis of eleven realistic planning instances, which have been made available over the Internet [9]. Table 1 displays several characteristic parameters of the constraint graphs associated with the planning instances. Notice the high average degrees of the vertices, i. e., the large numbers of transmitters that may directly be affected by the frequency assignment to one transmitter. Notice also the large maximum clique numbers, which in most cases proves directly that an interference free frequency assignment is impossible.

Table 1. Characteristics of constraint graphs of realistic planning instances [9]

	$\|V(G)\|$	density $\varrho(G)$ in %	avg. degree	max. degree $\Delta(G)$	max. clique $\omega(G)$	$\|\{e \in E : d(e) \neq 0\}\|$	$\|\{e \in E : c^{co}(e) \neq 0\}\|$	$\|\{e \in E : c^{ad}(e) \neq 0\}\|$	spectrum size
K	267	56.57	151.0	238	69	1053	19111	996	50
B[0]	1886	13.59	256.4	779	81	7288	234479	4263	75
B[1]	1971	13.46	265.3	805	84	7996	253441	4825	75
B[2]	2214	13.50	299.0	916	93	10284	320684	6871	75
B[4]	2775	13.44	373.0	1133	120	16663	500805	12524	75
B[10]	4145	13.41	555.9	1704	174	38234	1113850	33548	75
SIE1	930	9.03	84.0	209	52	6039	33002	9911	75
SIE2	977	49.17	480.4	877	182	17761	216912	25615	43
SIE3	1623	9.18	149.1	519	78	23093	97861	15069	76
SIE4	2785	10.50	292.3	752	100	27964	379052	26445	39
SW	310	8.29	25.7	94	21	3984	0	2075	3 + 49

4 Automatic Frequency Planning

In essence, the following observations are made [5, Chap. 5]. There is one particular strong combination of the fast heuristics presented: a self-tuning variant of the DSATUR WITH COSTS start heuristic, combined with the VDS improvement heuristic. This combination achieves a decent balance between solution quality and running times.

The resulting assignments are usually not much worse than those obtained by the elaborate THRESHOLD ACCEPTING method [4, Section 4.2.5]. With respect to the maximum incurred co- and adjacent channel interference, they are even sometimes better. The precise running times of THRESHOLD ACCEPTING are not public, but they are roughly one order of magnitude higher than those of the fast heuristic combinations. In case yet faster methods are needed, a combination of a self-tuning variant of T-COLORING with VDS or, even faster, with ITERATED 1-OPT may be attractive.

Interference plots are commonly used for frequency planning. They depict the (likely) occurrence of interference on the basis of the signal level predictions. Figure 3 contains two such plots, where the difference in dB between the serving sector's signal and the second strongest signal at the same frequency is color-coded. Clearly visible are the interference reductions achievable with the proposed methods in comparison to a formerly established, commercial routine. (This routine has meanwhile been replaced by the tool vendor.) In another example more than 96% of the interference could have been removed [5, Chap. 5].

Fig. 3. Interference plots: improvements from optimization

In order to rigorously assess the quality of the plans, lower bounds on the amount of unavoidable interference are in demand. By far the best lower bounds on the ***unavoidable co-channel interference*** are currently obtained from semidefinite programming. Significant bounds are given for the five scenarios K, B[4], B[10], SIE2, and SIE4 [5, Chap. 6]. The reported bounds LB yield quality gaps, computed as $1 - LB/UB$, between the provably unavoidable co-channel interference and a "good" heuristic solution with co- and adjacent channel interference totaling to UB. The gaps are 50% for K, 77% for B[4], 63% for B[10], 53% for SIE2, and 66% for SIE4. From the application point of view, these gaps may not be satisfying. Nevertheless, they are the first noteworthy bounds on the gap for large realistic instances.

The link between the bound on unavoidable co-channel interference in frequency planning and semidefinite programming is a semidefinite relaxation of the well-known graph minimum k-partition problem [5, Chaps. 7, 8]. These problems are obtained by relaxing the original frequency planning problems as follows. Each vertex may receive any of the k frequencies in the available spectrum; all separation requirements are reduced to at most 1; and the adjacent channel interference is ignored. A lower bound for the optimal k-partition, and hence a bound on the unavoidable co-channel interference, is computed by solving the semidefinite relaxation of the k-partition instances. The dual semidefinite programming solvers [3, 10] are used for this purpose.

5 Conclusions

Planning the use of frequencies is a central task in managing a GSM network. It is a cornerstone for providing the desired grade and quality of service. Three planning situations are distinguished.

- In the *relaxed* situation, a new frequency assignment is to be generated for a large network region. Many frequencies are available, and the objective is to minimize interference, thus providing radio service at high quality.
- In the *congested* situation, again a new plan for large network portions is to be produced, but the number of available frequencies hardly allows to provide the desired grade of service (at the least accepted level of quality).
- In the *adaption* case, the assignment shall be adapted locally to changes in the network.

Each of these situations seems to call for different planning methods. For surveys directed towards the "congested" case, see [11, 12, 15]. The "adaption" case has hardly been addressed explicitly yet. The focus here is on the "relaxed" planning situation.

The goal was to design algorithms for generating frequency plans quickly that incur as little interference as possible. They are particularly attractive for interactive planning processes, where alternative plans are produced for tentative network changes. Heuristic methods of small theoretical running times were proposed. Their computational behavior was analyzed on eleven realistic, publicly available scenarios.

Several of these methods are successfully used at E-Plus. Better frequency assignments are obtained much quicker than through the previous planning process. The software is also incorporated into a commercial GSM radio network planning tool as the standard frequency planning component.

The development of new planning methods seems to slow down lately. The lacking demand from major European GSM operators might be a reason. Market saturation is approached and the need for network expansions decreases. A more fundamental reason may lie in the difficulty to provide reliable interference predictions. These are the basis for the frequency plan optimization. At the current stage, the quality of a frequency assignment may depend more on the field strength prediction model used for interference prediction than on which modern planning heuristics is applied [4].

GSM is a second generation, digital system for mobile communication. The upcoming Universal Mobile Telecommunication System (UMTS) is a third generation, offering transmission rates up to 384 kbps. UMTS uses a fundamentally different way to support multiple radio links in parallel (code division multiple access). Frequency planning is no longer necessary with UMTS. A high price has to be paid for this convenience, however: provisioning coverage and capacity are tightly coupled. Base stations affect each other much more with respect to coverage and capacity. This spawns a new line of research, focusing on dimensioning UMTS radio networks [2, 8, 13, 14].

References

1. Aardal K.I., van Hoesel S.C.P.M., Koster A.M.C.A., Mannino C., Sassano A. (2001). Models and solution techniques for the frequency assignment problem. ZIB-report 01–40, Konrad-Zuse-Zentrum für Informationstechnik Berlin, Germany. URL `http://www.zib.de/PaperWeb/abstracts/ZR-01-40/`.
2. Amaldi E., Capone A., Malucelli F. (2002). Planning UMTS base station locations: Optimization models with power control and algorithms. *IEEE Transactions on Wireless Communications*, 1.
3. Burer S., Monteiro R.D., Zhang Y. (1999). Interior-point algorithms for semidefinite programming based on a nonlinear programming formulation. Tech. Rep. TR 99-27, Department of Computational and Applied Mathematics, Rice Unviversity.
4. Correia L.M. (ed.) (2001). *COST 259: Wireless Flexible Personalized Communications*. John Wiley & Sons Ltd.
5. Eisenblätter A. (2001). *Frequency Assignment in GSM Networks: Models, Heuristics, and Lower Bounds*. Cuvillier-Verlag. URL `ftp://ftp.zib.de/pub/zib-publications/books/PhD_eisenblaetter.ps.Z`.
6. Eisenblätter A., Grötschel M., Koster A.M.C.A. (2002). Frequency assignment and ramifications of coloring. *Discussiones Mathematicae Graph Theory*, 22:51–88. URL `http://www.zib.de/PaperWeb/abstracts/ZR-00-47/`.
7. Eisenblätter A., Grötschel M., Koster A.M.C.A. (2002). Frequenzplanung im Mobilfunk. *DMV-Mitteilungen*, (1):18–25. URL `http://www.zib.de/Paper%Web/abstracts/ZR-02-09/`. In German.
8. Eisenblätter A., Koch T., Martin A., Achterberg T., Fügenschuh A., Koster A., Wegel O., Wessäly R. (2002). Modelling feasible network configurations for UMTS. In *Telecommunications Network Design and Management*, pp. 1–24. Kluwer Academic Publishers.
9. FAP web (2000). *FAP web—A website about Frequency Assignment Problems*. Eisenblätter A., Koster A. URL `http://fap.zib.de/`.
10. Helmberg C. (2000). Semidefinite programming for combinatorial optimization. Habilitationsschrift. Technische Universität Berlin, Germany.
11. Jaumard B., Marcotte O., Meyer C. (1999). Mathematical models and exact methods for channel assignment in cellular networks. In Sansò B., Soriano P. (eds.), *Telecommunications Network Planning*, chap. 13, pp. 239–255. Kluwer Academic Publishers.
12. Koster A.M.C.A. (1999). *Frequency Assignment - Models and Algorithms*. Ph.D. thesis, Universiteit Maastricht, The Netherlands.
13. Mathar R., Schmeink M. (2000). Optimal base station positioning and channel assignment for 3G mobile networks by integer programming. Tech. rep., RWTH Aachen, Germany.
14. MOMENTUM (2001). Models and simulations for network planning and control of UMTS. URL `http://momentum.zib.de`. European Information Society Technologies (IST) project, IST-2000-28088.
15. Murphey R.A., Pardalos P.M., Resende M.G.C. (1999). Frequency assignment problems. In Du D.Z., Pardalos P.M. (eds.), *Handbook of Combinatorial Optimization*, Kluwer Academic Publishers.

Optimal Control of Methadone Treatment in Preventing Blood-Borne Disease*

Julia Almeder

Institute for Econometrics, Operations Research, and Systems Theory,
Vienna University of Technology, Argentinierstr. 8/119, A-1040 Vienna, Austria;
email: julia.almeder@rkag.at

Abstract. In this paper an optimal control model describing the influence of methadone maintenance treatment (MMT) on the spread of blood-borne diseases like Hepatitis C Virus (HCV) and Human Immunodeficiency Virus (HIV) among injection drug users is analyzed. The aim of the model is to find the optimal policy to minimize the discounted stream of the overall costs arising from MMT and the social costs caused by new infections from HIV and HCV.

1 Introduction

Human Immunodeficiency Virus (HIV) and Hepatitis C (HCV) are the most common diseases among injection drug users (IDUs). It is estimated that about 25% of the new HIV infections in the U.S. are a result of transmission from needle sharing among drug users [7]. HCV is less lethal than HIV and therefore the social costs (e.g. treatment, medication...) associated with one infected person are far below the costs caused by one IDU infected with HIV. But HCV, which has a prevalence of more than 50% in many populations [2], is far more widespread than HIV. The reason for that is a very high infectivity rate. (A study based on data of hospital workers in the U.S., who were exposed to HCV through accidents with infected needles, observed a infectivity rate between 3 and 9% [1].) It is estimated that the possibility to be infected with HCV is about four times higher than the infectivity rate of HIV. For this reason, treatment and prevention interventions have proven less successful in slowing HCV infection than they have in slowing the spread of HIV [1,2].

Syringe exchange programs (SEP) and methadone maintenance treatment (MMT) are two of the most important instruments which can slow down the disease spread. SEP have no impact on the frequency and duration of the clients' drug use, but they can reduce infectious disease spread through provision of sterile syringes. In contrast to that, MMT, which is more cost-intensive than SEP [3], tries to reduce the injection drug use. This has two

* This research was partly financed by the Austrian Science Foundation (FWF) under Contract No. P14060-OEK ("Dynamics and Control of Illicit Drug Consumption").

results: on the one hand, it lowers the spread of HCV and HIV because it lowers the participants of needle sharing groups, and on the other hand, it can also reduce side effects of illicit drug use like drug-related crime.

Pollack [8,9] provides an explicit epidemiological model to explore the impact of substance abuse treatment on the incidence and prevalence of HIV and HCV. Studying the cost-effectiveness of MMT, Pollack computes the (average and marginal) cost per infection averted. Although his approach is inter-temporal, the number of treatment slots, M, may not vary over time in his model. More precisely, he minimizes the cost per averted infection by comparing various values of M.

The purpose of the present paper is to extend Pollack's analysis to a dynamic cost-effectiveness analysis in which the number of treatment slots M acts as a time-dependent control variable. The performance functional contains (at least) two terms, i.e. the social costs (damage) created by the incidence of new HIV/HCV cases and the costs of MMT. Optimal control theory (see, e.g., [4]) provides a useful tool to investigate such inter-temporal trade-offs (between damages and costs).

2 Model Formulation

We start with presenting the dynamic equations Pollack used in his paper [8]. Let δ, μ, and γ represent the exit rate from the active IDU population, the exit rate from treatment, and the permanent cure rate of treatment, respectively. Then the dynamics of the IDU population ($N(t)$) are given by

$$\dot{N}(t) = \theta - (N(t) - M)\,\delta - M\mu\gamma, \tag{1}$$

with initial condition $N(0) = N_0$. θ is the exogenous inflow of drug users, M is the number of users under methadone treatment, $(N(t) - M)\,\delta$ is the outflow of users not in MMT, and $M\mu\gamma$ is the outflow of users getting MMT. Note that

$$\mu\gamma > \delta, \tag{2}$$

i.e. the desistance rate is higher for users in treatment.

According to Pollack [8], the number of infected IDUs, $I(t)$, evolves over time as follows (time argument t is omitted partly in what follows to increase readability):

$$\dot{I} = \left(1 - \frac{M}{N}\right)(-\delta I) + \kappa\lambda\left(1 - \frac{M}{N}\right)(\Omega N - I)\frac{I}{\Omega N} - \mu\gamma\frac{M}{N}I \tag{3}$$

with initial condition $I(0) = I_0$. Here, κ, δ, and Ω denote the infectivity, the arrival rate into shooting galleries, and the proportion of shooting gallery participants, respectively. (Shooting galleries are places where IDUs meet and share needles.) $\left(1 - \frac{M}{N}\right)(-\delta I)$ is the outflow of infected users getting no MMT. $\left(1 - \frac{M}{N}\right)(\Omega N - I)$ are the candidate users to be infected, while $\frac{I}{\Omega N}$

is the probability of sharing a needle with an infected user. Finally, $\mu\gamma\frac{M}{N}I$ is the outflow of infected users receiving MMT.

In this work we assume that $N = \bar{N}$ is constant while $M = M(t)$ is a time-dependent control variable. We change the state and control variables to proportions $J = \frac{I}{N}$ and $U = \frac{M}{N}$. Hence, the new state equation is

$$\dot{J} = J\left[-\delta - aU + (1-U)(b - fJ)\right], \tag{4}$$

where $a = \mu\gamma - \delta$, $b = \kappa\lambda$, and $f = \kappa\lambda/\Omega$.

The objective function has the form

$$\max_{U} -\bar{N}\int_0^\infty e^{-rt}\left[\bar{N}U^2 + \nu J(1-U)(b - fJ)\right]dt, \tag{5}$$

which represents the discounted stream of treatment costs and social costs of the new infected IDUs, with r and ν denoting the discount rate and the lifetime social cost per new infected user, respectively.

In order to solve this optimal control problem the maximum principle according to [4] is applied to the model equations.

3 Specification of Parameters

It is assumed that IDUs leave the population at random with the constant rate δ per person per time unit. Thus, the drug career length is exponentially distributed with mean duration $\frac{1}{\delta}$ of 4000 days (or about 11 years). This exit rate is also assumed to be independent of the disease status, which seems reasonable because of the long, largely asymptomatic post-infection period for both HIV and HCV.

To simplify the complex pattern of drug-using behavior, we assume that 30% of IDUs participate in needle-sharing ($\Omega = 0.3$). The remaining 70% do not face any disease risk. The shooting gallery participants have a constant arrival rate of $\lambda = 1/7$ days (i.e. once per week).

The infectivity rate κ, which describes the probability of a disease transmission when an uninfected person shares a needle with an infected IDU, is quite different for HIV and HCV. Hospital accidents data (cf. section 1) suggest that HCV is easily spread through syringes and other sources. In one of these studies ([1]), an infectivity rate between 3% and 9% was observed. Infectivity under realistic needle sharing conditions among IDUs is not known. Hence, κ is assumed to be 0.04. HIV has a much smaller infectivity rate of about 1%.

The mean treatment length $\frac{1}{\mu}$ is assumed to be 400 days (about 1.1 years). Furthermore, it is assumed that the exit rate from injection drug use due to treatment γ is 0.75. For the discount rate r we assume a value of 3% per year or $\frac{0.03}{365}$ per day.

To estimate the lifetime social costs of one new infection, ν, we use the lifetime expected treatment costs, which is clearly an underestimate. In the case of HIV, Holtgrave and Pinkerton [6] give a lifetime discounted present value of \$195000. Wong [10] puts the lifetime treatment costs of HCV at \$19900 for conservative treatment.

4 Results

Analyzing the canonical system resulting from the maximum principle with the parameters for HIV leads to the following result: Independent of the initial situation it is always optimal to approach a stable equilibrium with nearly no infected IDUs and about 38% of the IDU population under treatment. For details see figure 1.

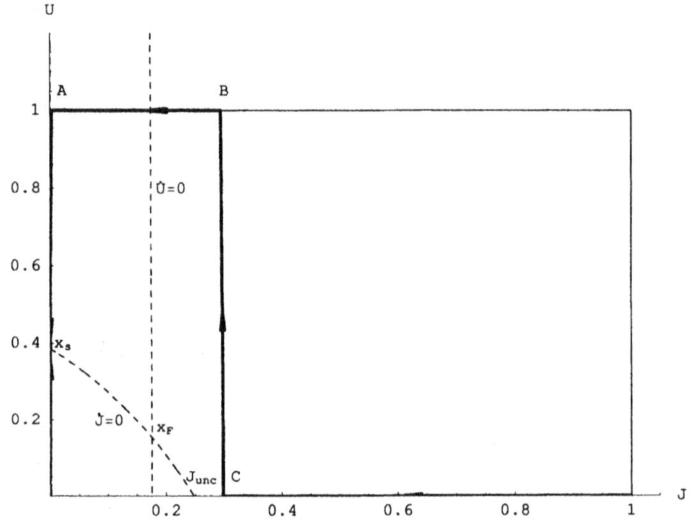

Fig. 1. Optimal control for HIV

In the case of HCV the results are similar, but the level of IDUs under treatment necessary to remain in the stable equilibrium is about 74%. This is due to the fact that HCV has a much higher infectivity rate.

Comparing the optimal dynamic control with the case of constant control which does not vary over time, the dynamic control reduces the total social costs significantly, both for HIV and HCV (see table 1).

The major outcome of this work can be summarized as follows:

- It is never optimal to completely eradicate the infection by applying full MMT all the time.
- In most cases it is optimal to move into a stable equilibrium with only a few infected people (usually below 0.01% of all IDUs).

Strategy	Costs HIV	Costs HCV
optimal control	$4774.35\bar{N}$	$8134.15\bar{N}$
no control	$138975\bar{N}$	$17113.7\bar{N}$
full control	$12163.4\bar{N}$	$12163.4\bar{N}$
optimal constant control	$12090.7\bar{N}$	$10946.3\bar{N}$

Table 1. Total costs for HIV and HCV for different strategies and initial value $J = 0.3$

- For a plausible range of the estimated parameters the optimal solution does not depend on the initial situation.
- It is very unlikely that no control at all is optimal. As the sensitivity analysis shows, the occurrence of a *Dechert-Nishimura-Skiba* (cf. [5] threshold is only possible for parameter values far from the original estimates (especially for the HIV parameters). Furthermore, if no control is applied, the number of infected persons tends to a very high level, at which nearly all shooting gallery participants are infected.
- A dynamic control where the number of treatment slots varies over time is much better than a constant level of the MMT program. A constant control at a high level has comparable effects on the number of infections, but the long run costs are significantly higher.

5 Conclusions and Extensions

This work uses a random-mixing epidemiological model, and it applies optimal control theory to derive the optimal path of services to maximize the net benefits associated with infectious disease control among injection drug users. Optimal control theory provides a useful tool to explicitly investigate the tradeoffs between protection and program costs.

Like any formal analysis, this model includes limitations. We use a random-mixing model most appropriate for populations with prevalent random sharing. This random-mixing model can be extended to overlapping subgroups or more complex compartment models. Segregated subgroups tend to depress disease incidence by reducing the proportion of "discordant" needle-sharing that matches infected and uninfected IDUs. More sociologically complex models provide a more sophisticated framework to examine the context of needle-sharing and infectious disease spread. However, mathematical models indicate that the random-mixing model provides a good approximation to non-random models when there is even a small degree of overlap across sharing networks.

This analysis focuses on an idealized harm reduction intervention. We do not consider many other benefits associated with harm reduction interventions. Methadone maintenance treatment and best-practice syringe exchange programs include diverse components to shorten drug-using careers and to otherwise halt or reduce individuals' injection drug use. Including these benefits in the intervention would increase our estimates of program effectiveness and might also alter our analysis of optimal policy.

There exist several other possible extensions of the model presented in this thesis. One of the main simplifications of the current model is the assumption of a constant size of the injection drug users' population. In particular, the optimal MMT program leading to the stable equilibrium takes several years, and during this time the IDU population can change very much, especially in the presence of a MMT program. A model taking this into account has already been presented in section 2.

We conclude this paper by pointing to another fact, which is completely neglected in the current model formulation. Definitely, it is not true that HIV and HCV occur separately. Hence, a model which combines both diseases would be more realistic. However, such a model would have to deal with at least four state variables, which describe the following four subgroups of the IDU population: infected with HIV and HCV, infected with HIV but not with HCV, infected with HCV but not with HIV, and not infected, respectively.

References

1. M. Alter and L. Moyer. The importance of preventing Hepatitis C Virus infection among injection drug users in the United States. *Journal of Acquired Immune Deficiency Syndromes and Human Retrovirology*, 18(S1):6–10, 1998.
2. CDC. Recommendations for prevention and control of Hepatitis C Virus (HCV) infection and HCV-related chronic disease. Technical Report 47 (RR 19), MMWR, 1998.
3. T. D'Aunno, T. Vaughn, and P. McElroy. An institutional analysis of HIV prevention efforts by the nations outpatient drug abuse treatment units. *Journal of Health and Social Behaviour*, 40(2):175–92, 1999.
4. G. Feichtinger and R.F. Hartl. *Optimale Kontrolle ökonomischer Prozesse.* Walter de Gruyter, Berlin, 1986.
5. G. Feichtinger and G. Tragler. Skiba thresholds in optimal control of illicit drug use. In G. Zaccour, editor, *Optimal Control and Differential Games.* Kluwer Academic Publisher, Boston, 2002.
6. D.R. Holtgrave and S.D. Pinkerton. Updates of cost of illness and quality of life estimates for use in economic evaluations of HIV prevention programs. *Journal of Acquired Immune Deficiency Syndromes and Human Retrovirology*, 16(1):54–62, 1997.
7. IOM. *No Time to Lose: Making the Most of HIV Prevention.* National Academy Press, Washington, DC, 2000.
8. H.A. Pollack. The cost-effectiveness of methadone in preventing blood-borne disease: A comparison of HIV and Hepatitis C. *Working Paper*, 2000.
9. H.A. Pollack. Controlling infectious diseases among injection drug users: Learning (the right) lessons from HIV. *Working Paper*, 2001.
10. J.B. Wong. Cost-effectiveness of treatments for chronic Hepatitis C. *American Journal of Medicine*, 107(6B):74–78, 1999.

Strategien zur Rüstzeitvermeidung in der Elektronikfertigung

Claus-Burkard Böhnlein

Lehrstuhl für BWL und Wirtschaftsinformatik, Universität Würzburg, Neubaustraße 66, D-97070 Würzburg, boehnlein@wiinf.uni-wuerzburg.de

1 Problemumfeld

Mit der Einführung der Surface Mount Technology wurde die Grundlage für eine hochautomatisierte Fertigungsabwicklung in der Elektronikindustrie geschaffen. Durch die Normierung und Miniaturisierung oberflächenmontierter Bauelemente (Surface Mounted Device, SMD) kann mit gängigen SMD-Bestückungssystemen eine Vielzahl unterschiedlicher SMD-Gehäuseformen verarbeitet werden (Pawlischek 1991). Dies führte zu einem Wechsel von der werkstatt- zur linienorientierten Fertigungssteuerung (Schweitzer 1994).

In den vergangenen Jahren konzentrierte sich die Massenproduktion elektronischer Güter in Asien. In den Hochlohnregionen Westeuropas wird dagegen zunehmend in Kleinserien bzw. kundenauftragsorientiert gefertigt (ZVEI 2001). Kleine Losgrößen und häufige Auftragswechsel führen aber dazu, dass SMD-Bestückungen wegen Umrüstung zu 30-40 % der Betriebszeit stillstehen und meist einen Engpass in der Elektronikfertigung darstellen. Mit den heute verfügbaren Planungssystemen kann dieses Problem nicht befriedigend gelöst werden (o.V. 2001).

2 Planung in SMD-Bestückungen

Der Bereich der Fertigungsplanung wird wegen seiner verschiedenartigen Problemstellungen in mehrere Planungsstufen aufgeteilt. In Anlehnung an FELDMANN, ROTH und ROTHHAUPT (Feldmann u. Roth 1991; Feldmann et al. 1992; Rothhaupt 1995) sowie GÜNTHER (Günther et al. 1996) lassen sich fünf Stufen unterscheiden, die nachfolgend beschrieben werden (vgl. Abb. 1):

1. Umrüstungsminimale Auftragseinlastung

Stehen für einen Bestückungsauftrag mit kleiner Losgröße verschiedene Bestückungslinien zur Auswahl, so wird er tendenziell so eingeplant, dass der geringste Umrüstungsaufwand und gleichzeitig eine Bestückung mit gutem Line-Balancing erreicht werden kann. Es handelt sich hier um ein Zuordnungsproblem, unter Einhaltung sowohl von Lieferterminen als auch von technischen und kapazitiven Restriktionen, mit dem Ziel die Leistung des Gesamtsystems zu maximieren bzw. eine möglichst günstige Kostenstruktur zu erreichen.

2. Umrüstungsminimale Auftragsreihenfolge vor einer Bestückungslinie
Durch die Bildung einer umrüstungsminimalen Reihenfolge für die vor der Bestückungslinie wartenden Aufträge werden Stillstandzeiten vermieden und die Systemleistung erhöht.

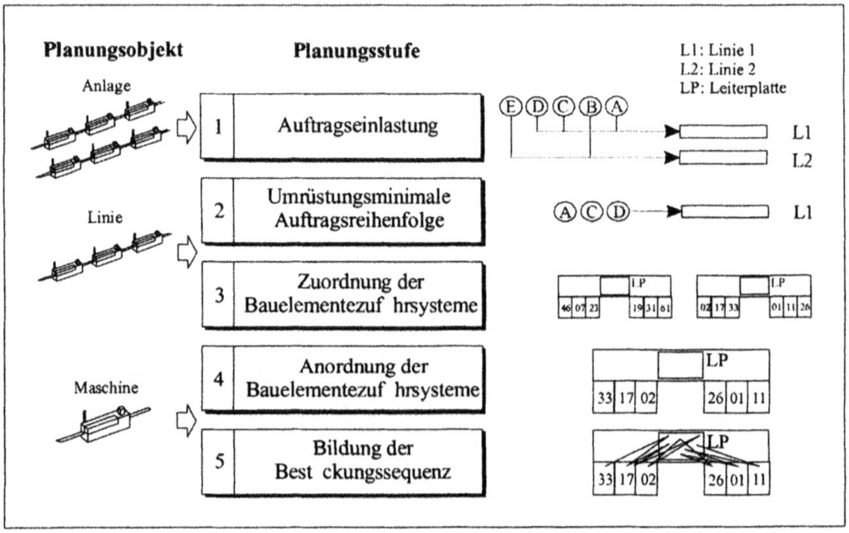

Abb. 1. Planungsstufen in SMD-Bestückungen

3. Optimale Zuordnung der Bauteilförderer zu den Bestückungssystemen
Sind in einer Bestückungslinie mehr als ein Bestückungssystem eingesetzt, dann muss der Auftrag aufgeteilt werden, indem die Bauteiltypen, die auf der Leiterplatte zu setzen sind, einzelnen Bestückungssystemen zugeordnet werden. Dabei sind einerseits technische Restriktionen zu beachten und im Hinblick auf eine umrüstungsminimale Zuordnung sollte andererseits die Ausgangsrüstung möglichst wenig verändert werden.

4. Anordnung der Bauteilförderer an den Bestückungssystemen
Die unter 3. getroffene Zuordnung wird hier für jedes Bestückungssystem weiter verfeinert. Stand bisher die Reduzierung des Rüstaufwands im Vordergrund, so wird jetzt versucht, durch eine günstige Platzierung der Bauteilförderer an den einzelnen Bestückungssystemen der Linie die Bestückungsdauer für den Auftrag zu minimieren. Dies geschieht in dem z. B. häufig benötigte Bauelemente in der Nähe des Bestückungstisches platziert werden, um kurze Verfahrwege des Bestückungskopfes zwischen der Abholposition eines Bauelements am Förderer und der Zielposition auf der Leiterplatte zu erreichen.

5. Optimierung der Bestückungssequenz
Auf der Basis der Planungsschritte 3. und 4. wird für jeden Auftrag eine günstigste Bestückungssequenz gebildet und als Bestückungsprogramm gespeichert. Ziel ist es für jeden Bestückungskopf einer Linie den zeitlich kürzesten Gesamtverfahrweg für den jeweiligen Auftrag zu ermitteln und so die Bestückungsdauer zu minimieren.

3 SIMOS Rüstkonzept für die Kleinserienfertigung

In der betrieblichen Praxis liegen die realen mittleren Bestückungsleistungen deutlich unter den Werten der technischen Systemspezifikation, insbesondere wenn kleine Losgrößen gefertigt werden. Die realen Leistungseinbußen sind im wesentlichen auf zwei Bereiche zurückzuführen, zum einen auf Stillstandszeiten wegen Umrüstung, Störungen etc. und zum anderen auf ein schlechtes Line-Balancing, d. h. eine ungleichmäßige Austaktung der Bestückungseinheiten. Dies hat unterschiedliche Ursachen:

- In der Praxis wird für jeden neuen Bestückungsauftrag die Rüstung der Bestückungslinie angepasst. Die Ziele Reduzierung des Rüstaufwands einerseits und Verbesserung des Line-Balancing und damit der Bestückungsdauer für den Auftrag andererseits sind dann gegenläufig. Als Konsequenz werden z. B. Rüstschritte zur Verbesserung des Line-Balancing bei kleinen Losgrößen nicht durchgeführt, wenn der dafür erforderliche Aufwand durch Einsparungen in der Bestückungsphase nicht kompensiert werden kann.
- Die gegenseitige Abhängigkeit der einzelnen Planungsstufen führt bei einer sequentiellen Bearbeitung der einzelnen Planungsschritte meist nicht zu einem Gesamtoptimum.
- Konzepte wie Fest- bzw. Mehrfachrüstungen einzelner Bauteiltypen sind in der Literatur zwar bekannt, werden wegen ungeeigneter Rüsttaktiken und Werkzeuge in der Variantenfertigung aber nicht konsequent eingesetzt.

Das SIMOS-Rüstkonzept (SImultane MatrixOrientierte Stücklistenauflösung) basiert auf der Grundüberlegung, die Einmaligkeit eines Fertigungsauftrags in der SMD-Bestückung in Frage zu stellen. Wird der Planungshorizont nur hinreichend erweitert, dann ergibt sich ein weitgehend stabiler Auftragsmix mit einem ebenso stabilen Bauteiltypenspektrum. Dies ist die Ausgangssituation für die Ermittlung dauerhafter Rüstungen, sog. Festrüstungen. In Praxistests haben sich Planungshorizonte und damit Lebensdauern der berechneten Festrüstungen von bis zu mehreren Monaten bewährt.

In SMD-Bestückungen mit Kleinserienfertigung gilt erfahrungsgemäß, dass ca. 80% aller gesetzten Bauelemente (nach Stückzahl) in weniger als 20% aller Leiterplattentypen verbaut werden. Zudem haben Analysen ergeben, dass einzelne Bauteiltypen in sehr großer Anzahl und in der Mehrheit aller Bestückungsprogramme benötigt werden. Als Faustregel der bisherigen Ergebnisse gilt, dass ca. 20% aller gesetzten Bauelemente einer Planungsperiode auf nur 1% des Bauteiltypenspektrums entfallen. Diese Bauteiltypen werden im SIMOS Planungswerkzeug gezielt ermittelt, mehrfach auf der Bestückungslinie gerüstet und durch Modifikation der Bestückungsprogramme für die Verbesserung des Line-Balancing genutzt.

Mit dem Ziel die Bestückungsleistung bei Kleinserienfertigung zu steigern, wurden im Forschungsprojekt SIMOS verschiedene Rüsttaktiken entwickelt, implementiert und getestet. Bei allen Taktiken wird durch geeignete Maßnahmen versucht die mittlere Rüstzeit pro Auftrag zu verkürzen ohne das Line-Balancing nennenswert zu verschlechtern. Als Konsequenz kann einerseits wegen reduzierter Stillstandszeiten die Systemleistung erhöht und andererseits die eingesparte Rüstzeit zur schnelleren Materialbereitstellung, Behebung von Störungen und für Wartungsarbeiten etc. genutzt werden. Dies führt zu einer Verbesserung der Produktionsabwicklung in der Bestückungslinie (Böhnlein 2001).

Für die Einführung des SIMOS Rüstkonzepts in SMD-Bestückungen wurde ein geeignetes Vorgehensmodell mit sechs Phasen entwickelt (vgl. Abb. 2).

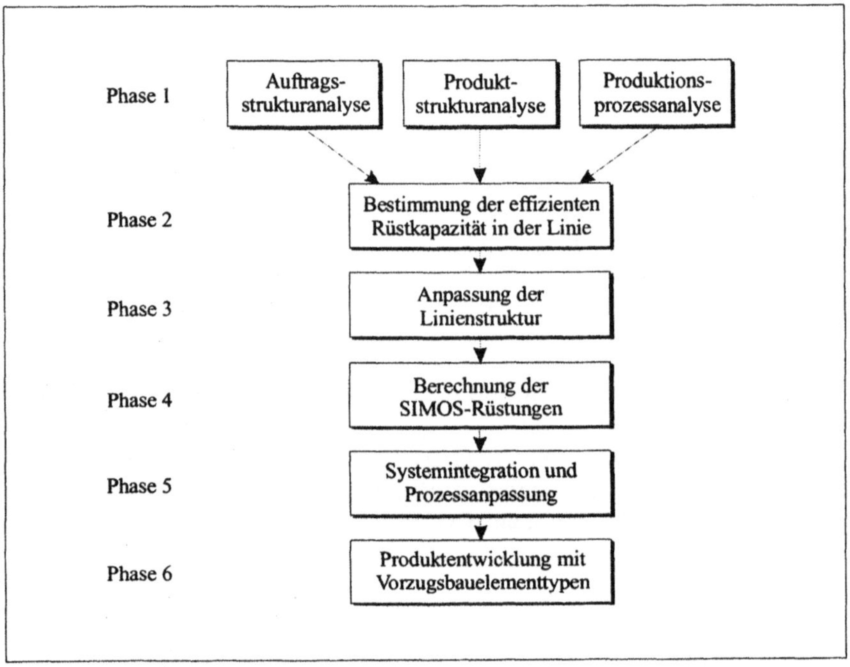

Abb. 2. Vorgehensmodell für die SIMOS-Einführung

In Phase 1 werden eine Auftrags-, Produktstruktur- und Produktionsprozessanalyse durchgeführt, um den Spielraum für Anpassungsmaßnahmen auszuloten. Auf der Basis der Analyseergebnisse werden in den Phasen 2 und 3 die Linienkonfiguration unter Rüstaspekten untersucht und angepasst. In Phase 4 werden nach Maßgabe der Auftrags- und Produktstruktur geeignete Rüstungen berechnet. Zum Abschluss der SIMOS-Einführung werden in der Phase 5 die SIMOS-Werkzeuge in die bestehende Systemwelt eingebunden. Die Phase 6 geht über eine reine Systemeinführung hinaus, denn durch die Versorgung der Abteilungen Auftragsvorbereitung, Entwicklung, Materialwirtschaft und Vertrieb mit SIMOS-Informationen werden wichtige Erkenntnisse aus der Fertigung in diesen Abteilungen nutzbringend für das Unternehmen eingesetzt.

4 Ergebnisse aus Praxiseinsätzen

Das SIMOS-Rüstkonzept wird seit 1998 in einer industriellen Fertigung produktiv zur Ermittlung von Festrüstungen eingesetzt und wurde im Zeitraum 07/1998 bis 03/2000 wissenschaftlich im Produktivbetrieb begleitet. Nach einer Anpassung des Linienlayouts (Phase 3: Erweiterung von 2 auf 3 Bestückungsautomaten pro Linie) konnten insbesondere in der Unterseitenbestückung deutliche Verbesserungen erzielt werden. Die Lebensdauer der bislang eingesetzten SIMOS Festrüstungen liegt zwischen 3 und 6 Monaten, wobei neue Leiterplattentypen in bestehende Festrüstungen aufgenommen werden können ohne bereits bestehende Rüstzuordnungen zu verändern. Von den pro Halbjahr produzierten Bestückungsprogrammvarianten können 50% in der Unterseitenbestückung in einer Festrüstung berücksichtigt werden und decken 90% aller auf dieser Linie bestückten Bauelemente ab (Böhnlein 1999).

In der Unterseitenbestückung konnte von Juli 1998 bis Juli 1999 die durchschnittliche, umrüstungsbedingte Stillstandszeit bei einem Auftragswechsel um 60% und in der Oberseitenbestückung im gleichen Zeitraum um mehr als 40% gesenkt werden. Dadurch wurden im 3-Schicht-Betrieb Rüstzeiten im Umfang von mehreren Stunden pro Tag und Bestückungslinie eingespart.

Im Beobachtungszeitraum hat sich durch die Einführung einer neuen Bauelementreihe das monatlich gefertigte Produktspektrum in der Unterseitenbestückung um 100 % und in der Oberseitenbestückung um 50% erhöht. Dadurch waren mehr Umrüstungen erforderlich, die die genannten Einsparungen weitgehend kompensierten und nur einen geringen Zuwachs der nominellen Bestückungsleistung ermöglichten. Insgesamt konnte aber die geplante Beschaffung einer weiteren Bestückungslinie um zwei Jahre verschoben werden (Böhnlein 2002).

Das Unternehmen setzt das SIMOS-Rüstkonzept auch zwei Jahre nach Abschluss des Kooperationsprojekts immer noch täglich produktiv ein.

Literatur

Böhnlein C (1999) Potential und Einsatz von SIMOS-Festrüstungen in der SMD-Bestückung. In: Schlecht M (Hrsg.) SMT Surface Mount Technology, EMP, Böblingen, S 22-28

Böhnlein C (2001) Verfahren zur Aufrüstung der Bestückungsautomaten einer Bestückungslinie für ein Mix unterschiedlicher Leiterplattentypen. Deutsches Patent und Markenamt, Patent DE198 34 620 vom 31.05.2001, München

Böhnlein C (2002) Rüstzeitvermeidung in der Elektronikfertigung. DUV, Wiesbaden

Feldmann K, Roth N (1991) Optimization of Set-up Strategies for Operating Automated SMT Assembly Lines. Annals of the CIRP, Vol. 40/1/1991

Feldmann K et al. (1992) Optimale Rüst- und Umrüststrategien steigern die Produktivität. In: Leiterplattentechnik (Supplement zu F&M, MO, QZ), 05/1992. Carl Hanser Verlag, LP46-LP50

Günther H-O et al. (1996) A Heuristic for Component Switching on SMD Placement Machines. Wirtschaftswissenschaftliche Dokumentation der TU Berlin, Fachbereich 14. Berlin

o. V. (2001) Effizientes SMD-Bestücken. Informationsabfrage vom 14.9.2001. In: http://www.mimot.de/produkte/wirtschaflich_konzept/wirtschaftlich/wirtschaftlich.htm

Pawlischek H (1991) SMT-Bestückungstechniken. In: Herrmann G (Hrsg.) Handbuch der Leiterplattentechnik, Neue Verfahren, Neue Technologien. Leuze, Saulgau, S 191-228

Rothhaupt A (1995) Modulares Planungssystem zur Optimierung der Elektronikfertigung. Carl Hanser Verlag, München

Schweitzer M (1994) Industriebetriebslehre – Das Wirtschaften in Industrieunternehmungen. 2. Aufl., Vahlen, München

ZVEI (2001) Elektroindustrie meldet zweistelliges Wachstum für 2000. Zentralverband Elektrotechnik- und Elektronikindustrie e. V. In: http: //www.zvei.de/news/Presseinformationen/2000-12/Pr115-2000.htm

Optimale Belegung von Stranggießanlagen mittels 2-dimensionaler Bin-Packing-Modelle

Thomas Spengler / Oliver Seefried[1]

Technische Universität Braunschweig, Institut für Wirtschaftswissenschaften, Abteilung BWL, insb. Produktionswirtschaft,
Katharinenstr. 3, 38106 Braunschweig,
Tel.: 0531/391-2201, Fax: 0531/391-2203,
e-mail: t.spengler@tu-bs.de, o.seefried@tu-bs.de

Abstract. In integrierten Hüttenwerken wird in der Produktionsstufe Stranggießen Rohstahl, der in der Konvertermetallurgie in deterministischen Pfannenmengen produziert wurde, zu kontinuierlichen Strängen vergossen. Im vorliegenden Beitrag wird das Problem der Belegung einer Stranggießanlage als zweidimensionales Bin-Packing-Problem mit variabler Stranglänge und variabler Strangbreite modelliert. Als zu minimierende Zielfunktion wird hierbei auf die Anlagenleistung/Gießzeit (und damit aufgrund der konstanten Gießgeschwindigkeit die Stranglänge) zurückgegriffen. Es wird ein Lösungsverfahren vorgestellt, das mit kommerzieller OR-Software auf einem Standard-PC implementiert in der Lage ist, für praxisrelevante Problemgrößen gute Lösungen im Minutenbereich zu generieren.

1 Einleitung und Problembeschreibung

Ein maßgeblicher Produktionsschritt bei der Herstellung von Rohstahl im Rahmen eines integrierten Hüttenwerksprozesses stellt das Stranggießen dar, bei dem flüssiger Rohstahl, der in der Konvertermetallurgie in einem Frischprozess auf die benötigten Qualitätseigenschaften eingestellt wurde, in den für die Weiterverarbeitung auf unterschiedlichen Walzwerkstufen notwendigen festen Aggregatszustand überführt wird. Ausführliche Prozessbeschreibungen des Stranggießens finden sich u.a. bei Tanner (1997) und V.d.E.H. (1999).

Nachdem der Gießvorgang abgeschlossen ist, wird der erstarrte Stahlstrang zunächst durch einen Guillotineschnitt quer in Sektionen der vollen Strangbreite unterteilt. In der Regel erfolgt in einem zweiten Schneidevorgang eine Teilung längs zur Flussrichtung, um die zuvor abgetrennten Abschnitte in Brammen unterschiedlicher Breite zu unterteilen. Die Produktionsleistung einer Stranggießanlage kann daran gemessen werden, welche Zeit benötigt wird, ein vorgegebenes Sorti-

[1] Die Autoren danken den beteiligten Mitarbeitern der Salzgitter Flachstahl GmbH für ihre Unterstützung bei der Durchführung der zugrunde liegenden Projektarbeiten.

ment an Fertigungsaufträgen abzugießen.[2] Da die Gießgeschwindigkeit als einer der darauf einwirkenden Einflussfaktoren in der Regel von den Analysespezifikationen der herzustellenden Güte beschränkt wird, kann eine Minimierung der benötigten Stranglänge als synonym zur Maximierung der Anlagenleistung angesehen werden. Hierbei wird die abgegossene Stranglänge aufgrund der deterministischen Schmelzengrößen in der Konvertermetallurgie und der über den gesamten Gießvorgang konstanten Stranghöhe durch die beiden Aktionsparameter Strangbreite und Anzahl der benötigten Schmelzen eindeutig beschrieben.

Während die maximale Gießbreite (bmax) der Stranggießanlage die zu bestimmende Strangbreite technisch nach oben begrenzt, ergibt sich eine untere Grenze in Abhängigkeit davon, mit welchen Schnittmustern der Strang in die einzelnen Brammen zerlegt wird. Je nach Verteilung der Brammen auf die Sektionen der Sequenz ergeben sich unterschiedliche Sektionsbreiten, aus deren Maximalwert sich die auftragsgebundene Breite (ba) der Sequenz errechnen lässt.

Dabei treten bei heterogenen Brammenabmessungen im Regelfall zwei Arten von Verschnitt auf:

- Zwischenverschnitt (ZV), der innerhalb der Packung der Brammen anfällt,
- „Schmelzenverschnitt“, der darauf zurückzuführen ist, dass die diskreten Schmelzengrößen in der Konvertermetallurgie nicht mit der Summe der Brammengewichte übereinstimmen, und der entweder in Form von Längenverschnitt (LV) oder Breitenverschnitt (BV) auftritt.

Zur Minimierung der abgegossenen Stranglänge und somit der benötigten Gießzeit bietet es sich an, die einzelnen Sequenzen möglichst breit abzugießen und einen etwaigen Längenverschnitt in einen Breitenverschnitt zu überführen. Mit dieser Strategie ist jedoch die Gefahr verbunden, zusätzliche Schmelzen zur Bereitstellung des Brammenkollektivs zu benötigen. Dieser Zielkonflikt und die Interdependenzen der Entscheidungsgrößen Stranglänge, Strangbreite und Anzahl n der vergossenen Schmelzen werden in Abbildung 1 veranschaulicht:

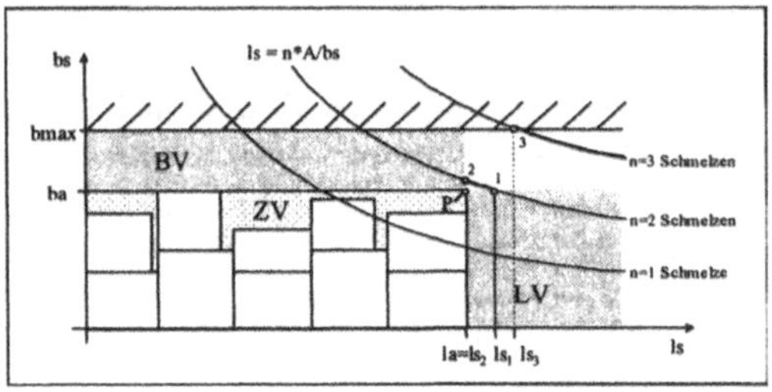

Abb. 1. Interdependenzen zwischen Brammenpositionierung und Sequenzparametern

Die Darstellung verdeutlicht, dass ein Abgießen mit maximaler Breite nicht automatisch zur Reduzierung der Sequenzlänge führen muss. Ausgehend von einer Brammenpositionierung, welche die auftragsgebundene Stranglänge la minimiert, beschränkt sich die Wahl der Sequenzparameter auf die weiße Fläche rechts ober-

[2] Stillstandszeiten durch Rüstvorgänge, die entweder durch Breiten- bzw. Qualitätsdifferenzen nacheinander abgegossener Sequenzen bedingt sein können (Spengler/Seefried/Köck (2001)), haben ebenfalls Einfluss auf die Anlagenleistung. Sie werden jedoch in den folgenden Ausführungen nicht weiter berücksichtigt.

halb des Punktes P im Diagramm. Es wird deutlich, dass für n = 1 Schmelze keine Breiten-Längen-Kombination existiert, mit der das geforderte Brammenkollektiv abgegossen werde könnte. Für n = 2 und n = 3 Schmelzen hingegen ergeben sich zulässige Kombinationsmöglichkeiten, von denen beispielhaft drei Randpunkte hervorgehoben sind:
(1) Wird mit der minimal möglichen, auftragsgebundenen Breite (ba) abgegossen, ergibt sich bei zwei Schmelzen ein Längenüberschuss.
(2) Eine geringfügige Erhöhung der Strangbreite führt zu einer Verringerung der Stranglänge auf ls_2, wobei der Längenüberschuss eliminiert wird und in einen Breitenüberschuss übergeht.
(3) Die Verbreiterung der Sequenz bis zu ihrer oberen Grenze führt zu einer Erhöhung der Gießzeit, da die Funktion für n = 2 Schmelzen unzulässig wird.

Wird beim betrachteten Planungsproblem auf eine differenzierte ökonomische Bewertung der verschiedenen Verschnittarten verzichtet und die Stranglänge als zu minimierende Ersatzzielfunktion betrachtet, kann es in die Typologie von Dyckhoff/Finke (1992) als zweidimensionales Bin-Packing-Problem eingeordnet werden, bei dem die zusätzlich zu berücksichtigenden technisch bedingten Restriktionen deterministischer Schmelzgrößen zu „Containern" mit variabler Seitenlänge und –breite, jedoch konstanter Grundfläche führen. Umfassende Überblicke über Forschungsarbeiten im Bereich der Verschnittminimierung finden sich u.a. bei Dyckhoff / Finke (1992) und Dyckhoff / Scheithauer / Terno (1995).

Zur Lösung soll im Folgenden ein Optimierungsmodell formuliert werden, das in der Lage ist, die folgenden Fragestellungen zu beantworten:

- An welcher Position liegen die Brammen innerhalb der Sequenz?
- Wie viele Schmelzen werden für das Abgießen einer Sequenz benötigt?
- Wie breit ist der Strang auf der Anlage einzustellen?

2 Modellformulierung und Lösungsansatz

Bei der Formulierung des Entscheidungsmodells werden folgende **Annahmen** zugrunde gelegt:

- Alle in der Planungsperiode herzustellenden Brammen einer Güte werden in nur einer Sequenz gefertigt.
- Die Strangbreite, mit welcher der Stahl abgegossen wird, soll innerhalb dieser Sequenz nicht variiert werden.
- Das Gewicht aller Schmelzen einer Güte ist identisch und kann aufgrund der konstanten Stranghöhe und Rohstahldichte in eine Fläche überführt werden.
- Anfallende Verschnittmengen werden nicht differenziert bewertet.
- Bei der Anordnung der Brammen können höchstens zwei Brammen nebeneinander in einer Sektion liegen[3].

[3] Diese einschränkende Annahme ist darauf zurückzuführen, dass im Referenzunternehmen Salzgitter Flachstahl GmbH aufgrund der Geometrien der nachfolgenden Walzwerkstufen Brammen nur in Breiten zwischen 1000 und 2600mm weiterverarbeitet werden (Preussag Stahl (Hrsg.) (1993)). Bei einer maximalen Gießbreite der untersuchten

Indizes:
i, p = 1, ..., m   Brammen

Variablen:

$$x_{ip} = \begin{cases} 1 & \text{falls Bramme i neben einer nicht längeren Bramme p angeordnet wird} \\ 0 & \text{sonst.} \end{cases}$$

Daten:
l: Längen der Brammen [mm]
b: Breiten der Brammen [mm]
bmax: maximale Gießbreite der Stranggießanlage [mm]

$$\text{Min} \quad la = \sum_{i=1}^{m} \sum_{p=1}^{m} x_{ip} \cdot l_i \qquad (1)$$

$$\text{u.d.N.} \quad \sum_{\substack{p=1 \\ p \neq i}}^{m} x_{ip} + \sum_{\substack{p=1 \\ p \neq i}}^{m} x_{pi} \geq 1 \qquad \text{für } i = 1, ..., m \qquad (2)$$

$$x_{ip} \cdot l_i \geq x_{ip} \cdot l_p \qquad \text{für } i = 1, ..., m;\ p = 1, ..., m \qquad (3)$$

$$x_{ip} \cdot (b_i + b_p) \leq b\max \qquad \text{für } i = 1, ..., m;\ p = 1, ..., m \qquad (4)$$

$$x_{ip} \in \{0, 1\} \qquad \text{für } i = 1, ..., m;\ p = 1, ..., m \qquad (5)$$

Damit ergibt sich für jede Güte/Sequenz in Abhängigkeit der darin enthaltenen Brammenzahl m ein lineares binäres Optimierungsproblem mit m^2 Entscheidungsvariablen und $2m^2+m$ (linearen) Nebenbedingungen, das mit dem OR-Softwarepaket LINGO 6.0 auf einem PC (Pentium III - Prozessor mit einer Taktfrequenz von 600 MHz) implementiert bis zu Größenordnungen von m = 80 im Sekunden- bzw. Minutenbereich lösbar ist.

Die Frage, ob mit der Minimierung der auftragsgebundenen Länge la auch die Lösung des eigentlichen Planungsproblems, eine Minimierung der Sequenzlänge ls, einhergeht, soll anhand von Abbildung 2 veranschaulicht werden.

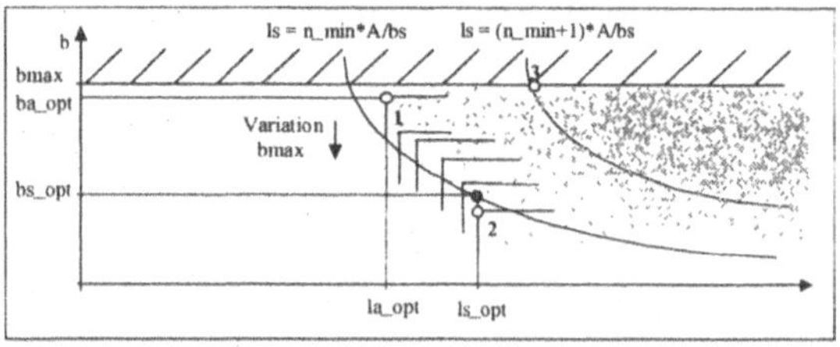

Abb. 2. Optimalitätsanalyse eines heuristischen Lösungsansatzes

Ausgehend von der technisch maximalen Gießbreite wird mittels des Optimierungsmodells (1) – (5) eine Brammenpositionierung mit minimaler auftragsgebundener Länge ermittelt (Punkt 1). Wird mit dieser Brammenanordnung die mi-

Anlage von 2650mm ist der Fall, mehr als zwei Brammen nebeneinander anzuordnen, somit ausgeschlossen.

nimale Schmelzenanzahl[4] n_min erreicht, ist eine optimale Lösung des Planungsproblems gefunden. Andernfalls kann durch eine Reduktion der maximalen Strangbreite der zulässige Lösungsraum rechts unterhalb des zuletzt gefundenen Ausgangspunkts sukzessive eingeschränkt werden. Gelingt es im Verlauf dieses iterativen Vorgehens, eine Brammenanordnung unterhalb der Isoquante für n_min Schmelzen zu finden (Punkt 2), wird die optimale Lösung durch vertikale Transformation auf die entsprechende Isoquante ermittelt. Ist dies nicht möglich, ergibt sich die optimale Lösung im Punkt 3.

Aufbauend auf diesen Überlegungen wird im Zusammenspiel einer Excel/VBA-Anwendung mit LINGO ein heuristischer Lösungsansatz implementiert. Falls das Verfahren bei der Minimierung der auftragsgebundenen Sequenzlänge aufgrund einer Rechenzeitüberschreitung[5] bzw. Speicherplatzmangels abbricht, ohne eine Lösung generiert zu haben, wird auf die Verwendung einer einfachen Prioritätsregel zurückgegriffen: Wähle die längste noch einzuplanende Bramme und weise ihr die längste Bramme der Liste als Partnerbramme zu, deren Breite die verbleibende Restbreite der Sektion nicht überschreitet.

3 Fallstudie

Zur Validierung des in Abschnitt 2 entwickelten Optimierungsmodells wird es auf einen Testdatensatz der Salzgitter Flachstahl GmbH angewendet, in dem sich 3243 Brammen (59 Güten) mit Sollbreiten zwischen 1000 und 2580mm und Längen von 9 bis 12m befinden[6]. Die Anzahl der pro Stahlgüte im Datensatz enthaltenen Brammen variiert dabei zwischen 8 und 714. Insgesamt sind im Testdatensatz ca. 1200 Güten-Abmessungs-Kombinationen zu finden, was für eine starke Heterogenität des Auftragsspektrums spricht.

Bei den Optimierungsrechnungen gelingt es, für 51 der 59 Sequenzen optimale Lösungen zu generieren, wobei bei 11 Sequenzen im Verlauf des Verfahrens eine Reduktion um eine Schmelze erzielt werden kann. Die gesamte Rechenzeit beläuft sich in etwa auf 8 Stunden, wobei die untersuchten 3243 Brammen einem zweiwöchigen Produktionsprogramm entsprechen, so dass das Verfahren im realen Einsatz als Zeit unkritisch angesehen werden kann. Insgesamt ergibt sich über alle Sequenzen ein Verschnittanfall von 15,1%, wovon 5,3% auf Zwischenverschnitt innerhalb der Packung der Brammen entfällt. Da der größere Teil als Längen- bzw. Breitenverschnitt anfällt, liegt die Vermutung nahe, bessere Ergebnisse erzielen zu können, wenn die Annahme der vollständigen Einbeziehung aller Brammen in eine einzige Sequenz aufgehoben wird.

---

4 Das theoretische Minimum an Schmelzen berechnet sich durch Aufrunden des Quotienten aus summierten Brammengewichten und mittlerer Schmelzmenge.

5 Diese wird bei den Testläufen auf 180 Sekunden festgelegt.

6 Um jeweils zwei Brammen einer Sektion zuweisen zu können, werden dem Testdatensatz für alle Güten Dummy-Brammen mit einer Länge und Breite von 0 mm hinzugefügt, so dass insgesamt 3504 Datensätze betrachtet werden.

4 Ausblick und weitere Vorgehensweise

Im vorliegenden Beitrag wird ein heuristisches Lösungsverfahren entwickelt, mit dessen Hilfe gute Produktionsprogramme für ein bestehendes Referenzaggregat geplant werden können.

Zukünftig bleibt zu überprüfen, welche Auswirkungen die zugrunde liegenden (teilweise recht restriktiven) Annahmen auf die Lösungsqualität haben. Neben der bereits erwähnten Möglichkeit, einzelne Brammen im Planungszeitraum unberücksichtigt zu lassen, sollte insbesondere eine Verallgemeinerung auf den Fall, dass aufgrund unterschiedlicher Anlagenkonfigurationen mehr als zwei Brammen parallel abgegossen werden können, erfolgen.

Darüber hinaus gilt es zu untersuchen, welche Rückkopplungen sich über die Integration von Stillstandszeiten aufgrund breitenbedingter Rüstvorgänge beim Sequenzübergang auf die bisher erzielten Ergebnisse einstellen.

Literatur

Dyckhoff, Finke (1992): „Cutting and Packing in Production and Distribution", Heidelberg.

Dyckhoff, Scheithauer, Terno (1997): "Cutting and Packing", in Dell'Amico, Maffioli, Martello, Annotated Bibliographies In Combinatorial Optimization, S. 393 – 413, Chichester.

PREUSSAG STAHL AG (Hrsg.) (1993): „Werkstoff Stahl – Herstellung und Verarbeitung", Salzgitter.

Spengler, Seefried, Köck (2001): „Planungsmodell zur kostenminimalen Brammenversorgung eines integrierten Hüttenwerks", in: Operations Research Proceedings 2000, S. 345 – 350.

Tanner (1997): „Revolution in der Stahlindustrie: Strangguss", Zürich.

VEREIN DEUTSCHER EISENHÜTTENLEUTE (Hrsg.) (1999): „Stahlfibel", Düsseldorf.

Short-term Capacity Planning in Manufacturing Companies with a Decentralized Organization

Peter Letmathe, Institut für Umwelt- und Technologiemanagement,
Universität Bayreuth,
D-95440 Bayreuth

Abstract: Bottleneck stations have a production rate lower than the demand rate for parts they are producing. Therefore they limit the production system's output rate. Short-term capacity planning's purpose is to reduce the bottlenecks' negative impact on the output rate. This paper discusses several opportunities to adjust capacity to the system's requirements in a decentralized production organization. Since these opportunities are related to additional costs, it is crucial to find a cost-efficient solution of this problem. For this purpose, a linear model is proposed which can be used to determine the optimal mix of short-term capacity adjustment measures. The model can be recalculated easily to take the actual output and capacity requirements into account if any deviations from the original plan occur.

1. Starting point

This paper considers a decentralized job shop production divided in several organizational units (production stations). Each station has a given workload of different tasks. Every task has to be performed within a defined time window which comprises the task's processing time including the average setup time per unit, and a time buffer. The size of the buffers determine the station's scope of changing the tasks order, of assigning them to different machines and scheduling them at different times. Since the master production schedule (MPS) might lead to a workload above the stations' standard capacity, the production stations are sometimes confronted with bottleneck situations leading to the need of exceeding standard capacity. For this purpose, the production stations can use several opportunities of short-term capacity adjustment combined with the linear planning model described in the following sections (see Letmathe, 2002).

2. Opportunities of short-term capacity adjustment

In case of bottleneck situations, production stations have several opportunities to adjust their capacity to actual demand (Gutenberg, 1983):

- The station may perform tasks in advance which means that parts are not produced in the same period they are needed.
- The station may work overtime which means that actual working hours are higher than standard working hours per week.
- The production rate of one ore more machines may be increased to perform more tasks than production with cost-optimal production rate.
- The number of machine setups may be reduced. The saved time is capacity which can be used to produce more parts.

- The station may solve the bottleneck problem in concert with other production stations. The mutual coordination of production stations may help relax temporary bottleneck situations by extending time buffers of bottleneck stations.
- Some of the tasks may be performed by external suppliers which leads to a reduced demand for a bottleneck station's capacity and therefore loosens the bottleneck's confining impact.

3. Linear model

All these measures to adjust capacity are related to additional costs which should be minimized by the production stations. Since this is a problem of high complexity, the stations need planning methods to find the cost-optimal solution. In many cases, an optimal solution indicates to use several measures of capacity adjustment simultaneously.

The goal function calculates direct capacity adjustment costs by multiplying direct adjustment cost per capacity unit of each measure with the total additional capacity gained through this measure. The sum of capacity adjustment costs of all measures leads to the total capacity adjustment costs during the planning horizon:

$$K^{Adj} = \sum_{n=1}^{N} \sum_{\tau=1}^{T} \left(c_n^A \cdot R_{n\tau}^A + \sum_{m=1}^{M_n} c_n^{I_m} \cdot R_{n\tau}^{I_m} + c_n^S \cdot R_{n\tau}^S \right) + \sum_{\tau=1}^{T} c^O \cdot R_\tau^O \quad \Rightarrow \min!$$

with:

K^{Adj}	Total capacity adjustment costs
c_n^A	Cost per additional capacity unit gained through performing task n in advance by one period
$R_{n\tau}^A$	Additional capacity through production in advance weighted with $t-\tau$ periods if task n is produced in period τ
$c_n^{I_m}$	Cost per additional capacity unit if task n is produced with production rate I_m [1]
$R_{n\tau}^{I_m}$	Additional capacity through performing task n with production rate I_m in period τ
c_n^S	Cost per additional capacity unit in case of external performance of task n
$R_{n\tau}^S$	Virtual capacity gain through external performance of task n in period τ
c^O	Costs per time unit above standard working time
R_τ^O	Additional capacity through overtime in period τ
τ	Period of production
t	Period of demand

[1] I_m is a production rate index and does not reflect the production rate itself.

T Planning horizon

This goal function includes only capacity adjustment measures which cause direct costs. Other measures like the reduction of the number of machine setups, shifting of time windows and informal agreements with other stations are considered in the model's constraints. E.g. the reduction of machine setups increases the capacity available to perform tasks in the affected periods. Coordination with other stations is reflected through shifted time windows for one or several tasks. This means that team members of production stations will make decisions in advance which are exogenous to the model. The result is a linear model without integer variables which can be solved easily. To find the optimal solution several constraints have to be taken into account:

1. Satisfaction of current demand

All tasks to be performed can be derived from the plant's master production schedule. To ensure that tasks assigned to the station are performed, the demand x_{nt} for task n in t has to be equal or smaller than the available amount of task n in t. The available amount of task n is calculated through the externally performed quantity y_{nt}^S of task n in t plus all quantities $y_{n\tau t}$ produced internally in the periods $\tau = 1,\ldots,t$ to serve demand in period t. This constraint has to be fulfilled in all periods $t = 1,\ldots,T$ and for all tasks $n = 1,\ldots,N$:

$$\sum_{\tau=1}^{t} y_{n\tau t} + y_{nt}^S \geq x_{nt} \qquad n = 1,\ldots,N,\ t = 1,\ldots,T$$

2. Calculation of production volume

The output quantity of task n produced in period τ to meet the demand in period t result from the sum of the performed tasks $y_{n\tau t}^{I_m}$ over all possible production rates I_m $(m = 1,\ldots,M_n)$ with whom task n can be produced:

$$y_{n\tau t} = \sum_{m=1}^{M_n} y_{n\tau t}^{I_m} \qquad n = 1,\ldots,N,\ \tau = 1,\ldots,T,\ t = 1,\ldots,T,\ \tau \leq t$$

The total amount of task n performed in period τ is calculated through the sum of $y_{n\tau t}$ over all periods $t = \tau,\ldots,T$.

$$y_{n\tau} = \sum_{t=\tau}^{T} y_{n\tau t} \qquad n = 1,\ldots,N,\ \tau = 1,\ldots,T$$

3. Capacity constraints

The model has to ensure that the available capacity of every period is sufficient to perform all tasks assigned to the bottleneck station. The available capacity results from the sum of the period's standard capacity R_τ plus the capacity gained through overtime R_τ^O and adjustment of production rates R_τ^I. The required capacity is calculated as the number of machine setups $z_{n\tau}$ in period τ multiplied with the time a_n^R to setup a machine to perform task n, plus the quantity $y_{n\tau}$ of task n in period τ times the capacity coefficient a_n of task n. Note, that the capacity coefficient reflects the standard processing time to perform task n. If the production rate is above the standard production rate the additional capacity is included in R_τ^I and therefore does not need to be taken into account in different capacity coefficients for one task. These considerations lead to the following constraints:

$$R_\tau + R_\tau^O + R_\tau^I \geq \sum_{n=1}^{N} a_n \cdot y_{n\tau} + \sum_{n=1}^{N} z_{n\tau} \cdot a_n^R \qquad \tau = 1,\ldots,T$$

4. Limitation of overtime

Overtime R_τ^O in period τ is restricted to an upper limit $\overline{R}_\tau^O$ which may be based on an agreement with the company's workers' council:

$$R_\tau^O \leq \overline{R}_\tau^O \qquad \tau = 1,\ldots,T$$

5. Capacity adjustment through variation of production rates

The additional capacity $R_{n\tau}^{I_m}$ gained through performing task n at production rate m in period τ reflects the quantity $y_{n\tau}^{I_m}$ times the difference of the capacity required for performing the task in standard processing time a_n minus performing it in reduced processing time $a_n^{I_m}$. This term is measured over all periods from τ to the end of the planning horizon T:

$$R_{n\tau}^{I_m} = \sum_{t=\tau}^{T} (a_n - a_n^{I_m}) \cdot y_{n\tau}^{I_m} \qquad n = 1,\ldots,N,\ \tau = 1,\ldots,T,\ m = 1,\ldots,M_n$$

The total additional capacity R_τ^I in period τ is then calculated as the sum of all $R_{n\tau}^{I_m}$ over all production rates m and all tasks n:

$$R_\tau^I = \sum_{m=1}^{M_n} \sum_{n=1}^{N} R_{n\tau}^{I_m} \qquad \tau = 1,\ldots,T$$

6. Capacity adjustment through production in advance

Capacity adjustment through production in advance does not yield additional capacity. Advanced production means to use capacity which is not required in former periods. Hence, production in advance is not included in the calculation of additional capacity. Production in advance is useful if a production station is confronted with temporary bottleneck situations which can be healed by shifting production to earlier periods. Production in advance causes holding costs of the part which have to be kept in stock between the production period τ and the period t when the part is required. In order to calculate the costs of production in advance in the goal function correctly, the performed tasks $y_{1\tau}$ have to be multiplied by the time difference of production in advance $t-\tau$ and the capacity coefficient a_n of task n. The sum over all periods $t=\tau+1,\ldots,T$ leads to the virtual capacity gain in period τ through performance of task n weighted with $t-\tau$:

$$R_{n\tau}^{A} = \sum_{t=\tau+1}^{T}(t-\tau)\cdot a_n \cdot y_{n\tau t} \qquad n=1,\ldots,N,\ \tau=1,\ldots,T$$

In the goal function, this weighted virtual capacity gain is multiplied by the holding cost c_n^A per capacity unit of task n.

7. Capacity adjustment through external task performance

If a station does not perform all tasks by itself it may be allowed to give orders to external suppliers. This leads to a further relaxation of the station's bottleneck situation. The plant management or contracts with external suppliers may restrict external task performance y_{nt}^S to an upper limit $\overline{y}_{nt}^S$ of task n in t. The required quantity of external parts is calculated through the difference of the demand x_{nt} of task n in t minus the number of parts $y_{n\tau t}$ of task n produced in $\tau=1,\ldots,t$ for period t. This difference has to be equal or lower than the upper limit $\overline{y}_{nt}^S$:

$$y_{nt}^S = x_{nt} - \sum_{\tau=1}^{t} y_{n\tau t} \le \overline{y}_{nt}^S \qquad n=1,\ldots,N,\ t=1,\ldots,T$$

The capacity relaxation through external task performance $R_{n\tau}^S$ of task n in t is measured as the quantity of externally purchased parts y_{nt}^S multiplied by the capacity coefficients a_n in case of internal performance at optimal production rate:

$$R_{n\tau}^S = a_{nl}\cdot y_{nt}^S \qquad n=1,\ldots,N,\tau=1,\ldots,T \quad \text{mit}:\ \tau=t$$

8. Production within the given time windows

All tasks have to be performed within the given time windows TW_n. Therefore, it is necessary to calculate the length of the time windows first. Here it is assumed that the length of the time windows depends on the processing time of each task. This means that the minimum length of a time window $a_n\cdot TB_n$ is calculated as a given percentage TB_n of the processing time a_n of task n. Since this model measures time in discrete units, the earliest start of performing a task is $t-[a_n\cdot TB_n]-1$. The brackets stand for the Gauss brackets. Production planning has to ensure that all preceding tasks are finished before this starting time. The following restrictions guarantee that $y_{n\tau t}$ is zero if τ lies before and is non-negative if τ lies after the earliest starting time:

$$y_{n\tau t} \begin{Bmatrix} = \\ \ge \end{Bmatrix} 0 \quad \begin{matrix} \text{if } \tau < t-TW_n = t-[a_n\cdot TB_n]-1 \\ \text{if } \tau \ge t-TW_n = t-[a_n\cdot TB_n]-1 \end{matrix} \qquad n=1,\ldots,N,\ \tau=1,\ldots,T,\ t=1,\ldots,T$$

9. Non-negative constraints

Finally, the non-negative constraints guarantee that the amounts of the performed tasks and the overtime working hours are non-negative:

$$y_{n\tau t}^{I_m} \geq 0 \qquad n = 1,\ldots,N,\ \tau = 1,\ldots,T,\ t = 1,\ldots,T,\ m = 1,\ldots,M_n$$

$$y_{nt}^{S} \geq 0 \qquad n = 1,\ldots,N,\ t = 1,\ldots,T$$

$$R_{\tau}^{O} \geq 0 \qquad \tau = 1,\ldots,T$$

The non-negativity of all other decision variables ($R_{n\tau}^{A}$, $R_{n\tau}^{I_m}$, R_{τ}^{I}, $R_{n\tau}^{S}$, $y_{n\tau t}$) follows automatically from the restrictions above.

4. Conclusions

The presented model allows bottleneck stations to quickly calculate an optimal capacity adjustment strategy. If it is programmed operator-friendly it offers the opportunity to adjust capacity to short term requirements cost efficiently. It is then suitable to support a decentralized manufacturing organization. Decisions requiring integer variables like the number of machine setups and the order sequencing are made by the station's workforce and are therefore exogenous to the model. Since no integer variables are included, the model can be recalculated easily according to production progress. All in all, the model enhances the company's flexibility to cope with bottleneck situations without compromising the advantages of decentralized organizations.

References

Gutenberg, E.: Grundlagen der Betriebswirtschaftslehre, Bd. 1, Die Produktion, Springer, Berlin / Heidelberg / New York, 24. Aufl. 1983.

Letmathe, P.: Flexible Standardisierung – Ein dezentrales Produktionsmanagement-Konzept für kleine und mittlere Unternehmen, Gabler, Wiesbaden 2002.

Performance Analysis of Make to Stock Supply Chains Using Discrete-Time Queueing Models

Sandeep Jain and N. R. Srinivasa Raghavan

Department of Management Studies, Indian Institute of Science, Bangalore, 560012, INDIA
email: {sandeep, raghavan}@mgmt.iisc.ernet.in

Abstract. Manufacturing supply chains are formed out of complex interconnections among several vendors, manufacturing facilities, warehouses, retailers, logistics providers. In this paper, we concern ourselves with a Supply Chain Network (SCN) consisting of single manufacturing plant, one warehouse and several retail outlets. We present a discrete-time queueing model which can be used for evaluating the performance of a given SCN which processes customer orders at discrete time intervals. It can also be used for determining the optimal inventory level at the warehouse that minimizes total cost of carrying inventory and back order cost associated with serving orders in the backlog queue, subject to service level constraint in terms of expected waiting time for customers. The model is analyzed by using discrete-time queues. We use CONWIP as the inventory control model.

1 Introduction

Supply Chain Management (SCM) focuses on the co-ordination among enterprises which work for procuring, producing, delivering products and services to customers who are at different geographical locations. Performance analysis of SCNs is very complex and difficult. In this paper, we present a new methodology to evaluate the performance of 3–stage SCN. This model can be used to formulate and solve optimization problem in the area of inventory control. We briefly survey the literature on mathematical models for inventory control in supply chains. For an overview of various inventory models, the reader may refer [9]. Kim and Tang [4] highlight the trade-off between manufacturing lead time and response time. Bertsimas and Paschalidis [5] develop a model for devising the optimal production policy to minimize the inventory cost. Buzacott [8] is the first researcher who proposes queueing models for production systems including kanban based system. Bruneel [7] elaborates about the general discrete-time queue (DTQ) for the single server and infinite waiting room case. Supply chains can be viewed as discrete-events dynamic state (DEDS) [6]- order arrivals, goods arrivals and departure represent the discrete events, themselves being dynamic. This paper is organized as follows: In section 2, we present the notation and the model. Some illustrations of performance analysis follow in section 3. We then present the optimization problem and results in section 4.

2 Notations and Model

2.1 Notations

Q : Number of units in one bucket
K : Total number of buckets at warehouse
λ : Demand arrival rate at warehouse
$I(t)$: Number of finished goods inventory at warehouse
$N(t)$: Number of orders in the factory being processed
$B(t)$: Number of back orders in the system at time t
h : Inventory holding cost ($ per unit)
b : Back order cost ($ per unit)
p : Probability that processing of a batch of products will finish in Q slots of time at the manufacturing plant
T : Waiting time for the orders placed by the retailers
$M_0(t)$: Probability that zero external arrivals occur within a time interval of duration t where time interval is placed at random in time
$M_n(t)$: Probability that n arrivals occur within a time interval of duration t where time interval is placed at random in time
$A(z)$: Probability generating function (pgf) of arrival distribution
$S(z)$: Pgf of service distribution

In this paper we analyze a 3–stage supply chain which consists of one manufacturing plant, one warehouse, and n identical retailers (see fig.1). We model the warehouse inventory as the input control mechanism for the manufacturing plant which itself is modeled as a single stage discrete time queueing system. The orders arrive at the warehouse from various retailers as a Poisson process with a constant rate λ. The warehouse keeps finished goods inventory in K buckets, each of which holds exactly Q units. Arriving orders from retailers deplete the on-hand inventory at warehouse, if any. Otherwise, (in a stock-out situation) the arriving orders have to wait to be fulfilled. We refer this situation as "back order". In our model, we assume that back orders can be infinite. The warehouse places orders to manufacturing plant when one bucket (Q units) is depleted. Thus, when the retail orders are Poisson, the inter-arrival time of the orders at the manufacturing plant is Erlang distributed with Q phases and rate λ/Q. We assume that manufacturing plant has infinite waiting line capacity. At the manufacturing plant, orders arrive in batches of Q units. The manufacturing plant processes orders at fixed discrete time slots. The time slots are decided appropriately. For instance, slots could be a couple of hours, one shift, a day or so on. In our model, the processing time of orders includes setup, manufacturing and logistics time. Orders arriving in between slots have to wait for the next slot for getting processed. The orders are processed based on the first-come-first-serve policy. We assume processing time is Negative binomially distributed with parameters p and Q. Our work represents the first attempt in terms of applying discrete time queueing models to supply chain analysis and design.

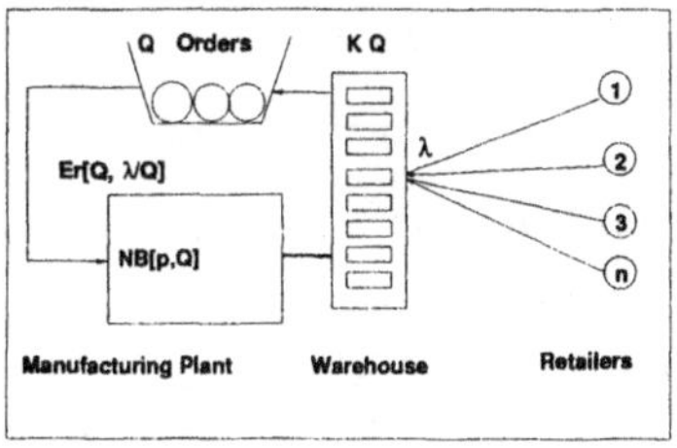

Fig. 1. 3-Stage Supply Chain Network

3 Analysis

At the warehouse there is some cost associated with keeping inventory and there is some backorder cost associated with orders in backlog. We assume that both inventory carrying cost and backorder cost are linear in nature. We would like to minimize the expected total cost at the warehouse.
Mathematically we write

$$Minimize \text{ Total Cost}_{\{KQ\}} = h\ E[I] + b\ E[B] \tag{1}$$

We also want to guarantee the retailers that their expected waiting time will not exceed c time units

$$E[T] \leq c \tag{2}$$

$$K, Q \in \mathbf{Z}^+ \tag{3}$$

We thus have formulated an optimization problem where (1) is the objective function, subject to constraints (2) and (3). In the manufacturing plant processing of orders starts at discrete time intervals; hence we can analyze the system as a discrete time queue. The pgf of number of units at the beginning of a random slot [7] can be given as

$$U(z) = [1 - A'(1)S'(1)]\left[\frac{(z-1)S(A(z))}{z - S(A(z))}\right] \tag{4}$$

For stability $A'(1)S'(1) < 1$
Let $V(z)$ be the pgf of system occupancy at random time points, then $V(z)$ can be given as

$$V(z) = \int_0^1 d\theta\ L(1 - X(\theta) + X(\theta)K(z))\ .\ U(z) \tag{5}$$

$$l(n) = \text{Prob}\,[n \text{ arrival instants in one slot}],\ n \geq 0$$

$$L(z) = \sum_{n=0}^{\infty} l(n) z^n$$

$$K(z) = z^Q$$
$$A(z) = L(K(z))$$

The arrival process when the time between two arrivals is Erlang [1] is

$$M_0(t) = e^{-Qt} \sum_{n=0}^{Q-1} \left[\left(1 - \frac{n}{Q}\right) \frac{(Qt)^n}{n!} \right] \tag{6}$$

$$M_n(t) = e^{-Qt} \sum_{s=o}^{Q-1} \left[\left(1 - \frac{s}{Q}\right) \frac{(Qt)^{nQ-s}}{(nQ-s)!} + \left(1 - \frac{s+1}{Q}\right) \frac{(Qt)^{nQ+s+1})}{(nQ+s+1)!} \right] \tag{7}$$

$$L(z) = M_0(t) + \sum_{n=1}^{\infty} M_n(t) z^n \tag{8}$$

$$X(\theta) = \frac{M_1(s) M_0(t-s)}{M_1(t)} \tag{9}$$

$$S(z) = \frac{(pz)^Q}{[1-(1-p)z]^Q} \tag{10}$$

For computing inventory and back-orders, we have to develop stochastic equations which capture the properties of the system [2].

$$I(t) = \{KQ - N(t)\}^+ \tag{11}$$

$$B(t) = \{N(t) - KQ\}^+ \tag{12}$$

$$E[I] = \sum_{n=0}^{KQ} (KQ-n) P\{N=n\} = \sum_{n=0}^{KQ} (KQ-n) \frac{1}{n!} \frac{d^n V(z)}{dz^n} |_{z=0} \tag{13}$$

$$E[B] = \sum_{n=KQ}^{\infty} (KQ-n) P\{N=n\} = \sum_{n=KQ}^{\infty} (n-KQ) \frac{1}{n!} \frac{d^n V(z)}{dz^n} |_{z=0} \tag{14}$$

For computing the expected waiting time (in discrete time slots), we can use Little's law, $E[T] = E[B]/\lambda$.

4 Implementation and Result

We have implemented our model using Mathematica package. The equations (13) and (14) are not possible to solve analytically in terms of variables like K and Q. So, we solve the expression by substituting numerical values of K, Q and observe the behavior of output parameters. It is not possible to compute exact pgf of the Erlang arrival distribution in closed form. So we truncate summation to 10 terms instead of infinity. We observed the results varying summation limits up to 100. The results were differing merely by 0.003%, so we found our approximation reasonable. For objective function

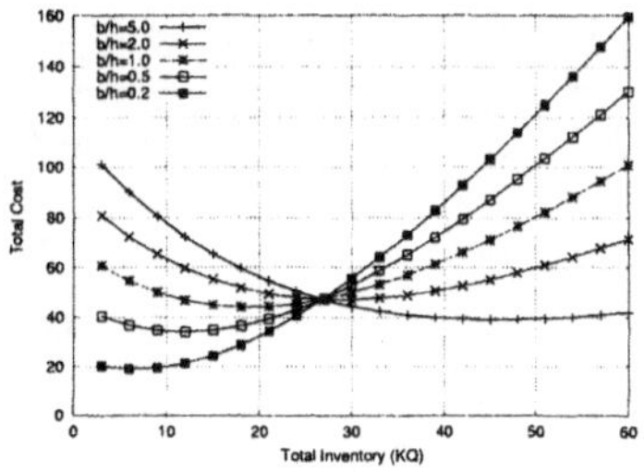

Fig. 2. Total Inventory (KQ) Total Cost when Q=3, λ=0.90, p=0.95

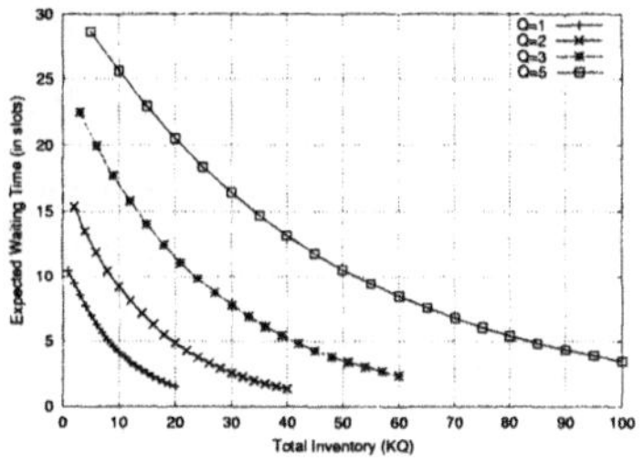

Fig. 3. Total Inventory (KQ) Vs Expected waiting time, λ=0.90, p=0.95

we plot graph (fig.2) for total inventory at warehouse (KQ) verses total cost. We keep Q constant and varying the ratio of back-orders and inventory (b/h) cost. Similarly for constraint we plot (fig.3) for total inventory verses expected waiting time which is in slots. After plotting these curves we try to fit some functions so that we can capture the nature of objective and constraint function in some form of eqns. In our model, for the objective function, we input the values of Q, K, b/h which gives the output value of total cost. Similarly for the constraint function, we input the value of Q which gives the expected waiting time (in slots). We solve the formulation as an Integer programming problem. We used branch and bound method [3]. We can see the results in table 1. The findings are as follows :

1. As we increase Q, for achieving same service level, required inventory at warehouse is higher which increases total inventory carrying cost for same b/h ratio.
2. As b/h ratio decreases, total inventory at warehouse is non-increasing.
3. At higher value of Q, b/h has no say in determining K^* (optimal K) .

5 Conclusion

In this paper, we have developed a model that examines the trade-off between total cost and total inventory at warehouse for the given customer service level. This model can be viewed as a starting point of modeling of SCN

Table 1. Results for λ=0.9, p=0.95, c $\leq$ 5

Q	b/h	K^*	Total Inventory	Total Cost (in $)
3	5.0	15	45	39.0725
3	2.0	14	42	52.6487
3	1.0	14	42	65.9768
3	0.5	14	42	79.3044
3	0.2	14	42	92.6322
5	5.0	17	70	60.072
5	2.0	17	70	94.0169
5	1.0	17	70	127.962
5	0.5	17	70	161.907
5	0.2	17	70	195.852

where service is provided at discrete time intervals. In the present work, we have considered a very basic model of inventory system. We assumed setup cost is zero. The model can be further analyzed assuming setup cost. We worked under the assumption that orders can be back ordered. The model can be extended for the lost sales case also.

References

1. M. M. Philip (1958) Queues, Inventories, Maintenance. John Wiley and Sons Inc.
2. Buzacott, J. A. and Shanthikumar, J. G. (1993) Stochastic Models of Manufacturing Systems. Prentice Hall, New Jersey
3. Rao, S. S. (1998) Engineering Optimization: Theory and Practice(Third Edition). New age International Ltd. Publishers
4. Kim, Illyung and S. Tang, Christopher (1997) Lead time and response time in a pull production control system. European Journal of Operational Research. **101**, 474–485
5. Bertsimas, D. and Ch. Paschalidis, Ioannis (2001) Probabilistic service level guarantees in make-to-stock manufacturing systems. Opers. Res. **49**, 0119–0133
6. Viswanadham, N. and Raghavan, N. R. Srinivasa (2000) Performance analysis and design of supply chains: a Petri net approach. Journal of the Operational Research. **51**, 1158–1169
7. Bruneel, Herwig (1993) Performance of discrete-time queueing systems. Computers Operations Research.**3**, 303–320
8. Buzacott J. A. (1989) Queueing models of kanban and MRP controlled production systems. Engineering Costs and Production Economics. **17**, 3–20
9. Hadley, G. and Whitin, T. M. (1963) Analysis of Inventory Systems.Printice-Hall Inc, Englewood Cliffs, NJ

A New Optimal Demand Forecast Model

Joachim Althaler, Herbert Jodlbauer

FH-Steyr, Wehrgrabengasse 1-5, A-4400 Steyr, Austria.
e-mail:joachim.althaler@fh-steyr.at, herbert.jodlbauer@fh-steyr.at

Abstract

This paper presents a model for determining an optimum demand plan which has been developed whereby both demand figures from the past as well as the results of market research concerning price elasticity are known.

The sales demand demonstrates a high seasonal fluctuation as well as a short-term fluctuation, which is also relevant. The production capacity is restricted.

The main idea of the model is first of all to approximate the past demand figures in order to predict future demand figures through extrapolation. In the second step, a non-linear production optimisation model is applied for calculation of the cumulative demand plan taking into account both the price elasticity and resource restrictions. The final predicted demand data is shown by an extrapolated curve corrected by a constant. The latter is determined by the fact that the integral of the demand curve has to be the cumulative demand plan.

In addition to this result, a beta level of service and consequently the level of the safety stock can be calculated by means of a curvilinear regression analysis.

1 Introduction

By using well-known methods in the field of non-linear regression analysis and non-linear production planning and combining these approaches, a new model for determining a sales forecast with maximum profit, high seasonal fluctuations and limited resources is introduced. The main idea is firstly, to analyse the past sales data structure in order to depict the basic shape of the forecast. In the second step, the total sales for the subsequent year are fixed by optimising the profit whereby resource-restrictions are taken into account. In the final step, the optimal sales curve is calculated by solving an integral equation describing the relationship between the total sales and the extrapolated sales curve.

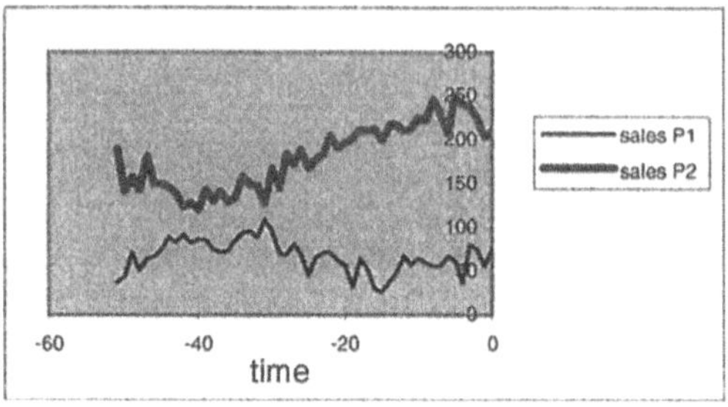

Fig. 1. Past sales data

The past sales data (t_l, x_l) whereby $l = -T+1,...,0$ is known. The values t_l are the sub time periods, for instance the past weeks and x_l the vector of total sales in the time period t_l. T is the number of observed time periods, for instance 52 for one year.

Typically, the sales data has seasonal fluctuations as well as short-term fluctuations as shown in Fig.1 for two products.

As a result of market research, the price-elasticity $a(p)$ describing the relationship between the price p and the possible total sales $a(p)$ for each product is given (see Fig. 2). In addition, lower and upper boundaries u and o for the sales are defined. For instance, the sales force knows that the maximum possible sales are limited by o or for strategic reasons a minimum u should be sold.

The required resources (employees, machines, ...) for producing the products are fixed in the matrix $M = (m_{ij})$ describing how many units of the resource i are necessary to produce one unit of the product j. All the resources, of course, are limited by the maximum available resources b. The vector $c = (c_i)$ describes the costs of one unit of the resource i.

The goal is to determine a forecast $x_{opt}(t)$ of the sales of the next T periods using all the given data above, especially the past sales data, the price elasticity and the resource restriction, whereby the contribution margin is maximized.

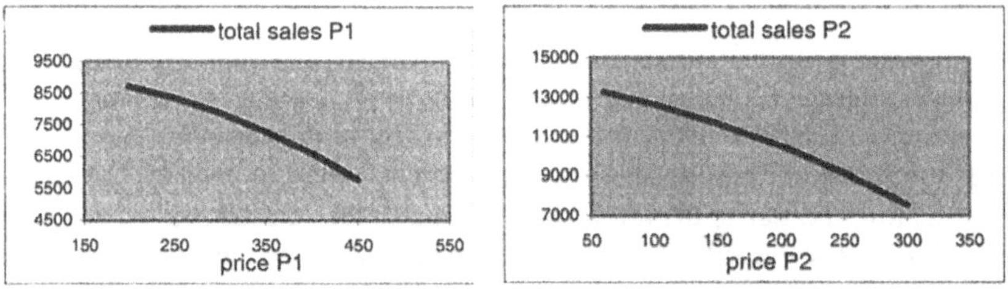

Fig. 2. Price elasticity

2 Mathematical Model

To begin with, a suitable parameterised function for each product $x_j(t)$ has to be introduced in order to approximate the past sales data. The sales data is characterized by seasonal fluctuations and a trend. The functions

$$x_j(t) = d_j + k_j t + A_j \cos\left(\frac{2\pi}{T}(t - \tau_j)\right) \tag{1}$$

can meet these requirements in a more generalized case than the functions proposed in [1]. The parameters d_j, k_j, A_j and τ_j are interpreted as follows:

- d_j constant summand
- k_j descent of the trend
- A_j amplitude of the seasonal fluctuations
- τ_j phase translation.

The shape of the curves are fixed by these parameters. If the quality of the seasonal fluctuations remains then the parameter τ_j will be independent in time. In addition, constant k_j and a constant ratio A_j of the amplitude of the seasonal fluctuations to the total sales are assumed. By varying the summand d_j the total sales can be changed.

The unknown parameters d_j, k_j, A_j and τ_j are calculated by applying a non-linear regression analysis. The non-linearity is caused by the parameter τ_j.

$$\text{minimise } \sigma_{xj} = \sqrt{\frac{1}{T-1} \sum_{i=-T+1}^{0} \left(x_j(t_i) - x_{ij}\right)^2} \tag{2}$$

- $x_j(t_i)$ approximated sales curve (product j)
- x_{ij} real past sales at time t_i (product j)
- σ_{xj} deviation (product j).

The idea in (2) is to minimise the deviation of the approximated sales function to the real past sales data as shown in Fig. 3 for two products.

It is not enough to extrapolate the approximated sales function to guarantee a maximised profit. Instead of this, a non-linear adjusted production planning model

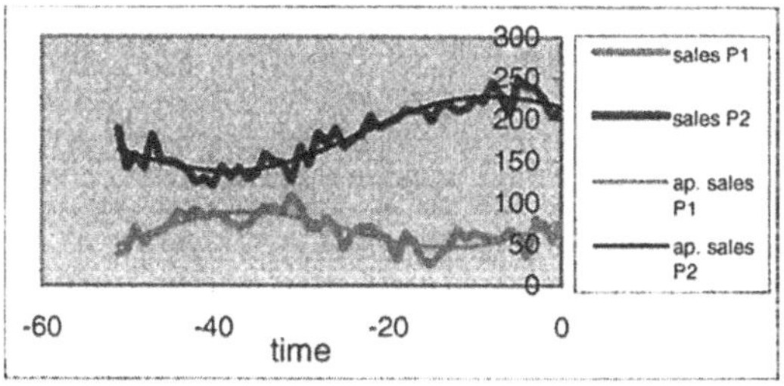

Fig. 3. Approximated past sales data

is applied, whereby the parameters τ_j and k_j are fixed by solving (2) and the two parameters A_j and d_j are not yet optimised. The model to be solved is defined by

$$\left(p^{\mathrm{T}}-c^{\mathrm{T}}M\right)a(p)\rightarrow \text{max. contribution margin} \tag{3}$$

whereby the following conditions must hold:

$$u \le a(p) \le o \quad \text{lower and upper boundaries for sales} \tag{4}$$

$$Ma(p) \le b \quad \text{limited resources} \tag{5}$$

$$\frac{A_j}{\int_{-T}^{0} x_j(t)dt}=\frac{A_j(p)}{\int_{0}^{T} y_j(t,p)dt} \quad \text{ratio amplitude to total sales remains} \tag{6}$$

$$a_j(p)=\int_0^T y_j(t,p)dt \quad \text{total sales} \tag{7}$$

with $$y_j(t,p)=d_j(p)+k_jt+A_j(p)\cos\left(\frac{2\pi}{T}\left(t-\tau_j\right)\right) \tag{8}$$

A_j amplitude of the approximated curve

$A_j(p)$ amplitude of the unknown optimal curve

$d_j(p)$ summand of the unknown optimal curve.

Solving problem (3) the optimal price p, the optimal amplitude A_j and the optimal constant summand d_j is determined by maximising the contribution margin. By using the price elasticity the optimal prize defines the optimal total sales. The amplitude A_j is calculated by ensuring a constant ratio amplitude to total sales. In order to determine the constant summand d_j the relationship between the total sales $a_j(p)$ and the sales function $y_j(t,p)$ has to be taken into account.

Because of having the chance to separate problem (3), the solution can be carried out in two steps. The first step is to solve the optimisation problem (3), s.t. (4) and (5).

This results in the optimal price p_{opt} and the optimal total sales $a(p_{opt})$.

The second and last step is to determine the optimal parameters $A_j(p_{opt})$, $d_j(p_{opt})$ and the optimal sales forecast $x_{opt}(t)$ calculating subsequently

$$A_j(p_{opt})=a_j(p_{opt})\frac{A_j}{\int_{-T}^{0} x_j(t)dt}, \tag{9}$$

$$d_j(p_{opt})=\frac{a_j(p_{opt})-\int_0^T k_jt+A_j(p_{opt})\cos\left(\frac{2\pi}{T}\left(t-\tau_j\right)\right)dt}{T} \tag{10}$$

and $x_{opt}(t)=y(t,p_{opt})$. (11)

Equations (9), (10) and (11) are a direct result of "the constant ratio amplitude to total sales equation" (6), "the total sales equation" (7), the definition of $y_j(t,p)$ in (8) and additionally of the definition of $x_j(t)$ in (1).

As Fig. 4 shows, the optimal forecast for the product P_2 is drastically below the extrapolated function. More contribution margin and profit is earned by reducing the total sales but by increasing the price. For the product P_1 there is an opposite result. Because of the increased total sales of product P_1 the amplitude of product P_1 changed significantly.

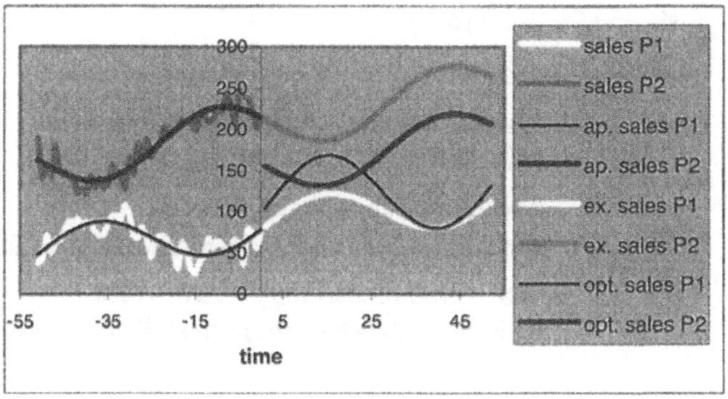

Fig. 4. Extrapolated sales and optimal sales

3 Additional Results

The short-term fluctuations cause a fluctuation in the production load as well as in the sourcing demand. One idea for handling the short-term fluctuation in the sourcing is to define a safety stock level (see for instance [2, 3]) for the sourcing warehouse. The safety stock level should guarantee to a high percentage β that both the forecasted sourcing demand within the sourcing time period and the additional sourcing demand caused by short-term fluctuations can be dealt with. For simplicity (12) is conceived only for one specific sourcing product:

$$p\left(\sum_{r=1}^{T_S} s_{i+r} \leq l_{\beta\text{-safety}}\right) = \beta \quad \text{availability condition} \tag{12}$$

whereby

$l_{\beta\text{-safety}}$	safety stock with safety-level β
$s_i = S\bar{x}_i \in$ □	sourcing demand caused by vector of real sales $\bar{x}_i$
S	sourcing matrix
T_S	sourcing time for a certain product (number of sub periods, for instance days)
$p(E)$	probability of some event E.

Formula (13) is an explicit expression for the safety stock and is the result of applying the normal distribution transformation on the availability condition in (12) and taking into account that the real demand s_i is a random number with deviation σ_s and mean $s(t_i)$.

$$l_{\beta\text{-safety}} = \sum_{j=1}^{T_S} s(t_{i+j}) + Q_{N(0,1)}(\beta)\sigma_S\sqrt{T_s} \qquad (13)$$

whereby

$s(t_i) = Sx(t_i) \in \square$ sourcing demand caused by the forecasted sales-vector $x(t_i)$, S sourcing matrix

$\sigma_S = \sqrt{S\sigma_x^{\ 2}}$ deviation of the sourcing demand

$\sigma_x = (\sigma_{x1} \ \cdots \ \sigma_{xn})^{\mathrm{T}}$ deviation vector of the approximated sales functions (for n products) to the past sales data

$Q_{N(0,1)}$ quantile of the standard normal distribution $N(0,1)$.

As shown in Fig. 5, the real sourcing demand within the sourcing time period is given by the area below the real sourcing data and between the two arrows. The safety stock level is defined by the area, which can be seen in Fig. 5 between the upper safety boundary curve and the forecasted sourcing demand curve.

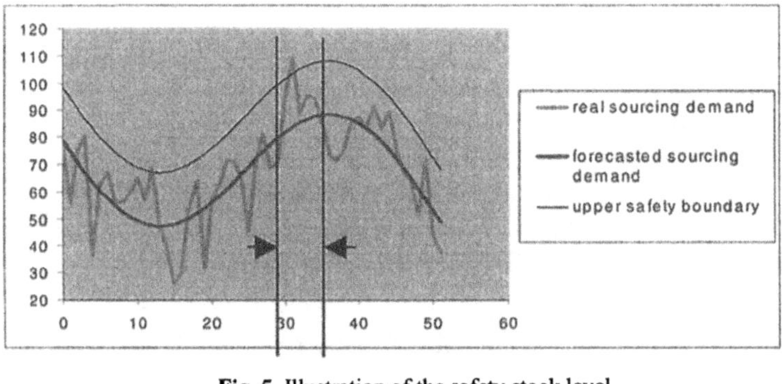

Fig. 5. Illustration of the safety stock level

References

1. Tempelmeier H (1999) Material-Logistik: Modelle und Algorithmen für die Produktionsplanung und –steuerung und das Supply-chain-Management. 4., überarb. u. erw. Aufl. Springer, Berlin Heidelberg New York, pp 88-90
2. Grubbström RW, Tang O (1999) Further Developments on Safety Stocks in an MRP System applying Laplace Transforms and Input-Output Analysis. In: International Journal of Production Economics, Vol. 60-61, pp 381-387
3. Minner S (Ed.) (2000) Strategic Safety Stocks in Supply Chains. In: Lecture Notes in Economics and Mathematical Systems, Vol. 490, Springer, Berlin Heidelberg New York, pp 33-53

A Multi-Product Batch-Available-to-Promise Model for Make-to-Stock Manufacturing

Richard Pibernik

Seminar für Logistik und Verkehr, Johann Wolfgang Goethe-Universität, Mertonstr. 17, 60054 Frankfurt am Main, pibernik@wiwi.uni-frankfurt.de

1 Introduction

Available-to-Promise (ATP) comprises of a variety of tools that enhance the responsiveness of order promising and the reliability of order fulfilment. Based on customer requests (i.e. requested product, order quantity and delivery time window) they support "order quantity" and "order due date quoting".
ATP is usually integrated in Enterprise Resource Planning (ERP) systems and Advanced Planning Systems (APS). Their functional scope can vary significantly. "Conventional" ATP, commonly implemented in ERP systems, merely determines the availability of finished goods at a certain point of time in the future. "Advanced" ATP provides a broader scope of functions. In this paper, Advanced ATP is considered as a decision-making system which simultaneously allocates available finished goods inventory to customer orders and quotes order due dates in a make-to-stock manufacturing environment.
The major goals pursued with the implementation of ATP are the improvement of on time delivery by generating reliable quotes, the reduction of the number of missed business opportunities by employing more effective methods for order promising and an increase of revenue and profitability by increasing the average sales price (see [4]).
The development of methods and their application to support order promising has primarily been driven by providers of ERP and APS (e.g. SAP, Manugistics, i2 Technologies). Up to now, a very limited number of theoretically founded contributions have been made. A variety of papers discuss the needs or propose features for ATP Systems (e.g. [3], [4] and [6]). Very few contributions provide quantitative models or algorithms for quantity and due date promising. Major contributions have been made by Chen/Zhao/Ball ([1],[2]). The authors develop a mixed-integer programming ATP model that allocates resources among customer orders that arrive within a pre-determined time interval (batching interval). Their contribution, however, is focused on a specific configure-to-order case. In this paper a batch ATP model for due date quoting on the basis of finished goods inventory is presented. This model can be customized in order to support company specific needs and is therefore applicable for any firm employing a make-to-stock manufacturing strategy.
The remainder of this paper is organized as follows: In section 2 we first describe the ATP scheduling problem and introduce the relevant notations. Thereafter we present a mixed integer programming formulation applicable for generating ATP

schedules and characterize major applications and extensions. The paper concludes with a summary of the findings.

2 The Batch Available to Promise Model

2.1 Problem Description and Notations

Consider a manufacturer of N different products n = 1,...,N. The manufacturer employs a make-to-stock production strategy. Let $[t_a, t_e]$ be the ATP planning horizon, consisting of T discrete time periods $[t_a, t_a+1],...,[t_e-1, t_e]$. The point of time when the model is being executed is denoted by t_a. Order promising decisions will be made for a batch of potential customer orders, collected within the time interval $[t_a - \tau, t_a]$ (batch-interval), where τ represents the length of the batch interval. $A(t_a)$ denotes the set of potential customer orders collected within the batch-interval. Every potential order $i \in A(t_a)$ can be characterized by a triple $(\mathbf{d}_i, z_i^u, z_i^o)$, with $\mathbf{d}_i = (d_i^1,...,d_i^N)$ the vector of order quantities for the N products, $z_i^u \in \{t_a+1,...,t_e\}$ the earliest date of delivery and $z_i^o \in \{t_a+1,...,t_e\}$, $z_i^o \geq z_i^u$ the latest date of delivery.[1] The interval $[z_i^u, z_i^o]$ represents the customer's delivery time window, with $z_i^u = z_i^o$ if the customer insists on a fixed delivery date.

We assume, the manufacturing schedule is fixed for the ATP planning horizon $[t_a, t_e]$. At every point of time t ($t = t_a+1,...,t_e$) a given quantity of q_t^n units of product n are produced and put into stock. For simplicity and without loss of generality, we further assume that the q_t^n units can be delivered to a customer at the point of time t and that the delivery time is zero. Let b_t^n denote the inventory on hand at point of time t ($t = t_a+1,...,t_e$). We assume, the inventory on hand b_{t_a+1} at t_a+1 is determined by the ATP run executed at point of time $t_a - \tau$. The inventory on hand b_t^n and the production quantity q_t^n determine the quantity of product n, "available to promise" to a customer at point of time t.

We assume, an order can, in certain cases, also be fulfilled by means of substitute products.[2] Let M(n) denote the set of ordered tuples of substitution possibilities for product n:

---

[1] For simplicity, we assume that the customer specifies the earliest and latest due date for all ordered products. Considering different due dates, however, is not difficult: The orders can be divided into sub orders with different due dates which are then subject to order promising.

[2] A manufacturer of hard disks, for example, may deliver drives with a higher capacity if he runs out of drives with lower capacity.

$$M(n) := \left\{ (n, \bar{n}) \middle| \begin{array}{l} \text{product } \bar{n} \text{ can be employed as a substitute for } n \\ \bar{n} \in \{1, \ldots, N\}; \bar{n} \neq n \end{array} \right\}.^{3} \qquad (1)$$

Let M be the set union of all sets M(n) (n = 1,…,N).

Apparently, the firm has to solve an assignment problem. The potential customer orders, collected within the batch-interval have to be assigned to the quantities of the N products, available to promise during the time interval $[t_a + 1, t_e]$.[4] We will call a (feasible) solution of this assignment problem an "ATP schedule". This schedule specifies, which orders are fulfilled by which quantities of the N products at which point of time within the ATP planning horizon.

In order to determine an ATP schedule with the batch model, presented in the following section, we define the following variables:

$x_i^n(t)$: Delivered quantity of product n at point of time t in order to fulfil customer order i.

$x_i^{n,\bar{n}}(t)$: Delivered quantity of product $\bar{n}$ as substitute for product n at point of time t in order to fulfil customer order i.

$$r_i^{n,\bar{n}} = \begin{cases} 1, \text{if product } n \text{ is substituted by product } \bar{n} \text{ in order to fulfil order } i \\ 0, \text{else} \end{cases}$$

$$u_i(t) = \begin{cases} 1, \text{if due date of order } i \text{ is } t \\ 0, \text{else} \end{cases}$$

$$v_i = \begin{cases} 1, \text{if order } i \text{ is fulfilled within } [z_i^u, z_i^o] \\ 0, \text{else} \end{cases}$$

With the decision variables $x_i^n(t)$ and $x_i^{n,\bar{n}}(t)$ an ATP schedule can be represented by an $(n(A(t_a))) \times T$-matrix **x**.[5] The components of **x** are the ordered tuples $(\mathbf{x}_i(t), \bar{\mathbf{x}}_i(t))$ $(i \in A(t_a), t = t_a + 1, \ldots, t_e)$ with

$$\mathbf{x}_i(t) := (x_i^1(t), \ldots, x_i^N(t)) \text{ and } \bar{\mathbf{x}}_i(t) := \begin{pmatrix} x_i^{1,1}(t) = 0 & \cdots & x_i^{N,1}(t) \\ \vdots & & \vdots \\ x_i^{1,N}(t) & \cdots & x_i^{N,N}(t) = 0 \end{pmatrix}. \qquad (2)$$

It is the firm's objective to determine an ATP schedule which maximizes overall net profit, denoted by P, within the time interval $[t_a + 1, t_e]$. In order to calculate the overall profit, associated with an ATP schedule, we define:

p_i^n: profit contribution of one unit of product n with regard to order i.[6]

---

[3] M(n) can be defined individually for every customer order $i \in A(t_a)$ or for different types of customer orders. For simplicity we assume, M(n) is applied to every potential customer order.

[4] The first delivery can take place at point of time $t_a + 1$, i.e. at the point of time after executing the Model.

[5] $n(A(t_a))$ denotes the number of Elements in the set $A(t_a)$.

[6] We assume that the profit contribution does not include holding costs.

$p_i^{n,\bar{n}}$: profit contribution of one unit of product $\bar{n}$ when used as substitute for n with regard to order i ($p_i^{n,\bar{n}} < p_i^n$).[7]

hc^n : holding costs for one unit of product n.

dp_i : penalty associated with the denial of customer order i.[8]

2.2 Model Formulation

To determine an ATP schedule, which maximizes the overall profit P, we can employ the following mixed-integer-programming model:

$$\max P = \sum_{t=t_a+1}^{t_e} \sum_{i\in A(t_a)} \sum_{n=1}^{N} p_i^n \cdot x_i^n(t) + \sum_{t=t_a+1}^{t_e} \sum_{i\in A(t_a)} \sum_{n=1}^{N} \sum_{(n,\bar{n})\in M(n)} p_i^{n,\bar{n}} \cdot x_i^{n,\bar{n}}(t) - \sum_{t=t_a+1}^{t_e} \sum_{n=1}^{N} hc^n \cdot b_t^n - \sum_{i\in A(t_a)} dp_i \cdot (1 - v_i) \quad (3)$$

s.t.

$$x_i^n(t) + \sum_{(n,\bar{n})\in M(n)} x_i^{n,\bar{n}}(t) = d_i^n \cdot u_i(t) \quad (4)$$

for all $i \in A(t_a); n = 1,\dots,N;\ t = t_a + 1,\dots,t_e$

$$b_t^n = b_{t-1}^n + q_t^n - \sum_{i\in A(t_a)} x_i^n(t) - \sum_{i\in A(t_a)} \sum_{(\bar{n},n)\in M} x_i^{\bar{n},n}(t) \quad (5)$$

$n = 1,\dots,N;\ t = t_a + 1,\dots,t_e$

$$\sum_{t=z_i^u}^{z_i^o} u_i(t) = v_i \qquad \text{for all } i \in A(t_a) \quad (6)$$

$$\sum_{(n,\bar{n})\in M(n)} r_i^{n,\bar{n}} \le \bar{r} \qquad \text{for all } i \in A(t_a); n = 1,\dots,N \quad (7)$$

$$x_i^{n,\bar{n}}(t) \le r_i^{n,\bar{n}} \cdot d_i^n \qquad \text{for all } i \in A(t_a), (n,\bar{n}) \in M(n); n = 1,\dots,N \quad (8)$$

$$x_i^n(t) \ge 0 \qquad \text{for all } i \in A(t_a); n = 1,\dots,N \quad (9)$$

$$x_i^{n,\bar{n}}(t) \ge 0 \qquad \text{for all } i \in A(t_a), (n,\bar{n}) \in M(n); n = 1,\dots,N \quad (10)$$

---

[7] If a substitute product is used to fulfil an order, it has to be considered that the customer will usually not pay the regular price for the substitute, but rather the lower price for the product, originally ordered.

[8] dp_i has to account for contract penalties, loss of profits associated with order i and loss of future profits, if the customer switches to a different supplier.

$$b_t^n \geq 0 \qquad n = 1,...,N; t = t_a + 1...,t_e \qquad (11)$$

$$u_i(t) \in \{0,1\} \qquad \text{for all } i \in A(t_a); z_i^u \leq t \leq z_i^o \qquad (12)$$

$$v_i \in \{0,1\} \qquad \text{for all } i \in A(t_a) \qquad (13)$$

$$r_i^{n,\bar{n}} \in \{0,1\} \qquad \text{for all } i \in A(t_a), (n,\bar{n}) \in M(n); n = 1,...,N \qquad (14)$$

The objective function (3) accounts for both tangible and intangible terms. The tangible terms include profit from promised orders (taking into account the delivery of originally ordered products and substitutes) as well as inventory costs for finished goods. The intangible term represents penalties associated with order denial. A feasible solution (ATP schedule) is subject to the constraints (4)-(14). Constraint (4) ensures that the delivered quantity is equal to the ordered quantity for every accepted order. Balance of finished goods inventory is provided by (5). Note that the last term in (5) accounts for the quantity of product n, used for substitution. Constraint (5) and constraint (11), which models non-negativity of inventory on hand, ensure that only quantities "available to promise" are actually assigned to customer orders. Constraint (6) ensures delivery within the delivery time window $[z_i^u, z_i^o]$ and determines the value of the binary variable v_i. Constraint (7) limits the number of products, employed for the substitution of a product n to $\bar{r}$. We assume, the customers will not accept the delivery of a number of substitute products greater $\bar{r}$. Constraint (8) determines the value of the binary variable $r_i^{n,\bar{n}}$. Constraints (9), (10) and (11) assure nonnegativity of delivered quantities and inventory on hand. Constraint (12) defines the binary variable $u_i(t)$ within the delivery time window $[z_i^u, z_i^o]$.

2.3 Applications and Extensions

The model presented in the previous section is suitable for implementation in a software application, which supports order promising on a daily basis. As a component of a supply chain management system it can then obtain the relevant data, needed to solve the presented scheduling problem, from ERP and APS. Apart from application in day-to-day order promising, the model can also be employed in order to support tactical decisions, e.g. concerning the length of the batch interval (see [2]) and the consideration of customer specific order priorities.

The model can easily be enhanced to account for a variety of additional functionalities. It can, for example, be modified in order to allow for partial deliveries. If the order quantity is not available within the given delivery time window, the customer order can be fulfilled with two or more partial deliveries, whereas the first partial delivery is carried out within the given time window. The model then determines the quantities and delivery dates for each partial delivery (see [5]).

The model can also be applied to a distribution network, rather than only to a single location. It then has to be modified to account for different manufacturing and transportation lead times and costs, depending on the considered locations.

3 Summary

In this paper we presented a basic batch ATP model, which can be employed to support day-to-day order promising of a firm, applying a make-to-stock manufacturing strategy. Based on potential orders with specific customer requests (requested product, order quantity and delivery time window) and the product quantities „available to promise", the model generates an ATP schedule for a predefined planning horizon. This schedule specifies order due dates as well as product quantities and product types, used to fulfil the set of potential customer orders. The model can be customized in order to account for firm specific needs. It is suitable for implementation in a software application to support order promising, which can be integrated in ERP and APS.

References

[1] Chen C-Y, Zhao Z-Y, Ball MO (2000) A Model for Batch Advanced Available-To-Promise. To appear in: Production and Operations Management, available Internet: http://bmgt1-notes.umd.edu/faculty/km/papers.nsf.

[2] Chen C-Y, Zhao Z-Y, Ball MO (2001) Quantity and Due Date Quoting Available To Promise. Information Systems Frontiers 3:4: 477-488.

[3] Fischer ME (2001) „Available to Promise": Aufgaben und Verfahren im Rahmen des Supply Chain Management. Regensburg.

[4] Kilger C, Schneeweiss L (2000) Demand Fulfilment and ATP. In: Stadtler H, Kilger C (Eds) Supply Chain Management and Advanced Planning. Berlin et al., pp 79-95.

[5] Pibernik R (2002) Advanced Available to Promise: Models and Algorithms for Order Promising and Fulfilment. Working Paper Lehrstuhl Logistik, Universität Frankfurt.

[6] Robinson A, Dilts DM (1999) OR and ERP: Can operations research play a role in fast-growing, enterprise-wide information systems?. OR/MS Today 26/3: 30-37.

Scheduling of Rolling Ingots Production

Christoph Schwindt and Norbert Trautmann

Universität Karlsruhe, Institut für Wirtschaftstheorie und Operations Research, D-76128 Karlsruhe, Germany,
e-mail: {schwindt,trautmann}@wior.uni-karlsruhe.de

Abstract. We consider a real-world scheduling problem arising in the context of a rolling ingots production. We review the production process and discuss peculiarities that have to be observed when scheduling a given set of production orders on the production facilities. We then describe a model for this scheduling problem using prescribed time lags between operations, different kinds of resources, and sequence-dependent changeovers. The basic principle of the solution procedure is to relax the resource constraints by assuming infinite resource availability. Resulting resource conflicts are then stepwise resolved by introducing precedence relationships among operations competing for the same resources. The algorithm has been implemented as a beam search heuristic enumerating alternative sets of precedence relationships.

1 Introduction

This paper deals with a scheduling problem arising in aluminium industry. We consider the production of rolling ingots, i.e. ingots of a certain aluminium alloy in rectangular form. These ingots are the starting material for the rolling of sheet, strip, and foil. The production flow is as follows (cf. Kammer 1999, Section 1.4, and Figure 1).

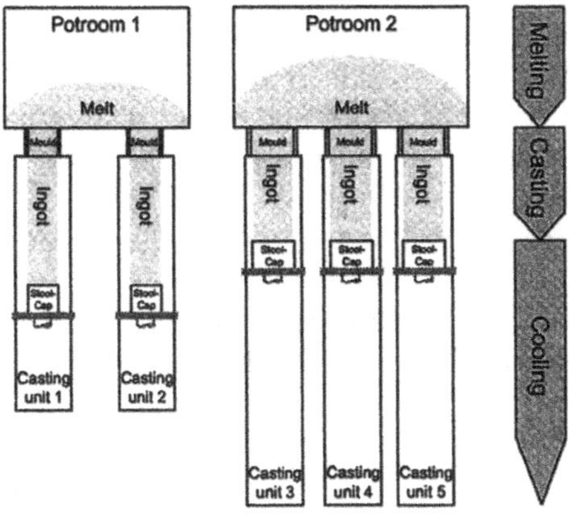

Fig. 1. Production flow chart

In a melting furnace called potroom, the ingredients composing the alloy are smelted in an electrolytical process. In general, several alternative potrooms are available. One or several casting units belong to each potroom. A casting unit consists of a holding fixture for a mould, a retractable hydraulic cylinder named stool, and a so-called stool-cap, which closes the bottom of the mould at the start of the casting process. The melt is cast through the mould, which determines the cross-section of the ingot.

As soon as the metal in the mould begins to solidify, the stool is lowered and the resulting ingot is cooled by spraying water. The maximum stroke of the stool determines the maximum cast length, which implies that not every ingot can be produced on each casting unit. However, all casting units belonging to one potroom are of the same height. After the completion of the casting process, the ingot stays some time in the casting unit for cooling. All ingots produced within one casting are of the same alloy and same length. The casting has to be started and completed at the same time at all casting units of a potroom. It is not necessary to use all casting units during a casting process nor to use their full length. When passing to the casting of an ingot with a different cross-section, the mould of the casting unit has to be changed. The changeover can only be performed when no casting is in process. Moreover, only one mould per potroom can be exchanged at a time.

The production scheduling problem is as follows. Given a set of production orders for ingots characterized by their size and alloy, the problem consists of computing a feasible production schedule with minimum makespan. To the best of our knowledge, there is no scheduling procedure known from literature that can be applied to this scheduling problem. Some approaches have been proposed for related problems (cf. Harjunkoski and Grossmann 2001, Kempf et al. 1998, and Fleischmann and Jess 1985).

2 Temporal and resource constraints

In this section we discuss the set of operations to be scheduled, the prescribed temporal relationships between individual operations, and the different types of resource constraints. For a formal statement of the production scheduling problem we refer to Schwindt and Trautmann (2002).

Each production order for an individual ingot corresponds to a job consisting of the three **operations** melting, changeover of the mould, and casting and cooling. The duration of a melting operation equals the melting plus the casting time because the potroom is occupied up to the end of the casting. The duration of a casting and cooling operation corresponds to the time needed for casting and cooling the ingot. The duration of a changeover operation depends on the sequence in which the jobs are processed on the casting units. If two consecutive jobs require different mould types, the duration of the changeover operation equals the setup time for installing the mould belonging to the second job. Otherwise, the mould need not be replaced, and the duration is equal to zero.

Since the melting operations may be performed in alternative potrooms, we define a set of alternative **execution modes** for each job, where each job has to be carried out in exactly one mode.

The following **temporal constraints** arise from technological requirements:

1. The mould has to be installed before the casting, i.e., there is a minimum time lag of 0 between the completion of the changeover and the start of the casting operation.

2. Since the potroom remains occupied up to the end of the casting, melting and casting must be completed simultaneously. Thus, there is a minimum and a maximum time lag equal to the cooling time between the completion of the melting and the completion of the casting operation.

Basically, all production facilities (potrooms, moulds, and casting units) correspond to **renewable resources** (cf. Brucker et al. 1999). An operation takes up a mode-dependent number of units of each resource. For each renewable resource, no more units than available can be used simultaneously.

In a potroom, only one alloy can be melted at a time. This condition can be taken into account by using the concept of **batching machines** and incompatible job families discussed in literature on machine scheduling (cf. Potts and Kovalyov 2000, Uzsoy 1995). A batching machine has to be operated in a way that operations running in parallel must be started at the same time. In case of incompatible job families, those operations must also belong to the same batching type. Each potroom is modelled as a batching machine whose capacity equals the number of casting units that belong to the potroom. The batching type of a job identifies the alloy and length of the corresponding ingot.

Obviously, it is not possible to use the installed mould for processing another ingot before the casting has been completed. To model the latter requirement, we introduce a set of **allocatable resources**. The units of an allocatable resource used for executing an operation remain occupied from the start of a given *allocating operation* up to the completion of the operation. For each allocatable resource, at no point in time more units than available can be allocated. Moulds of the same type form an allocatable resource whose capacity is equal to the number of moulds of that type. A mould gets allocated to a casting operation at the start of the corresponding changeover operation.

We associate each group of casting units belonging to one potroom with a **changeover resource** (cf. Neumann et al. 2001) whose capacity equals the number of casting units in the group. The changeover times on this resource coincide with the durations of the respective changeover operations.

Finally, only one of the casting units belonging to a potroom can be changed over at a time. Moreover, a mould cannot be installed on any casting unit during a casting operation. We take those requirements into account by defining an extra renewable resource for each potroom, where again the capacity is chosen to be equal to the number of casting units of the potroom. The full capacity of this resource is taken up by the changeover and one unit by the casting and cooling operations.

3 Solution method

The solution procedure is based on a schedule-generation scheme for resource-constrained project scheduling presented in Neumann et al. (2001). The constraints that make the problem intractable are

1. the necessity to select a mode for each job,
2. the limited capacity of the renewable resources,
3. the requirement that operations running in parallel on batching machines must be of the same batching type and must be started jointly, and
4. the need for changeover operations with sequence-dependent duration.

If we relax all those constraints, the remaining problem consist of scheduling all operations subject to the temporal constraints. This *temporal scheduling problem* can efficiently be solved by standard network-flow algorithms (cf. Ahuja et al. 1993).

An optimal solution to the temporal scheduling problem may be infeasible for the production scheduling problem due to two reasons: (i) there may be jobs for which no mode has been selected so far, or (ii) one of the resource constraints may not be met. In case (i), we select some mode for a job whose mode has not been fixed so far. In case (ii), the violation of some resource constraint at a given time is resolved by introducing appropriate time lags between operations competing for the respective resource (cf. Schwindt and Trautmann 2002). By assigning modes to jobs and adding time lags, we obtain a new scheduling problem with a refined relaxation, which belongs to a reduced set of feasible modes and to an expanded set of temporal constraints. The selection of modes and the addition of arcs is continued until either a mode has been selected for each job and the solution to the temporal scheduling problem provides a feasible schedule or the temporal scheduling problem is unsolvable. Figure 2 summarizes this schedule-generation scheme.

Due to the sequence-dependent need for changeovers, the resource demands over time arising from the changeover operations are not known until the job sequences on the changeover resources have been established. That

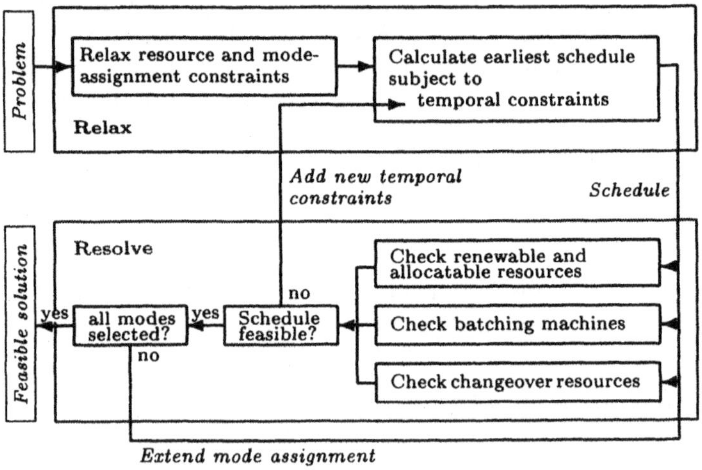

Fig. 2. Schedule-generation scheme

is why we follow a heuristic two-phase approach where in the first phase, while observing the changeover times, we assume that changeover operations do not require resources for their execution. In the terms of our production scheduling problem the latter assumption means that we allow for parallel changeovers on casting units belonging to one and the same potroom. When we have obtained a feasible schedule for this subproblem, we fix the job sequences on the changeover resources (and thus the durations of the changeover operations) by introducing a precedence constraint between any two consecutive operations. We then proceed with the second phase, where we apply the branching process subject to the time lags introduced thus far.

4 Computational results

In this subsection we return to the example from Section 1, which serves as a basis to benchmark problems for testing the efficiency of the scheduling procedure proposed. Ingots can be cast in three different alloys, two different lengths, and four different cross sections. For each of the four mould types, two moulds are available. Long ingots can only be produced in potroom 2. Table 1 gives the alloy, the length, the mould type corresponding to the cross-section, and the durations of the individual production steps for the different ingot types. We assume that installing the mould takes 10 minutes.

Based on the schedule-generation scheme described in Section 3, we have implemented a depth-first search branch-and-bound procedure enumerating alternative mode assignments and alternative sets of minimum time lags between operations. We have evaluated the performance of a truncated version of the branch-and-bound algorithm (a filtered beam search) using 15 instances which have been generated by varying the number of production orders for the different ingot types of our example. For each instance, we have imposed a CPU time limit of one minute to the scheduling procedure running on a Pentium-800 PC. The makespans obtained for the 15 instances are shown in Table 2. The results for instances 6 to 10 indicate that the algorithm scales well. In particular, the makespan-per-order ratio seems to be independent of the problem size, though the CPU time limit has not been varied.

Table 1. Ingot data

Ingot Type	1	2	3	4	5	6	7	8	9	10
Alloy ID	1	1	1	1	2	2	2	3	3	3
Length (s: short, l: long)	s	l	s	l	s	l	l	s	s	l
Mould Type	1	2	3	4	1	1	2	3	4	4
Duration Melting [min]	56	56	56	56	73	73	73	57	57	57
Duration Casting [min]	60	80	60	80	60	80	80	60	60	80
Duration Cooling [min]	45	45	45	45	70	70	70	75	75	75

Table 2. Results

Ingot Type	1	2	3	4	5	6	7	8	9	10	Makespan
Orders Instance 1	3	3	3	3	3	3	0	0	0	0	816
Orders Instance 2	0	3	3	3	3	3	3	0	0	0	834
Orders Instance 3	0	0	3	3	3	3	3	3	0	0	848
Orders Instance 4	0	0	0	3	3	3	3	3	3	0	816
Orders Instance 5	0	0	0	0	3	3	3	3	3	3	917
Orders Instance 6	1	1	1	1	1	1	1	1	1	1	485
Orders Instance 7	2	2	2	2	2	2	2	2	2	2	829
Orders Instance 8	3	3	3	3	3	3	3	3	3	3	1421
Orders Instance 9	4	4	4	4	4	4	4	4	4	4	1664
Orders Instance 10	5	5	5	5	5	5	5	5	5	5	2131
Orders Instance 11	3	6	9	12	15	3	6	9	12	15	5121
Orders Instance 12	6	9	12	15	3	6	9	12	15	3	5446
Orders Instance 13	9	12	15	3	6	9	12	15	3	6	4112
Orders Instance 14	12	15	3	6	9	12	15	3	6	9	4731
Orders Instance 15	15	3	6	9	12	15	3	6	9	12	5626

References

1. Ahuja, R., Magnanti, T., Orlin, J. (1993): Network Flows. Prentice Hall, Englewood Cliffs
2. Brucker, P., Drexl, A., Möhring, R., Neumann, K., Pesch, E. (1999): Resource-constrained project scheduling: notation, classification, models, and methods. European Journal of Operational Research **112**, 3–41
3. Fleischmann, B., Jess, H. (1985): Erfahrungen mit einem Simulationsmodell in einer Aluminiumgießerei. OR Spektrum **7**, 175–185
4. Harjunkoski, I., Grossmann, I. (2001): A decomposition approach for the scheduling of a steel plant production. Computers & Chemical Engineering **25**, 1647–1660
5. Kammer, C. (1999): Aluminium Handbook Vol. 1: Fundamentals and Materials. Aluminium, Dsseldorf
6. Kempf, K.G., Uzsoy, R., Wang, C.S. (1998): Scheduling a single batch processing machine with secondary resource constraints. Journal of Manufacturing Systems **17**, 37–51
7. Neumann, K., Schwindt, C., Zimmermann, J. (2001): Project Scheduling with Time Windows and Scarce Resources. Springer, Berlin
8. Potts, C., Kovalyov, M. (2000): Scheduling with batching: A review. European Journal of Operational Research **120**, 228–249
9. Schwindt, C., Trautmann, N. (2002): Production Scheduling with Batching Resources: Industrial Context, Model, and Solution Method. Report WIOR-618, University of Karlsruhe
10. Uzsoy, R. (1995): Scheduling batch processing machines with incompatible job families. International Journal of Production Research **33**, 2685–2708

Determination of Economic Production Quantity for a Multi-Stage Production System with Limited Storage Capacity

U. Buscher[1] and G. Lindner[2]

[1] Dresden University of Technology, „Friedrich List"-Faculty of Transportation Sciences, D-01062 Dresden
[2] Otto-von-Guericke University Magdeburg, Faculty of Economics and Management, D-39016 Magdeburg

This paper describes a model for a multi-stage production system in which a uniform lot size is produced through all stages with a single setup and without interruption at each stage. Transportation of partial lots, called batches, is allowed between stages before the whole lot is completed. Although the batch size must be equal at any particular stage, the optimal number of equal-sized batches may differ across stages. Additionally, we assume that inventories between consecutive stages are constrained by a limited storage capacity. Considering setup costs, inventory holding costs, and transportation costs, an optimisation method is developed to determine the economic production quantity and the optimal batch sizes for each stage.

1 Introduction

The production of parts for a complex product often involves a series of operations (stages). Although the assembly of the final product is frequently continuous over time, the fabrication of parts is performed in lots since the rhythm of the assembly line is seldom synchronised with part-manufacturing. Because of this, items facing a continuous demand are manufactured intermittently. Moreover, in many cases production rates differ between adjacent stages which leads in effect to corresponding work-in-process inventories.

In such a context, a fundamental shortcoming of most conventional lot size models is that they pay no attention to the fact that produced items have to be transported from one stage to the next. Implicitly, they restrict the number of conveyances per lot to only two alternatives. Either complete lots are transferred or each item is shipped immediately after its completion. It seems to be more realistic also to allow the transport of partial lots (batches) larger than one and smaller than the entire production quantity (see e.g. [1, 4, 5]).

Furthermore, in practice the storage space for each work-in-process inventory is never unlimited. Thereby, the usable total capacity of the before mentioned inventories is influenced by the storage method selected. In this paper, we assume that each item has a fixed storage bin dedicated solely to it. Including restricted storage capacity in lot-sizing models is well known. Nevertheless, in the overwhelming number of cases, systems are examined which consist of only one inventory.

In the remainder of this article, we consider a multi-stage production system having multiple inventories with constrained storage capacities. The authors know of only one other approach in this area of investigation (see [2]). By modelling explicitly the influence of transportation frequencies on the maximum process inventory, a suitable formulation of the storage capacity restriction will be achieved. In consequence the range of feasible stage-specific lot sizes expands in comparison to [2]. This, in turn, opens up possibilities for cost-savings.

2 The Model and the Cost Function

This section deals with a single-item multi-stage Economic Production Quantity (EPQ) model. The analysed serial system consists of an unrestricted number of S manufacturing stages, s = 1, 2, ..., S. The stage that meets the demand for the finished product is stage S + 1. The main characteristics of our model can be summarised as follows.

The production process of the considered item is comprised of a fixed sequence of operations to be carried out. At each stage one facility is available for performing the respective operation. The corresponding production rates P_s are assumed to be finite, deterministic, and constant. Thereby, a uniform lot size Q is produced through all stages of the underlying system. This quantity is fabricated without interruption and with a single setup at each stage. A setup causes a stage-specific fixed setup cost S_s. Before the whole lot is finished, equal sized batches q_s can be transported from one stage to the next. For every shipment a fixed stage-specific transportation cost T_s is charged. Moreover, a different (integer) number $m_s = Q/q_s$ of batches can be realised at the various stages.

By calculating the EPQ we have to ensure that the maximum process inventory at stage s does not exceed the related storage capacity L_s. All other capacity constraints of the production system will hold. The cost of carrying one unit of physical inventory over one unit of time at a particular stage is represented by c_s. Additionally, the units of the item shall be infinitely divisible, setup and transportation time are negligible small and no backlogging is permitted. Finally, the demand rate D for the item is deterministic and constant over an infinite time horizon.

To derive the optimal lot-sizing policy for the entire production system we use the criterion of minimisation of total relevant costs per unit time. The cost function C(Q, M) where $M = (m_1, m_2, \ldots, m_S)$ is composed of inventory holding costs (first and second term), setup costs (third term), and transportation costs (last term) as follows (see [2, 5]):

$$C(Q,M) = Q \cdot \sum_{s=1}^{S} \frac{\alpha_s}{m_s} \cdot D + Q \cdot \beta \cdot D + \gamma \cdot \frac{D}{Q} + \sum_{s=1}^{S} m_s \cdot T_s \cdot \frac{D}{Q} \tag{1}$$

$$\alpha_s = \frac{c_s}{\max[P_s;P_{s+1}]};\ \beta = \sum_{s=1}^{S} \frac{1}{2} \cdot \left| \frac{1}{P_s} - \frac{1}{P_{s+1}} \right| \cdot c_s;\ \gamma = \sum_{s=1}^{S} S_s$$

Furthermore, we have to respect the limited storage capacity. In order to formulate the corresponding constraint, it is helpful to look at the development of the total inventory per production lot between two consecutive stages s and s + 1. As easily can be shown, the maximum inventory level is $Q \cdot [1+P_{s+1}/P_s \cdot (1/m_s-1)]$, when $P_s \geq P_{s+1}$, and is not allowed to exceed L_s. Since the maximum inventory level is given by $Q \cdot [1+P_s/P_{s+1} \cdot (1/m_s-1)]$, when $P_{s+1} > P_s$, the general expression for the storage capacity restriction is:

$$Q \leq L_s / [1 + \lambda_s \cdot (1/m_s - 1)] \quad \text{where} \quad \lambda_s = \min[P_s / P_{s+1}; P_{s+1} / P_s] \tag{2}$$

Note, the total cost function C(Q, M) is convex for all positive real values of Q and $(m_1; m_2; \ldots; m_S)$ of the vector M. To derive the optimal lot-sizing policy it is necessary to minimise Eq. (1) subject to (2).

3 Feasible and Optimal Transportation Frequencies

The total cost function depends on both the lot size as well as the transportation frequencies. For the time being, let us ignore the storage capacity constraints. Obviously, a certain integer m_s-value is only optimal for a particular range of lot sizes. In order to determine the latter range, one has to look at that part $H_s(Q, m_s)$ of the total cost function which is influenced by m_s. Clearly, for a given Q it is advantageous to choose the integer shipment frequency m_s^i instead of $m_s^i + 1$ if:

$$H_s(Q, m_s^i) \leq H_s(Q, m_s^i + 1) \tag{3}$$

Inequality (3) can be used to calculate the critical value of the lot size Q for which the optimal integer m_s-value changes from m_s^i to $m_s^i + 1$. Exactly at that point will both transportation frequencies result in the same cost and (3) is fulfilled as an equation. Now, solving this equation for Q gives:

$$Q_s^v(m_s^i; m_s^i + 1) = \sqrt{T_s \cdot m_s^i \cdot (m_s^i + 1) / \alpha_s} \tag{4}$$

Since the same consideration can be done with m_s^i - 1 and m_s^i, m_s^i is the optimal integer number of transports if Q lies in the range:

$$Q_s^v(m_s^i - 1; m_s^i) < Q \leq Q_s^v(m_s^i; m_s^i + 1) \tag{5}$$

So far we have ignored constraint (2). Taking this restriction into account, the largest feasible lot size $Q_s^r(m_s)$ for a particular m_s is:

$$Q_s^r(m_s) = L_s / [(1 - \lambda_s) + \lambda_s / m_s] \tag{6}$$

As can be seen, the denominator in Eq. (6) is always larger than zero. Hence, for each number of transports a feasible value $Q_s^r(m_s)$ exists. Moreover, $Q_s^r(m_s)$ becomes larger the higher the considered m_s-values are. Consequently, a specific shipment frequency m_s^i gives the lowest possible costs according to $H_s(Q, m_s)$

only up to $Q_s^w(m_s^i; m_s^i+1) = \min\{Q_s^v(m_s^i; m_s^i+1), Q_s^r(m_s^i)\}$. Following this line of argument, we can deduce that by incorporating the capacity constraint, the modified range of Q-values for which m_s^i is optimal can be described as follows:

$$Q_s^w(m_s^i-1; m_s^i) < Q \leq Q_s^w(m_s^i; m_s^i+1) \tag{7}$$

4 Optimal Lot Size Policy

Subsequently we carry out the optimisation for the modelled planning problem in two phases. In the first phase an initial feasible solution is generated. Afterwards, this solution serves as starting point for the second phase in which we obtain the optimal solution.

Now, we turn to the first phase. For the time being let us ignore the restriction (2) as well as the integrality constraints on the m_s-values in order to determine the global optimum for Q. For this, it is helpful to use a cost function which represents a lower bound on the total costs for given production lot sizes. Formulating the mentioned function requires solving the extremal equations $\partial C(Q, M)/\partial m_s = 0$ for each m_s and inserting the resulting expressions into Eq. (1). After some simplifications we obtain the following cost function depending only on Q:

$$C(Q) = 2 \cdot D \cdot \sum_{s=1}^{S} \sqrt{\alpha_s \cdot T_s} + Q \cdot \beta \cdot D + \gamma \cdot \frac{D}{Q} \tag{8}$$

Thus, the global optimal lot size is given by $Q^{**} = (\gamma/\beta)^{1/2}$. The next step consists of calculating the unconstrained optimal stage-specific integer transportation frequencies m_s^* for Q^{**}. In order to achieve this, it is necessary to go back to (5), to solve at first the left hand side of this inequality for m_s^i separately. After that, we solve for the right hand side. Combining those expressions results for a given lot size, e.g. Q^{**}, in (For a detailed derivation see [2]):

$$m_s^*(Q) = (e)\updownarrow = \sqrt{Q^2 \cdot (\alpha_s / T_s) + 0{,}25}\updownarrow \tag{9}$$

The up- and down-arrow in (9) denotes the rounding to the closest integer to e. Note, the transportation frequencies $m_s^*(Q^{**})$ for $s = 1, 2, \ldots, S$, denoted as M^{**}, minimise the total cost function given Q^{**}. However, usually Q^{**} is not the optimal lot size for M^{**}. Given a vector M, the corresponding optimal lot size is obtained by solving the extremal equation $\partial C(Q, M)/\partial Q = 0$ for Q:

$$Q^*(M) = \sqrt{\left(\gamma + \sum_{s=1}^{S} m_s \cdot T_s\right) \Bigg/ \left(\sum_{s=1}^{S} \frac{\alpha_s}{m_s} + \beta\right)} \tag{10}$$

Inserting M^{**} in Eq. (10) gives $Q^*(M^{**})$. But, $Q^*(M^{**})$ does not necessarily represent a feasible solution. If $Q^*(M^{**})$ is smaller than $Q^r(M^{**}) = \min\{Q_1^r(m_1^{**}); Q_2^r(m_2^{**}); \ldots; Q_S^r(m_S^{**})\}$ then $Q^*(M^{**})$, defined as Q^a, constitutes the initial feasible solution, otherwise it is $Q^r(M^{**}) = Q^a$. The initial feasible solution leads to costs in size of $C^a = C(Q^a, M^{**})$.

After we have found this solution, the second phase of our optimisation procedure begins. Clearly, an upper bound on the cost of the optimal solution is given by C^a. Moreover, using Eq. (8) we can establish a lower bound (Q_l) and an upper bound (Q_u) within which the optimal lot size lies. This is achieved by finding those Q-values yielding a cost equal to C^a. We choose Q_l as the starting point in searching for the optimal lot size. Inserting Q_l in Eq. (9) we receive the corresponding unconstrained stage-specific optimal transportation frequencies. Note, any pair of Q_l and $m_s^*(Q_l)$ has to satisfy restriction (2). In case that (2) is violated, we have to search for any stage the smallest integer m_s-value which guarantees the adherence of (2). For this, assume constraint (2) is binding. Now, solving this equation for m_s and rounding the right hand side up to the next integer value gives:

$$m_s^r(Q_l) = \lceil (\lambda_s \cdot Q_l)/[L_s - Q_l \cdot (1-\lambda_s)] \rceil \tag{11}$$

Since $m_s^r(Q_l)$ represents the minimum number of transports regarding Q_l, the optimal feasible values of the stage-specific shipment frequencies are given by:

$$m_s^+(Q_l) = \max \{m_s^*(Q_l) ; m_s^r(Q_l)\} \tag{12}$$

Recall the fact that for a given vector M only the lot size $Q^*(M)$ minimises the total cost function. However, due to the storage capacity constraint a realisable lot size can not be larger than $Q^r(M)$. Hence, after obtaining the vector $M^+(Q_l) = (m_1^+(Q_l); m_2^+(Q_l); \ldots; m_S^+(Q_l))$, the restricted optimal lot size is $Q^+ = \min \{Q^*(M^+); Q^r(M^+)\}$. The corresponding costs are $C(Q^+; M^+(Q_l))$. This costs have to be compared with those of the initial feasible solution. If $C(Q^+; M^+(Q_l)) \geq C^a$ we retain our existing temporary optimal solution. Otherwise, our actual solution becomes the new temporary optimal solution.

Nevertheless, it is possible that higher lot size solutions exist leading to lower costs than our temporary optimal solution. To obtain the critical Q-value for which the vector M^+ is no longer optimal, we must determine Q_s^w for each element m_s^+ of the vector M^+. Note, an additional shipment between two adjacent stages can result in lower costs $H_s(Q, m_s^i)$ only if the production lot size is at least equal to Q_s^w. For the smallest Q_s^w-value the corresponding shipment frequency m_s^+ has to be increased by one while all other elements of the vector M^+ remain at the same value. By using Eq. (10), we calculate now the optimal lot size for the modified vector M_{mod}^+. To evaluate the new solution we have to compute the resulting costs according to Eq. (1) and compare them with those of our temporary optimal solution. Obviously, if the latter costs are larger than the costs of the actual solution we define the new solution as our new temporary optimal solution. Now, an iterative procedure follows. The steps which we had described for M^+ must be repeated for M_{mod}^+. The algorithm ends if a vector M is found whose optimal lot size $Q^*(M)$ is equal or greater than the upper bound Q_u. Hence, the best solution so far is also the optimal solution for the given planning problem.

5 Conclusions

The objective of this paper was to determine the Economic Production Quantity for a multi-stage production system with limited storage capacity. For this purpose a two-phase algorithm was developed which minimises the sum of setup costs, inventory holding costs, and transportation costs. In contrast to most known approaches we have taken into account the transportation processes between consecutive stages explicitly. The proposed procedure determines not only the optimal production lot size but also the stage-specific shipment frequencies. Due to the formulation of the storage capacity restriction we were able to realise possibilities for considerable cost-savings compared with a conventional formulation of the restriction which ignores the influence of transportation activities.

References

1. Bogaschewsky R, Buscher U and Lindner G (2001) Optimizing Multi-Stage Production with Constant Lot Size and Varying Number of Unequal Sized Batches. Omega 29:183–191
2. Buscher U and Lindner G (2002) Simultane Bestimmung von Fertigungs- und Transportlosgrößen bei beschränkter Lagerkapazität – Planung stufenbezogener Transporthäufigkeiten. Working Paper, Bayerische Julius-Maximilians University Würzburg, Germany
3. Drezner Z, Szendrovits AZ and Wesolowsky GO (1984) Multi-Stage Production with Variable Lot Sizes and Transportation of Partial Lots. European Journal of Operational Research 17:227–237
4. Szendrovits AZ (1975) Manufacturing Cycle Time Determination for a Multi-Stage Economic Production Quantity Model. Management Science 22: 298–308
5. Szendrovits AZ and Drezner Z (1980) Optimizing Multi-Stage Production with Constant Lot Size and Varying Number of Batches. Omega 8:623–629

A Reverse Logistics Model with Integer Setup Numbers

Knut Richter[1], Imre Dobos[2]

[1] European University Viadrina, Frankfurt (Oder), Grosse Scharrnstr. 59, D-15230 Frankfurt (Oder), Germany, richter@euv-frankfurt-o.de - corresponding author

[2] Budapest University of Economics and Public Administration, Department of Business Economics, H-1053 Budapest, Veres Pálné u. 36, Hungary, idobos@mercur.bke.hu

Abstract. A production-recycling system is investigated. A constant demand can be satisfied by production and recycling. The used items are bought back and then recycled. The not recycled products are disposed off. A model is examined with EOQ-type inventory holding costs and linear waste disposal, recycling, production and buyback costs. It will be shown that under these circumstances the mixed strategies are dominated by the pure strategies. The paper generalizes a former model proposed by the authors for the case of integer recycling and production batch.

Keywords: EOQ model, Reverse Logistics, Production, Recycling, Waste disposal, Cost minimization

1. Introduction

A producer serves a stationary product demand occurring at the rate $D > 0$. This demand is served by producing new items as well as by recycling some part $0 \leq \delta \leq 1$ of the used products coming back to the producer at a constant return rate $d = \alpha D$, $0 \leq \alpha \leq 1$. It is assumed that the producer is in the situation to buy back all used product to recycle and/or to dispose off them. The parameters δ and α are called *marginal use rate* and *marginal buyback (return) rate*, respectively. The remaining part of the non-serviceable products $(1-\delta)d$ will be disposed off. $(1-\delta)$ is called *marginal disposal rate.* The length of the production and recycling is T.

The inventory stocks for *serviceable* products from the *production and recycling processes* (PRP) and for the *non-serviceable* items are determined. The following costs inputs will be used: The setup costs of recycling S_R, the setup costs of production S_P, the waste disposal cost for $(1-\delta)\cdot\alpha D\cdot T$ is C_w, the linear production cost for $(1-\delta\alpha)D\cdot T$ is C_P, the linear recycling cost for $\delta\cdot\alpha d\cdot T$ is C_R, the buyback cost for $\alpha\cdot D\cdot T$ is C_B, holding cost of serviceable items h_s and holding cost of non-serviceable items h_n. All of these parameters are positive. The following notations are applied to formulate the models: the number of recycling lots, positive integer m, the number of production lots, positive integer n and the lot size $x = D\cdot T$. If these variables are fixed then the demand is satisfied by recycling $\alpha\delta\cdot x$ units in m lots of size $\dfrac{\alpha\delta\cdot x}{m}$ and by producing $(1-\alpha\delta)\cdot x$ units of new items in n lots of size $\dfrac{(1-\alpha\delta)\cdot x}{n}$. The production and recycling rates are $1/\beta\cdot D$ and $1/\gamma\cdot D$.

First, only the EOQ-related setup cost and the holding cost parameters are considered. The overall cost for a production and recycling cycle T is [12]

$$C_O=(S_R\cdot m+S_P\cdot n)+\frac{h_s+h_n}{2\cdot D}\cdot(1-\gamma)\cdot\frac{(\alpha\delta x)^2}{m}+\frac{h_s}{2\cdot D}\cdot(1-\beta)\cdot\frac{[(1-\alpha\delta)\cdot x]^2}{n}+\frac{h_n}{2\cdot D}\cdot\frac{1-\alpha}{\alpha}\cdot(\alpha\delta x)^2. \quad (1)$$

The per time unit cost is then

$$C_A(x,m,n,\alpha,\delta)=\frac{C_O}{T}=D\cdot\frac{S_R\cdot m+S_P\cdot n}{x}+\frac{x}{2}\cdot V(m,n,\alpha,\delta) \quad (2)$$

with $V(m,n,\alpha,\delta)=(h_s+h_n)(1-\gamma)\alpha^2\delta^2\cdot\frac{1}{m}+h_s(1-\beta)(1-\alpha\delta)^2\cdot\frac{1}{n}+h_n\alpha(1-\alpha)\delta^2$ (3)

as per unit total holding cost. The formula (3) can be also presented as

$$V(m,n,\alpha,\delta)=H_R(\alpha,\delta)\cdot\frac{1}{m}+H_P(\alpha,\delta)\cdot\frac{1}{n}+H_n(\alpha,\delta), \quad (4)$$

where the factors express three different types of inventory holding cost occurring in our system: those induced by collecting and recycling used items, those induced by the production of new items, and those induced by collecting those items which were used for the first time. Later we need the relationships

$$M(\alpha,\delta)=\frac{H_R(\alpha,\delta)}{H_R(\alpha,\delta)+H_n(\alpha,\delta)} \text{ and } N(\alpha,\delta)=\frac{H_P(\alpha,\delta)}{H_P(\alpha,\delta)+H_n(\alpha,\delta)}, \quad (5)$$

Let now the non EOQ-related cost inputs be included. The sum of linear waste disposal cost, recycling cost, production cost, and buyback cost per unit time is given by the function [12]

$$C_N(\alpha,\delta)=C_W\cdot(1-\delta)\alpha D+C_R\cdot\delta\alpha D+C_P\cdot(1-\delta\alpha)D+C_B\cdot\alpha D.$$

Hence, the overall per time unit cost is $C(x,m,n,\alpha,\delta)=C_A(x,m,n,\alpha,\delta)+C_N(\alpha,\delta)$ and the corresponding optimal decision (solution), i.e. the lot size $x>0$, the setup numbers $m, n\in\{1,2,\ldots\}$, buyback rate α and waste disposal rate δ, have to be determined.

If the cost function $C_A(x,m,n,\alpha,\delta)$, which is obviously convex in x, is to be minimized in $x>0$ for fixed $m, n\geq 1$ and fixed α and δ, then the cost minimal lot sizes $x(m,n,\alpha,\delta)$ can be derived as $x(m,n,\alpha,\delta)=\sqrt{\frac{2D(S_R\cdot m+S_P\cdot n)}{V(m,n,\alpha,\delta)}}$ and the minimal cost is $\tilde{C}_A(m,n,\alpha,\delta)=\sqrt{2D\cdot(S_R\cdot m+S_P\cdot n)\cdot V(m,n,\alpha,\delta)}$.

These results contain the pure strategies, i.e. no recycling ($\alpha=\delta=0$) and no production ($\alpha=\delta=1$). The optimal costs can be written for these cases, as

$$\begin{aligned}\tilde{C}_P&=\tilde{C}_A(0,1,0,0)=\sqrt{2DS_P\cdot h_s(1-\beta)}, & \tilde{x}_P&=\sqrt{2DS_P/h_s(1-\beta)}\\ \tilde{C}_R&=\tilde{C}_A(1,0,1,1)=\sqrt{2DS_R\cdot(h_s+h_n)(1-\gamma)}, & \tilde{x}_R&=\sqrt{2DS_R/(h_s+h_n)(1-\gamma)}\end{aligned}, \quad (6)$$

where parameter $\tilde{C}_P$ is the minimal cost for the pure strategy production, and $\tilde{C}_R$ for recycling.

The two problems to be studied are the integer program

$$\hat{C}_A(\alpha,\delta)=\min\{\tilde{C}_A(m,n,\alpha,\delta): m,n\in\{1,2,\ldots\}\} \quad (7)$$

and by this way $\widetilde{C}(\alpha,\delta)=\min\{C(m,n,\alpha,\delta): m,n\in\{1,2,\ldots\}\}$, and the continuous program $\min\{\widetilde{C}(\alpha,\delta): \alpha\in[0,1], \delta\in[0,1]\}$. (8)

EOQ type reverse logistics problem was extensively studied in the literature [1-10]. The model discussed in this paper was introduced in paper [11]. The extended model [12] has examined the continuous case of setup numbers.

The following paper is organized as follows. The next section investigates the model for integer setup numbers (1). In section 3 the optimal buyback and recycling strategy is provided for the integer problem (2), and the last section summarizes the results of this paper.

2. The case of integer numbers of lots for production and recycling

Now we will minimize the cost function $\widetilde{C}_A(m,n,\alpha,\delta)$ in order to determine the optimal number of lots. After some calculation this cost function can be written in the following the form [12]

$$\widetilde{C}_A(m,n,\alpha,\delta)=\sqrt{2D\cdot\left[A(\alpha,\delta)\cdot\frac{m}{n}+B(\alpha,\delta)\cdot\frac{n}{m}+C(\alpha,\delta)\cdot m+D(\alpha,\delta)\cdot n+E(\alpha,\delta)\right]} \quad (9)$$

where

$$A(\alpha,\delta)=S_R\cdot H_P(\alpha,\delta),\quad B(\alpha,\delta)=S_P\cdot H_R(\alpha,\delta),$$

$$C(\alpha,\delta)=S_R\cdot H_n(\alpha,\delta),\quad D(\alpha,\delta)=S_P\cdot H_n(\alpha,\delta),\quad E(\alpha,\delta)=S_R\cdot H_R(\alpha,\delta)+S_P\cdot H_P(\alpha,\delta).$$

To solve this problem we can introduce a relaxed auxiliary problem (meta-model) ([5]). The continuous solution for the lot numbers $m(\alpha,\delta)$ and $n(\alpha,\delta)$ are

$$(m^o(\alpha,\delta),n^o(\alpha,\delta))=\begin{cases}\left(1,\dfrac{1-\alpha\delta}{\alpha\delta}\cdot\dfrac{\widetilde{x}_R}{\widetilde{x}_P}\cdot\sqrt{M(\alpha,\delta)}\right) & (\alpha,\delta)\in I\\ (1,1) & (\alpha,\delta)\in J\\ \left(\dfrac{\alpha\delta}{1-\alpha\delta}\cdot\dfrac{\widetilde{x}_P}{\widetilde{x}_R}\cdot\sqrt{N(\alpha,\delta)},1\right) & (\alpha,\delta)\in K\end{cases}$$

where

$$I=\{(\alpha,\delta)|\ \delta\le\delta_1(\alpha), 0\le\alpha\le1, 0\le\delta\le1\},\quad J=\{(\alpha,\delta)|\ \delta_1(\alpha)\le\delta\le\delta_2(\alpha), 0\le\alpha\le1, 0\le\delta\le1\},$$

$$K=\{(\alpha,\delta)|\ \delta\ge\delta_2(\alpha), 0\le\alpha\le1, 0\le\delta\le1\},\ \text{and}\ \ \delta_1(\alpha)=\frac{1}{\alpha}\cdot\frac{\widetilde{x}_R}{\widetilde{x}_R+\widetilde{x}_P/\sqrt{M(\alpha,\delta)}},$$

$$\delta_2(\alpha)=\frac{1}{\alpha}\cdot\frac{\widetilde{x}_R}{\widetilde{x}_R+\widetilde{x}_P\sqrt{\dfrac{(S_P+S_R)\cdot M(\alpha,\delta)-S_R}{S_P\cdot M(\alpha,\delta)}}}.$$

The regions I, J and K were determined in paper [12]. The continuous optimal inventory cost function after eliminating the setup numbers is

$$C_I(\alpha,\delta)=\begin{cases}(1-\alpha\delta)\cdot\widetilde{C}_P+\alpha\delta\cdot\widetilde{C}_R/\sqrt{M(\alpha,\delta)} & (\alpha,\delta)\in I\\ \sqrt{2D(S_R+S_P)\cdot V(1,1,\alpha,\delta)} & (\alpha,\delta)\in J\\ \alpha\delta\cdot\widetilde{C}_R+(1-\alpha\delta)\cdot\widetilde{C}_P/\sqrt{N(\alpha,\delta)} & (\alpha,\delta)\in K\end{cases}.$$

The optimal solution is automatically integer on set J. In other cases the optimal integer solution is not necessarily reached on $m=1$ or $n=1$, i.e. not necessarily on the boundary of $(m,n)\ge(1,1)$. Let therefore a feasible solution with $m=1$ or $n=1$ be called *boundary*. The next example shows such a case.

Example 1. Let $D=1{,}000$, $\beta = \gamma = 2/3$, $\alpha = 9/10$, $\delta = 1/2$, $S_P = 1{,}350$, $S_R = 440$, $h_s = 850$ and $h_n = 80$. Then the optimal continuous solution for *(m,n)* is *(1.484,1)* with inventory costs *35,547.4*, while the optimal integer solution is *(3,2)* with cost value *35,794.3*. The values of the cost function at points *(2,1)* and *(1,1)* are of *35,896.6* and *36,157.6*. The costs of the optimal integer boundary solution *(2,1)* are with 0.3 percent higher than those of the optimal solution *(3,2)*.

The same situation occurs in other models of reverse logistics, for example in the model of Teunter [9]. This model is similar to that proposed in our article, with the difference that this article investigates a situation with infinite production and recycling rates and the inventory holding costs for serviceable items are different for remanufactured and manufactured items, i.e. the unit holding costs for manufactured products are higher than those of remanufactured items. Let us show an example in this model that the manufacturing and remanufacturing lot numbers can be strictly greater than one in an optimal integer solution. We take the notations and numerical example of article [9].

Example 2. Let demand $\lambda=1{,}000$, the manufacturing and remanufacturing setup costs $K_m = 750$, $K_r = 100$, the unit holding costs for manufactured, remanufactured and non-serviceable items $h_m = 200$, $h_r = 50$ and $h_n = 20$, the return rate $r = 0.9$ and the reuse rate $u = 0.48$. Let us assume that the remanufacturing lot number is *R* and for manufacturing *M*. Then the optimal continuous solution for *(R,M)* is *(1.489, 1)* with inventory costs $AC^C = 10{,}845.2$, while the optimal integer solution is *(3,2)* with cost value $AC^I = 10{,}887.6$. This situation is displayed in fig. 1. The lines are the isocost lines for the optimal continuous AC^C and integer solutions AC^I. The values of the cost function at *(1,1)* and *(2,1)* are of *10,964.7* and *10,910.8*. The costs of the optimal integer boundary solution *(2,1)* are with 0.213 percent higher than those of the optimal solution *(3,2)*.

The structure of the optimal boundary solution is provided by

Theorem 1 [5]. The optimal boundary solution is $(m^b(\alpha,\delta),\ n^b(\alpha,\delta))$ with

$$\text{(i)}\ (\alpha,\delta)\in I \Rightarrow m^b(\alpha,\delta)=1,\ n^b(\alpha,\delta)=\left\lfloor \sqrt{\frac{(1-\alpha\delta)^2}{\alpha^2\delta^2}\cdot\frac{\tilde{x}_R^2}{\tilde{x}_P^2}\cdot M(\alpha,\delta)+\frac{1}{4}}+\frac{1}{2}\right\rfloor,$$

$$\text{(iii)}\ (\alpha,\delta)\in K \Rightarrow m^b(\alpha,\delta)=\left\lfloor \sqrt{\frac{\alpha^2\delta^2}{(1-\alpha\delta)^2}\cdot\frac{\tilde{x}_P^2}{\tilde{x}_R^2}\cdot N(\alpha,\delta)+\frac{1}{4}}+\frac{1}{2}\right\rfloor,\ n^b(\alpha,\delta)=1,$$

where $\lfloor x \rfloor$ denotes the maximal integer not greater than *x*.

The optimal boundary inventory cost function is

$$\hat{C}_A^b(\alpha,\delta)=\sqrt{A(\alpha,\delta)\cdot\frac{m^b(\alpha,\delta)}{n^b(\alpha,\delta)}+B(\alpha,\delta)\cdot\frac{n^b(\alpha,\delta)}{m^b(\alpha,\delta)}+C(\alpha,\delta)\cdot m^b(\alpha,\delta)+D(\alpha,\delta)\cdot n^b(\alpha,\delta)+E(\alpha,\delta)}.$$

Figure 1. An optimal not boundary solution in the model [9]

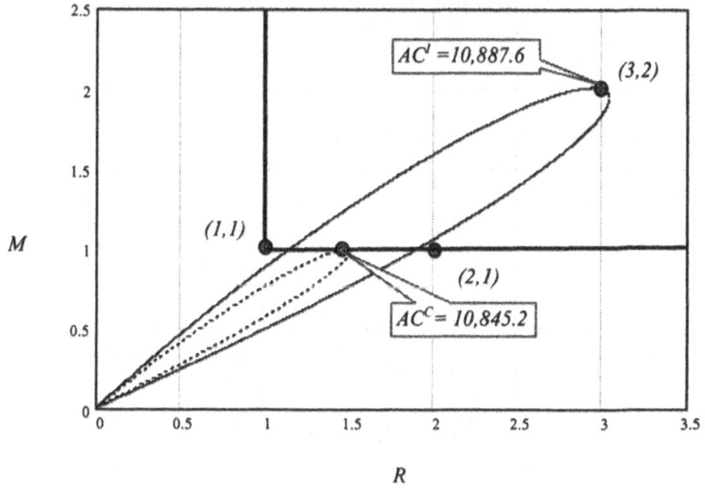

For a wide range of situations the optimal boundary solutions are really optimal. However, as we have seen, instances can be found, where such solutions are not optimal. The upper bound for the deviation from the optimum is given by

Lemma 1 [5]. Let $\hat{C}_A^b(\alpha,\delta)$ denote the minimal value for the optimal boundary solution and let $\hat{C}_A^o(\alpha,\delta)$ be the global minimum. Then the relation $\frac{\hat{C}_A^b(\alpha,\delta)-\hat{C}_A^o(\alpha,\delta)}{\hat{C}_A^o(\alpha,\delta)} \le \frac{1}{48}$ holds, i.e. relative error is less than 2.1 %.

If the continuous optimal setup numbers are integer for values ($m(\alpha,\delta)$, $n(\alpha,\delta)$), then the relative error is zero. Let us call this points (α,δ) as *switching points*. Below these points (lines) are determined.

Lemma 2 [5]. If the boundary property holds there are two sets X and Y of switching points (α,δ) with the properties

(i) $X=\{(\alpha,\delta): A(\alpha,\delta)=n(n+1)\cdot(B(\alpha,\delta)+D(\alpha,\delta)),\ n=1,2,\dots\}$ and
$\hat{C}_A^o(\alpha,\delta)=\tilde{C}_A(1,n^b(\alpha,\delta),\alpha,\delta)=\tilde{C}_A(1,n^b(\alpha,\delta)+1,\alpha,\delta)$ for $(\alpha,\delta)\in X$ and

(iii) $Y=\{(\alpha,\delta): B(\alpha,\delta)=m(m+1)\cdot(A(\alpha,\delta)+C(\alpha,\delta)),\ m=1,2,\dots\}$ and
$\hat{C}_A^o(\alpha,\delta)=\tilde{C}_A(m^b(\alpha,\delta),1,\alpha,\delta)=\tilde{C}_A(m^b(\alpha,\delta)+1,1,\alpha,\delta)$ for $(\alpha,\delta)\in Y$.

Remark. The sets X and Y separate the sets I and K into such subsets of identical optimal setup numbers $m^b(\alpha,\delta)$ and $n^b(\alpha,\delta)$. Then $I=\bigcup_{n=1}^{\infty} I_n$ and $K=\bigcup_{m=1}^{\infty} K_m$. For example, the sets X and Y consist of the union of the following lines

$$X=\bigcup_{n=1}^{\infty}\{(\alpha,\delta)\mid \delta=\delta^n(\alpha),0\le\alpha\le 1,0\le\delta\le 1\},\text{ where }\delta^n(\alpha)=\frac{1}{\alpha}\cdot\frac{\tilde{x}_R}{\tilde{x}_R+\tilde{x}_P\cdot\sqrt{n(n+1)}/\sqrt{M(\alpha,\delta)}},$$

and $Y=\bigcup_{m=1}^{\infty}\{(\alpha,\delta)\mid \delta=\delta^m(\alpha),0\le\alpha\le 1,0\le\delta\le 1\}$, where

$$\delta^m(\alpha)=\frac{1}{\alpha}\cdot\frac{\tilde{x}_R}{\tilde{x}_R+\tilde{x}_P\cdot\sqrt{\dfrac{1}{m(m+1)}-\dfrac{S_R}{S_P}\cdot\dfrac{1-M(\alpha,\delta)}{M(\alpha,\delta)}}}.$$

3. The dominance of pure strategies

In paper [12] we have shown that the EOQ-related and the sum of EOQ-related and non EOQ-related costs are dominated by the pure strategies for the case of continuous production and recycling setup numbers, i.e. for the inventory cost $C_I(\alpha,\delta)\geq\min\{\tilde{C}_P;\tilde{C}_R\}$ and for the total cost model $C(x,m,n,\alpha,\delta)\geq\min\{\tilde{C}_P+D\cdot C_P,\tilde{C}_R+D\cdot(C_B+C_R)\}$.

It is obvious that the continuous solutions to these models have lower costs than the boundary integer and the global integer solutions: $\hat{C}_A^b(\alpha,\delta)\geq\hat{C}_A^o(\alpha,\delta)\geq C_I(\alpha,\delta)\geq\min\{\tilde{C}_P;\tilde{C}_R\}$, and for the linear costs

$$\hat{C}_A^b(\alpha,\delta)+C_N(\alpha,\delta)\geq\hat{C}_A^o(\alpha,\delta)+C_N(\alpha,\delta)\geq C_I(\alpha,\delta)+C_N(\alpha,\delta)\geq$$
$$\geq\min\{\tilde{C}_P+D\cdot C_P,\tilde{C}_R+D\cdot(C_B+C_R)\}$$

By this last inequality a proof is given for the

Theorem 2 [12]. The optimal inventory holding and production-recycling strategy in this production-recycling model is a pure strategy: either to produce to meet the demand ($\alpha^o=\delta^o=0$) or to buy back and to recycle all used product without production ($\alpha^o=\delta^o=1$). The optimal pure strategy can be simply found by comparing the pure strategies.

Example 3. Let $D=1{,}000$, $\beta=\gamma=2/3$, $S_P=1{,}350$, $S_R=440$, $h_s=850$ and $h_n=80$. Then the inventory holding costs of recycling is *16,516.7* and those of production *33,326.7*. It is economical to recycle with buyback of all used items: $\alpha^o=\delta^o=1$.

Example 4. Let $D=1{,}000$, $\beta=2/5$ $\gamma=2/3$, $S_P=360$, $S_R=440$, $h_s=85$ and $h_n=80$. Then the inventory holding costs of production is *6,059.7* and those of recycling *6,957.01*. It is more effective to produce and not to recycle: $\alpha^o=\delta^o=1$.

4. Conclusions and further research

In this paper we have investigated a production-recycling model. By minimizing the inventory holding costs it was shown that one of the pure strategies (to produce or to recycle all products) is optimal. A similar proposition can be obtained minimizing the total EOQ and non-EOQ related costs. A similar result was obtained by Richter [3] in a waste disposal model with remanufacturing and by Dobos and Richter [11, 12] in a production and recycling model.

Probably these pure strategies are technologically not feasible and some used products will not return or even more as the sold ones will come back, and some of them will be not recycleable. This kind of generalization of this basic model could be the introduction of an upper bound on the buyback rate which is strictly smaller than one. In such a case a mixed strategy would be economical compared to the pure strategy "production". The answer to this question is left to a next paper.

References

[1] Richter, K. (1996): The EOQ repair and waste disposal model with variable set-up numbers, European Journal of Operational Research 96, 313-324

[2] Richter, K. (1996): The extended EOQ repair and waste disposal model, International Journal of Production Economics 45, 443-447

[3] Richter, K. (1997): Pure and mixed strategies for the EOQ repair and waste disposal problem, OR Spektrum 19, 123-129

[4] Richter, K. Dobos, I. (1999): Analysis of the EOQ repair and waste disposal model with integer setup numbers, International Journal of Production Economics 59, 463-467

[5] Dobos, I., Richter, K. (2000): The integer EOQ repair and waste disposal model – further analysis. Central European Journal of Operations Research 8, 173-194

[6] Schrady, D.A. (1967): A deterministic inventory model for repairable items. Naval Research Logistic Quarterly 14, 391-398

[7] Nahmias, N., Rivera, H: (1979): A deterministic model for repairable item inventory system with a finite repair rate. International Journal of Production Research 17(3), 215-221

[8] Mabini, M.C., Pintelon, L.M., Gelders, L.F. (1998): EOQ type formulation for controlling repairable inventories. International Journal of Production Economics 54, 173-192

[9] Teunter, R.H. (2001): Economic Ordering Quantities for Recoverable Item Inventory Systems, Naval Research Logistics, Vol. 48, 484-495

[10] Dobos, I., Richter, K. (1999): Comparison of Deterministic One-Product Reverse Logistics Models, in: Hill, R., Smith, D. (Eds.): Inventory Modelling: A Selection of Research Papers Presented at the Fourth ISIR Summer School (1999), Exeter 1999, 69-78

[11] Dobos, I., Richter, K. (2002): A production/recycling model with stationary demand and return rates, Central European Journal of Operations Research, to appear

[12] Dobos, I., Richter, K. (2002): An extended production/recycling model with stationary demand and return rates, , Proceedings of the 12th International Working Seminar on Production Economics, Vol. 1, 47-60, Igls/Innsbruck

Lotsizing in a Production System with Rework and Product Deterioration

K. Inderfurth, G. Lindner and N.P. Rahaniotis

Otto-von-Guericke University Magdeburg, Faculty of Economics and Management, D-39016 Magdeburg

Abstract. Producing new or recovering defective products often takes place on a common facility, with these activities carried out in lots. Consequently, there is a necessity to coordinate the production and rework activities with respect to the timing of operations and also with regard to appropriate lot sizes for both processes while completely satisfying a given demand. Thereby, it has to be taken into account whether the state of defective items that await rework worsens in the course of time or not. In this paper we present an EPQ model which addresses all of these aspects. Considering set-up and inventory holding costs, optimization algorithms are developed covering different planning situations.

1 Introduction

For manufacturing purposes most often a quite sophisticated equipment is used. However, some defective units are fabricated even by production systems comprised of high quality facilities. In many cases defective items incorporate substantial values and hence there is an economical incentive to rework those products into an "as new" condition. Besides this motive, rework activities can also be mandated by existing laws or due to the green image a company will present.

Often rework takes place on the same machinery used for producing new items. Moreover, in such a context, a company has two sources for satisfying the demand of a certain product. Namely supplying reworked units or new produced items. Consequently, there is a necessity to coordinate the production and rework activities with respect to the timing of operations as well as with regard to appropriate lot sizes for both processes. Thereby, it has to be taken into account whether the state of defective items that await rework deteriorates in the course of time or not. Perishable reworkable defectives occur e.g. in the pharmaceutical industry [1].

Although determining lot sizes in the presence of deteriorating items has already received a lot of attention, all approaches dealing with this problem assume (implicitly) that defective items have to be disposed of (see e.g. [2, 6]). Likewise, joint planning of production and rework lot sizes has attracted many researchers (see e.g. [4, 5]). Nevertheless, the application of the available models is restricted to situations where deterioration has no influence on lotsizing decisions.

To the best of our knowledge, the only approach which integrates both mentioned streams of research was suggested in [7]. There a single-facility manufacturing system is considered where items are produced and, in case of insufficient quality, reworked in subsequent lots. Due to time-based deterioration of recoverables it is assumed that rework processing time and costs per item increase linearly with the time span that an item is held in stock and awaits rework. Defects occur stochastically, and the objective is to determine the production lot size in such a way that the expected system profit, given that all serviceable products can be sold at a given price, is maximized. In [7] a closed- form expression for the objective function is given which can be used to determine the optimal production lot size numerically.

In contrast to this approach we address a more complicated cost minimization problem with additional feasibility constraints. By restricting defects to be caused by a deterministic process we are able to develop closed-form expressions not only for the objective function but also for the optimal production lot size. These formulas can be used for an easy computation of the optimal lot size as well as for gaining general insights into how this lot size depends on the various parameters.

The remainder of this paper is organized as follows. In section 2 we give a detailed description of the considered planning situation. Subsequently we analyze the so-called basic approach which deals with non-deteriorating reworkable items. In section 3 we extend the basic approach by incorporating different aspects of deterioration. Finally, conclusions are outlined in section 4.

2 Fundamentals of the Model

2.1 The Planning Situation

In this paper we present a static deterministic lot-size model for a single product, focusing on a two-stage manufacturing system within which production as well as rework activities are carried out. The main characteristics of our model can be summarized as follows.

The considered manufacturing system faces a constant and continuous demand D over an infinite time horizon. Moreover, the production and rework activities are carried out in lots of size Q_p (Q_r) at the same facility and with finite rates P_p (P_r). Hence, we have to ensure that the production line is used for only one process at a time while satisfying the entire demand without backlogging. Each changeover from production to rework and vice versa causes a set-up cost (set-up time) of S_p or S_r (t_{pr} or t_{rp}), respectively.

According to our coordination policy, one inventory cycle starts with fabricating a production lot. Due to an imperfect underlying process this quantity contains only a fraction α ($0 < \alpha < 1$) of serviceable units. A portion ($1-\alpha$) of units are defectives and will be reworked. After reprocessing these items are always as good as new. Disposal of defectives is not considered. During production, all serviceable (reworkable) items are transported continuously to the so-called serviceables (recoverables) inventory. Demand is satisfied from serviceables inventory. For every unit at the corresponding inventory a holding cost rate of h_s (h_r) per unit per unit time is charged.

Defective units stay in the recoverables inventory until the rework process is carried out. This takes place directly after the preceding production run and the necessary set-up are finished. Once reprocessing of the rework lot, containing $Q_r = (1-\alpha)\cdot Q_p$ units, has started, items are shipped according to a parallel movement of units to the serviceables inventory as long as all units of the actual lot are completed. Afterwards, the next inventory cycle is initiated such that the first good quality item out of one production lot arrives at the serviceables inventory exactly at the moment where all previously delivered products are used up.

Finally, for ascertaining the optimal production and rework lot sizes we use the criterion of minimization of total relevant cost per unit time. For assuring feasible solutions, the relation $\alpha\cdot P_p > D$ must always hold.

2.2 The Basic Model

The basic approach is featured by the property that the state of reworkable defectives does not change in the course of time while they wait to be reworked. Hence, the model considers the case of non-deteriorating items. In order to determine the optimal solution, we have to solve the minimization problem as stated below.

$$C(Q_p) = 0{,}5 \cdot Q_p \cdot H \cdot D + F \cdot D / Q_p \rightarrow \text{Min!} \qquad (1)$$

Subject to:

$$Q_p \geq Q_p^{+} = (t_{rp} + t_{pr}) \cdot D / [1 - D / P_p - (1-\alpha) \cdot D / P_r] \qquad (2)$$

$$Q_p \geq Q_p^{++} = t_{pr} \cdot D / (\alpha - D / P_p) \qquad (3)$$

$$Q_p > 0 \qquad (4)$$

where:

$$H = (1-\alpha) \cdot [(1-\alpha) / P_r + 1 / P_p] \cdot (h_r - h_s) + (1/D - 1/P_p) \cdot h_s$$

$$F = S_p + S_r$$

The convex objective function (1) is comprised of inventory holding costs per unit time related to the recoverables as well as serviceables inventory (first term) and the second term represents the fixed set-up costs per unit time. Constraint (2) guarantees that the time span between two subsequent production runs is large enough to set-up the system for the rework as well as the production lot and to process both corresponding quantities. From (2) it can easily be seen that P_r can take on values smaller than D. However, a feasible solution exists if and only if the rework rate fulfills the condition: $P_r > (1-\alpha)\cdot D/(1-D/P_p)$.

Moreover, restriction (3) ensures in case of $P_r > D$ that the time between the end of production and the moment where the corresponding serviceables inventory level hits zero, is sufficiently large in order to prepare the facility for the rework run. In fact, if $P_r \leq D$ occurs, constraint (3) is always looser than (2).

If we first ignore constraints (2) and (3), the optimal production lot size can be obtained by solving the extremal equation $dC(Q_p)/dQ_p = 0$ for Q_p.

$$Q_p^* = \sqrt{2 \cdot F/H} \tag{5}$$

Finally, because of convexity of total cost function, the optimal solution of the constrained planning problem is given by:

$$Q_p^{opt} = \max\{Q_p^*; Q_p^+; Q_p^{++}\} \tag{6}$$

3 Modeling Aspects of Deterioration

So far we have restricted our attention to the case of non-deteriorating defectives. However, in several practical situations this assumption does definitely not hold. Consequently, if the rate of deterioration is not sufficiently low or if there is no deterioration at all, its impact on modeling the considered decision problem must not be neglected. In order to study the impact of deteriorating reworkables on ascertaining the optimal lot size policy, we will extend the planning situation – as already described – by the following assumptions.

Since the state of defective units worsens in the course of time while they wait to be reworked, there will be differences in deterioration among items coming out of one production lot. However, in our model we do not consider each item individually, but use an average waiting time approach. Furthermore, we presume that the deterioration of recoverables exhibits a significant dependence on the production lot size, since with increasing this quantity the average waiting time of a defective item will increase. Due to the deterioration caused by waiting, the cost and time of rework operations necessary to bring the defective item into an acceptable state also will increase.

Now, let us turn to the first modified planning situation, namely where the rework costs c_r per unit increase linearly with lot size Q_p. Specifically, we assume: $c_r = c_{rb} + c_{ra} \cdot Q_p$. In this relation c_{rb} represents the basic rework cost per unit whereas c_{ra} denotes the additional rework cost per unit per unit incorporated in one production lot. Determining the optimal solution regarding the considered problem is quite easy. One simply has to replace H in the basic approach by $H_I = H + 2 \cdot (1-\alpha) \cdot c_{ra}$ and the solution procedure, discussed in the previous subchapter, can be applied. Of course, since H_I is larger than H, the unrestricted optimal production lot size is smaller than its counterpart of the basic approach.

Subsequently we take a closer look at the more realistic case where besides the rework cost per unit also the rework processing time per unit is influenced by the deterioration of recoverables. Equivalently to c_r we assume a linear dependence of the rework processing time per unit on Q_p. Thereby, the rework rate is given by the relation: $P_r = 1/(\tau_{rb} + \tau_{ra} \cdot Q_p)$. The symbol τ_{rb} denotes the basic rework time per unit and τ_{ra} stands for the additional rework time per unit for each unit included in one production lot of size Q_p. By substituting the expression for P_r into Eq. (1) the modified total cost function is represented by:

$$C(Q_p) = 0{,}5 \cdot Q_p^2 \cdot T \cdot D + 0{,}5 \cdot H_{II} \cdot D + F \cdot D / Q_p \rightarrow \text{Min!} \tag{10}$$

where:

$T = 0{,}5 \cdot (1-\alpha)^2 \cdot \tau_{ra} \cdot (h_r - h_s)$ and $H_{II} = H_I$ with $P_r = 1/\tau_{rb}$

Moreover, constraints (2) and (3) change to:

$$\max\{Q_{pu}; Q_p^{++}\} \leq Q_p \leq Q_{po} \tag{11}$$

with:

$$Q_{pu} / Q_{po} = Z -/+ \sqrt{Z^2 - (t_{rp} + t_{pr})/[(1-\alpha) \cdot \tau_1]} \tag{12}$$

where:

$$Z = 0{,}5 / \tau_1 \cdot [(1/D - 1/P_p) \cdot 1/(1-\alpha) - \tau_0]$$

Note, for the planning situation considered here, a feasible solution exists if and only if the expression in the square root in Eq. (12) is larger than or equal to zero and if the Q_p-value where (3) is binding is smaller than or equal to Q_{po}.

For determining the optimal solution regarding the minimization problem defined just now we have to distinguish three different cases depending on the relation between the holding cost rates h_r and h_s.

If $h_r = h_s$ the optimal unrestricted production lot size can be obtained by substituting H in Eq. (5) by H_{II}. Since the cost function is convex in this case, the constrained optimal solution is given by:

$$Q_p^{opt} = \begin{cases} \max\{Q_{pu}; Q_p^{++}\} & \text{if} \quad Q_p^* < \max\{Q_{pu}; Q_p^{++}\} \\ Q_p^* & \text{if} \quad \max\{Q_{pu}; Q_p^{++}\} \leq Q_p^* \leq Q_{po} \\ Q_{po} & \text{if} \quad Q_p^* > Q_{po} \end{cases} \tag{13}$$

It is worthwhile to mention that in case of $h_r = h_s$ neither α nor P_r influence Eq. (10). Therefore, the fraction of defectives as well as the deterioration of recoverables have only an implicit impact on total costs via the restrictions.

When h_r is larger than h_s Eq. (10) remains convex. However, solving the extremal equation $dC(Q_p)/dQ_p = 0$ for Q_p in order to ascertain the unconstrained optimal production quantity does now require to solving a cubic equation. This can be done using available standard methods like e.g. the Newton-Raphson-method.

The last case to be analyzed is that of $h_r < h_s$. Unfortunately, if this relation occurs, the objective function is no longer convex. In this case, for $Q_p \geq 0$, the cost function can be shown to be either monotonously decreasing or to have a single local minimum followed by a single local maximum. Consequently, determining the constrained optimal solution can be done in the following way. First of all, it has to be checked whether there is a local minimum which fulfills (11) or not. If such a value exists, we have to compare its total costs with those related to the lower and the upper bound where (11) is binding in order to define the optimal solution. Otherwise only the two mentioned boundary values must be tested for selecting the cost minimizing solution. (For a detailed treatment of the planning problems discussed in this paper see [3]).

4 Conclusions

In this paper we have presented a model which allows the simultaneous determination of production and rework lot sizes in the presence of deteriorating reworkable defectives. To the best of our knowledge, there is no further lotsizing approach available in this area of investigation.

Of course, the optimality of the solution depends on the assumptions made. The relaxation of some of them allows for possible extensions of the model. For instance alternative forms of deterioration, more general rules for coordinating production and rework lots, and the inclusion of a disposal option could be analyzed.

References

1. Flapper SDP, Fransoo JC, Broekmeulen RACM, and Inderfurth K (2002) Planning and control of rework in the process industries: a review. Production Planning & Control 13: 26-34
2. Goyal SK and Giri BC (2001) Recent trends in modeling of deteriorating inventory. European Journal of Operational Research 134: 1-16
3. Inderfurth K, Lindner G and Rahaniotis NP (2002) Lotsizing in a Production System with Rework and Product Deterioration. Working Paper, Otto-von-Guericke-University Magdeburg, Germany
4. Lindner G and Buscher U (2002) An Optimal Lot and Batch Size Policy for a Single Item Produced and Remanufactured on One Machine in the Presence of Limitations on the Manufacturing and Handling Capacity. Working Paper, Otto-von-Guericke-University Magdeburg, Germany
5. Minner S (2001) Economic Production and Remanufacturing Lot-sizing Under Constant Demands and Returns. In: B. Fleischmann et al. (eds) Operations Research Proceedings 2000. Springer, Berlin et al., pp 328-332
6. Misra RB (1975) Optimum production lot size for a system with deteriorating inventory. International Journal of Production Research 13: 495-505
7. Teunter RH and Flapper SDP (2003) Lot-sizing for a single-stage single-product production system with rework of perishable rework defectives. OR Spectrum forthcoming

Mathematical Programming Models for Strategic Supply Chain Planning and Design

J. Kalcsics, M. T. Melo, and S. Nickel

Fraunhofer Institut für Techno- und Wirtschaftsmathematik,
Gottlieb-Daimler-Str. 49, D-67663 Kaiserslautern, Germany

Abstract. Although facility location and configuration of global manufacturing and distribution networks have been studied for many years, a number of important real world issues has not received adequate attention in the literature. In this paper we propose realistic models for the strategic planning and design of global supply chains. Our main concern is to provide comprehensive models which explicitly capture the essential elements of many industrial environments. A generic mathematical programming model is described in detail which became part of *Supply Chain Design*, a module within mySAP™ Supply Chain Management, developed by SAP AG. Moreover, extensions to the dynamic situation are discussed.

1 Introduction

Structuring global supply chain networks is a complex decision making process. The typical inputs to such a process consist of a set of customer zones to serve, demand projections for the different customer zones, a set of products to be manufactured and sold, and information about future conditions and costs (e.g. transportation and production). Given the above inputs, companies have to decide, among other things, where to locate new service facilities (e.g. plants, warehouses), how to allocate procurement and production activities to the various manufacturing facilities of the network, and how to manage the distribution of products.

Although facility location and configuration of global manufacturing and distribution networks have been studied for many years, a number of important real world issues have not received adequate attention in the literature (see Bender *et al.* [1] for a review of models, company-specific case studies, and decision support systems). For example, the supply chain structure typically consists of arborescent structures usually limited to at most two echelons (e.g. plants and warehouses), a system of distribution channels between these echelons, and a relatively simple cost structure. Excluding a few cases, the primary shortcoming of such models arises from the fact that their results cannot be extrapolated to real world supply chains. Our objective is to fill this gap by proposing realistic models for the strategic planning and design of global supply chains. Our models are part of the software tool *Supply Chain Design* which assists decisions-makers in the design of supply chain networks. *Supply Chain Design* is a module within the software mySAP™ Supply

Chain Management, developed by SAP AG, which supports the organization and operation of supply chain networks (see e.g. Bender *et al.* [1]). Model extensions to dynamic planning horizons are also discussed.

2 Strategic Supply Chain Planning

We consider a general supply chain network where different products are delivered to satisfy the requirements of several customer zones. Since a strict categorization of facilities into echelons is not required, the network may accommodate any number of different types of facilities (e.g. plants, central and regional warehouses), which already operate at the beginning of the planning project. No restrictions are imposed on the number of different facility types and on the transportation channels used by the company for shipping its products. In other words, commodities can be transported between any type of facility. This allows, for example, the return of pallets or empties (e.g. bottles, printer cartridges) to be modeled.

Figure 1 shows a supply chain network with suppliers, plants, distribution centers (DCs) and customers. The arrows indicate the transportation channels that are available to ship the products. In addition to inbound and outbound transportation, products may flow between facilities of the same type and from customer locations to other facilities. The latter flows describe product recovery activities for the purposes of recycling, remanufacturing and re-use.

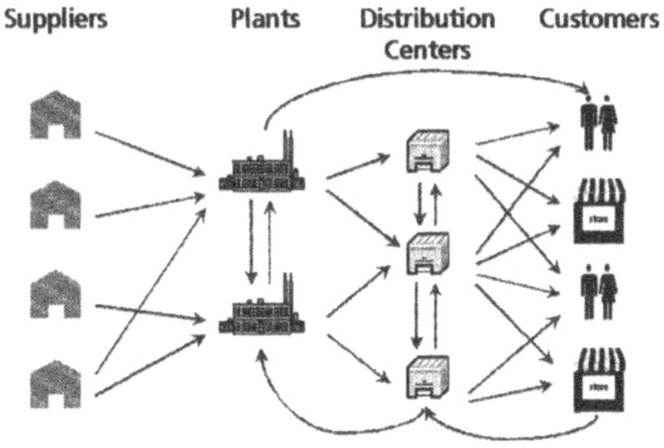

Fig. 1. Example of a supply chain network

For strategic supply chain planning we consider general networks such as that of Figure 1 in order to capture important practical aspects. Continuous as well as discrete mathematical models are useful to improve the performance of a supply chain. In particular, models should focus on the analysis and redesign of an existing network structure with respect to facility location, procurement, production, distribution and transportation decisions (see Kalcsics *et al.* [3]).

While classical facility location models mainly focus on location-allocation decisions, we take the whole supply chain into account. This means that our models deal with the planning of supply chain activities such as those described above, and thus generalize many well-known location models.

3 A General Location Model for Strategic Supply Chain Planning

Locations for new facilities are selected among a pre-defined set of candidate sites. Moreover, location planning may be carried out for different types of facilities simultaneously (e.g. plants and DCs). Regarding the production structure, each end product may have a bill of materials which describes the requirements for components, subassemblies and raw materials. In addition to production, commodities may be purchased from outside suppliers. The objective is to determine the optimal number of operating facilities and their locations, as well as the amount of products to procure, the amount of products to manufacture and the flow of products throughout the network in such a way that the overall costs are minimized subject to various side-constraints.

Before introducing a formulation of the problem, we first define the set L of facilities. This set includes the so-called *selectable* and *non-selectable* facilities. Selectable facilities form the set S, a subset of L, and correspond to existing facilities as well as potential sites for establishing new facilities. The second category of facilities, the so-called non-selectable, forms the set $L \setminus S$ and includes all facilities that already exist prior to the planning period and which must remain in operation. Note that a non-selectable facility may also have demand requirements, that is, it may correspond to a customer zone. In addition, we also define the set P of all product types ranging from raw materials and semi-finished items to end products. Further notation is introduced next.

Inputs

$BC_{\ell,p}$: cost of purchasing one unit of product $p \in P$ in facility $\ell \in L$ from an outside supplier

$a_{\ell,r,p}$: number of units of product $r \in P$ required to manufacture one unit of product $p \in P$ in facility $\ell \in L$

$PC_{\ell,p}$: cost of manufacturing one unit of product $p \in P$ in facility $\ell \in L$

$TC_{\ell,\ell',p}$: cost of shipping one unit of product $p \in P$ from facility $\ell \in L$ to facility $\ell' \in L$ $(\ell \neq \ell')$

SC_ℓ : costs derived from closing an existing facility $\ell \in S$

FC_ℓ : fixed costs for setting up a new facility at site $\ell \in S$

$d_{\ell,p}$: demand for product $p \in P$ at facility $\ell \in L$

Δ_1 : minimum number of *selectable* facilities to be operated

Δ_2 : maximum number of *selectable* facilities to be operated

Decision variables

$b_{\ell,p}$ = amount of product $p \in P$ purchased from an outside supplier at facility $\ell \in L$

$h_{\ell,p}$ = amount of product $p \in P$ manufactured in facility $\ell \in L$

$v_{\ell,\ell',p}$ = amount of product $p \in P$ shipped from facility $\ell \in L$ to facility $\ell' \in L$ $(\ell \neq \ell')$

$$\delta_\ell = \begin{cases} 1 \text{ if facility } \ell \text{ is operating, } \ell \in S \\ 0 \text{ otherwise} \end{cases}$$

Assuming that all inputs are nonnegative and no capacity limits are set, our formulation is as follows:

$$\text{MIN} \sum_{\ell \in L} \sum_{p \in P} BC_{\ell,p}\, b_{\ell,p} + \sum_{\ell \in L} \sum_{p \in P} PC_{\ell,p}\, h_{\ell,p}$$

$$+ \sum_{\ell \in L} \sum_{\ell' \in L \setminus \{\ell\}} \sum_{p \in P} TC_{\ell,\ell',p}\, v_{\ell,\ell',p} + \sum_{\ell \in S} FC_\ell\, \delta_\ell + \sum_{\ell \in S} SC_\ell\, (1 - \delta_\ell) \quad (1)$$

s. t.

$$b_{\ell,p} + \sum_{\ell' \in L \setminus \{\ell\}} v_{\ell',\ell,p} + h_{\ell,p} =$$

$$\sum_{\ell' \in L \setminus \{\ell\}} v_{\ell,\ell',p} + \sum_{r \in P} a_{\ell,p,r}\, h_{\ell,r} + d_{\ell,p}, \quad \ell \in L, p \in P \quad (2)$$

$$\sum_{\ell' \in L \setminus \{\ell\}} \sum_{p \in P} v_{\ell,\ell',p} \leq M\, \delta_\ell, \quad \ell \in S \quad (3)$$

$$\sum_{\ell \in L \setminus S} \sum_{p \in P} v_{\ell,\ell',p} \leq M\, \delta_{\ell'}, \quad \ell' \in S \quad (4)$$

$$\sum_{p \in P} h_{\ell,p} \leq M\, \delta_\ell, \quad \ell \in S \quad (5)$$

$$\sum_{p \in P} b_{\ell,p} \leq M\, \delta_\ell, \quad \ell \in S \quad (6)$$

$$\Delta_1 \leq \sum_{\ell \in S} \delta_\ell \leq \Delta_2 \quad (7)$$

$$b_{\ell,p} \geq 0, \quad h_{\ell,p} \geq 0, \quad v_{\ell,\ell',p} \geq 0, \quad \ell \in L, \ell' \in L \setminus \{\ell\}, p \in P \quad (8)$$

$$\delta_\ell \in \{0, 1\}, \quad \ell \in S \quad (9)$$

The total costs to be minimized are described in (1). These range from procurement, production, and transportation costs to fixed charges for opening and closing facilities. Note that the latter costs may also include operating costs. Furthermore, handling costs for incoming and outgoing goods in a given facility can easily be incorporated in (1). Constraints (2) are conservation of flow equations, where the inbound flow to a facility ℓ with respect to some product p results from procurement and production activities at the facility plus the total product flow from other facilities $\ell' \neq \ell$. The outbound flow in equations (2) consists of the product demand to be satisfied at facility ℓ, the production of new materials using product p as raw material, and the total product flow to other facilities $\ell' \neq \ell$. Inequalities (3), (4), (5) and (6) ensure that transportation, production and procurement activities are only allowed in operating facilities. The fixed term M denotes the maximum amount that may be required by any facility with respect to all product types. The calculation of M takes not only the external demand $d_{\ell,p}$ into account, but also the secondary requirements for a product p that may arise due to production activities where p is used as a raw material or intermediate product. Finally, constraints (7) impose a lower and upper limit on the total number of facilities to be selected from the set S.

The above formulation describes a large scale Mixed Integer Problem (MIP) which is solved for reasonable size problems with the commercial package ILOG CPLEX 7.0 using the C++ modeling library ILOG Concert Technology 1.0 (ILOG [2]).

A natural extension of the above MIP formulation is to consider capacity constraints. These are modeled in terms of consumption of resources required for manufacturing and handling commodities. Resources can be site and product independent meaning that a given (production/handling) resource may be used by different products in different facilities. This generalizes the classical way capacity availability is modeled in facility location problems. In addition, overtime work can be considered by allowing extension of the available resources at the expense of additional costs. The decision maker is also given the possibility to allow external demand not to be completely met. In the case that part or all of the demand of a customer zone for a given product is not satisfied, a penalty cost is charged. All these extensions are included in the software module *Supply Chain Design.*

4 Outlook: Extension to Dynamic Situations

Since the establishment of facilities is a comprehensive project requiring substantial investment capital and coordination of all operational aspects involved in setup activities, it is often necessary to smooth the relocation process and eventual phase-out of facilities in a dynamic multiperiod process. A gradual relocation of facilities in terms of shifting capacity from an existing site to the site of a newly established facility has the advantage of not disrupting supply chain activities and allows relocation plans to be carried out smoothly. For example, in some cases it is not possible to relocate a large facility such as a plant in one step, both from an investment and operational point of view. Instead, it is preferable to gradually transfer production capacity from the current location to a new site. Such a smooth transition also allows a better management of the required investment capital. Instead of investing a substantial amount at once and thus putting great financial strain on the company, capital expenditures can be distributed among several time periods. Another practical situation to which a relocation scenario applies concerns the merging of companies. In this case, formerly separated supply chains are consolidated and a joint system structure needs to be planned. This entails shutting down some of the existing facilities and concentrating capacities in new locations.

Currently, we are focusing on the extension of our static models to the case of dynamic relocation of facilities (see Melo *et al.* [4]). The new mathematical programming models capture a number of important practical aspects and focus on the following strategic issues:

- Where and when should a partial or total relocation of existing facilities take place?
- Using the available budget, how gradually should the capacity of existing facilities be reduced and when should the transition to newly established facilities be made without disrupting the supply chain activities?
- How should the supply chain be operated in each planning period with respect to the provision of materials, storage and shipment of products so that demand requirements are satisfied at least cost?

References

1. T. Bender, H. Hennes, J. Kalcsics, M. T. Melo, and S. Nickel. Location Software and Interface with GIS and Supply Chain Management. In Z. Drezner and H. W. Hamacher, editors, *Facility Location: Applications and Theory*, chapter 8, pages 233–274. Springer-Verlag, Berlin, Heidelberg, 2002.
2. ILOG Optimization Suite. ILOG, Inc., Incline Village, Nevada, 2000. http://www.ilog.com/products/optimization.
3. J. Kalcsics, T. Melo, S. Nickel, and V. Schmid-Lutz. Facility Location Decisions in Supply Chain Management. In K. Inderfurth, G. Schwödiauer, W. Domschke, F. Juhnke, P. Kleinschmidt, and G. Wäscher, editors, *Operations Research Proceedings 1999*, pages 467–472. Springer-Verlag, Berlin, 2000.
4. M. T. Melo, S. Nickel, and F. Saldanha da Gama. Dynamic multi-commodity capacitated facility location: A mathematical modeling framework for strategic supply chain planning. Technical report, Fraunhofer Institut für Techno- und Wirtschaftsmathematik (ITWM), Kaiserslautern, Germany, 2002. In preparation.

Ein dynamisches Verhandlungsmodell des Supply Chain Management

Eric Sucky

Seminar für Logistik und Verkehr, Johann Wolfgang Goethe-Universität, Mertonstr. 17, 60054 Frankfurt am Main, esucky@wiwi.uni-frankfurt.de

1 Einleitung

Im Fokus von Supply Chain Management (SCM) steht die unternehmensübergreifende Koordination der Material- und Informationsflüsse über den gesamten Wertschöpfungsprozess, von der Rohstoffgewinnung über die einzelnen Veredelungsstufen bis hin zum Endkunden (vgl. [8], S. 8). Die Koordination in der Supply Chain (SC) kann generell nach dem hierarchischen oder heterarchischen Prinzip erfolgen: Während das hierarchische Koordinationsprinzip dadurch gekennzeichnet ist, dass eine übergeordnete Planungsebene Rahmenpläne entwirft, die untergeordneten Planungsebenen als Vorgaben dienen, treffen nach dem heterarchischen Prinzip die Akteure in der SC ihre Entscheidungen durch gegenseitige Übereinkunft (vgl. [10], S. 5). Das hierarchische Koordinationsprinzip ist im Rahmen von SCM kritisch zu beurteilen: „Most supply chains are composed of independent agents with individual preferences. These agents could be distinct firms or they could even be managers within a single firm. (...) It is also reasonable to assume that each agent will attempt to optimize his own preference, knowing that all of the other agents will do the same.“ ([2], S. 113). Im Rahmen einer unternehmensübergreifenden Koordination der Wertschöpfungsprozesse sind Planungsprobleme mit interpersonellen Mehrfachzielen zu lösen. Die Koordination erfolgt i. d. R. nach dem heterarchischen Prinzip, d. h., es erfolgt eine dezentrale Abstimmung der Akteure, die ihre interdependenten Entscheidungen durch unmittelbare Interaktion im Rahmen von Verhandlungen treffen (vgl. [3], S. 55-59).

Gegenstand dieses Beitrags ist das Planungsproblem der Bestimmung integrierter Bestell- und Produktionspolitiken in der SC. Aus der Verknüpfung der durch die Bestell- und Produktionspolitiken der SC-Partner festgelegten Beschaffungs- und Produktionsprozesse resultiert der Materialfluss durch die gesamte SC. Während produktionswirtschaftliche und logistische Ansätze zur Planung integrierter Materialflüsse durch eine weitgehende Vernachlässigung von Wettbewerbsaspekten und der Dominanz hierarchischer Koordinationsprinzipien gekennzeichnet sind, eignen sich spieltheoretische Ansätze, die strategische Interaktion zwischen Unternehmen zu analysieren und eine dezentral abgestimmte Planung zu unterstützen (vgl. [6], S. 307). Ziel des Beitrags ist die Entwicklung eines Verhandlungsmodells zur Bestimmung integrierter Bestell- und Produktionspolitiken.

2 Das Verhandlungsproblem

Gemäß eines Rahmenvertrags zwischen einem Abnehmer (A) und einem Hersteller (P) wird der bekannte Periodenbedarf b [ME/PE] durch Lieferungen der Quantität x [ME] gedeckt. Unter dem Ziel der Minimierung ihrer relevanten Kosten je Periode bestimmen (A) und (P) ihre individuell optimale Bestellmenge bzw. Losgröße auf Basis der klassischen Losgrößenformel. (P) disponiert seine Produktionspolitik dabei auf einer Lot-for-Lot-Basis, d. h., die Liefermenge entspricht der Losgröße. Die Bestellmenge von (A) entspricht ebenfalls der Liefermenge, sodass die Bestellmenge von (A) gleich der Losgröße von (P) sein muss (vgl. [9], S. 965). Die streng konvexen Kostenfunktionen von (A) und (P) lauten:

$$z_A(x) = B \cdot \frac{b}{x} + \frac{x}{2} \cdot h_A \;;\; z_P(x) = R \cdot \frac{b}{x} + \frac{x}{2} \cdot \frac{b}{d} \cdot h_P . \tag{1}$$

b : Periodenbedarf von (A) [ME/PE]
R : Rüstkosten von (P) [GE]
B : Bestellkosten von (A) [GE]
d : Produktionsrate von (P) [ME/PE]
h_P: Lagerkosten von (P) [GE/ME PE]
h_A: Lagerkosten von (A) [GE/ME PE]

Für beide Zielgrößen gilt „Minimierung" als Ausprägung der Höhenpräferenzrelation. Das Planungsproblem der Ermittlung der integrierten Bestell- und Produktionspolitik kann als Vektorminimumproblem formuliert werden:

$$\text{"min"}\left\{ \mathbf{z} = \begin{pmatrix} z_A(x) \\ z_P(x) \end{pmatrix} = \begin{pmatrix} B \cdot b & \frac{1}{2} \cdot h_A \\ R \cdot b & \frac{1}{2} \cdot \frac{b}{d} \cdot h_P \end{pmatrix} \cdot \begin{pmatrix} \frac{1}{x} \\ x \end{pmatrix} \middle| \, x \in R_+ \right\}. \tag{2}$$

Existiert eine Bestell- und Produktionspolitik $\hat{x}$, für die beide Zielfunktionen in (1) ihre individuellen Minima erreichen, so wird diese Lösung als perfekte Lösung von (2) bzw. als perfekte Bestell- und Produktionspolitik bezeichnet. Die individuell optimalen Bestell- und Produktionspolitiken von (A) und (P) lauten:

$$x_A^* = \sqrt{\frac{2 \cdot B \cdot b}{h_A}} \;;\; x_P^* = \sqrt{\frac{2 \cdot B \cdot d}{h_P}} . \tag{3}$$

Es existiert nur dann eine perfekte Lösung von (2), wenn die individuell optimale Bestellmenge und die individuell optimale Losgröße identisch sind. In diesem Fall gilt $\hat{x} = x_A^* = x_P^*$ und es ergibt sich der ideale Zielfunktionswertvektor:

$$\hat{\mathbf{z}}' = \left(z_A(\hat{x}), z_P(\hat{x})\right) = \left(\sqrt{2 \cdot B \cdot b \cdot h_A}, b \cdot \sqrt{\frac{2 \cdot R \cdot h_P}{d}} \right). \tag{4}$$

In realen Planungssituationen wird die individuell optimale Bestellmenge nicht der individuell optimalen Losgröße entsprechen. Im Allgemeinen gilt daher $x_A^* \neq x_P^*$ und (2) besitzt keine perfekte Lösung. Die Akteure müssen eine Kompromisslösung finden, d. h. sie müssen sich im Rahmen von Verhandlungen über die zu realisierende Bestell- und Produktionspolitik einigen. Kandidaten für eine Verhandlungslösung sind Bestell- und Produktionspolitiken, die bezüglich der

Zielfunktionen in (1) funktional-effizient sind. Die Menge der funktional-effizienten Bestell- und Produktionspolitiken ist:

$$X^{eff} = \left\{ x \in R_{+} \middle| x \in [x_A^*, x_P^*] \right\}. \tag{5}$$

Für jede Bestell- und Produktionspolitik $x \notin X^{eff}$ kann mindestens eine Bestell- und Produktionspolitik $x \in X^{eff}$ angegeben werden, für die (A) und (P) niedrigere Kosten realisieren. Nicht funktional-effiziente Alternativen sind keine potenziellen integrierten Bestell- und Produktionspolitiken, ihre Wahl verstößt gegen den Grundsatz der formalen Rationalität (vgl. [7], S. 134). Für $x_A^* \neq x_P^*$ existiert ein Zielkonflikt, da ausgehend von z. B. der individuell optimalen Bestellmenge von (A) eine Bestell- und Produktionspolitik x, mit $x \neq x_A^*$ und $x \in X^{eff}$, zu höheren Kosten bei (A) führt, während (P) niedrigere Kosten realisiert. Es gilt:

$$z_A(x_P^*) > z_A(x) > z_A(x_A^*) \text{ und } z_P(x_A^*) > z_P(x) > z_P(x_P^*) \forall x \in]x_A^*, x_P^*[\tag{6}$$

Durch die Ermittlung von (5) erfolgt eine Komplexitätsreduktion des dem Vektorminimumproblem (2) zu Grunde liegenden Planungsproblems. Eine endgültige Auswahl der integrierten Bestell- und Produktionspolitik gelingt aber nicht. Es existiert der Interessengegensatz, dass (A) und (P) ihre jeweils individuell optimalen Lösungen realisieren möchten. Verhandlungsprobleme mit der Möglichkeit der Schließung bindender Verträge sind Gegenstand der kooperativen Spieltheorie, deren Lösungskonzepte in axiomatische Ansätze und Ansätze zur Modellierung konkreter Verhandlungsprozesse klassifiziert werden können (vgl. [5], S. 24-25). Charakteristisch für axiomatische Ansätze ist, dass sie Verhandlungen nicht als zeitliche Abfolge von Geboten der Akteure abbilden, sondern Verhandlungsprobleme dadurch zu lösen versuchen, dass sie den Akteuren eine Lösung anbieten, die als „fair" zu beurteilen ist. (vgl. [1], S. 150). Im Rahmen dieses Beitrags wird der Verhandlungsprozess von (A) und (P) jedoch explizit modelliert.

3 Die Verhandlungslösung

Es wird ein Verhandlungsmodell entwickelt, mit dem es gelingt, wechselseitige Angebote und Konzessionen abzubilden und die integrierte Bestell- und Produktionspolitik als Ergebnis eines dynamischen Verhandlungsprozesses zu ermitteln. Das Ergebnis von Verhandlungen ist abhängig von den Machtbeziehungen zwischen den Verhandlungsparteien. „A firm's power reflects its potential for influence on the decision making and behavior of another firm." ([4], S. 58). Es kann zwischen asymmetrischer und symmetrischer Machtverteilung unterschieden werden. Im Rahmen der hier zu Grunde liegenden Problemstellung bedeutet eine asymmetrische Machtverteilung, dass sich ein Akteur in der Machtposition befindet, seine individuell optimale Lösung als integrierte Bestell- und Produktionspolitik zu realisieren. Im Weiteren soll von einer symmetrischen Machtverteilung ausgegangen werden, sodass kein Akteur seine individuell optimale Lösung gegen den Willen des anderen realisieren kann.

Im Rahmen des zu analysierenden Verhandlungsprozesses haben beide Akteure die Möglichkeit zu den Zeitpunkten $t = 0,1,2,...,T$ ein Angebot zu unterbreiten.

Dabei bezeichnet x_i^t die vom Akteur $i \in \{A, P\}$ zum Zeitpunkt t angebotene Bestell- und Produktionspolitik. Es ist $\mathbf{z}(x_i^t)' = (z_A^t(x_i^t), z_P^t(x_i^t))$ der aus dem Angebot x_i^t des Akteurs $i \in \{A, P\}$ zum Zeitpunkt t resultierende Zielfunktionswertvektor, mit $z_A^t(x_i^t)$ und $z_P^t(x_i^t)$ als die entsprechenden Zielfunktionswerte von (A) und (P). Die Akteure verhalten sich individuell rational im Sinne ihrer Zielsetzungen. Ein Angebot x_i^t ist individuell rational, wenn gilt:

$$z_A^t(x_A^t) \leq z_A^{t-1}(x_P^{t-1}) \text{ bzw. } z_P^t(x_P^t) \leq z_P^{t-1}(x_A^{t-1}) \text{ für } t = 1,2,...,T\,. \tag{7}$$

Ein Angebot eines Akteurs $i \in \{A, P\}$ heisst Konzession, wenn gilt:

$$z_P^{t+1}(x_A^{t+1}) < z_P^t(x_A^t) \text{ bzw. } z_A^{t+1}(x_P^{t+1}) < z_A^t(x_P^t) \text{ für } t = 0,1,2,...,T-1\,. \tag{8}$$

Eine Konzession heisst volle Konzession wenn ein Akteur ein Angebot annimmt:

$$x_A^{t+1} = x_P^t \text{ bzw. } x_P^{t+1} = x_A^t \text{ für } t = 0,1,2...,T-1\,. \tag{9}$$

Eine Konzession ist partiell, wenn sie (8) erfüllt aber keine volle Konzession ist. Zur Analyse des Verhandlungsprozesses wird angenommen, dass beide Akteure zu jedem Verhandlungszeitpunkt t ein Angebot abgeben. Auch wird angenommen, dass beide Akteure ein starkes Interesse an einer Einigung haben, d. h. ein Scheiterns der Verhandlungen induziert bei beiden Akteuren höhere Kosten als die Annahme des ungünstigsten Angebots. Verhalten sich die Akteure individuell rational im Sinne ihrer Zielsetzungen, so bieten sie zum Zeitpunkt $t = 0$ ihre jeweils individuell optimalen Lösungen als Bestell- und Produktionspolitik an. Es folgt $x_A^0 = x_A^*$ und $x_P^0 = x_P^*$. Im Fall von $x_A^* \neq x_P^*$ ergibt sich keine Einigung und zum Zeitpunkt $t = 1$ müssen weitere Angebote abgegeben werden. Zum Zeitpunkt $t = 1$ hat beispielsweise (A) folgende Alternativen:

- (A) wiederholt sein Angebot: $x_A^1 = x_A^0 = x_A^*$,
- (A) macht eine volle Konzession: $x_A^1 = x_P^0 = x_P^*$ oder
- (A) macht eine partielle Konzession: $x_A^1 \neq x_P^0 = x_P^*$,

  mit $z_P^1(x_A^1) < z_P^0(x_A^0)$ und $z_A^1(x_A^1) \leq z_A^0(x_P^0)$.

Die Wiederholung des Angebots aus Zeitpunkt $t = 0$ birgt die Gefahr, dass (P) die Verhandlungen aufgrund mangelndem Entgegenkommens abbricht, sie somit scheitern. Eine volle Konzession wiederum ist nicht individuell rational, da für alle $x \in [x_A^*, x_P^*[$ auch $z_A(x) < z_A(x_P^*)$ gilt. Einzige individuell rationale Alternative für (A) zum Zeitpunkt $t = 1$ ist die Abgabe einer partiellen Konzession. Analoges gilt für (P). Zum Zeitpunkt $t = 1$ geben somit beide Akteure eine individuell rationale, partielle Konzession ab, welche die Bedingungen (7) und (8) erfüllen. Zur Abgabe dieser Konzessionen benötigen die Akteure keine Informationen über die Kostenfunktionen des anderen, d. h. der Verlauf des Verhandlungsprozesses ist unabhängig vom Informationsstand der Beteiligten. Nur auf Basis der Angebote zum Zeitpunkt $t = 0$ können partielle, individuell rationale Konzessionen abgegeben werden. Zum Zeitpunkt $t = 1$ ergeben sich folgende Konzessionen:

$$x_A^1 \in]x_A^0, x_P^0[=]x_A^*, x_P^*[\ ; \ x_P^1 \in]x_A^0, x_P^0[=]x_A^*, x_P^*[\ . \tag{10}$$

Für Konzessionen zu den Folgezeitpunkten ergibt sich analog:

$$x_A^t \in]x_A^{t-1}, x_P^{t-1}[\ , \ x_P^t \in]x_A^{t-1}, x_P^{t-1}[\ \text{für } t=2,3,...,T. \tag{11}$$

Die Verhandlungen enden, wenn ein Akteur eine volle Konzession macht. Zur Ermittlung der idealtypischen Verhandlungslösung muss bestimmt werden, wie lange Konzessionen möglich sind. Für eine mögliche Konzessionen von (A) gilt:

$$\left(z_A(x_A^{t+1}) - z_A(x_P^*)\right) \cdot \left(z_P(x_A^{t+1}) - z_P(x_A^*)\right) > \left(z_A(x_A^t) - z_A(x_P^*)\right) \cdot \left(z_P(x_A^t) - z_P(x_A^*)\right). \tag{12}$$

Analoges gilt für (P). Die Verhandlungen enden, wenn keine Konzessionen mehr möglich sind, die Bedingung (12) erfüllen. Dies gilt für die Bestell- und Produktionspolitik $x \in]x_A^*, x_P^*[$, die $\tilde{z} = \left(z_A(x) - z_A(x_P^*)\right) \cdot \left(z_P(x) - z_P(x_A^*)\right)$ maximiert. Der idealtypische Verhandlungsprozess wird an einem Beispiel illustriert.

4 Ein Beispiel

Die dem Beispiel zu Grunde liegenden relevanten Daten zeigt Tabelle 1.

Tabelle 1. Daten des Beispiels

Abnehmer (A)	Hersteller (P)
b=10000 [ME/PE	d=16000 [ME/PE]
B=100 [GE]	R=240 [GE]
h_A=50 [GE/ME PE]	h_P=48 [GE/ME PE]

Die Akteure wissen nicht , ob $x_A^* = x_P^*$, $x_A^* > x_P^*$ oder $x_A^* < x_P^*$ gilt. Die einzigen individuell rationalen Angebote zum Zeitpunkt $t = 0$ sind die individuell optimalen Lösungen: $x_A^0 = x_A^* = 200$ und $x_P^0 = x_P^* = 400$. Die Angebote sind nicht kompatibel, sodass beide Akteure zum Zeitpunkt $t = 1$ eine partielle, individuell rationale Konzession tätigen, mit $x_A^1, x_P^1 \in]200,400[$, z. B. $x_A^1 = 250$ und $x_P^1 = 350$. Die Verhandlungen werden solange fortgesetzt, bis einer der Akteure ein Angebot annimmt oder bis keine partiellen Konzessionen mehr möglich sind. In Tabelle 2 ist der mögliche Verhandlungsverlauf bis zur idealtypischen Verhandlungslösung $\tilde{x} = 282{,}84$ dargestellt.

Tabelle 2. Verlauf des Verhandlungsprozesses

t	x_A^t	$z_A(x_A^t)$	$z_P(x_A^t)$	$\tilde{z}$	x_P^t	$z_A(x_P^t)$	$z_P(x_P^t)$	$\tilde{z}$
0	200	10000	15000	0	400	12500	12000	0
1	250	10250	13350	3712500	350	11607,1	12107,1	2582919
2	270	10453,7	12938,9	4217669,9	300	10833,3	12500	4166675
3	280	10571,4	12771,4	4298077,9	290	10698,3	12625,9	4277535,5
⋮	⋮	⋮	⋮	⋮	⋮	⋮	⋮	⋮
T	282,84	10606,6	12727,9	4301994,1	282,84	10606,6	12727,9	4301994,1

5 Schlussbetrachtung

In einer SC agieren rechtlich und wirtschaftlich selbständige Unternehmen, sodass sich eine Koordination nach dem heterarchischen Prinzip ergibt. Unter der Annahme einer symmetrischen Machtverteilung konnte der Verhandlungsprozess über die zu realisierende Bestell- und Produktionspolitik modelliert und die Verhandlungslösung ermittelt werden. Der dargestellte Verhandlungsprozess führt zu einer Win-Win-Situation, da beide Akteure, gegenüber dem aus ihrer Sicht ungünstigsten Einigungsfall, niedrigere Kosten realisieren. Die Instrumente und Methoden der Spieltheorie stellen geeignete Ansätze zur Analyse der Interaktion der SC-Partner sowie zur Unterstützung dezentraler Planungsansätze in der SC dar.

Literatur

[1] Berninghaus SK, Erhart KM, Güth W (2002) Strategische Spiele - Eine Einführung in die Spieltheorie. Berlin Heidelberg New York

[2] Cachon GP (1999) Competitive Supply Chain Inventory Management. In: Tayur S, Ganeshan R, Magazine M (eds) Quantitative Models For Supply Chain Management. Boston Dordrecht London, S 111-145

[3] Corsten H, Gössinger R (2001) Einführung in das Supply Chain Management. München Wien

[4] Frazier GL, Spekman RE, O'Neal CR (1988) Just-In-Time Exchange Relationships in Industrial Markets. Journal of Marketing 52: S 52-67

[5] Holler MJ, Illing G (2000) Einführung in die Spieltheorie. Berlin Heidelberg New York

[6] Inderfurth K, Minner S (2001) Produktion und Logistik. In: Jost PJ (Hrsg) Die Spieltheorie in der Betriebswirtschaftslehre. Stuttgart, S 307-349

[7] Isermann H (1974) Lineare Vektoroptimierung. Dissertation, Regensburg

[8] Scholz-Reiter B, Jakobza J (1999) Supply Chain Management - Überblick und Konzeption. HMD Praxis der Wirtschaftsinformatik 207: S 7-15

[9] Toporowski W (1999) Unternehmensübergreifende Optimierung der Bestellpolitik - Das JELS-Modell mit einem Intermediär. ZfbF 10: S 963-989

[10] Zäpfel G (2000) Supply Chain Management. In: Baumgarten H, Wiendahl HP, Zentes J (Hrsg) Logistik-Management. Berlin Heidelberg New York, S 1-32

Modeling the Interaction between Operational and Financial Decisions in the Inventory Pooling of Repairable Spare Parts Problem

Hartanto Wong, Dirk Cattrysse, Dirk Van Oudheusden

Centre for Industrial Management, Katholieke Universiteit Leuven
Celestijnenlaan 300 A, B-3001 Leuven, Belgium
E-mail: Hartanto.Wong@cib.kuleuven.ac.be

1 Introduction

Equipment-intensive industries such as airlines, nuclear power plants, various process and manufacturing plants using complex machineries are often confronted with the difficult task of maintaining a high system performance while controlling their inventory holding cost. An important type of components in such industries are called repairable items. The typical problem is to determine the optimal stocking level of spare parts. An insufficient stock of spare parts can lead to an excessive downtime cost. On the other hand, maintaining an excessive number of spare parts increases the cost of tying up capital in non-revenue-generating spare parts inventories.

Inventory pooling, an 'inter-company' cooperation where the cooperating companies share their inventories, is an effective way to improve a company's logistical performance without requiring any additional cost. Cooperation usually takes the form of lateral transshipments from a location with a surplus of on-hand inventory to a location that faces a stock-out.

Previous research in the area of inventory pooling for non-repairable items [1, 6, 7, 9, 15, 16] and for repairable items [4, 8, 14, 18] has mainly focused on cooperation among 'bases' which are assumed to be owned by a single 'parent-firm' whose objective is to maximize some measures of aggregate performance. Little research has been done in the context where the pooling members are independent companies and hence, there exists a competitive behavior in the sense that all members have interests in maximizing their own benefits from pooling. To model the competitive behavior in a decision making process, classical optimization concepts may no longer be applicable. Instead, game theoretic models should be used. Several studies [9, 10, 12, 17] have been done to model the inventory problem using game theory. They consider a competitive version of the newsboy problem. Anupindi et al. [2, 3] develop a "co-opetitive" framework for the analysis of decentralized distribution systems. They consider the sequential decisions of inventory ordering and allocation of revenues/costs. For the cooperative allocation decision, they use the concept of core and develop sufficient conditions for the existence of the core. They develop conditions for the existence of a pure strategy Nash Equilibrium for the inventory decision.

This paper is in the spirit of the work of Anupindi et al. Two sets of decisions are made when dealing with inventory pooling of repairable spare parts. The first decisions are the operational decisions dealing with: optimal number of spare parts; and critical inventory level for allowing a lateral transshipment. In relation to the critical inventory level, we have complete and partial pooling when the critical inventory level is zero and positive, respectively. The second decision is the financial decision dealing with allocation of pooling costs/benefits to every pooling member. In this paper, our objective is to give an analysis on the interaction between those two decisions.

2 Problem description

Without loss of generality, we will use the airline company's repairable spare parts inventory problem to describe our problem. We use an example case of two cooperating airline companies to describe our analysis. Below are all the parameters for both companies.

- number of airplanes operated: $M_1 = 15$; $M_2 = 10$;
- failure rate of the repairable item: $\lambda = 0.002$/day;
- average repair time: $1/\mu = 50$ days;
- inventory holding cost: $h_1 = h_2 = 10000$ \$/unit/year;
- AOG (Aircraft On Ground) unit cost: $g_1 = g_2 = 1000$ \$/grounded-airplane/day;
- transshipment cost: $t = 3000$ \$/transshipment.

3 Operational and financial decisions

3.1 Operational decisions

In this section, we consider the operational decisions without considering the allocation of cost or profit to each pooling member. In this setting, we treat all members as 'bases' and cooperation among them takes place within a single company.

The problem can be formulated as follows. Each company wants to find the number of spare parts, S_i, and the critical inventory level, k_i, to minimize the total system cost TC (see Eq. 1). In this equation, AOG_i is the expected number of AOG at company i while Tr_i is the expected number of lateral transshipments received by company i.

$$TC = \sum_{i=1}^{2} h_i S_i + g_i AOG_i + tTr_i \tag{1}$$

In this paper, the machine-repair queuing model is chosen to solve the problem. The more detailed description of the model is given in [18]. After investigating a number of values of S_i and k_i, we obtain the optimal solutions: S_1=3; k_1=0; S_2=3;

k_2=0; with the total cost TC = 72210 \$. Without inventory pooling, the optimal solutions for the two companies are: S_1=4 and S_2=3, and the resulted costs are C_1 = 48390 \$ and C_2 = 38075 \$. A significant cost saving (14255 \$) is achieved when the two companies pool their spare parts inventories.

3.2 Financial decision

The main motivating factor behind the members' desire to cooperate is the potential cost savings for all pooling members. Thus, it is imperative that the allocated cost is lower than the cost that each member can guarantee itself by acting individually. Otherwise, the members will have no incentive to participate in this cooperation. The concept of the "core" from cooperative game theory is used to find "fair" cost allocation [5]. Let us define X_i as the cost allocated to company i. Using the optimal result from the previous section, the core in our example case can be simply defined by: $X_1 + X_2 = 72210$; $X_1 \leq 48390$; and $X_2 \leq 38075$. Determining which point to choose within the core has been the subject of considerable study within the game theory and cost accounting literature [11]. No attempt is made here to find the "best" cost allocation method.

At a glimpse, one can see that the operational and financial decisions are made sequentially and they can be treated separately. First, optimal operational decisions are determined and secondly the resulted cost is allocated. In fact, this treatment is only valid in the situation where no competitive behavior is considered. The existence of competitive behavior of the pooling members needs a different view of the decision processes. For any agreement on financial decision, the process of making operational decisions involves strategic interaction, i.e., each pooling member reacts optimally to the other's decisions. In the next section, we discuss the interaction between the two decisions. For this purpose, four cost allocation methods are considered.

4 The interaction

We will now look at the Nash equlibria of the games in our example. In the first part, all costs and parameters are treated as common knowledge in the system. Consequently, this competitive situation can be considered as a two-player game with complete information. In the second part, we extend our analysis by considering the game with incomplete information. All costs and parameters, except for AOG unit cost g_i, remain as common knowledge.

4.1 Games with complete information

In this part we consider four games corresponding to the four selected allocation methods. The description of each allocation method is given below.

- Cost allocation method 1: inventory holding cost is allocated based on the number of spare parts provided by each company; AOG cost is allocated based on individual AOG; transshipment cost is paid by the "receiver" company.

- Cost allocation method 2: inventory holding cost and transshipment cost are allocated based on the number of airplanes operated by each company; AOG cost is allocated based on individual AOG.
- Cost allocation method 3: the total cost is allocated based on the demand rate of each company (the number of airplanes operated by each company).
- Cost allocation method 4: Shapley value (the formula to calculate Shapley value can be seen in [11]).

In each game, the set of operational decisions $(S_1,k_1)^*$ and $(S_2,k_2)^*$ is a Nash equilibrium if $(S_1,k_1)^*$ is the best response for company 1 to $(S_2,k_2)^*$ and $(S_2,k_2)^*$ is the best response for company 2 to $(S_1,k_1)^*$. In the first and second games, however, the Nash equilibria are not optimal as in the other two games. The optimal solutions for the system are S_1=3; k_1=0; S_2=3; k_2=0 which are the optimal decisions when the cooperation is treated as if occuring inside a single company. Cost allocation methods 1 and 2 do not give enough incentives for both companies to cooperate. The Nash equilibria in these two games are S_1=4; k_1=4; S_2=3; k_2=3. That means no pooling applied in these equilibria. The presence of individual payments of some cost components in the first two methods causes the two companies getting incentives to lower their individual costs by unilaterally deviating from any cooperative solutions. These examples clearly show that the selection of operational decisions is influenced by the agreement on cost allocation (financial decision).

4.2 Games with incomplete information

Since many individual parameters are more likely to be private rather than common knowledge, cost allocation methods based on the total joint cost (like methods 3 and 4 in our example) are not easy to implement. To make this type of cost allocation methods applicable, all pooling members should be willing to open all their private information. This is not easy especially when the pooling members are competitors to each other. Moreover, the possibility of making profits by providing untrue information makes the problem more complicated. In this part, we analyze how the game will be played when not all cost parameters are opened as public information. We choose AOG unit cost as private information because among the three components, AOG unit cost is the most difficult to quantify and can be very subjective. We assume that the other two cost components, i.e., inventory holding and transshipment costs are still traceable. We use an example case where AOG unit costs of both companies are private information and their true values are g_1=g_2=2000. Let us define g_i' as the submitted value of AOG unit cost by company i. To simplify the problem, we assume that g_i' can only have a value which is a multiple of 500 from a minimum of 1000 and up to a maximum of 3000.

In this example, cost allocation method 4 (Shapley value) is applied. The game is played as follows. First, company 1 and 2 submit the values of g_1' and g_2'. Based on these values, they make operational decisions to minimize their individual costs. Note that we assume the following settlement is used to conclude the final cost allocation. First, company i claims its payment as:

$$TCC_i = h_i S_i + g_i' AOG_i + tTr_i \,. \tag{2}$$

The Shapley value method gives the cost allocation X_i' for company i. Thus, to reach the final value X_i', company i has to pay his partner X_i'-TCC_i if X_i' is higher than TCC_i. Each company will finally, of course, consider its actual cost, X_i, as the basis of its decisions. Eq. 3 is used to calculate X_i.

$$X_i = h_i S_i + g_i AOG_i + tTr_i + X_i' - TCC_i \tag{3}$$

It can be shown that the Nash equilibrium of the considered game is g_1'=1000; g_2'=1000; with the operational decisions S_1=3; k_1=0; S_2=3; k_2=0. The resulted costs are 45454 \$ and 33716 \$ for company 1 and 2, respectively. This gives the total cost of 79170 \$ which is 2286 \$ higher than the optimal total cost if both companies reveal the true values of their AOG unit costs (the optimal operational decisions for g_1=g_2=2000: S_1=4; k_1=0; S_2=3; k_2=0; total cost = 76884 \$). This example case shows that given the condition of incomplete information, there is an incentive for each pooling member to reveal the untrue information as the optimal reaction to the other member's revelation.

5 Conclusions

In this paper, we present the use of game theoretic models to analyze the decision processes in the problem of inventory pooling of repairable spare parts. The analysis considers both cooperative and competitive behavior of the pooling members in making their operational and financial decisions. We show the interaction between the two decisions. The financial decision on cost allocation gives the influence on whether or not the pooling members have any incentives to cooperate by sharing their spare parts inventories. We also extend our analysis to the situation with incomplete information. The AOG unit cost is assumed to be private rather than common knowledge. We show that there is an incentive for each pooling member to reveal the untrue information as the optimal reaction to the other member's revelation and this can lead to the pooling members playing the game with a non-optimal solution.

Two related issues are of interest for further research. The first issue is to extend the problem to the games with more than two players. The second issue that could be of importance is to find a general framework for the design of cost allocation policies such that all pooling members are motivated to fully cooperate and reveal the true information ending up with the equilibrium which gives the optimal solution.

References

1. Alfredsson P, Verrijdt J (1999) Modeling emergency supply flexibility in a two-echelon inventory system. Management Science 45: 1416–1431
2. Anupindi R, Bassok Y, Zemel E (1999) Study of decentralized distribution systems: part I – general framework. Working paper, Kellog Graduate School of Management, Northwestern University, Evanston, IL

3. Anupindi R, Bassok Y, Zemel E (1999) Study of decentralized distribution systems: part II – applications. Working paper, Kellog Graduate School of Management, Northwestern University, Evanston, IL
4. Axsater S (1990) Modelling emergency lateral transshipments in inventory systems. Management Science 36 : 1329–1338
5. Gerchak Y, Gupta D (1991) On apportioning costs to customers in centralized continuous review inventory systems. Journal of operations Management 10: 546-551
6. Grahovac J, Chakravarty A (2001) Sharing and lateral transshipment of inventory in a supply chain with expensive low-demand items. Management Science 47: 579–594
7. Kukreja A, Schmidt CP, Miller DM (2001) Stocking decisions for low-usage items in a multilocation inventory system. Management Science 47: 1371–1383
8. Lee HL (1987) A multi-echelon inventory model for repairable items with emergency lateral transshipments. Management Science 33: 1302–1316
9. Lippman SA, McCardle KF (1997) The competitive newsboy. Operations Research 45: 54–65
10. Parlar M (1988) Game theoretic analysis of the substitutable product inventory problem with random demands. Naval Research Logistics 35: 397–409
11. Robinson LW (1993) A comment on Gerchak and Gupta's "On apportioning costs to customers in centralized continuous review inventory systems". Journal of Operations Management 11: 99–102
12. Rudi N, Kapur S, Pyke DF (2001) A two-location inventory model with transshipment and local decision making. Management Science 47: 1668–1680
13. Sherbrooke CC (1968) METRIC: A multi-echelon technique for recoverable item control. Operations Research 16: 122–141
14. Sherbrooke CC (1992) Multi-echelon inventory systems with lateral supply. Naval Research Logistics 39: 29–40
15. Tagaras G (1999) Pooling in multi-location periodic inventory distribution systems. Omega International Journal of Management Science 27: 39–59
16. Tagaras G, Cohen MA (1992) Pooling in two-location inventory systems with non-negligible replenishment lead times. Management Science 38: 1067–1083
17. Wang Q, Parlar M (1994) A three-person game theory model arising in stochastic inventory control theory. European Journal of Operational Research 76: 83–97
18. Yanagi S, Sasaki M (1992) An approximation method for the problem of a repairable-item inventory system with lateral supply. IMA Journal of Mathematics Applied in Business and Industry 3: 305–314.

Die Schätzung von Markentreue, Nichtkäuferanteil und Marktpotenzial aus Handelspaneldaten

Heribert Reisinger, Udo Wagner, Matthias Schuster

Institut für Betriebswirtschaftslehre, Universität Wien, Brünner Straße 72, 1210 Wien, e-mail: {heribert.reisinger | udo.wagner | matthias.schuster}@univie.ac.at

1 Problemstellung

Eine Kernaufgabe des Marketings ist die Identifizierung der relevanten Kundenbedürfnisse und die dementsprechende, spezifische Ausrichtung des Angebots. Häufig ist es dabei nicht möglich, den gesamten Markt in derselben Weise zu bearbeiten und so bedient man sich geeigneter Segmentierungstechniken. Eine wichtige Art der Segmentierung auf Konsumgütermärkten bezieht sich auf verhaltensbezogene Variablen und hier spielen wiederum Kaufhäufigkeit sowie Markentreue eine wesentliche Rolle. Die dafür benötigten Informationen liefert der Marktforscher in der Regel auf der Grundlage von Haushaltspaneldaten oder von eigens dafür durchgeführten Primärerhebungen. Die vorliegende Arbeit beschäftigt sich mit der Ermittlung der Anteile an loyalen Kunden, Markenwechslern und Nichtkäufern, beschreitet zu deren Bestimmung aber einen anderen Weg, indem sie vorschlägt, aus Handelspaneldaten, also *nichtpersonenbezogenen* Angaben, *individuelles* Nachfrageverhalten abzuleiten. Der Vorteil einer solchen Datenquelle liegt darin, dass sie regelmäßig (zumeist mit Hilfe der modernen Scannertechnologie) erhoben, insbesondere für Logistikzwecke eingesetzt wird und daher häufig ohne Zusatzkosten zur Verfügung steht.

Die Grundideen der zur Diskussion gestellten Methodik wurden mehreren Forschungsgebieten entlehnt. Zunächst stammen aus dem *Marketing* etablierte Erkenntnisse über Regelmäßigkeiten zwischen Marktanteilen, Markentreue und Markenwechsel für häufig gekaufte Güter des täglichen Bedarfs. Diese Muster fanden Berücksichtigung in verschiedenen Modellen (insbesondere Ehrenberg 1988; Colombo, Morrison 1989), die ihrerseits hier als Ausgangspunkt dienen. Weiters liefert die *ökologische Inferenz* (King 1997) das Prinzip, dass Kennzahlen, die sich auf Querschnitte von Personen beziehen und durch Aggregation individueller Daten gewonnen wurden, tendenziell noch immer die charakteristischen Symptome der Mitglieder dieser Gruppen widerspiegeln. Solche Merkmale können daher mittels einer statistischen Analyse erfasst werden. Das bekannteste Beispiel hierfür sind vermutlich Wählerstromanalysen auf der Grundlage der Ergebnisse von zwei aufeinander folgenden Wahlgängen in einer großen Anzahl von Wahlbezirken oder -sprengeln. Schließlich kommt aus der *Ökonometrie* die ‚Maximum Entropie Schätzmethode' (Jaynes 1957), die vorteilhafter Weise vor allem dann eingesetzt wird, wenn über einen Sachverhalt nur unvollständiges Wissen vorhanden ist und der Untersuchende von subjektiven Einflüssen möglichst unberührte Ergebnisse erhalten möchte. Eine derartige Schätzmethode scheint für die vorliegende Aufgabenstellung besonders zweckmäßig zu sein, weil es sich dabei einerseits um eine Art von Disaggregation handelt, andererseits über Nichtkäufer überhaupt keine unmittelbar aus den Daten ableitbare Informationen vorhanden sind.

2 Lösungsansatz

2.1 Modellierung von Markentreue und Wechselkäufen

Den vorstehenden Ausführungen gemäß postulieren wir in diesem Abschnitt zunächst gewisse Zusammenhänge zwischen den Anteilen der loyalen Kunden und der Markenwechsler, wobei wir unsere Betrachtungen auf die zwei aufeinander folgenden Perioden t und $(t+1)$ konzentrieren.

$$p_{j|i} = \rho_i \cdot m_{j(t+1)} \quad \forall j \neq i \tag{1}$$

$$p_{j|j} = 1 - \rho_j \cdot \left(1 - m_{j(t+1)}\right) \quad \forall j \tag{2}$$

Mit:

$p_{j\|i}$	bedingte Wahrscheinlichkeit zu Marke j zu wechseln, nachdem zuvor Marke i gekauft wurde
$p_{j\|j}$	bedingte Wahrscheinlichkeit Marke j wieder zu kaufen
m_{jt}	Marktanteil von Marke j zum Zeitpunkt t
ρ_j	von Marke j abhängiger Parameter; $0 \leq \rho_j \leq 1$
i, j	Markenindex, $(i, j = 1,...,I)$
I	Anzahl der berücksichtigten Marken
t	Zeitindex

Die inhaltliche Begründung für diesen Ansatz liefert die Erkenntnis, dass auf vielen Konsumgütermärkten die Anteile der zwischen zwei Marken wechselnden Käufer proportional zu dem Produkt ihrer Marktanteile schwanken. Beschränkt man sein Augenmerk auf die Kunden einer Marke i, so erhält man (1), berücksichtigt man ferner die Summenbedingung (logische Konsistenz) für die $p_{j|i}$ so erhält man (2). Der Parameter ρ_j hat unter dem Namen ‚Switching Constant' Eingang in die Literatur gefunden und wird dort ausführlich diskutiert (für eine diesbezügliche Darstellung in einem sehr ähnlichem Kontext siehe Wagner et al. 2001). Anschaulich gesprochen beschreibt ρ_j das Ausmaß der Fluktuation am betrachteten Markt: je kleiner ρ_j ist, desto höher ist der Anteil der loyalen Konsumenten von Marke j; ein Wert nahe Eins hingegen steht für ausgeprägtes Wechselwahlverhalten. In einem formal identischen Modell für Investitionsgüter bzw. Dienstleistungen unterscheiden Colombo, Morrison (1989) zwischen jedenfalls treuen Kunden (‚hard-core loyal') und potenziellen Wechselkäufern; $(1 - \rho_j)$ beschreibt dann in ihrem Szenario den Anteil ersterer.

2.2 Parameterschätzung auf Basis von Querschnittsdaten

Wie in der Problemstellung dargelegt stehen für die Schätzung der Parameter ρ_j querschnittsbezogene Informationen, in diesem Falle die Absätze q_{jkt} und $q_{jk(t+1)}$ für alle Marken j, in $k = 1,...,K$ Querschnitten (z.B. Geschäften, geographischen Bezirken) in zwei aufeinander folgenden Perioden t und $(t+1)$ zur Verfügung. Dem Prinzip der ökologischen Inferenz gemäß postulieren wir über die Querschnitte konstante Parameter und formulieren unser Schätzproblem als Mehrgleichungsregression.

$$q_{jk(t+1)} = \sum_{i=1}^{I} p_{j|i} \cdot q_{ikt} \quad \forall j,k \tag{3}$$

Zur Bestimmung der Parameter ersetzen wir die $p_{j|i}$ durch (1) bzw. (2). Wir wählen ein verallgemeinertes ‚Kleinste Quadrate' Kriterium, wobei wir als Gewichte die reziproken Querschnittsgrößen verwenden und erhalten ein Optimierungsproblem unter Nebenbedingungen, die die logische Konsistenz der Schätzwerte sicherstellen (für die formalen Details siehe Wagner et al. 2001).

2.3 Das Problem der Nichtkäufer

Die Identifikation der für unseren Ansatz zentralen Regelmäßigkeit (1) und (2) erfolgte durch die Untersuchung individueller Kaufgeschichten. Das heißt, man analysierte das Markenwahlverhalten bei aufeinander folgenden Kaufakten, wobei die Zeitspannen dazwischen unterschiedliche Längen haben können. Stehen nun Handelspaneldaten zur Verfügung, so ist der Beobachtungsrhythmus vorab festgelegt (etwa wöchentlich oder monatlich) und individuelle Verbrauchszyklen bleiben daher unberücksichtigt. Bei einer ausreichend großen Stichprobe kann argumentiert werden, dass in den beiden Perioden zwar nicht die Absätze *derselben* Haushalte aufgezeichnet werden, sich aber in Summe jene, die entweder nur im ersten oder nur im zweiten Beobachtungszeitraum nachfragen, sowohl bezüglich des Mengen- als auch bezüglich des Markenwahlverhaltens die Waage halten (‚left censoring' und ‚right censoring' Probleme gleichen einander aus). Dies setzt allerdings im Wesentlichen stationäre Verhältnisse und damit einen gleichbleibenden Einsatz der Marketinganstrengungen der Anbieter voraus. Insbesondere kurzfristig wirkende Aktionen an den Einkaufsstätten können aber sehr wohl einen Einfluss auf die Wahl des Einkaufszeitpunktes haben (man denke beispielsweise an Vorziehkäufe auf Grund von Preisnachlässen). Das bereits erwähnte, zu (3) formal sehr ähnliche, Modell der Wählerstromanalyse braucht sich zwar nicht mit der Problematik der Zwischeneinkaufszeiten zu befassen – die Wahltermine sind ja festgelegt – berücksichtigt in manchen Versionen aber sehr wohl die ‚Nichtwähler'.

Wir erweitern daher unseren Ansatz und führen einen zusätzlichen, mit $(I+1)$ bezeichneten, Zustand ‚Nichtkauf' ein. Sowohl der Wertebereich der Laufindizes i und j als auch der Summationsbereich in (3) wird entsprechend vergrößert. Aus Sicht des Kaufverhaltens bedeutet dies, dass zwischen der Marken- und der Einkaufszeitpunktswahl modellmäßig kein Unterschied gemacht wird. Es wäre daher mit dem Ansatz unvereinbar, wenn Käufer, die eine gewisse Marke präferieren, häufiger oder seltener als andere nachfragen. Dies ist natürlich eine einschränkende Annahme, sie wird jedoch üblicherweise in der Literatur getroffen und auch hier vor allem aus pragmatischen Gründen postuliert.

Während die Berücksichtigung der Nichtkäufer im Modell vergleichsweise einfach durchzuführen ist, stellt dies für den Dateninput ein fundamentales Problem dar. Zum einen sind die $q_{(I+1)kt}$ und $q_{(I+1)k(t+1)}$ nun nicht mehr Absätze, sondern die Anzahlen der Nichtkäufer in den einzelnen Querschnitten, zum anderen liegen darüber keine Angaben in Handelspaneldaten vor. Der erste Aspekt wird dadurch relativiert, dass auch bei der Ableitung der Regelmäßigkeiten (1) und (2) das Mengenwahlverhalten außer Acht blieb, welches sich aber sehr wohl in den Absätzen aus dem Handelspanel niederschlägt. Allerdings zeigen Forschungsergebnisse (bspw. Ehrenberg 1988), dass solche Unterschiede auf individueller Ebene durch Aggregation weitgehend ausgeglichen werden. Aus diesem Grund scheint eine derartige Vereinfachung – ebenfalls aus pragmatischen Gründen – gerechtfertigt und dann ist auch der Unterschied zwischen der Anzahl an Kaufakten (anstelle von Absätzen) und jener an Nichtkäufern nicht mehr gravierend. Der zweite Gesichtspunkt erfordert allerdings, die $q_{(I+1)kt}$ und $q_{(I+1)k(t+1)}$ zu schätzen.

Wir schlagen dafür zwei unterschiedliche Vorgangsweisen vor, wobei wir zunächst jeweils annehmen, dass das Marktpotenzial Q bzw. die Gesamtzahl aller Käufer am untersuchten Markt bekannt und über die beiden Beobachtungsperioden konstant ist. Letzteres dürfte im Lichte der betrachteten Zeiträume unproblematisch sein. Eine einfache Möglichkeit stellt dann die proportionale Aufteilung von Q auf die einzelnen Querschnitte ihrer Größe gemäß dar (analog für $(t+1)$).

$$q_{(I+1)kt} = Q / n_{kt} \quad \forall k \tag{4}$$

Mit:

$n_{kt} = \sum_i q_{ikt}$ Größe des Querschnitts (bezogen auf den Absatz am untersuchten Markt)

Eine andere Variante ergibt sich aus einem Optimierungsproblem, welches jenem aus Abschnitt 2.2 sehr ähnlich ist.

$$Min \sum_{k=1}^{K} \sum_{j=1}^{I+1} \left(q_{jk(t+1)} - \sum_{i=1}^{I+1} p_{j|i} \cdot q_{ikt} \right)^2 / n_{kt} \tag{5}$$

$$\sum_{k=1}^{K} \sum_{j=1}^{I+1} q_{jk(t+1)} = \sum_{k=1}^{K} \sum_{i=1}^{I+1} q_{ikt} = Q, \quad q_{(I+1)kt}, q_{(I+1)k(t+1)} \geq 0 \quad \forall k \tag{6}$$

Die Zielfunktion ist wieder das verallgemeinerte ‚Kleinste Quadrate' Kriterium, die gesuchten Variablen sind aber die $q_{(I+1)kt}$ und $q_{(I+1)k(t+1)}$; die Nebenbedingungen (6) stellen die logische Konsistenz sicher. (5) setzt voraus, dass eine erste Schätzung der $p_{j|i}$ vorliegt. Dies kann etwa durch Anwendung des Hendry Modells (Vanhonacker 1980) auf aggregierter Ebene, mit für alle Marken sowie dem Zustand Nichtkauf konstantem ρ, erfolgen. Mittels der so bestimmten $q_{(I+1)kt}$ und $q_{(I+1)k(t+1)}$ erhält man durch Anwendung von (3) revidierte $p_{j|i}$ bzw. ρ_j. Unterscheiden sich die beiden Schätzungen für die ρ_j sehr stark voneinander, sollten die obigen zwei Rechenschritte iterativ wiederholt werden, bis Konvergenz eintritt.

2.4 Die Bestimmung des Marktpotenzials

Im vorangehenden Abschnitt wurde angenommen, dass das Marktpotenzial Q bekannt ist. Dies ist für die Anwendung des Ansatzes bei der Wählerstromanalyse erfüllt, weil die Anzahl der Wahlberechtigten amtlich festgelegt wird. Im Marketing ist die Sachlage schwieriger, da Q nur in Ausnahmefällen bekannt sein wird und ein Handelspanel keine unmittelbar verwertbare Information darüber bereitstellt. Wir schlagen daher einen Maximum Entropie Ansatz vor.

$$H = -\left(\sum_{i=1}^{I+1}\sum_{j=1}^{I+1} p_{j|i} \cdot \ln p_{j|i}\right) \tag{7}$$

(7) wird dabei als eindimensionale Funktion in Q aufgefasst. Ein wesentliches Argument für die Wahl dieser Methode wurde bereits im Abschnitt 1 gegeben. Ein anderes resultiert daraus, dass damit ein Kompromiss zu erwarten ist zwischen der generellen Tendenz von Maximum Entropie Schätzungen annähernd gleichverteilte Wahrscheinlichkeiten zu liefern und der Modellformulierung (1) und (2), die in der Regel hauptdiagonal-dominante Übergangsmatrizen erzeugt.

Zusammenfassend lautet demnach die Vorgangsweise zur Lösung des dargelegten Problems folgendermaßen:

1) Für gegebenes Q bestimmt man zunächst die Gesamtzahl der Nichtkäufer in t und $(t+1)$.
2) Auf aggregierter Ebene berechnet man eine Näherung von ρ mit Hilfe des Hendry Modells.
3) (4) oder (5) und (6) liefern Schätzwerte für die Anzahl der Nichtkäufer je Querschnitt.
4) Das mit (3) korrespondierende quadratische Optimierungsproblem ergibt verbesserte Schätzwerte ρ_j.
5) Wenn erforderlich werden die Schritte 3) und 4) iteriert.
6) Der Funktionswert für H – vgl. (7) – wird berechnet, Q verändert und die Schritte 1) bis 5) werden wiederholt.
7) Jene Parameter werden letztlich gewählt, für die H maximal wird.

3 Zusammenfassung und Ausblick

In diesem Beitrag haben wir eine Vorgangsweise dargelegt, mit deren Hilfe die Anteile loyaler Kunden, an Wechsel- und Nichtkäufern sowie das Marktpotenzial aus Handelspaneldaten geschätzt werden können. Ein derartiger Algorithmus wurde unseres Wissens nach noch nicht in der Literatur vorgestellt. In einem weiteren Schritt muss freilich seine Leistungsfähigkeit überprüft werden. Dabei stößt man allerdings auf eine zusätzliche, der Problemformulierung inhärente, Schwierigkeit. Eine *stringente* Validierung an Hand empirischer Daten ist deswegen nicht möglich, weil ein Handelspanel eben definitionsgemäß keine personenbezogenen Informationen liefert und ein Haushaltspanel – über den selben Markt – mit ersterem nicht vollständig kompatibel ist (vgl. bspw. Nenning et al. 1979). Die Autoren haben dennoch einige empirische Untersuchungen durchgeführt, die durchaus vielversprechend verlaufen sind.

Das Instrumentarium der Simulation bietet hier Abhilfe. Zunächst werden dabei individuelle Kaufakte generiert, die interessierenden Kennzahlen (Markentreue etc.) berechnet und die Aggregation auf Querschnittsebene vorgenommen. Die so erhaltenen Daten dienen als Input für den vorgeschlagenen Algorithmus und die geschätzten Parameter können mit den ‚wahren‘ Werten verglichen werden. Das Design des Simulationsexperiments sollte überdies die potenziell wichtigsten Einflussfaktoren auf die Leistungsfähigkeit des Modells (wie Anzahl der Nichtkäufer, der Marken und der Querschnitte, die Stichprobengröße etc.) berücksichtigen. Auch hier liefern erste Berechnungen zufriedenstellende Ergebnisse.

Literatur

Colombo RA, Morrison DG (1989) A Brand Switching Model with Implications for Marketing Strategies. Marketing Science 8 (1), 89-99

Ehrenberg ASC (1988) Repeat Buying: Facts, Theory and Applications. Charles Griffin & Company LTD, London

Jaynes ET (1957) Information Theory and Statistical Mechanics. Physical Review 106, 620-630

King (1997) A Solution to the Ecological Inference Problem: Reconstructing Individual Behavior from Aggregate Data. Princeton University Press, Princeton

Nenning M, Topritzhofer E, Wagner U (1979) Zur Kompatibilität alternativer kommerziell verfügbarer Datenquellen für die Marktreaktionsmodellierung: Die Verwendung von Prewhitening-Filtern und Kreuzspektralanalyse sowie ihre Konsequenzen für die Analyse betriebswirtschaftlicher Daten. Zeitschrift für Betriebswirtschaft 49 (9), 281-297

Vanhonacker WR (1980) The Hendry partitioning methodology: Theory and applications, Research Working Paper No. 288A, Graduate School of Business, Columbia University

Wagner U, Reisinger H, Gausterer K (2001) Die Bestimmung des Markenwechselverhaltens mit Hilfe von Querschnittsdaten. Zeitschrift für Betriebswirtschaft 71 (10), 1113-1130

A Conjugate Direction Frank-Wolfe Method with Applications to the Traffic Assignment Problem

Maria Daneva and Per Olov Lindberg

Linköping University, Department of Mathematics, SE-581 83 Linköping, Sweden,
e-mails: madan, polin@mai.liu.se

Abstract. We present a version of the Frank-Wolfe method for linearly constrained convex programs, in which consecutive search direction are made conjugate to each other. We also present preliminary computational studies in a MATLAB environment. In these we apply the pure Frank-Wolfe, the Conjugate Direction Frank-Wolfe (CDFW) and the "partanized" Frank-Wolfe to some classical Traffic Assignment Problems. CDFW compares favorably to the other methods in this study.

1 Introduction

The Frank-Wolfe (FW) method [4] was originally suggested for quadratic programming problems, but the original paper also noted that it could be applied to linearly constrained convex programs. The main usage of the FW method has been in routing problems in the telecom and traffic areas, where it is usually attributed to [5] and [8], respectively; see however [2] also. It still is the most popular method in these areas. According to the survey [13], "The reason for this is twofold. The method easily generates feasible solutions from shortest path calculations and enjoys fast convergence in the early iterations". However, to cite [14], "The unsatisfactory performance of the Frank-Wolfe algorithm, in particular at the vicinity of an optimal solution, was observed quite early." Since this behavior was observed early, many attempts have been made to improve upon the FW method, in particular finding better search directions, (e.g. [6], [9], [11]). The main proponent of these attempts is the PARTAN approach [9], where one gets better directions by taking PARTAN steps (e.g. [10], p. 255-257, [16]) between pure Frank-Wolfe steps. An intrinsic difficulty with the PARTAN technique in the Frank-Wolfe context is the determination of appropriate bounds on the step lengths to stay feasible. In the early papers these bounds were determined individually for the first few steps. Later, [1] and [3] determined analytical recursion formulas for these bounds. In spite of improved convergence characteristics, the PARTAN approach has not superseded the Frank-Wolfe method in the telecom and transportation areas.

In the current paper we suggest a simple improvement of the FW directions: making them conjugate. Every textbook on nonlinear programming advocates conjugate gradients over pure gradient directions (i.e. steepest descent). PARTAN owns some of its motivation to conjugate directions. In fact, applied to an unconstrained convex quadratic problem, PARTAN will generate conjugate directions ([10], p.255-257).

In the next section we review the FW method and its "PARTANized" version. Section 3 is devoted to conjugate direction techniques and our own conjugate direction FW method (CDFW). In Section 4 we apply MATLAB versions of pure FW, partanized FW (PFW) and CDFW to a set of classical traffic assignment problems. Measuring performance per major iteration, it is demonstrated that FW is clearly outperformed by PFW, which in turn is clearly outperformed by CDFW.

2 The Frank-Wolfe Method and Modifications

Let $f : \mathbb{R}^n \mapsto \mathbb{R}$ be a twice continuously differentiable convex function and let $X \subset \mathbb{R}^n$ be a compact convex polyhedral set. The methods discussed in this section apply to the problem of minimizing f over X:

$$\text{(P)} \qquad f^* \triangleq \min_{\mathbf{x}\in X} f(\mathbf{x}).$$

By continuity of f, (P) has a solution, unique if f is strictly convex.

The problem (P) is in principle well-behaved, but in the traffic and telecom applications we aim at, it is a convex cost multicommodity flow problem, which typically becomes very large. Thus, one needs to utilize structure. The Frank-Wolfe algorithm allows efficient utilization of the multicommodity structure of (P). We briefly discuss it in the next section.

2.1 The Frank-Wolfe Method

The conventional FW algorithm [4] was introduced for solving general linearly constrained quadratic and convex problems. At iteration k, FW approximates f by linearizing at the current point $\mathbf{x}_k$, giving an affine minorant f_k to f:

$$f_k(\mathbf{x}) \triangleq f(\mathbf{x}_k) + \nabla f(\mathbf{x}_k)^T(\mathbf{x} - \mathbf{x}_k).$$

The first step in FW method is the *determination of the search direction*, which is done by minimizing f_k over X. This is a linear program

$$(\text{LP}_k) \qquad f_k^* \triangleq \min_{\mathbf{x}\in X} f_k(\mathbf{x}),$$

the solution of which we denote by $\mathbf{x}_k^{LP}$. In the telecom and traffic applications, however, (LP_k) decomposes into a set of shortest path problems, (e.g. [14], p.97). Note that f_k^* is a *lower bound*, LBD_k, to f^*, a fact which may be used in termination criteria.

The second step of the FW method is to perform a *line search*, i.e. a one-dimensional minimization of $f(\mathbf{x})$, along the line segment between the current solution $\mathbf{x}_k$ and the solution $\mathbf{x}_k^{LP}$. The point where this minimum is achieved (at least approximately) is chosen as the next iterate $\mathbf{x}_{k+1}$.

As already mentioned this algorithm is very easy to implement, but due to the slow asymptotic convergence, many modifications have been suggested. The line search can e.g. be modified in order to take predetermined [15] or longer steps [17] or the search direction can be combined with previous ones [6], [9], [11]. In other modifications the linear subproblem is modified in order to avoid generating extreme point solutions [7].

2.2 The PARTAN Technique

The method of parallel tangents (PARTAN) originates in attempts to avoid zig-zagging in gradient based methods for unconstrained optimization ([10], p.254, [16]), by performing an extra line search at each iteration. In the PARTAN technique for unconstrained optimization, at the iterate $\mathbf{x}_k$, one first performs a line search in the gradient direction, giving the intermediate point $\mathbf{y}_k$. Then one applies an extra line search from $\mathbf{y}_k$ in the direction $\mathbf{d}_k = \mathbf{y}_k - \mathbf{x}_{k-1}$, giving the new iterate $\mathbf{x}_{k+1}$.

In the partanized FW method, the line search in the gradient direction is replaced by one in the direction towards the solution, $\mathbf{x}_k^{LP}$, of the linearized problem.

3 The Conjugate Direction Frank-Wolfe Method

In this section we describe our modification of the FW method. In conjugate directions methods (e.g. [10], Ch.8) for unconstrained convex quadratic optimization, one performs line searches consecutively in a set of directions, $\mathbf{d}_1, \cdots, \mathbf{d}_n$, mutually conjugate with respect to the hessian H of the objective (i.e. fulfilling $\mathbf{d}_i^T H \mathbf{d}_j = 0$ for $i \neq j$). In $\mathbb{R}^n$ the optimum then is identified after n line searches ([10], p.241, Expanding Subspace Theorem).

In conjugate gradient methods, one obtains conjugate directions by "conjugatizing" the gradient direction with respect to the previous search direction; i.e. $\mathbf{d}_k \triangleq \nabla f(\mathbf{x}_k) + \beta_k \mathbf{d}_{k-1}$, with β_k chosen so that $\mathbf{d}_k$ is conjugate to $\mathbf{d}_{k-1}$. In the quadratic case, $\mathbf{d}_k$ then in fact becomes conjugate to all previous directions $\mathbf{d}_1, \cdots, \mathbf{d}_{k-1}$ (e.g. [10], p. 245, Conjugate Gradient Theorem).

The same trick can be applied to the FW method: Let $\mathbf{x}_k$ be the current point (see Figure 1), found by a line search in the direction $\mathbf{d}_{k-1}$, from $\mathbf{x}_{k-1}$ towards $\mathbf{w}_{k-1} \in X$. At $\mathbf{x}_k$ we solve the linearized problem (LP_k) giving $\mathbf{y}_k = \mathbf{x}_k^{LP} \in X$. In the FW method the new search direction is taken as $\tilde{\mathbf{d}}_k = \mathbf{y}_k - \mathbf{x}_k$. In the *conjugate direction FW method* (CDFW), we choose the search direction as

$$\mathbf{d}_k = \tilde{\mathbf{d}}_k + \beta_k \mathbf{d}_{k-1}, \tag{1}$$

where β_k is chosen to make $\mathbf{d_k}$ conjugate to $\mathbf{d}_{k-1}$ with respect to the hessian H_k of f at $\mathbf{x}_k$, which can be done cheaply, since H_k is diagonal. Equivalently (up to scaling) we choose $\mathbf{d}_k = \mathbf{w}_k - \mathbf{x}_k$, where $\mathbf{w}_k$ is a convex combination of $\mathbf{y}_k$ and $\mathbf{w}_{k-1}$, both belonging to X.

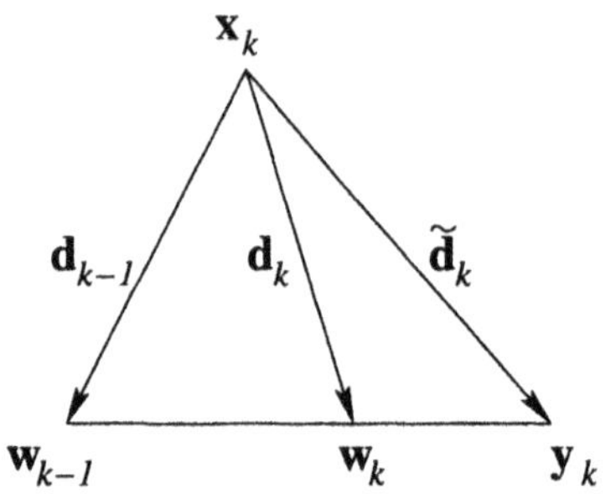

Fig. 1. Determination of the search direction for the CDFW algorithm

4 Computational Experiments on Traffic Assignment Problems

As mentioned, the main areas of applications of the Frank-Wolfe method and its modifications are traffic assignment (e.g.[14]) and telecom traffic routing (e.g.[13]). In this paper we consider the traffic assignment problem (TAP). This problem can be formulated as a mathematical program with nonlinear objective and linear constraints ([14], Ch.2). Thus we can use the methods proposed in this paper to solve the TAP.

The three methods discussed have been implemented in MATLAB [12] and run on a 480 MHz Sun Sparc 10 Station. Two networks, frequently used to test methods for problems in transportation planning, are used in our experiments (see Table 1).

In the experiments we start in an all-or-nothing assignment using free flow costs. The solver for the shortest path subproblems is implemented in C, using a MEX interface for communication between MATLAB and C. In the line search, we take a single Newton step per line search, computed cheaply by the diagonality of H_k. In figure 2 we display the relative errors of all

Table 1. Test networks.

Network	nodes	arcs	commodities
Sioux Falls	24	76	528
Winnipeg	1052	2836	4345

three methods (in log-log scale diagrams) against the iteration count, for the Sioux Falls and Winnipeg cases. The relative error RE_k is defined as $RE_k \triangleq$

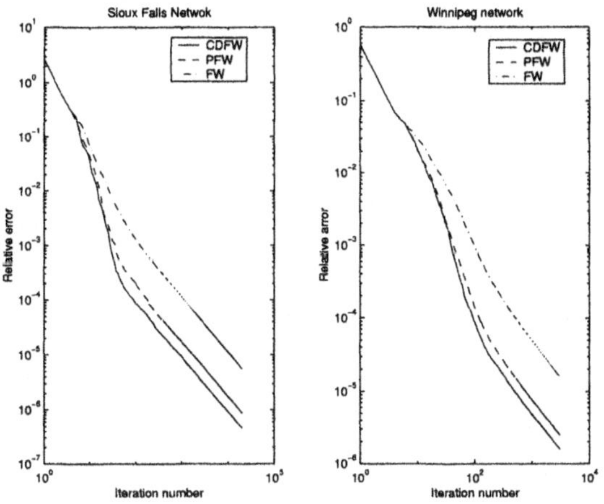

Fig. 2. Comparison between the FW, PFW and CDFW .

$(f(\mathbf{x}_k) - LBD)/LBD$, where LBD is the best known lower bound (which in the Sioux Falls case is the full precision optimal value). The iterations are continued much longer than what is typically done in practice, to show the asymptotic behavior, which is linear with slope 1 in the log-log diagram, corresponding to a relative error, RE_k, of the form

$$RE_k = const./k. \tag{2}$$

The seemingly small difference between PFW and CDFW is illusive. Dis-

Fig. 3. The ratio between the relative error for CDFW and PFW

playing the ratio of the relative errors of PFW and CDFW in the Sioux Falls and Winnipeg cases (figure 3), we see that the ratio stabilizes rather soon around 2 and 1.5 respectively. In view of (2), this implies that PFW asymptotically needs to take around 100% and 50% more iterations respectively to achieve the same accuracy as CDFW. In the same way comparing CDFW with FW, the needed number of iterations of FW is around ten times that of CDFW.

References

1. Arezki, Y. and van Vliet, D. (1990) A full analytical implementation of the PARTAN/Frank-Wolfe algorithm for equilibrium assignment. Transportation Sci. **24**(1), 58–62.
2. Bruynooghe, M., Gibert, A. and Sakarovitch, M. (1969) Une méthode d'affectation du trafic. Proceedings of the 4th International Symposium on the Theory of Road Traffic Flow, Bundesminister für Verkehr, Bonn, 198–204.
3. Florian, M., Guelat, J. and Spiess, H. (1987) An efficient implementation of the "PARTAN" variant of the linear approximation method for the network equilibrium problem. Networks **17**, 319–339.
4. Frank, M. and Wolfe, P. (1956) An algorithm for quadratic programming. Naval Res. Logist. Quart. **3**, 95–110.
5. Fratta, L., Gerla, M. and Kleinrock, L. (1973) The flow deviation method: An approach to store-and-forward communication network design. Networks **3**, 97–133.
6. Fukushima, M. (1984) A modified Frank-Wolfe algorithm for solving the traffic assignment problem. Transportation Res. Part B **18**(2), 169–177.
7. Larsson, T., Patriksson, M. and Rydergren, C. (1997) Applications of simplicial decomposition with nonlinear column generation to nonlinear network flows. Network optimization, 346–373. Springer, Berlin.
8. LeBlanc, L. J. (1973) Mathematical programming algorithms for large scale network equilibrium and network design problems. PhD thesis, IE/MS Dept, Northwestern University, Evanston IL.
9. LeBlanc, L., Helgason, R. and Boyce, D. (1985) Improved efficiency of the Frank-Wolfe algorithm for convex network programs. Transportation Sci. **19**(4), 445–462.
10. Luenberger, D. G. (1984) Linear and Nonlinear Programming. Addison-Wesley, Reading, MA.
11. Lupi, M. (1986) Convergence of the Frank-Wolfe algorithm in transportation network. Civil Engineering Systems **3**, 7–15.
12. Matlab Reference Guide (1996), The MathWorks, Mass.
13. Ouorou, A., Mahey, P. and Vial, J.P. (2000) A survey of algorithms for convex multicommodity flow problems. Management Sci. **46**(1), 126–147.
14. Patriksson, M. (1994) The Traffic Assignment Problem - Models and Methods. VSP, Utrecht, The Netherlands.
15. Powell, W. and Sheffi, Y. (1982) The convergence of equilibrium algorithms with predetermined step sizes. Transportation Sci. **16**(1), 45–55.
16. Shah, B., Buehler, R. and Kempthorne, O. (1964) Some algorithms for minimizing a function of several variables. J. Soc. Indust. Appl. Math. **12**, 74–92.
17. Weintraub, A., Ortiz, C. and González, J. (1985) Accelerating convergence of the Frank-Wolfe algorithm. Transportation Res. Part B **19**(2), 113–122.

Online-Algorithmus zur Steuerung von Verkehrslichtsignalanlagen

Klaus Ladner[1]

Institut für Statistik und Operations Research, Karl-Franzens Universität Graz, Universitätsstraße 15, 8010 Graz, Österreich

Abstract. Es wird ein Algorithmus entworfen, der laufend verkehrsabhängige Daten einer Kreuzung und strategische Vorgaben des Verkehrsrechners verarbeitet, um den jeweils nächsten Umlauf einer Ampelanlage zu planen.

Dabei werden die Vorstellungen des Verkehrsplaners berücksichtigt. Es können Zeitfenster definiert werden, innerhalb denen Verkehrsströme entweder keinesfalls Grün haben dürfen oder unbedingt Grün haben müssen, um die Koordinierung mit den umliegenden Kreuzungen ermöglichen zu können.

Es werden Kombinationen von Verkehrsströmen der betrachteten Kreuzung so zu Phasen zusammengefasst, dass dabei sinnvolle Phasenwechselschemata erstellt werden können. Aufgrund einer definierten Zielfunktion und mehreren Nebenbedingungen werden zulässige Kombinationen von Phasen gesucht, welche die "Leistungsfähigkeit" der Steuerung maximieren. Dazu wird das Problem in einen Zustands-/Schichtengraph übergeführt, in dem ein längster zulässiger Weg gesucht wird.

Da der Algorithmus auf Kreuzungsrechnern laufen soll, muss er extrem schnell und mit bescheidenem Speicheraufwand gute Ergebnisse liefern können.

1 Einleitung

Die Steuerung von Verkehrslichtsignalanlagen ist eine komplexe Aufgabe, die versucht wird, modular zu lösen. In diesem Beitrag soll die Aufgabe eines dieser Module dargelegt und ein Lösungsansatz, aufbauend auf das Modul, das in [3] beschrieben wird, skizziert werden.

Die Thematik der Lichtsignalsteuerung wurde schon in einigen wissenschaftlichen Arbeiten untersucht (siehe z. B. [4] oder [1]). Riedel fasst in [2] die Steuerung als Regelkreis auf und versucht die Problematik sogar als Art Dynamische Programmierung optimal zu lösen.

Betrachtet wird hier eine Situation, in der ein Kreuzungsrechner mit einem Verkehrsrechner verbunden ist. Der Verkehrsrechner liefert Informationen die Koordinierung der Verkehrslichtsignalanlage mit umliegenden Anlagen betreffend. Der Kreuzungsrechner wertet ständig aktuelle Verkehrsdaten aus. Diese Verkehrsdaten bestehen z. B. aus Meldungen über herannahende öffentliche Verkehrsmittel, die bevorzugt über die Kreuzung geführt werden sollen. Problematisch dabei ist, dass es zu Zielkonflikten kommen kann, wenn sowohl eine Koordinierung aufrecht erhalten werden soll als auch auf das tatsächlich vorhandene Verkehrsgeschehen eingegangen werden soll.

Gesucht ist eine Grobplanung für die Ein- und Ausschaltzeitpunkte der einzelnen Verkehrsströme für den nächsten Umlauf unter Berücksichtigung dieser Informationen und weiteren notwendigen weiter unten genannter Nebenbedingungen. Unter Grobplanung wird hier verstanden, dass die ungefähren Zeitpunkte bestimmt werden, die exakte sekundengenaue Festlegung, sowie die Optimierung der Phasenübergänge ist an dieser Stelle nicht nötig, da diese Aufgabe von einem anderen Modul erledigt wird.

2 Aufgabenstellung

Es gebe n Verkehrsströme und eine Umlaufzeit (T_{u}). Jeder Verkehrsstrom habe eine maximale Wartezeit ($tw^{\max}$), eine minimale Grünzeit ($tg^{\min}$), eine minimale Rotzeit ($tr^{\min}$), und eine maximale Grünzeit ($tg^{\max}$). Verkehrsströme sind entweder verträglich oder nicht verträglich. Für je zwei nicht verträgliche Verkehrsströme i und j gebe es eine Zwischenzeit (tz_{ij}), die angibt, wie viele Sekunden nach dem Ende des Grüns von i vergehen müssen, bevor j Grün bekommen darf.

Für jeden Verkehrsstrom kann ein Grünband definiert sein, innerhalb dessen dieser Verkehrsstrom jedenfalls Grün haben muss (Grünband-Anfang BA_{i} bzw. Grünband-Ende BE_{i}).

Gegeben sei weiter für jeden Verkehrsstrom eine Funktion ($f_i(t)$), die in Abhängigkeit von der Zeit angibt, welchen Beitrag der betrachtete Verkehrsstrom zum Zielfunktionswert leistet. In diese Funktionen fließen alle Vorstellungen und Gewichtungen des Betreibers der Verkehrslichtsignalanlage bezüglich der Wartezeit und Anzahl der Halte eines Verkehrsstromes, der Priorität der Verkehrsströme untereinander, sowie statistischer Verteilungen der Verkehrsflüsse ein.

Gesucht sind Ein- und Ausschaltzeitpunkte (ein_i bzw. aus_i) für alle Verkehrsströme derart, dass die Summe ihrer Flächen maximal wird. Vereinfacht um die Problematik des Umbruchs am Ende der Umlaufzeit kann das Problem wie folgt formuliert werden:

$$\max \sum_{i=1}^{n} \int_{ein_{\mathrm{i}}}^{aus_{\mathrm{i}}} f_{\mathrm{i}}(t)dt$$

$$\begin{aligned} \text{bez.} \qquad ein_{\mathrm{i}} - aus_{\mathrm{i}} &\leq tw_{\mathrm{i}}^{\max} \\ aus_{\mathrm{i}} - ein_{\mathrm{i}} &\geq tg_{\mathrm{i}}^{\min} \\ ein_{\mathrm{i}} - aus_{\mathrm{i}} &\geq tr_{\mathrm{i}}^{\min} \\ aus_{\mathrm{i}} - ein_{\mathrm{i}} &\leq tg_{\mathrm{i}}^{\max} \\ ein_{\mathrm{i}} - aus_{\mathrm{j}} &\geq tz_{\mathrm{ji}} \\ ein_{\mathrm{i}} &\leq T_{\mathrm{u}} \\ aus_{\mathrm{i}} &\leq T_{\mathrm{u}} \\ aus_{\mathrm{i}}, ein_{\mathrm{i}} &\geq 0 \end{aligned}$$

3 Bilder

Ein Verkehrsstrom kann entweder ein- oder ausgeschaltet sein. Es gibt somit 2^n theoretische Zustände, in der sich die Verkehrslichtsignalanlage befinden kann. Tatsächlich sind es weniger, da nicht verträgliche Verkehrsströme nicht gleichzeitig eingeschaltet sein können.

Da in der Praxis aber nicht alle Zustände Verwendung finden werden, wird ein Algorithmus entwickelt, der eine Menge von "'sinnvollen"' Zuständen ermittelt. Der Zustand einer Verkehrslichtsignalanlage kann durch die Menge der eingeschalteten Verkehrsströme beschrieben werden. Eine solche Menge wird als Bild bezeichnet.

Als erstes werden Maximalbilder gesucht. Maximalbilder seien Bilder, die hinsichtlich der Eigenschaft, dass es nicht möglich ist, einen weiteren Verkehrsstrom hinzuzufügen, maximal sind.

Gesucht sind Maximalbilder und Untermengen von diesen, sodass ausgehend von einem beliebigen Maximalbild durch Kombination von Untermengen anderer Maximalbilder jeder Verkehrsstrom genau einmal Grün hat.

Gegeben sei die Menge aller Verkehrsströme $B^0 = \{VS_\mathrm{i} \mid i = 1 \dots n\}$. R sei eine Verträglichkeitsrelation. Ein Maximalbild sei die Menge $B \subseteq B^0$, für die gilt: $\forall\ i, j$ mit $VS_\mathrm{i}, VS_\mathrm{j} \in B : \ VS_\mathrm{i}\ \mathbf{R}\ VS_\mathrm{j}$ und $\forall i$ mit $VS_\mathrm{i} \notin B$, $\exists j$ mit $VS_\mathrm{j} \in B : \ VS_\mathrm{i}\ \neg\mathbf{R}\ VS_\mathrm{j}$.

Die Anzahl der Maximalbilder sei v. Ein Maximalbild werde mit B_i für $i \in \{1, \dots, v\}$ bezeichnet. Die Restmengen seien: $B^\mathrm{i} = B^0 \backslash B_\mathrm{i} \ \ i = 1 \dots v$.

Das k-te Maximalbild zweiter Ebene, ausgehend von der i-ten Restmenge, sei die Menge $B_\mathrm{ik} \subseteq B^\mathrm{i}$, für die gilt: $\forall\ j, r$ mit $VS_\mathrm{j}, VS_\mathrm{r} \in B_\mathrm{ik} : \ VS_\mathrm{j}\ \mathbf{R}\ VS_\mathrm{r}$ und $\forall j$ mit $VS_\mathrm{j} \in B_\mathrm{i} \backslash B_\mathrm{ik}$, $\exists r$ mit $VS_\mathrm{r} \in B_\mathrm{ik} : \ VS_\mathrm{j}\ \neg\mathbf{R}\ VS_\mathrm{r}$.

Diese Zerlegung wird analog solange durchgeführt bis die sich ergebenden Restmengen zulässig sind, d. h. Maximalbilder sind und daher nicht weiter zerlegt werden können.

Die Anzahl der so erhaltenen Bilder sei m. Jedes Bild wird zu jenen Zeitpunkten als gesperrt markiert, zu denen ein Verkehrsstrom, den das Bild enthält gesperrt ist oder ein Verkehrsstrom, der nicht im Bild enthalten ist, ein Grünband besitzt.

4 Zustands-/Schichtengraph

Die Zeitachse werde diskretisiert. Es werden Intervalle mit der Länge Δ gebildet. Dadurch entsteht ein Zustands-/Schichtengraph mit m Zuständen und k Schichten, wobei $k = 1 + \mathrm{int}\left(\frac{T_\mathrm{u}-1}{\Delta}\right)$. Theoretisch könnte es somit $m \cdot k$ Knoten geben. Tatsächlich fällt die Anzahl der Knoten geringer aus, da an jenen Stellen im Zustands-/Schichtengraph keine Knoten vergeben werden, wo es im zugeordneten Bild im entsprechenden Zeitintervall Sperrbereiche gibt.

Zwischen je zwei Knoten i und j gibt es genau dann einen Pfeil von i nach j, wenn es möglich ist, zu jenem Zeitpunkt, der dem Ende des Knotens

i zugeordnet ist, in das Bild des Knotens j zu wechseln, sodass aufgrund eines potentiellen Zwischenzeitenlaufs zu jenem Zeitpunkt das Bild des Knotens j eingeschaltet werden kann, der dem Anfang des Knotens j entspricht.

Es wird eine Quelle eingeführt, die mit allen Knoten der ersten Schicht, und eine Senke, die mit allen Knoten der letzten Schicht adjazent ist.

Jeder Knoten erhält eine Bewertung, nämlich die Summe der Integrale über die zugeordnete Zeit der eingeschalteten Verkehrsströme.

Gesucht ist ein zulässiger Pfad durch diesen Graph, der die Summe der Knotenbewertungen maximiert. Da die Rechenzeit für eine vollständige Enumeration nicht zur Verfügung steht, wird auf ein heuristisches Verfahren zurückgegriffen, das den Rahmen dieses Beitrags sprengen würde (siehe dazu auch Fig. 1).

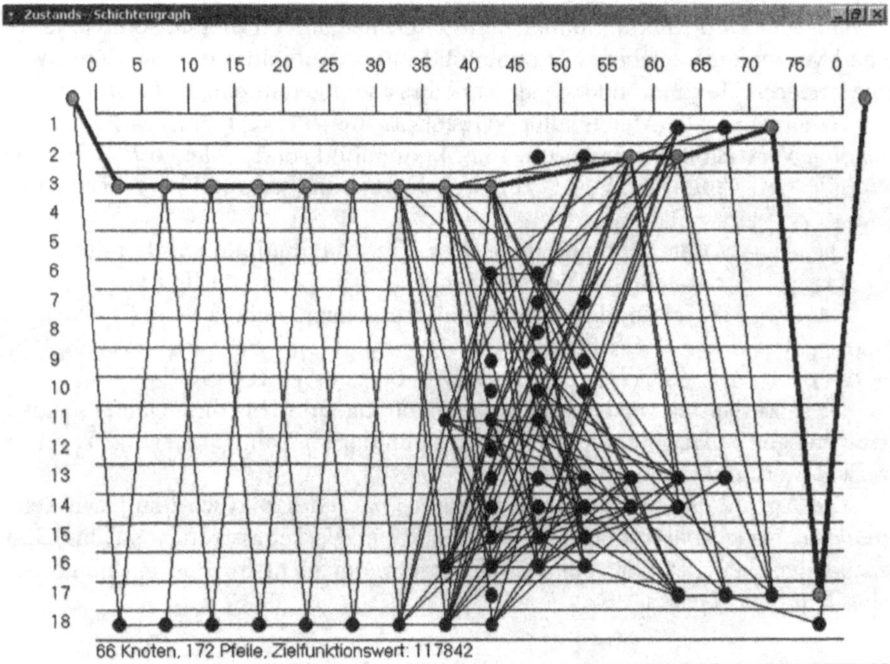

Fig. 1. Zustands-/Schichtengraph mit Lösungsvektor

5 Resultat

Der Lösungsvektor wird anschließend dahingehend transformiert, dass jedem Verkehrsstrom jene Grünzeiten zugeordnet werden, die den Zeitintervallen der Knoten im Lösungsvektor entsprechen, in denen der Verkehrsstrom vorkommt (siehe dazu auch Fig. 2).

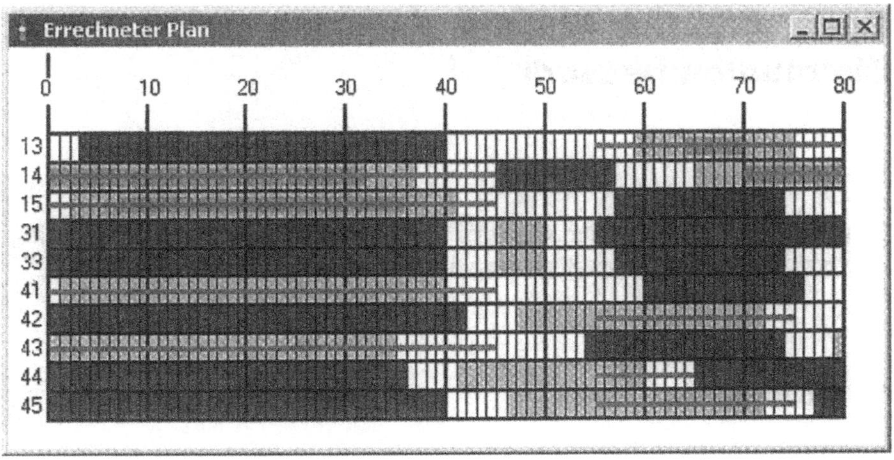

Fig. 2. Plan mit transformierten Grünzeiten

References

1. Mirchandani, P., Knyazyan, A., Head, L., Wu, W. (2001) An Approach Towards the Integration of Bus Priority, Traffic Adaptive Signal Control, and Bus Information/Scheduling Systems. In: Voß, S., Daduna, J. R. (ed.): Computer-Aided Scheduling of Public Transport, Springer, Heidelberg, 319–334
2. Riedel, Th. (1994) Regelung in Verkehrssystemen, Zrich
3. Rudolf Keller AG (1993) VS-PLUS Beschreibung, Muttenz
4. Schnabel, W., Lohse, D. (1980) Grundlagen der Straßenverkehrstechnik und der Straßenverkehrsplanung Band 1, Verlag für Bauwesen, Berlin

Optimal Sorting Machine Allocation in the Postal Distribution Network

Jaroslav Janáček

Department of Transportation Networks, Faculty of Management and Informatics, University of Žilina, Slovak Republic, 010 26, e-mail: jardo@frdsa.fri.utc.sk

1. Introduction

The postal distribution system constitutes many-to-many distribution system with several transshipments [1], [3]. At the beginning, consignments are transported by vehicles from post offices to transit centers. Another fleet continuous the transport from transit centers. This one finishes at so-called sorting center, which serve a given cluster of the transit centers. Rough sorting of the consignments coming from the individual transit centers of the cluster is done in the sorting center to separate the consignments according to destination sorting centers and each class is divided according to the destination transit centers. Having completed the rough sorting, each bulk of consignments is transported to the destination-sorting center. Fine sorting is performed there to separate the consignments in accordance to destination post offices. The fine sorting is completed after all bulks have been sorted. Then the sorted consignments are loaded on vehicles and transported to destination transit centers to be there up to the time, when distribution to the destination post offices starts. It can be noticed that the earliest time of consignment bulk departure from its original transit center results from given times of departures from the transit centers. The similar situation occurs at the opposite end of the transport chain. The arrival time of a bulk at the destination transit center is fixed for distribution fleet to be able to complete its work in time. It follows that the time of bulk departure from destination sorting center is given as well.

This way, each pair of origin and destination sorting centers is associated with pair of times, where the first time is the earliest possible end of rough sorting at the original sorting center and the second one gives the latest possible time, when fine sorting at the destination center must start. Any pair of times constitutes time window, in which transport of the associated bulk must be performed. Necessary transport time for any pair of sorting centers is fixed, but the end and starting times of sorting depend on numbers of rough and fine sorting machines respectively. The higher number of rough sorting machines, the earlier end time of rough sorting is reached and similarly the higher number of fine sorting machines, the latter starting time of fine sorting is allowed. The objective of the problem is to minimize the total cost of allocated rough and fine sorting machines under the above-mentioned constraints.

In the following chapters we give the mathematical description of the time constraints, establish a mathematical programming model of the problem and outline a special exact approach to solve the problem.

2. Time constraints of sorting machine allocation problem

As mentioned in the introduction, the end time of rough sorting process is influenced by arrival times of bulks to the original sorting center. Let t_1, t_2, ..., t_q be all the arrival times ordered in increasing way. Let p_1, p_2, ..., p_q be sizes of the bulks associated with the arrival times. If r denotes time necessary for one rough sorting machine to process one consignment and w denotes number of rough sorting machines allocated at the sorting center, then w/r is rate of sorting at the center. The end of rough sorting can be determined as resulting value T of this algorithm [2]:

Step 1. Set $T = t_1+p_1(r/w)$ and $j = 1$.

Step2. Repeat $j = j+1$, $T = max\ \{T, t_j\}+ p_j(r/w)$ until $j=q$.

demonstrate this algorithm, we consider an example with three arrival times t_1, t_2 and t_3, which are equal to 450, 470 and 500 respectively and which are associated with numbers 6000, 6000 and 3600 of consignments. For $r = 0.01$ and $w = 2$, we get current number of unsorted consignments as saw-shaped function of time (see Fig. 1) and $T = 528$.

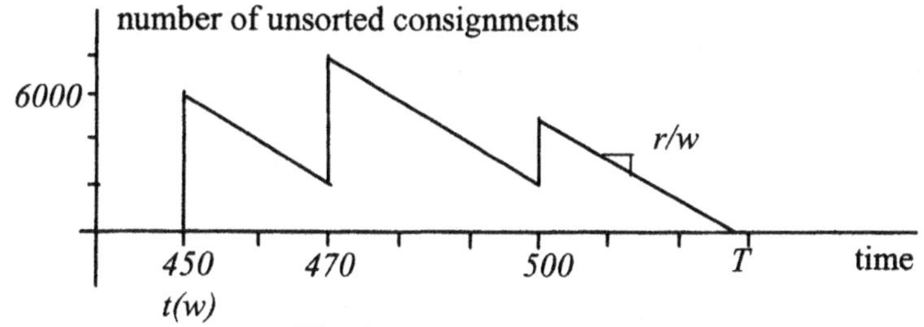

Fig. 1

When another example is considered with the same t_j, p_j and r, but with different number of machines $w = 3$, we get function of unsorted consignments in Fig 2, where $T = 512$.

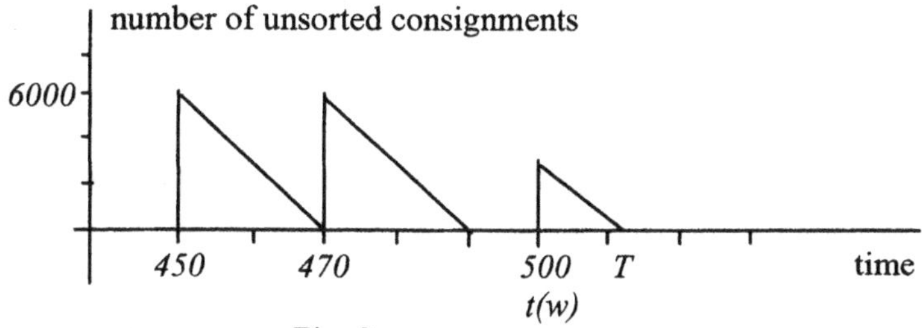

Fig. 2

It can be noticed that the end of rough sorting is given by arrival time $t(w)$ following the last idle interval of the sorting process and by the number of consignments $p(w)$ delivered after this time. Then resulting time $T^R(w)$ takes value of expression $t(w)+p(w)r$. Let us imagine for a moment that w increases continuously. Until other idle time emerges after former $t(w)$, $p(w)$ stays constant and $T^R(w)$ will decrease like hyperbola. Whenever new idle time after former $t(w)$ occurs, new $t(w)$ takes higher value and associated $p(w)$ shrinks. Then $T^R(w)$ decreases with hyperbola again but it decreases slower. We conclude that $T^R(w)$ is decreasing convex function of w.

Similar way we analyze the starting time $T^F(\underline{w})$ of fine sorting at a destination sorting center. If we denote $\underline{t}(\underline{w})$ the last departure before the first idle interval of the sorting process and $\underline{p}(\underline{w})$ the number of consignments, which is to be taken from the sorting center up to $\underline{t}(\underline{w})$, then resulting time $T^F(\underline{w})$ takes value of $\underline{t}(\underline{w}) - \underline{p}(\underline{w})\underline{r}$, when $\underline{r}$ denotes time of one fine sorting machine necessary for processing of one consignment. Thus function $T^F(\underline{w})$ is concave increasing function of $\underline{w}$.

Now, let us consider, for each sorting center $s \in S$, the above mentioned functions $T_s^R(w_s)$ and $T_s^F(\underline{w}_s)$, which provide the earliest end of rough sorting and the latest starting time of fine sorting for integer values of w_s and $\underline{w}_s$.

Let v_s^R, $v_s^R+m(s)$ denote lower and upper bounds on number of rough sorting machines, which is possible to allocate at sorting center s and let v_s^F, $v_s^F+p(s)$ denote lower and upper bounds on number of fine sorting machines. Then variables u_s^R and u_s^F may be introduced by equations $u_s^R = w_s - v_s^R$ and $u_s^F = \underline{w}_s - v_s^F$. Functions R_s, F_s can be obtained by substitution for w_s and $\underline{w}_s$ into $T_s^R(w_s)$ and $T_s^F(\underline{w}_s)$: $R_s(u_s^R) = T_s^R(v_s^R + u_s^R)$ and $F_s(u_s^F) = T_s^F(v_s^F + u_s^F)$ for $u_s^R \in \langle 0, m(s)\rangle$ and $u_s^F \in \langle 0, p(s)\rangle$. Furthermore, let t_{rs} denote time necessary to transport sorted consignments from original sorting center r to destination sorting center s. After this preliminaries the time constraints of the problem can be designed as follows:

$$R_r(u_r^R) + t_{rs} \le F_s(u_s^F) \quad for\ r, s \in S \tag{1}$$

$$u_s^R \in \{0, 1, \dots, m(s)\}\ for,\ s \in S \tag{2}$$

$$u_s^F \in \{0, 1, \dots, p(s)\}\ for,\ s \in S \tag{3}$$

These time constraints determine set of feasible solutions of convex nonlinear discrete problem. Let us denote e^R and e^F costs of one rough and fine sorting machines respectively. Then the whole model can be stated as:

$$Minimize \sum_{s \in S}\left(e^R u_s^R + e^F u_s^F\right). \tag{4}$$

Subject to (1), (2) and (3).

3. Integer linear program of machine allocation problem

We make use of the fact that LP-relaxation of model (4), (1), (2), (3) is convex and separable and replace it by piece wise linear model. Let us define the following constants for each $s \in S$:

$a_{sk} = R_s(k-1) - R_s(k)$, $d_{sk} = e^R/a_{sk}$ for $k = 1, 2, \dots, m(s)$,
$b_{st} = F_s(t) - F_s(t-1)$, $e_{st} = e^F/b_{st}$ for $t = 1, 2, \dots, p(s)$.

Due the fact that R_s is a convex decreasing function, inequalities $a_{s1} > a_{s2} > a_{s3} > \dots > a_{sm(s)} > 0$ and therefore $0 < d_{s1} < d_{s2} < d_{s3} < \dots < d_{sm(s)}$. must hold. Considering that F_s is concave increasing function, we get similarly $b_{s1} > b_{s2} > b_{s3} > \dots > b_{sp(s)} > 0$ and therefore $0 < e_{s1} < e_{s2} < e_{s3} < \dots < e_{sp(s)}$. Let us introduce two sequences of real variables u_{sk} and v_{st} for $s \in S$ and $k = 1, 2, \dots, m(s)$ and $t = 1, 2, \dots, p(s)$. Now, LP-relaxation of model (4), (1), (2), (3) can be replaced by the following linear model, which corresponds to piece wise linear approximation of the LP-relaxation.

$$Minimize \sum_{s \in S} \left(\sum_{k=1}^{m(s)} d_{sk} u_{sk} + \sum_{t=1}^{p(s)} e_{st} v_{st} \right). \tag{5}$$

$$Subject\ to \quad \sum_{k=1}^{m(r)} u_{rk} + \sum_{t=1}^{p(s)} v_{st} \geq R_r(0) + t_{rs} - F_s(0) \quad for\ r,\ s \in S, \tag{6}$$

$$u_{sk} \leq a_{sk} \quad for\ s \in S\ and\ k = 1,\ \dots\ m(s), \tag{7}$$

$$v_{st} \leq b_{st} \quad for\ s \in S\ and\ t = 1,\ \dots\ p(s), \tag{8}$$

$$u_{sk},\ v_{st} \geq 0 \quad for\ s \in S\ and\ k = 1,\ \dots\ m(s)\ and\ t = 1,\ \dots\ p(s). \tag{9}$$

The relations between former variables u_s^R, u_s^F and latter variables u_{sk}, v_{st} are given by equations $u_s^R = \sum_{k=1}^{m(s)} u_{sk} / a_{sk}$ for $s \in S$ and $u_s^F = \sum_{t=1}^{p(s)} v_{st} / b_{st}$ for $s \in S$.

The above-mentioned property of coefficients d_{sk} and e_{st} ensures that optimal values of u_{sk} and v_{st} constitute connected area of real numbers originated at zero. Another optimal solution property of (5)-(9) is that whenever all variables u_{sk} and v_{st} take values from sets $\{0, a_{sk}\}$ and $\{0, b_{st}\}$ respectively, then the optimal solution of (5)-(9) constitutes an optimal solution of (4), (1), (2), (3).

Furthermore, it follows from the introduction of constants a_{sk}, d_{sk}, b_{st} and e_{st} that they can be arbitrary enlarged by multiplication of (5) and (6) by a positive number. This enables to approximate the values of the constants by positive integer numbers with demanded precision. That is why only integer constants are considered in the next sections.

4. Branch and bound technique

To obtain an optimal integer solution of the problem (5)-(9) branch and bound method with depth-first searching scheme was employed. The branching is performed by addition one of constraints (a) or (b) to the model, which describes a processed node of the searching tree.

(a) Let $u_{sq} = 0$ for all $q \in \{k, k+1, \dots, m(s)\}$ for given s and k.

(b) Let $u_{sq} = a_{sq}$ for all $q \in \{1, 2, \dots, k\}$ for given s and k.

The same constraints can be constructed for variables v_{st} and constants b_{st}, $p(s)$. The core of the methods is an algorithm, which finds optimal non-integer solution of problem (5)-(9). Objective function value of optimal solution yields lower bound of an inspected node of the searching tree and rounding the values of non-zero u_{sk} or v_{st} up to corresponding a_{sk} or b_{st} gives feasible integer solution, which can be used to update the current best solution. Objective function value of the current best solution constitutes upper bound used in the searching scheme. To obtain the optimal solution of problem (5)-(9), any linear programming algorithm may be used, but we developed a special primal-dual algorithm to accelerate computation. This algorithm comes out of the primal model:

$$\text{Minimize} \sum_{s \in S} \left(\sum_{k=lk(s)+1}^{hk(s)} d_{sk} u_{sk} + \sum_{t=lt(s)+1}^{ht(s)} e_{st} v_{st} \right). \tag{10}$$

$$\text{Subject to} \quad \sum_{k=lk(r)+1}^{hk(r)} u_{rk} + \sum_{t=lt(s)+1}^{ht(s)} v_{st} \geq c_{rs} \quad \text{for } r, s \in S, \tag{11}$$

$$u_{sk} \leq a_{sk} \quad \text{for } s \in S \text{ and } k = lk(s)+1, \ldots, hk(s), \tag{12}$$

$$v_{st} \leq b_{st} \quad \text{for } s \in S \text{ and } t = lt(s)+1, \ldots, ht(s), \tag{13}$$

$$u_{sk}, \; v_{st} \geq 0 \quad \text{for } s \in S \text{ and } k = lk(s)+1, \ldots, kt(s) \text{ and } t = lt(s)+1, \ldots, ht(s). \tag{14}$$

In this model, constants $lk(s)$, $hk(s)+1$ denote the highest index of variables u_{sk} fixed at value a_{sk} and the lowest index of the variables fixed at 0. Constants $lt(s)$ and $ht(s)+1$ have the same meaning for variables v_{st}. It is considered that $lk(s) \leq hk(s)$ and $lt(s) \leq ht(s)$ hold. Constant c_{rs} in (11) denotes following expression:

$$R_r(0) + t_{rs} - F_s(0) - \sum_{k=1}^{lk(r)} a_{rk} - \sum_{t=1}^{lt(s)} b_{st}$$

Having introduce nonnegative dual variables x_{rs} for constraints (11), y_{sk} for constraints (12) and z_{st} for constraints (13), the dual model can be stated as:

$$\text{Maximize} \sum_{r \in S} \sum_{s \in S} c_{rs} x_{rs} - \sum_{s \in S} \sum_{k=lk(s)+1}^{hk(s)} a_{sk} y_{sk} - \sum_{s \in S} \sum_{t=lt(s)+1}^{ht(s)} b_{st} z_{st}. \tag{15}$$

$$\text{Subject to} \sum_{s \in S} x_{rs} - y_{rk} \leq d_{rk} \quad \text{for } r \in S, \text{ and } k = lk(r)+1, \ldots hk(r), \tag{16}$$

$$\sum_{r \in S} x_{rs} - z_{st} \leq e_{st} \quad \text{for } s \in S \text{ and } t = lt(s)+1, \ldots ht(s), \tag{17}$$

$$x_{rs,} \; y_{sk}, \; z_{st} \geq 0 \quad \text{for } r, s \in S, \; k = lk(s)+1, \ldots hk(s), \; t = lt(s)+1, \ldots, ht(s). \tag{18}$$

This primal-dual system of equations slightly remembers primal-dual model of transportation problem. It has lead to the idea to develop an algorithm like Hungarian method. Developed algorithm starts with feasible solution of (10)-(14), which ensures all complementary conditions with exception of those, which are formed by constraints (16) and by associated variables u_{sk}. Then a bipartite graph is considered with vertex subsets $r \in S$ and $s \in S$ and set of arcs, where each arc corresponds to pair of vertices r, s, for which associated constraint (11) is satisfied as equality. On this graph an augmenting path is searched starting from vertex r, for which (16) is satisfied by the current solution as inequality. If an augmenting path is found, the current solution of (15)-(18) is improved. Otherwise, a structure like Hungarian tree arises on the bipartite graph and it is used to improve the current solution of (10-(14). The process is repeated until all complementary conditions hold.

5. Conclusions

The above-described approach to the optimal sorting machine allocation problem was implemented as a part of decision supporting system for postal network design [1], [2]. This system was used for analysis of current state and possible variants of the Czech postal network. During these experiments, which were done on the network with 69 possible sorting center locations, the algorithm has proved to be safe and fast enough to solve problems of this size.

Furthermore, I was found that the primal-dual algorithm often produces integer solution of the problem unless branching process had to be employed.

References

1. Jablonský J, Lauber J (1999) A Time-Cost Optimization of the National Postal Distribution Network. Journal of Multi-Criteria Decision Analysis 8: 51-56
2. Janáček J (1977) Location and Capacity determination of Postal Sorting Centres (in Czech). Proceedings if the international conference POST POINT'97, University of Žilina, pp 77-81
3. Lauber J, Jablonský J (1977) The Modeling of the Postal Distribution Network. Proceedings if the international conference MOSIS'97, vol 3, Ostrava, pp 183-188

Acknowledgement

This work has been supported by research projects CETRA and VEGA 1/7211/20.

A Combined Approach to Solve the Pickup and Delivery Selection Problem

Jörn Schönberger, Herbert Kopfer, and Dirk C. Mattfeld

University of Bremen, Faculty of Economics, Chair of Logistics, D-28334 Bremen, Germany, {sberger,kopfer,dirk}@logistik.uni-bremen.de

Abstract. In this article we propose a model which decides upon the execution of transportation requests in order to maximize their contribution to the overall profit. Positive contributions of requests are obtained by composing appropriate routes for the vehicles involved. The composition of routes is hindered by limited capacity of vehicles and time windows associated to requests. A hybrid approach consisting of a parallel path construction heuristic which seeds a Genetic Algorithm is presented. We assess its capability for a set of suitable benchmark instances for pickup and delivery problems with time windows.

1 Introduction

In a pickup and delivery problem vehicles have to visit a set of locations in order to transport goods from origins to destinations. A request specifies the pickup and the delivery location and the volume respectively weight of goods to be transported. Each request must be executed by one vehicle. The minimization of execution costs is impeded by maximal vehicle capacities and often it is compromised by time windows.

In this contribution the well-known Pickup and Delivery Problem with Time Windows (PDPTW) [4] is extended by a decision concerning the acceptance of orders. It is aimed at identifying a selection of requests that leads to a maximal profit contribution, defined as the difference between the revenues and costs in monetary units. The revenue of a single request typically does not cover the costs for its execution so that the combination of several requests in a route is necessary to achieve positive contributions. The fulfillment of requests with negative contributions to a route is allowed to be rejected.

The incorporation of the selection decision into the route generation task has received only minor attention so far. Some contributions deal with delivery or collection problems with one [5,6] or several vehicles [3]. For pickup and delivery problems only one single vehicle case is investigated [9].

Next, an exact description of the investigated problem is given followed by the presentation of a hybrid Genetic Algorithm. Its capability is assessed by means of suitable benchmark instances.

2 The Pickup and Delivery Selection Problem

To fulfill the available customer requests a collection of routes for the m vehicles within a fleet $\mathcal{F}$ is to be determined.
Every customer request r_i is specified by a quadruple (PU_i, DL_i, c_i, REV_i). The pickup activity PU_i takes place at p_i^+ whereas the delivery activity DL_i is demanded at location p_i^-. A time window $[t_{min}, t_{max}]$ is specified for each activity. Load of volume c_i is to be picked up at p_i^+ and to be delivered to p_i^-. A revenue REV_i is associated with r_i and the costs of moving along one distance unit is one money unit.
The available customer requests form the set $\mathcal{R} := \{r_i \mid i = 1, \ldots, n\}$. The set of all involved locations is denoted by $\mathcal{V} := \mathcal{I} \cup \mathcal{T} \cup \{p_1^+, \ldots, p_n^+\} \cup \{p_1^-, \ldots, p_n^-\}$, where $\mathcal{I}$ consists of the initial positions of the vehicles and $\mathcal{T}$ of the depots.
Feasible Routes. An operation is a triple $\pi := (p, a(p), e(p))$, where p represents the location of a pickup, a delivery, a start or a stop activity. $a(p)$ carries the arrival time of the vehicle at location p and $e(p)$ the leaving time from p. If the vehicle arrives at p before the associated time window has been opened it has to wait until the earliest allowed operation time $t_{min}(\pi)$ for π has passed. The leaving time of π has to precede the latest allowed operation time $t_{max}(\pi)$.
A sequence of operations $\Pi = (\pi_1, \ldots, \pi_{n_\Pi})$ is called a route. In the remainder of this article the first component of π_i is denoted by p_i. Each route includes exactly one temporally unrestricted dummy request of zero capacity and zero revenue, that describes the initial and the final operation of a vehicle. The route Π includes N_Π requests.
Let $t_{p_i,p_{i+1}}$ be the travel time between p_i and p_{i+1}. The arrival and the leaving times are calculated recursively. We have $a(p_1) = e(p_1) := 0$ and $a(p_i) := e(p_{i-1}) + t_{p_{i-1},p_i}$. The leaving time is achieved by $e(p_i) := \max\{a(p_i), t_{min}(\pi_i)\}$. The vector $\boldsymbol{q}^\Pi = (q_1^\Pi, \ldots, q_{N_\Pi}^\Pi)$ describes the volumes that are collected along the route Π. For a pickup operation at p_i it is $q_i \geq 0$ and for the associated delivery operation at p_j we define $q_j := -q_i$. The capacity usage along Π is determined recursively: $c_1^\Pi := 0$ and $c_i^\Pi := c_{i-1}^\Pi + q_{i-1}$.
The route Π is called pd-path for vehicle v if it the following restrictions hold [8]. Either both operations of r_i or none of them are contained in Π (*pairing*), a pickup operation precedes its associated delivery operation (*precedence*), the maximal load is not exceeded for all i: $c_i^\Pi \leq C_v^{max}$ (*capacity*) and the leaving time for operation π_i lies in the specified time window: $t_{min}(p_i) \leq e(p_i) \leq t_{max}(p_i)$ (*time window*).
For the fulfillment of Π one monetary unit per distance unit driven is taken into account leading to costs C_Π for route Π.
The total received revenues E_Π for Π are given by $E_\Pi := \sum_{j=1}^{N_\Pi} REV_{i_j}$.
A pd-schedule S is a set of pd-paths $\Pi^1, \ldots, \Pi^L$ so that each customer request is assigned to at most one path.
Optimization Problem. For each vehicle k the set $\mathcal{S}_k$ of possible pd-paths is finite so that a set partition formulation [10] of the problem at hand is

appropriate. From the set $\mathcal{S} := \cup_{k\in\mathcal{F}}\mathcal{S}_k$ of pd-paths we select the routes that forms the schedule S. The following binary decision variables are introduced. Let $X_i = 1 \Leftrightarrow$ route i is in the schedule S and $Y_{ik} = 1 \Leftrightarrow$ route i is served by vehicle k. For the binary constants z_{ij} is assumed that $z_{ij} = 1 \Leftrightarrow$ route i contains customer request j. The Pickup and Delivery Selection Problem (PDSP) is than defined as shown in (1a)-(1f).

$$\max \sum_{i\in\mathcal{S}} E_i \cdot X_i - C_i \cdot X_i \tag{1a}$$

$$\sum_{i\in\mathcal{S}_k} Y_{ik} = 1 \; \forall k \in \mathcal{F} \tag{1b}$$

$$\sum_{i\in\mathcal{S}\setminus\mathcal{S}_k} Y_{ik} = 0 \; \forall k \in \mathcal{F} \tag{1c}$$

$$\sum_{k\in\mathcal{F}} Y_{ik} = X_i \; \forall i \in \mathcal{S} \tag{1d}$$

$$\sum_{i\in\mathcal{S}} z_{ij} \cdot X_i \leq 1 \; \forall \, j \in \mathcal{R} \tag{1e}$$

$$Y_{ik}, X_i \in \{0,1\}. \tag{1f}$$

Each vehicle k is assigned to exactly one pd-path (1b) so that the maximal capacity of k is not exceeded (1c), each pd-path is selected at most once (1d) and no request is contained in more than one selected pd-path (1e).

Benchmark Problems. A revenue is added to the PDPTW-instances introduced in [7]. Let $\Pi^1, \ldots, \Pi^m$ be an optimal or near optimal solution of such an instance. Let $r_{i_1}, \ldots, r_{i_{n_k}}$ the requests assigned to vehicle k. Then $\rho_{i_j} := C_{\Pi_k}/(d(p^+_{i_1}, p^-_{i_1}) + \ldots + d(p^+_{i_{n_k}}, p^-_{i_{n_k}}))$ for all $j = 1, \ldots, n_k$ is the charge for moving along one distance unit with vehicle k. The function $d(\cdot,\cdot)$ describes the distance between two locations. The total revenue REV_i is then determined as $REV_i := \rho_i \cdot c_i \cdot d(p^+_i, p^-_i)$. The solution of the PDPTW-instance is then evaluated as a PDSP solution calculating the overall profit contribution. The achieved value serves as a reference value.

We derive PDSP problems from 40 different PDPTW-instances and generate several classes of instances determining the maximal capacity of the vehicles with $C^{max} = 20, 40, \ldots, 200$. Now, $40 \cdot 10 = 400$ benchmark instances are available.

3 Hybrid Algorithm

To solve the PDSP we propose a Genetic Algorithm (GA) that is seeded by a parallel route construction heuristic.

Schedule Construction. The schedule construction heuristic consists of three phases. At first the operations are ordered along the time axis, then they are distributed among the vehicles and finally requests that violate the capacity or the time window restriction are removed from the routes.

Phase 1 (Slot-Assignment): The relevant part of the time axis is partitioned into $s > 1$ equidistant slots. Each operation is assigned to the slot in which its latest permissible leaving time falls. It is ensured that each pickup operation precedes its corresponding delivery operation.

Phase 2 (Routing): The sequences in which the requests are taken and the order in which the vehicles are checked are not externally determined. Let $\boldsymbol{R}$ be a request sequence and $\boldsymbol{V}$ be a vehicle sequence. Each request from $\boldsymbol{R}$ is assigned to the first vehicle from $\boldsymbol{V}$ with free pickup and delivery slot. A request that cannot be assigned to any vehicle remains unserved. The depot situated as near as possible to the last delivery operation of a vehicle is selected as its terminating point. The obtained routes $\Pi^1, \ldots, \Pi^m$ hold for the pairing and the precedence condition. The load of the vehicles and the leaving times of the operations are determined as described in section 2.

Phase 3 (Obtaining Feasibility): Feasibility concerning capacity and time windows is obtained greedily. Along the time axis each scheduled operation is checked. If a capacity or a time window constraint violation is asserted the currently considered request is removed from the route.

Adaptive Improvement. A Genetic Algorithm [1] is configured to improve the pd-schedules generated from the heuristic described above.

Initial Population. The application of the schedule generation heuristic with randomly generated request and vehicle permutations produces an initial population of feasible pd-schedules.

Schedule Encoding. A schedule is represented in a string $(\boldsymbol{p}, \boldsymbol{d}, \boldsymbol{v})$ of length $2n + m + m$. $\boldsymbol{p}$ carries a permutation of all customer locations. The i-th component of $\boldsymbol{d}$ carries the number of the terminating depot assigned to vehicle i. The i-th component v_i of $\boldsymbol{v}$ contains the number of requests assigned to vehicle i.

Schedule Decoding. The first v_1 pickup operations in $\boldsymbol{p}$ and the corresponding v_1 delivery operations form the route of the first vehicle. For the second vehicle the next v_2 pickup operations and their corresponding delivery operations are selected and so on. Operations of rejected requests are carried in the last permutation positions. Conditions (1b)-(1e) are kept syntactically. Following the above representation each individual represents a pd-schedule if it can be ensured that the routes obtained from $\boldsymbol{p}$ are pd-paths. Therefore, each route is greedily checked for capacity and time window constraint violations. Requests that cause constraint violations are removed from the routes. The genetic information is immediately updated and fed back into the population.

Cross-Over. Applying the cross-over operator PPX [2], a new permutation of operations is generated from two parental permutations. This leads to a new collection of routes, each satisfying the pairing and the precedence restriction. From the same parents a new depot sequence is achieved applying a standard uniform cross-over operator to the parental depot sequences. To derive a new

request assignment sequence we assign a modified uniform-cross-over operator to the last segment of both parental individuals. The modification ensures that at most n requests are distributed among the vehicles. Cross-over of the third segment changes the number of accepted customer requests.
Mutation. Two arbitrarily chosen alleles are swapped within each segment. Pairing and precedence feasibility are preserved. Additionally, the number of requests within the routes is increased. Therefore, the so far unselected requests are randomly distributed among the vehicles.
Selection. A $\mu + \lambda$ scheme [1] is used to derive a new population of individuals. The best individuals from the union of the offspring set and original population form the new population.
Evaluation. Since all individuals are feasible with respect to (1b)-(1f) a suitable fitness value is the objective value obtained by (1a).

4 Computational Investigations

Two experiments have been performed to compare the capability of the adaptive approach with a non-adaptive biased random search (BRS) heuristic.
In the BRS experiment a request permutation in which the most promising requests are averagely found in the first components is instantiated as follows. For each request r_i the standardized revenue ρ_i (cf. the benchmark paragraph) is calculated. Then, the components are successively determined by a roulette-wheel-selection, which prefers requests with large ρ_i. A selected request is not considered in subsequent selection steps. Altogether 2000 independent permutations have been generated. Since all vehicles offer the same capacity a random vehicle permutation is used.
The results of BRS are compared to those that are obtained from our hybrid genetic algorithm (HGA) seeded with a set of pd-schedules obtained from the BRS-experiments. The GA is configured for 100 individuals and generates 200 generations within each run. For each PDSP-instance ten independent runs have been executed. Best results are obtained for a cross-over probability of 0.6 and a mutation probability of 0.3 that means three of ten offsprings are mutated.

Table 1. results obtained for different capacities

	capacity	20	40	60	80	100	120	140	160	180	200
BRS	best. obj	0.25	0.40	0.46	0.49	0.52	0.53	0.54	0.54	0.55	0.55
	best. acc	0.34	0.44	0.49	0.50	0.52	0.54	0.54	0.54	0.54	0.54
HGA	avg. obj	0.32	0.52	0.56	0.60	0.62	0.64	0.65	0.65	0.65	0.65
	avg. acc	0.38	0.47	0.52	0.53	0.55	0.56	0.56	0.57	0.57	0.56

Tab. 1 shows the obtained results consolidated for different capacities. The values are compared to those derived from the solutions of the PDPTW in-

stances from [7]. For each capacity the averagely best solutions found are shown together with the corresponding ratio of accepted requests. For increasing vehicle capacities the objective value is doubled up to a capacity of 100. For larger resources no significant improvements are observed.
The results from BRS are outperformed by those obtained from HGA. The objective value is doubled up to capacity of 120. Further capacity enlargements do not lead to significantly higher values. The number of selected requests is slightly enlarged in the HGA results. In both experiments the quota of selected requests remains averagely unimproved if the capacity is enlarged to more than 120 units.

5 Conclusion and Future Work

In this contribution a formal model for the selection of the requests with maximal positive contributions to the overall profit has been proposed. The obtained results from a permutation controlled parallel construction heuristic can be improved by a Genetic Algorithm.
Future research will be dedicated to the identification of promising request permutations to generate pd-schedules of improved quality.
Acknowledgement: This work was supported by the BMBF Project SpiW, No. 01HT0144.

References

1. Bäck, T., Fogel, D.B., Michalewicz, Z. (2000) Evolutionary Computation 1 - Basic Algorithms and Operators, Institute of Physics Publishing
2. Bierwirth, C., Mattfeld, D.C., Kopfer, H. (1996) On Permutation Representations for Scheduling Problems, Proc. of PPSN IV, Springer, 310-318
3. Butt, S.E., Ryan, D.M. (1999) An optimal Solution procedure for the multiple tour maximum collection problem using column generation, Computer & Operations Research 26, 427-441
4. Dumas. Y., Desrosiers, J., Soumis, F. (1991) The pickup and delivery problem with time windows, European Journal of Operational Research 54, 7-22
5. Feillet, D. (2001) Traveling Salesman problems with profits: an overview, Proceedings of OPR^3, Paris
6. Laporte, G., Martello, S. (1990) The Selective Travelling Salesman Problem, Discrete Applied Mathematics 26, 193-207
7. Nanry, W.P., Barnes, J.W. (2000) Solving the pickup and delivery problem with time windows using reactive tabu search, Transportation Research Part B 34, 107-121
8. Savelsbergh, M.W.P., Sol, M. (1995) The General Pickup and Delivery Problem, Transportation Science 29, 17-29
9. Verweij, B. Aardal, K. (2000) The Merchant Subtour Problem, technical report UU-CS-2000-25, Universiteit Utrecht
10. Williams H.P. (1999) Model Building in Mathematical Programming, 4th Edition, Wiley, Chichester

VRP with Interdependent Time Windows - A Case Study for the Austrian Red Cross Blood Program

Karl Doerner[1], Manfred Gronalt[2], Richard F. Hartl[1], Marc Reimann[1], and Kerstin Zisser[1]

[1] University of Vienna, Department of Management Science, Bruenner Strasse 72, A-1210 Vienna

[2] University of Agricultural Sciences Vienna, Department of Forestry and Timber Industry, Gregor Mendel Strasse 33, A-1190 Vienna

Abstract. This work shows our endeavors to assist the fleet management division of the Austrian Red Cross blood programme by developing an algorithm for the VRP with interdependent time windows. Usually, blood donation campaigns are operating all day long and the collected blood must be transported from campaign locations to a central blood processing facility, where it has to be further processed within a certain amount of time. By considering this constraint, each location must be visited for a number of times during the course of a day. Certain vehicles are scheduled for picking up the stored blood donations and delivering it to the processing facility within the given time window. Time windows for each location show a dynamic behaviour as they depend on previous pickup times. In order to solve the problem we develop a savings-based constructive procedure for the VRP with interdependent time window constraints. First, our algorithm calculates the minimum number of pickups for each location and then it determines possible combinations when constructing a tour. Preliminary results on real life data show that we can easily decrease overall tour lengths.

1 Introduction

"Everybody knows giving blood saves lives [1]". But before blood reaches vulnerable patients suffering from cancer or injuries from an accident, it passes a highly complex screening, processing, testing, transporting and monitoring process. The blood collection process of the Austrian Red Cross blood program is divided up regionally, where each region plans and schedules its blood donation campaign independently of the others.

For collecting the donations a number of districts and locations usually spread in the area of Eastern Austria is selected. To each of the locations a team is assigned for a specific daily duration to organize the blood donation. They collect the donations and provide for a proper storage of these. A blood donation campaign usually takes place at the weekend.

Collected blood is transported from the campaign locations to a central blood processing facility, where it has to be further processed and separated into its component parts (Red Blood Cells, Platelets, and Plasma) within about 300 minutes.

Currently, the blood transport is manually scheduled and the resulting tours are registered on a sequential list. By doing this, possible combinations of tours are only found if they obviously can be put together, e.g. locations are resided along the same highway etc. In our work we both integrate a GIS which is already in use for locating blood donators in case of emergencies and develop a procedure for finding a minimum travel cost solution for the present VRP.

The problem characteristics can be stated as follows:

Any blood donated must arrive at the processing facility at most 300 minutes after donation. Campaign durations are typically longer than 300 minutes. Thus, each campaign needs to be visited more than once.
For each visit a time window applies, which is reset every time a vehicle leaves a particular location because of time window interdependence.
The last donations are taken back to the processing facility by the people organizing the campaign.
The goal is to find minimum cost tours and to allocate the appropriate transport devices to take back all the blood.
For calculated solutions an eligibility check with the people at the control station is done. Further, rendezvous options are considered in order to improve the solution found.

We develop a savings-based constructive solution procedure for the VRP with interdependent time windows. In the initialization phase the minimum number of pickups and the current slack time are determined for each location. This leads to a multiplication of locations and their respective time windows in the used construction graph. In the construction phase locations are added to a tour according to the savings values and the current slack times.

The problem considered in this paper is related to the periodic vehicle routing problem with time windows. Cordeau et al. ([5]) apply the unified tabu search heuristic to solve this problem, while Baptista et al. ([6]) apply a heuristic approach to a real world period vehicle routing problem for the collection of recycling paper containers.

For the problem class of the period travelling salesperson problem, where over a period of time each customer must be visited at least once with some customers requiring several visits, Chao et al. ([2]) and Paletta ([3]) developed heuristic approaches, while Drummond et al. ([4]) solve this problem class by using a parallel genetic algorithm and local search heuristics. These problems do not have time window and maximum tour lengths constraints.

Yi and Scheller-Wolf ([7]) design an exact solution procedure based on integer programming for a vehicle routing problem with time windows faced by the American Red Cross. In their approach the objective is to bring a predetermined amount of blood from the campaign locations back to the processing facility at minimum cost, where the amount collected at each campaign depends on the time of the visit. They consider only a single visit per campaign location thus eliminating interdependence between time windows. Still, a maximum tour length exists due to perishability of blood.

2 Model formulation

We use the following data and indices for our model:

i,j...	blood donation locations (BDL)
0...	blood bank (BB)
d_{i0}...	distance from BDL i to BB
d_{ij}...	distance from BDL i to BDL j
α...	one BDL or an already assigned feasible combination of BDLs
s_0...	service time at BB
s_i...	service time at BDL i
a_i, e_i...	starting/finishing time of the blood donation event at BDL i
τ...	max. time span between blood donation and processing at BB

Based on these data we can compute the following values:

$\tau'_i = \tau - s_i - d_{i0} - s_0$	maximum time period between two collections of blood donations at BDL i
$\beta_i = e_i - \tau'_i - a_i$	relevant time span to collect the donations at BDL i taking into account that the donated blood at the end of the blood donation event is returned by the organizing team of the event, which also returns to the blood bank.
$m_i = \lceil \beta_i / \tau'_i \rceil$	number of required blood pick-ups at BDL i
$p_i = m_i \cdot \tau'_i - \beta_i$	slack time at BDL i; results from the required integer number for the blood pick-ups
$t_{i1}^{max} = a_i + \tau'_i$	latest arrival time for the first blood pickup
$t_{ik}^{max} = t_{i,k-1}^{max} + \tau'_i$	latest arrival time for the k-th blood pickup
$t_{ik}^{min} = t_{ik}^{max} - p_i$	earliest arrival time for the k-th visit at BDL i

To ensure feasibility three constraints are particularly important:

1. Constraint (Detour Constraint):

$$p_\alpha \geq -d_{\alpha 0} + d_{\alpha m} + s_m + d_{m0} \tag{1}$$

Adding a pickup at BDL m to partial tour α is possible only, if the slack time of the partial tour is larger than the time of the detour associated with adding BDL m.

2. Constraint (Reachability Constraint):

$$t_\alpha^{min} + s_\alpha + d_{\alpha m} \leq t_{mn}^{max} \tag{2}$$

Adding the n-th pickup at BDL m to a partial tour α is possible only, if the campaign m can be reached on time.

3. Constraint (Punctuality Constraint):

$$t_{\alpha}^{max} + s_{\alpha} + d_{\alpha 0} - s_m - d_{m0} \geq t_{mn}^{min} \tag{3}$$

Adding the n-th pickup at campaign m to a partial tour α is possible only, if all the blood collected on the partial tour so far still arrives at the processing facility on time.

Note, that the detour constraint (constraint 1) and the punctuality constraint (constraint 3) are both necessary, as the detour constraint evaluates the detour, whereas the punctuality constraint ensures that donated blood already picked up does not perish as a cause of the detour.

3 Heuristic solution procedure

Let us now turn to the approach we propose to solve the problem with interdependent time windows. We will also discuss implementation issues.

- Step I: Read input data and initialize system
  Initially the blood donation campaign information and the distance information are read. The distance information is computed by the used GIS. For each BDL all the relevant information (e. g. the relevant time span to collect the donations, the number of required blood pickups, the slack time,...) is computed.
- Step II: Compute Savings matrix
  By using the distance information the Savings values between all BDLs are computed. The Savings algorithm, proposed in [8] is the basis of most commercial software tools for solving VRPs in industrial applications. It is initialized with the assignment of each BDL to a separate tour. After that for each pair of BDLs i and j the following savings measure is calculated:

  $$s_{ij} = d_{i0} + d_{0j} - d_{ij} \tag{4}$$

  where d_{ij} denotes the distance between BDLs i and j and the index 0 denotes the BB. Thus, the values s_{ij} contain the savings of combining two pickups at BDL i and BDL j on one tour as opposed to serving them on two different tours.
- Step III: Determine feasible combinations
  Having computed the Savings values for each BDL, we check the constraints (detour constraint, reachability constraint, punctuality constraint). Thus, we can check if there are any feasible combinations between two BDLs. Note, that at each BDL a number of visits may be necessary, such that combinations between all of these visits have to be checked. A combination is only feasible if all the constraints are satisfied.

- Step IV: Find combinations
  In the iterative phase of step IV pickups at BDLs are combined. We start to combine a feasible combination of two BDLs according to the earliest possible pickup time. Feasible combinations with a Savings value below a certain threshold are not considered. After the realisation of the combination of BDL i and BDL j, the time windows for the next pickups of these two BDLs and also the list of the remaining possible combinations are updated. This step is repeated until no combination with a savings value greater than a given threshold exists.
- Step V: Report results
  The combinations of BDLs, the number of required pickups and the tourlength are reported. The tours are visualized by our commercial desktop GIS.

4 Numerical example

In the following section, we present a numerical study that applies our approach described previously by using real world data from the Austrian Red Cross. Our example considers a typical day with 13 BDLs in the eastern part of Austria. The required number of pickups at one BDL varies between zero and three. The daily duration of the BDLs are given.

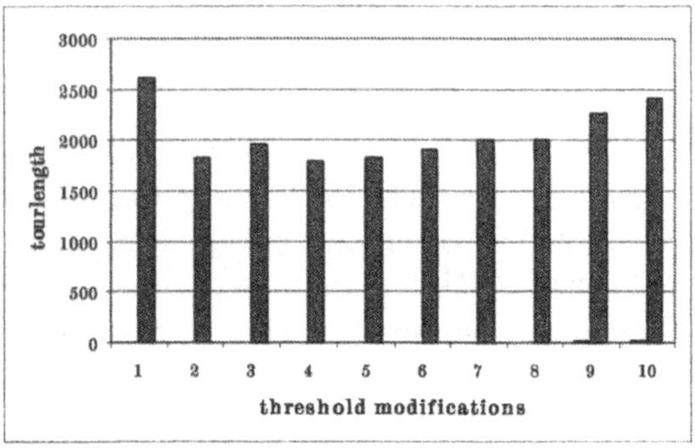

Fig. 1. Numerical Example

In our algorithm only combinations of two BDLs are considered. In Figure 1. the solution quality obtained by using different thresholds ranging from 10 (column 2) to 180 minutes (column 10) is shown. The tour length by assigning each pickup at a BDL on a separate tour is 2606 minutes (column 1). The minimum tour length in this example is achieved using a threshold between 30 and 70 (column 4) and equals 1795.23 minutes.

By choosing a threshold that is too small we can see that the solution quality decreases also. The reason is that the realisation of combinations with an early pickup time and a small Savings value prevents the realisation of other promising combinations.

On the other hand, by choosing a threshold that is too large, we reject too many promising combinations and thus solution quality decreases.

Note, that we can also decrease the costs in comparison with the real costs but due to contractual agreements we are not allowed to report the solution quality of the Austrian Red Cross.

5 Conclusions

In this paper we have proposed a model and a heuristic solution approach for a real world scheduling problem faced by the Austrian Red Cross. The approach can be used to instantly a candidate solution. By running the algorithm a few times with different thresholds, the user is provided with different solutions to choose from.

Starting from our initial approach, we are now examining strategies based on relaxing the assumption to visit each campaign the minimum number of times. Further, we plan to integrate a rendezvous option where two vehicles can exchange loads in order to save unnecessary travelling.

Acknowledgments

Financial support by grant #8630 from the Oesterreichische Nationalbank (OeNB) is gratefully acknowledged.

References

1. www.redcross.org/services/biomed/blood/learn/process.html
2. Chao, I.-M., Golden, B. L. and Wasil, E. A., A new heuristic for the period traveling salesman problem, Computers & Operations Research 22 (1995), pp. 553-565.
3. Paletta, G., The period traveling salesman problem: a new heuristic algorithm, Computers & Operations Research 29 (2002), pp. 1343-1352.
4. Drummond, L. M. A., Ochi, L. S., Vianna, D. S., An asynchronous parallel metaheuristic for the period vehicle routing problem, Future Generation Computer Systems 17 (2001), pp. 379-386.
5. Cordeau J.-F.; Laporte G.; Mercier A., A unified tabu search heuristic for vehicle routing problems with time windows, Journal of the Operational Research Society 52 (2001), pp. 928 - 936.
6. Baptista, S., Oliveira, R. C. and Zuquete E., A period vehicle routing case study, European Journal of Operational Research 139 (2002), pp. 220-229.
7. Yi, J. and Scheller-Wolf, A., A Vehicle Routing Problem with Time Window Constraints and Variable Rewards: An Application in the American Red Cross. Research Report.
8. Clarke, G. and Wright,J. W., Scheduling of vehicles from a central depot to a number of delivery points. Operations Research 12 (1964), pp. 568-581.

Incident Management Based on Real Time Simulation

Jürgen Zajicek, Martin Linauer, Katja Schechtner

arsenal research
Austrian Research and Testcentrum Arsenal Ges.m.b.H.
Business Area Transport Technologies
A-1030 Vienna, Faradaygasse 3
phone: +43-(0)50550-6610, fax: +43-(0)50550-6613
e-mail: juergen.zajicek@arsenal.ac.at

Abstract

The main function of a traffic simulation which is used to evaluate the current traffic situation is reflected in the spatio-temporal interpolation of the situation between the stationary detector points of any road network.

Today the traffic information systems are mainly based on traffic data from fixed metering points using technologies like induction loops, radar or infra red sensors, etc. Due to this static information current traffic information systems feature no or only meager spatio-temporal portability.

The following figure will show how simulation helps to fill up the scopes without traffic information using the **Floating Car Data (FCD)** method to get area-wide traffic information.

To be able to employ a dynamic traffic metering method a redesign of simulation algorithms has to take place in the near future. The Floating Car Data method will for the first time enable the creation of area-wide traffic information without expensive infrastructure and allow a rapid reaction onto current incidents on the road network regarded.

But this method presupposes simulation algorithms using variable time steps instead of constant time steps like the prevalent software offered today. Only on-line simulation software (macro simulation software) can evaluate traffic data sent by moving vehicles at different times and positions.

This ability of simulation software should not be reserved to the macro layer because of the great influence of incidents onto the traffic situation that can only be registered in the micro layer.

To prepossess this requirement interfaces between both layers must be realised to enable a bi-directional data exchange. However there is a lot of research work required to solve these specified problems.

This lecture should help to understand the existing problems and show the possibilities opening by using this new simulation method in combination with Floating Car Data in incident management.

Introduction

Considering the daily situation on the road in and around the congested city areas you will recognise that the traffic jams are getting worse and worse each day. To anticipate the impending collapse of traffic systems, strategies to manage the emerging problems have to be developed. But before starting a management system you first have to know what the current situation on the road network looks like.

Today the traffic information systems are mainly based on traffic data from fixed metering points using technologies like induction loops, radar or infrared sensors, etc. In Austria there are only 156 official points placed on the main routes of the road network comparing to about 8000 stations in Germany.

Due to this static information current traffic information systems feature no or only very limited spatio-temporal portability. A possible way out of this dilemma may be the use of a traffic simulation to fill "unknown" areas with information about the situation on the relevant roads [3].

However, a clincher of this idea must receive attention: with increasing distance to the metering station also the accuracy of the simulation of the traffic situation of the area between two metering stations will decrease to an unacceptable level. A relief to this problem will be the use of data from "Floating Car Data" vehicles to "feed" the simulation supplementary.

Principle of Floating Car Data (FCD) Acquisition

In floating car data applications (FCD services), the vehicle speed, in conjunction with the vehicle position and time, is recorded and transmitted to a service center in defined time intervals in order to generate traffic information (see figure 1). The time for transmitting the FCD data to the service center can be event controlled, or sent by a fixed time interval (e.g. 1 minute). Combining floating car data from different vehicles, the service center draws conclusions depending on the kind of FCD service, and sends these conclusions back to the service subscribers. Typical services are traffic information, like detection of traffic jams, information on traffic flow (level of service) and the information on the travel time. FCD services are expected to become a mass market in the future, resulting in a very high number of cars taking part.

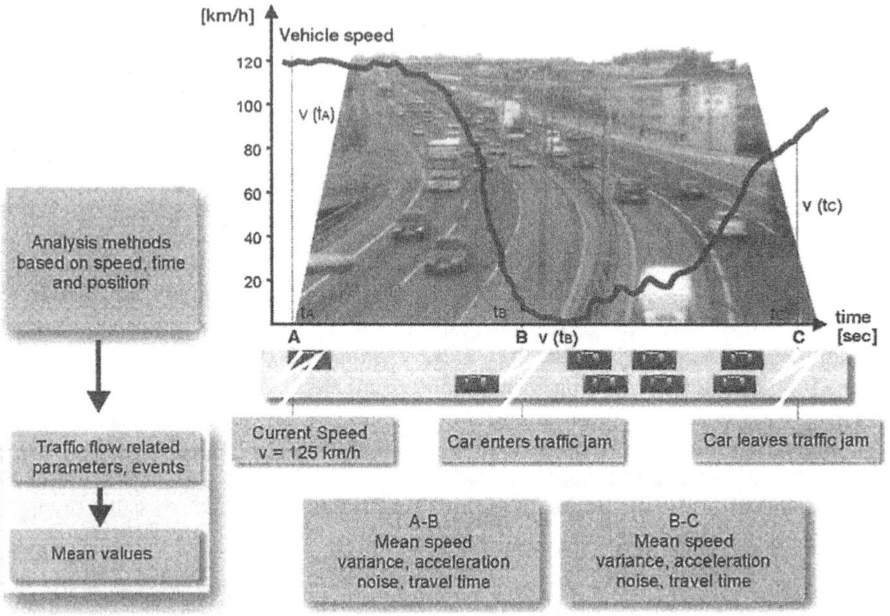

Figure 1: Principle of Floating Car Data Acquisition [1]

The floating car data applications are not only based on satellite navigation like GPS (EGNOS, GALILEO). It is also possible to generate position and time by a GSM Network by tracking the Cell ID i.e. to generate the Floating Car Data. In Austria the mobile penetration is at 84%. This means that in most cars the driver will have a cell phone. This way of localisation does not have the same quality as GPS but arsenal research develops some FCD models which are designed for using mobile Cell ID [2].

Integrate FCD in Real Time Simulation Models

Implementing this new kind of measuring traffic situations the next problem appears in the fact that the data do not arrive in fixed time steps in the traffic center and also is from different places. To use both kinds of data sources a special data management system must be established. The conventional data will be handled like in the past. The floating car data has to be assigned to road sections belonging to the near array of the co-ordinates of the current data point (Figure 2).

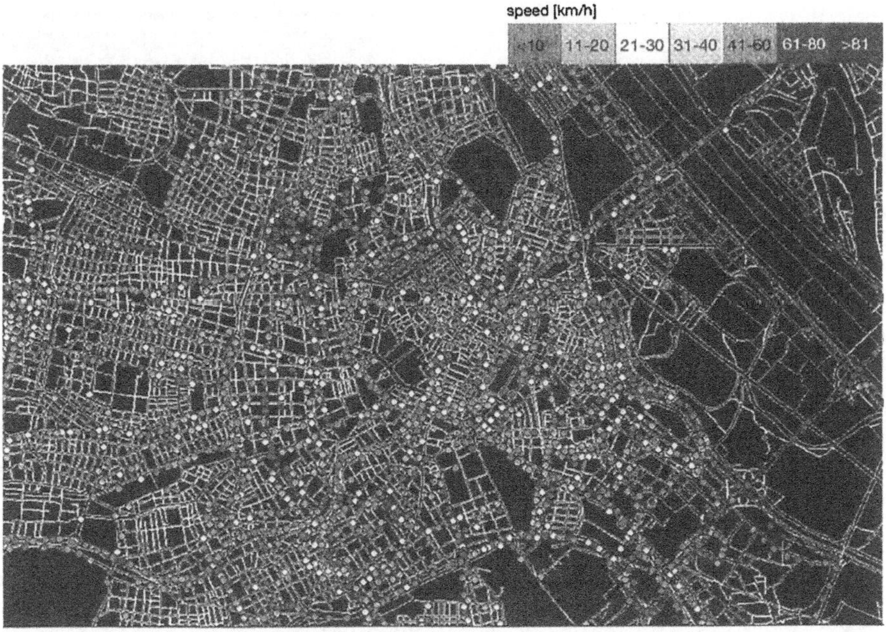

Figure 2. Positions of FCD-vehicles and their current speed

The position of each observed FCD-vehicle will be updated in fixed time steps. With the distance and the time gap between two following positions the speed of the vehicle can be calculated. Figure 3 shows the dedicated velocities to road sections of FCD vehicles in the urban area of Vienna.

Some critics may ask in view of FCD "Why do we need traffic simulation?". Today there are only a few cars equipped with FCD-devices. Due to this circumstance a fast and reliable traffic simulation is needed to manage different traffic situations. Based on the combined data of the different sources a meaningful traffic simulation has to be able to give prognoses about possible situations in the near future.

Metering stations can only detect some incidents on the road network when the congestion reaches back to its site. The network of metering stations in the area of Vienna is not able to detect incidents in proper time to set reactions to manage the changed situation because of the large distances between them. This will be possible if traffic simulation algorithms can handle vehicle volume (veh/h or veh/15min, etc.) and speed values dedicating to different road sections at the same time.

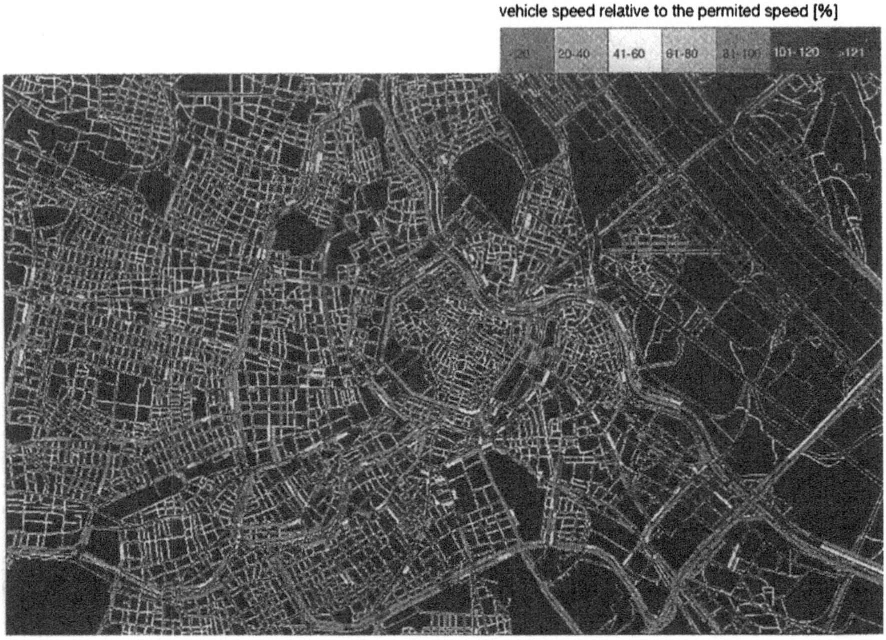

Figure 3: Dedicated velocities of FCD-vehicles in the urban area of Vienna

Only macroscopic simulation algorithms using variable time step will be able to achieve these postulations because they are able to evaluate traffic data sent by moving vehicles at different times and positions. This ability of simulation software should not be reserved to the macroscopic layer due to the fact of the massive influence of incidents onto the traffic situation that can only be registered in the micro layer.

To fulfill this requirement interfaces between both layers must be realised to enable a bi-directional data exchange. Known companies offering traffic simulation software plan to reorganise their algorithms from fixed to variable time steps. However, there is a lot of research work and man power required to solve these specified problems. Due to this fact there will be some products available in two to three years.

A fast and reliable evaluation of the current traffic situation is the presupposition of an efficient traffic information system to re-route vehicles on congested roads. This would mean that the vehicle, which would otherwise be stationary, or driving slowly through traffic jams, can better be distributed onto the low level road network.

In case of this the environment will not be so heavily bonded in future like today.

CONCLUSION

The potential to improve the quality of Real Time Simulation and Incident Management by combining data from fixed metering points (inductive loops, radar, laser or infrared sensors) and Floating Car Data is enormous.

On the one hand conventional metering points are very important for long and short term traffic prognoses who needs data from the vehicle volume. On the other hand Floating Car Data is the best technology for locating incidents and traffic jams in the road network. The possibility to accurately predict journey time is also an integral part of FCD application.

REFERENCES

[1] Huber, W.: XFCD and Local Danger Warning. Paper N°. 2032, 6th Intelligent Transport Systems World Congress. Toronto, Canada 1999

[2] Linauer, M.: Verkehrssteuerung per Handy – Kombination von Telekommunikation und Informatik eröffnet neue Wege der Verkehrssteuerung, Schriftenreihe Forschungsnews 3/2001, arsenal research

[3] Zajicek, J., ONLINE-Simulation–Eine Chance für den innerstädtischen Verkehr, IT`S T.I.M.E., Heft 1, ARC Austrian Research Centers GmbH, Wien 2001

[4] Daduna, J. R., Voß St., Informationsmanagement im Verkehr, Physica Verlag, Heidelberg 2000

[5] OSSA Consortium, Open Framework for Simulation of Transport Strategies and Assessment – User and Policy Requirement, GROWTH PROJECT, 2000

Online-Dispatching of Automobile Service Units

Martin Grötschel[1], Sven O. Krumke[1], Jörg Rambau[1], and Luis M. Torres[1]

Konrad-Zuse-Zentrum für Informantionstechnik, Berlin (ZIB)
Takustraße 7, 14195 Berlin-Dahlem, Germany
e-mail: {groetschel,krumke,rambau,torres}@zib.de

Abstract. We present an online algorithm for a real-world vehicle dispatching problem at ADAC, the German Automobile Association.

1 Problem Description

The German Automobile Association *ADAC* (*Allgemeiner Deutscher Automobil-Club*), the second largest such organization in the world, surpassed only by the American Automobile Association, maintains a heterogeneous fleet of over 1,600 service vehicles in order to help people whose cars break down on the road. All ADAC service vehicles (called *units* in the following, for short) are equipped with a GPS system which allows to locate their positions at any time. In five help centers (*Pannenhilfezentralen*) distributed throughout Germany, human dispatchers have to reply to incoming help requests (*events*) instantly. Their task is to assign a unit to serve each customer and to predict the estimated time of arrival at the customer's location. In addition to the ADAC fleet, about 5,000 units operated by service *contractors* can be employed to cover events that otherwise could not be served in time.

There is no unique objective. The goals are high-quality service (e.g., short waiting times for the customers) and low operational costs (e.g., short tour lengths and small overtime costs). With increasing costs and request volume, ADAC's dispatching system has come under stress. The task is to design an automatic system that guarantees small waiting times for events and keeps operational costs low. Such a system must address *two* different issues:

Realtime-Problem Given a "snapshot" of the situation at some moment in time, compute an optimal dispatch for attending all pending events with the available units and (if required) contractors. Since it contains the classic vehicle-routing problem as a special case, this problem is NP-hard. The difficulty is that such a dispatch has to be returned in a very short time, usually no more than 15 seconds for a system load of about 200 events and 100 units.

Online-Problem Once an algorithm for the first task has been found, design a good strategy for embedding it within the constantly running dynamic planning process, i.e., decide how often a new dispatch should be computed in response to incoming events, and/or how far changes should be admitted in a previous computed dispatch. The main challenge lies in the impossibility to predict if, where and when events in the near future will take place.

The modeling of the first task, which we shall in the following call the *vehicle dispatching problem* VDP, involves many technical and organizational side constraints—some hard, others soft—, and it takes some time to figure out which restrictions and objectives really count. As one example, we have considered constraints arising from a management decision: ADAC's imposition of a soft deadline on the service time of an event, which may be missed at the cost of a linearly increasing lateness penalty (*soft time windows*).

Using an approach based on column generation, it was possible to design an optimization algorithm for the VDP capable of finding optimal or near-optimal solutions for real-world instances and which was compliant with the real-time requirements of the problem. In section 3, we give a brief description of the algorithm (see [5] for the details). The main purpose of this paper is to present our first results concerning the *online-problem.* In section 2, we describe the criteria we used for evaluating different possible online solution schemes and introduce the strategy that we have chosen. Experimental competitive results for the application of this strategy on real-world instances are reported in section 4. Section 5 summarizes the key points of this paper.

2 Online Strategy

We postpone until the next section the discussion about *how* the VDP instances corresponding to snapshot situations are solved and focus first on the more subtle online-problem: the search for a strategy to carry out the actual planning without knowing where future events will pop up. A decision that is "optimal" at some point in time can prove later to have been unwise. In particular, even if we were able to compute locally optimal dispatches for any snapshot situation this does *not* mean that we obtain a dispatch which is (in hindsight) optimal for the whole planning period.

A by now standard tool for measuring the "goodness" of an online algorithm is *competitive analysis* [6]. Basically, one compares the solution provided by the algorithm for an instance of the online problem with the solution provided by a "hindsight" adversary, which solves the same problem instance but with knowledge of the whole input data in advance. It is usually almost impossible to obtain theoretical proofs of (useful) competitive results, except for elementary problems. Nevertheless, the concepts arising from competitive analysis can still be used in practice, in the form of experimental a-posteriori analysis of online strategies (see, e.g., [2] for some real-world examples), provided a measured hindsight adversary can be found.

In our case, this a-posteriori analysis was carried out over input data collected during one (resp. two) hour(s) of operation of the current system at

ADAC. This data was fed to our algorithm in a way that simulated how it would occur in practice: each event was labeled by a time-stamp that indicated the moment at which it became known to the system. Only after that time the algorithm was allowed to incorporate information from this event in a solution. After having briefly considered several simple online heuristics, we took the decision to follow a REPLAN strategy.

This strategy consists of recomputing from scratch, at certain moments in time, a completely new dispatch for the current VDP snapshot (the so-called *offline* problem). REPLAN assumes that an algorithm for this offline problem is at hand, which is capable of delivering solutions with certain guaranteed quality under real-time conditions (see [4] for more details). Fortunately, this is the case for the instances arising at ADAC, as we shall see in the next section.

3 Auxiliary Optimization Problem

In the following, we briefly specify the form of VDP that is tackled by our algorithm. More details concerning both the model as the solution method can be found in [5].

An instance of the VDP consists of a set of units, a set of contractors, and a set of events. Each unit u has a current position o_u, a home position d_u, a logon time t_u^{start}, a shift end time t_u^{end}, and a set of capabilities F_u. Moreover, the costs related to using this unit are specified by values for costs per time unit for each of the following actions: driving c_u^{drv}, serving c_u^{svc}, and overtime c_u^{ot}. Each contractor v has a home position d_v and a set of capabilities F_v. The costs for booking the contractor are specified by a fixed value per service c_v^{svc}. Each event e has a position a_e, a release time θ_e^{r}, a deadline θ_e^{d}, a service time δ_e, and a set of required capabilities F_e. Moreover, missing the deadline of an event means incurring in a lateness cost equal to the delay times the value of a lateness coefficient c_e^{late}.

A feasible solution of the VDP (a *dispatch*) is an assignment of events to units and contractors capable of serving them, as well as a tour for each unit such that all events are assigned, the service of events does not start before their release times (waiting of units is allowed at no extra-costs), and all tours for all units start at their current positions not before their logon times and end at their home positions. The costs of a dispatch are the sum of all unit costs, contractor costs, and event costs.

Following a common approach in the vehicle routing literature (see, e.g., [1] and the references therein), we state our model using binary tour variables. The VDP can then be formulated as a set partitioning problem, where the ground set of events and units has to be partitioned using a family of subsets that represent unit tours (plus some special subsets to account for the alternative of assigning events to contractors). The advantage of such a model lies in its flexibility to incorporate complicated technical and organizational side constraints within the individual tours and the possibility to tweak the cost function to achieve certain desired properties of the online behaviour (e.g., non-linear lateness penalty). This set partitioning model may be written as a huge integer linear program where the 0/1 restriction matrix contains one row for each event and unit and one column for each tour. To solve this problem, we use a column generation approach. Starting from some initial columns produced heuristically, the main iteration of our algorithm consists of adding new columns and resolving the linear relaxation of the problem until a certain stopping criterion is met. Whenever a new integral solution is found we output the corresponding dispatch.

The search for columns is done by enumeration in a *depth-first-search branch&bound tree* (*search tree*, for short) for each unit. To prune the search tree, a lower bound on the cost of a tour is used, which is based on the dual prices of events and units obtained from the previous solution of the linear problem. Our main contribution lies in what we call *dynamic pricing control*. The depth and degree of the search tree, as well as the value of a (negative) acceptance threshold imposed on the reduced cost of new columns are adjusted at each iteration according to the number of columns produced in the previous one. This ensures that (i) the effort of finding new columns is small in the beginning, when the dual variables are not yet in good shape, (ii) the dual variables are updated often in the beginning, (iii) this update is fast since the number of columns in the LP is still small, (iv) we can enforce the output of a feasible integer solution early, and (v) the search is exact later in the run when the dual information is reliable.

Our algorithm ZIBDIP turned out to be very efficient on real-world snapshot-instances provided by ADAC. In all tested cases, provably optimal or near-optimal (<1%) solutions were found in less than five seconds on state-of-the-art personal computers, even for high load situations containing about 200 events and 100 units. This figures ensure compliance with the real-time requirements of the application. The behavior of the algorithm remained stable for problem instances with (artificially augmented) extreme load: ZIBDIP found a dispatch to attend 770 events with 200 units whose quality was within 12% from optimum after 5 seconds, within 5% from optimum after 15 seconds, and within 2% after one minute.

Besides of solving the real-time "snapshot" problems, our algorithm is also used for the evaluation of online strategies: by running it a-posteriori on jobs collected during one/two hours, we found lower bounds on the value of the hindsight-optimum discussed in section 2.

4 Computational Results on Real-World-Data

The input data for the online-tests consisted of 68 *one-hour instances* and 68 *two-hour instances* that were extracted from accumulated whole-day datasets provided by ADAC. The events occurring in these instances were labeled with time-stamps that represent the moment in time when they pop up. Using this information, a simulation was run on the ADOptCmd implementation of our algorithm to test six variants of the REPLAN strategy mentioned in section 2. The total cost of the dispatch produced during the whole hour (resp. two hours) was compared to a lower bound on the hindsight optimum, which was obtained by solving to optimality the linear programming relaxation of the offline problem (i.e., the problem of finding an optimal dispatch if all events are known from the beginning). In all cases, the online algorithms were required to deliver their (partial) solutions within 15 seconds of running time.

Both for the one-hour as for the two-hours instances, the relative error of the online solution achieved by ADOptCmd for all six alternatives is well below 50% on the average with very rare substantial deviations of up to 230.71%, obtained for the setting `int120` (replan every 120 seconds) in one data set. There is a difference in the performances of the six different settings in favour of `newjob` (replan with every occurrence of a new job) and `int60` (replan every 60 seconds). This difference is, however, small enough to ensure that there will be no serious performance problems in waiting say 60 seconds for an optimization run to finish.

Figure 1 shows the distribution of the relative errors obtained for the best two settings `newjob` and `int60` on the one-hour instances. Setting `newjob` achieves the best results on average (a relative error of 40.21% against 44.33% for `int60`) and produces the least deviation (22.65 against 33.64). This trend continues in the two-hour instances.

(a) **`newjob`**: $\mu = 40.21$, $\sigma = 22.65$ (b) **`int60`**: $\mu = 44.33$, $\sigma = 33.64$

Fig. 1. Relative errors in the solutions produced for the one-hour instances.

Another fact to be noticed is that there is no substantial degradation in the performance of the algorithm when switching from the one-hour to the two-hours instances. In fact, the relative error obtained by `newjob` was slightly better in the second case. We expect the same to hold also for larger intervals of time. The computation of the required lower bounds, however, becomes then technically unmanageable.

5 Conclusion

We have developed a specialized column generation algorithm ZIBDIP that solves a real-world large scale vehicle dispatching problem with soft time windows under realtime requirements. The problem arises as a subproblem in an online-dispatching task that was proposed to us by the German Automobile Association (ADAC). A key concept behind the algorithm is the Dynamic Pricing Control, which can significantly speed up convergence of the column generation process, thereby making a method that has proven to be effective for large scale offline problems ready for the use in online-algorithms under realtime requirements. A further advantage of the column generation approach is its flexibility to incorporate complicated restrictions: we are planning to use non-linear lateness penalties in future tests.

Employing a-posteriori analysis on real-world problem instances provided by ADAC, we tested the performance of several settings of the REPLAN strategy (i.e., we determined a kind of *experimental competitivity*). The best results were obtained for the case when a new dispatch was computed either each time a new event was issued or at a fixed frequency of 60 seconds.

Although the development of the final version of the online algorithm is still under way, the first results are promising: On average, the online costs are within 50% above a lower bound on the "hindsight" (offline) adversary—not too bad, in our experience, for an online algorithm. We hope, however, to improve on these figures by utilizing estimates of future events in the snapshot-dispatches. This is work in progress and leads to most interesting questions as to how knowledge about the future can be exploited in combinatorial online optimization.

References

1. Jacques Desrosiers, Yvan Dumas, Marius M. Solomon, and Francois Soumis, *Time constrained routing and scheduling*, Handbooks in Operations Research and Management Science, vol. 8, Elsevier Science B.V., Amsterdam, 1995, pp. 35–139.
2. Martin Grötschel, Sven O. Krumke, and Jörg Rambau, *Online optimization of complex transportation systems*, in *Online Optimization of Large Scale Systems* [3], pp. 714–739.
3. Martin Grötschel, Sven O. Krumke, and Jörg Rambau (eds.), *Online optimization of large scale systems*, Springer Verlag, 2001.
4. Martin Grötschel, Sven O. Krumke, Jörg Rambau, Thomas Winter, and Uwe Zimmermann, *Combinatorial online optimization in real time*, in Grötschel et al. [3], pp. 687–713.
5. Sven O. Krumke, Jörg Rambau, and Luis M. Torres, *Real-time dispatching of guided and unguided automobile service units with soft time windows*, Preprint SC 01-22, Konrad-Zuse-Zentrum Berlin, Berlin, 2001.
6. D.D. Sleator and R.E. Tarjan, *Amortized efficiency of list update and paging rules*, Communications of ACM **28** (1985), 202–208.

Multi-Class User Equilibria under Social Marginal Cost Pricing

Leonid Engelson[1], Per Olov Lindberg[2], and Maria Daneva[2]

[1] Inregia, Box 12519, SE-10229, Stockholm,
[2] Linköping University, Dept. of Mathematics, SE-581 83 Linköping,
e-mails: lee@inregia.se, polin@mai.liu.se, madan@mai.liu.se

Abstract. In the congested cities of today, congestion pricing is a tempting alternative. With a single user class, already Beckmann et al. showed that "system optimal" traffic flows can be achieved by social marginal cost (SMC) pricing. However, different user classes can have wildly differing time values. Hence, when introducing tolls, one should consider multi-class user equilibria, where the classes have different time values. With SMC pricing, Netter claims that multi-class equilibrium problems cannot be stated as an optimization problems. We show that, depending on the formulation, the multi-class SMC-pricing equilibrium problem (with different time values) can be stated either as an asymmetric or as a symmetric equilibrium problem. In the latter case, the corresponding optimization problem is in general non-convex. For this non-convex problem, we devise descent methods of Frank-Wolfe type. We apply the methods and study a synthetic case based on Sioux Falls.

1 Overview

Traffic in major cities has become one of the biggest problems for society. It has become a common standpoint among transportation planners that charging some kind of fees from the users of the road network is necessary.

The classical social marginal cost pricing theory ([1], Ch. 4) states that if each user is charged a toll equal to the total value of time loss incurred on other users of the network, this will induce an equilibrium that is system optimal, assuming that all users have a fixed and identical time value.

However, different traveller groups may have grossly different time values, in Stockholm, e.g., ranging by a factor of more than 17. In such situations, computed tolled equilibria need to recognize the variation of time values between user groups, thus leading to multi-class user equilibria, (e.g., [2]). When the Jacobian of the multi-class link travel costs is symmetric and positive definite, such equilibria are unique and can be found by solving a corresponding optimization problem.

Netter [5], however, argues that when link travel times depend on class specific link volumes and are different for different user classes, the user equilibrium problem is not generally symmetric even in toll-free networks or in networks with fixed tolls.

The subject of this paper is social marginal cost (SMC) pricing in networks with several user classes that differ only by their value of time. We show in particular that the variational inequality defining the equilibrium can be stated in different forms, asymmetric or symmetric, hence not having or having a corresponding optimization formulation. We further see that this optimization problem in general is nonconvex. We conclude by stating an algorithm of Frank-Wolfe type for the SMC toll case, and applying it to the classical Sioux Falls network.

2 Multi-Class User Equilibria

Consider a road network consisting of nodes $n \in N$ and directed links $a \in A$. Let $P \subset N \times N$ be the set of OD pairs. Assume that OD demands q_p^k between the OD pairs $p \in P$ for each user class $k \in K$ are given and that to each link a are associated continuously differentiable cost functions $c_a^k : \Re_+^K \to \Re_+$, that give the cost of traversing link a for a user in class k dependent on the class specific volumes on the link. For the time being, c_a^k is a general function, but it will be endowed with a special structure in the next section.

Let R_p be the set of routes connecting OD pair p and $R = \cup_{p \in P} R_p$. Let $H = \{h \in \Re_+^{R \times K} : \sum_{r \in R_p} h_r^k = q_p^k, \forall p \in P, k \in K\}$ denote the set of feasible route flow vectors, and let
$F = \{f \in \Re_+^{A \times K} : f_a^k = \sum_{r \in R} \delta_{ra} h_r^k, \forall a \in A, k \in K, h \in H\}$, be the set of feasible link flows, where δ_{ra} is 1 if route r traverses link a, 0 otherwise.

Definition 1. (extended Wardrop) $\hat{h} \in H$ is a multi-class equilibrium flow if for any OD-pair p and user class k, the class specific costs of routes actually used (i.e. having $\hat{h}_r^k > 0$), are equal and not larger than those of any unused routes.

Similarly to the single user class case (see [6], Thm 3.14), the equilibrium condition can be written as a variational inequality in the set of link flows

$$\langle c(\hat{f}), f - \hat{f} \rangle \geq 0 \quad \forall f \in F \tag{1}$$

where $c = (c_a^k)_{a \in A, k \in K}$.

Variational inequality problems (VIP:s) are usually solved by reduction to optimization problems (or series of such). When the link specific cost functions are symmetric, i.e., with $f_a = (f_a^k)_{k \in K}$, when

$$\frac{\partial c_a^k(f_a)}{\partial f_a^l} = \frac{\partial c_a^l(f_a)}{\partial f_a^k} \quad \forall a \in A, k, l \in K, \tag{2}$$

then $c(f) = \nabla I(f)$, the gradient of some differentiable function $I : \Re_+^{A \times K} \to \Re$. In this case, all local minima of I restricted to F (and maybe some other points) are solutions to the VIP (1). If, in addition, the function I is convex, then the VIP (1) is equivalent to the optimization problem:

$$\min I(f) \text{ s.t. } f \in F. \tag{3}$$

If moreover, I is strictly convex, uniqueness of the solution to (3) and hence of the equilibrium is guaranteed (see, e.g., [4]).

3 Equilibria under Social Marginal Cost Pricing

For space reasons this section is quite terse; for more details, see [3]. In the remainder of the paper, it is assumed that the drivers' perceived link travel cost consists of two components: the toll p_a and the travel time t_a, which is a twice differentiable function of the total link volume $f_a^{tot} = \sum_{k\in K} f_a^k$. Using class specific *time values* $v_k > 0$, the perceived travel cost can be expressed either in time or in monetary terms. Thus we define *generalized time* $\bar{t}_a^k$ and *generalized cost* $\bar{c}_a^k$ of link a for class k, respectively, as

$$\bar{t}_a^k(f_a) = t_a(f_a^{tot}) + p_a/v_k, \tag{4a}$$

$$\bar{c}_a^k(f_a) = v_k t_a(f_a^{tot}) + p_a. \tag{4b}$$

Under SMC pricing, the users have to pay for the delays they incur on other users and one is interested in the traffic volumes that are established in the network and the corresponding toll values. The link toll then is the sum of all delay values for the users of the link caused by a marginal user, i.e.,

$$p_a = p_a(f) = t_a'(f_a^{tot}) \sum_{k\in K} v_k f_a^k. \tag{5}$$

The SMC equilibrium problem can now be formulated as the VIP (1) with link cost functions defined by inserting (5) into (4a) or (4b). However, using the costs $c_a^k = \bar{t}_a^k$ the symmetry condition (2) is not satisfied, precluding an optimization formulation. On the other hand, using the costs $c_a^k = \bar{c}_a^k$, we get

$$\frac{\partial \bar{c}_a^k(f_a)}{\partial f_a^l} = v_k t_a'(f_a^{tot}) + t_a''(f_a^{tot}) \sum_{n\in K} v_n f_a^n + t_a'(f_a^{tot}) v_l$$

which somewhat astonishingly satisfies (2). Integration yields

$$I(f) = \sum_{a\in A} t_a(f_a^{tot}) \sum_{k\in K} v_k f_a^k \tag{6}$$

with the property that $\nabla I(f) = \bar{c}(f)$. Note that (6) is the overall travel cost in the network as perceived by society under assumption that travel time of individual travellers is perceived as valuable as the travellers perceive it themselves.

In general, $I(f)$ is not convex and multiple SMC equilibria corresponding to different values of I can exist (see example below).

4 A Two-Link Example

Consider a network consisting of a single OD pair p connected by two parallel links a and b with identical travel time functions $t_a(u) = t_b(u) = u$. There are two traveller classes with time values $v_1 = 1$ and $v_2 = 5$, and with travel demand $q_p^1 = q_p^2 = 100$. The feasible set is $F = \{f = (f_a^1, f_a^2, f_b^1, f_b^2) \in \Re_+^4 : f_a^1 + f_b^1 = f_a^2 + f_b^2 = 100\}$. Without tolls, there is a continuous set of user equilibria $\{\hat{f}\}$, namely $\hat{f}_a^1 = \hat{f}_b^2 = \lambda, \hat{f}_b^1 = \hat{f}_a^2 = 100 - \lambda$ for $0 \leq \lambda \leq 100$. Note that travel time and total volume on each link is constant across $\{\hat{f}\}$.

Introduction of marginal congestion pricing leads to the VIP

$$\begin{array}{l}(2f_a^1 + 6f_a^2)(g_a^1 - f_a^1) + (6f_a^1 + 10f_a^2)(g_a^2 - f_a^2) + \\ (2f_b^1 + 6f_b^2)(g_b^1 - f_b^1) + (6f_b^1 + 10f_b^2)(g_b^2 - f_b^2) \geq 0, \ \forall g \in F\end{array} \tag{7}$$

Considering different cases, one can see that the equilibria are $f^{(1)} = (0, 80, 100, 20)$, $f^{(2)} = (100, 20, 0, 80)$ and $f^{(3)} = (50, 50, 50, 50)$ with corresponding tolls $p^{(1)} = (400, 200), p^{(2)} = (200, 400)$ and $p^{(3)} = (300, 300)$, respectively. Whereas $f^{(1)}$ and $f^{(2)}$ are local (and global) minima of the overall network cost function $I(f) = (f_a^1 + f_a^2)(f_a^1 + 5f_a^2) + (f_b^1 + f_b^2)(f_b^1 + 5f_b^2)$ on F, with $I(f^{(1)}) = I(f^{(2)}) = 56000$, $f^{(3)}$ is a saddle point with $I(f^{(3)}) = 60000$ (see Figure 1). When the tolls $p^{(1)}$ or $p^{(2)}$ are enforced, the only existing

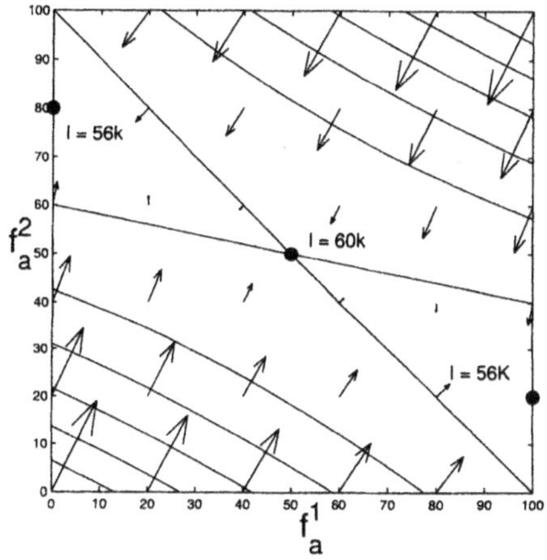

Fig. 1. Equilibria (•) and level curves and negative gradients of the function I.

user equilibria are $f^{(1)}$ and $f^{(2)}$, respectively. Implementation of the tolls $p^{(3)}$ does not affect the route choice, whence there is the same set of user equilibria $\{\hat{f}\}$ as in the toll free situation. Thus the equilibrium flow pattern with fixed equilibrium tolls $p^{(3)}$ need not coincide with $f^{(3)}$. However, these flow patterns are equivalent both from the individual and the societal points of view, since total flow and travel time along each link, as well as the total perceived travel cost at any of $\{\hat{f}\}$ is the same as at $f^{(3)}$.

5 A Frank-Wolfe Algorithm for the SMC Equilibrium

It follows from the results of sections 2 and 3 that the SMC equilibria can be determined by solving the optimisation problem (3) with the objective I of (6). To this end one can use (an adaption of) the Frank-Wolfe method.

Linearizing the objective, (3) decomposes into a set of shortest path problems; one for each user class and OD-pair. Assigning all demand to these shortest paths gives an extreme point feasible solution to (3). Performing a line search from the current solution towards this extreme point solution, one arrives at a better point, unless the current point is already an equilibrium. The process can be initialized in, e.g., the free flow extreme point solution.

6 Some Experimental Results

6.1 The Two Link Network

The algorithm has first been applied to the two link network in the example above, although with a different but qualitatively equivalent cost function. Due to the symmetric network structure, the algorithm, when started with the free flow conditions, quickly reaches the saddle point equilibrium and gets stuck there. When the algorithm is started from another feasible link flow, it reaches one of the local optima, although experiencing a lot of zig-zagging.

6.2 Sioux Falls Network

We have also applied the algorithm to the classical Sioux Falls network. For each OD-pair, we subdivided demand into three user classes (work, business and other trips) with class fractions equal to the Stockholm case $(.754, .036, .210)$. We also set the time values to their Stockholm values, $v = (.98, 3.30, .19)$, but normalized to average time value equal to 1. Note, that since the problem is nonconvex, we do not get an underestimate of the optimal value, when we solve the linearized problem, but it is reasonable to assume that we get a local lower bound in the limit. In order to get an accurate estimate of one local optimal value, we performed as many as 10000 iterations (which would give a relative error of 10^{-6} in the single class case).

Starting in the free flow solution, we get an objective value of 71.09, compared to the single class system optimum 71.94 and the systems cost of the single class user equilibrium at 74.80. The iteration history of the relative error $(I(f^{(i)}) - LBD)/LBD$ (where LBD is the optimal value of the linearized problem at iteration $i = 10000$) is displayed in a log-log diagram in figure 2a, together with the iteration history for the single class user optimum case (with LBD equal to the full precision optimal value), given for reference. As can be seen, apart from the first 10 iterations the progress in the multi-class case is comparable to that of the single class case.

In figure 2b we display iteration histories for the accurate free flow run together with three runs starting at random extreme points (with LBD from the free flow run). The behavior of the random runs are in line with the accurate one, making it probable that there is only one local optimum.

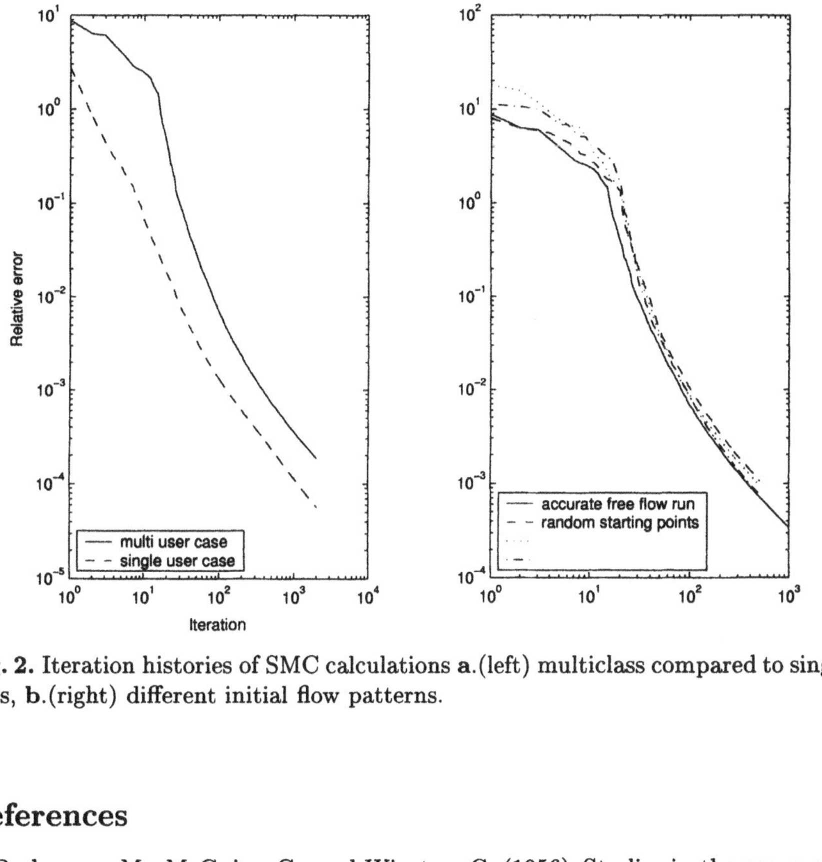

Fig. 2. Iteration histories of SMC calculations **a.**(left) multiclass compared to single class, **b.**(right) different initial flow patterns.

References

1. Beckmann, M., McGuire, C., and Winsten, C. (1956) Studies in the economics of transportation. Yale University Press, New Haven.
2. Dafermos, S.(1973) Toll patterns for multiclass-user transportation networks. Transportation Sci. **7**, 211–223.
3. Engelson, L., Lindberg, P.O., and Daneva, M. (2002) Congestion Pricing of Road Network Users with Different Time Values, Technical Report LiTH-MAT-R-2002-10, Linköping University, forthcoming.
4. Nagurney, A. (1993) Network economics: a variational inequality approach. Kluwer Academic Publishers Group, Dordrecht.
5. Netter, M. (1972) Equilibrium and marginal cost pricing on a road network with several traffic flow types. Proceedings of the 5th International Symposium on the Theory of Traffic flow and Transportation, G.F. Newell, Elsevier, 155–163.
6. Patriksson, M. (1994) The Traffic Assignment Problem - Models and Methods. VSP, Utrecht, The Netherlands.

School Bus Routing and Scheduling Problem

Michela Spada, Michel Bierlaire, and Thomas M. Liebling

École Polytechnique Fédérale de Lausanne, Institute of Mathematics, Switzerland.

Abstract. We consider the school bus routing and scheduling problem, where transportation demand is known and bus scheduling can be planned in advance. We first propose a modeling framework which aims to optimize a level of service for a given number of buses. Then, we describe a procedure building first a feasible solution, and subsequently improving it, using a heuristic. After the performance analysis of three types of heuristics on real and synthetic data, we recommend the use of simulated annealing exploring infeasible solutions, which performs slightly better than all others. More importantly, we find that the performance of all heuristics is not globally affected by the choice of the parameters. This is important from a practitioner viewpoint, as the fine tuning of algorithm parameters is not critical for its performance.

1 Introduction

Busing systems are common in countries where school assignment is imposed by home location, as in Switzerland, France, and Italy for instance. We propose a modeling framework, where the concept of level of service is captured by two different objective functions (Section 3). We propose and analyze automatic procedures to build a solution, all of which start by generating an initial feasible solution (see Section 4). In Section 5, we then propose three automatic procedures to improve the quality of the solution: a tabu search heuristic, and two variants of simulated annealing. We have tested several instances of each proposed heuristic on a set of 30 problems. The results, presented in Section 6, suggest that the simulated annealing heuristic exploring infeasible solutions is the most efficient out of the three, irrespectively of the choice of parameters.

2 Literature review

Various approaches are described in the literature for the school bus routing and scheduling problem. They differ in the way of the problem decomposition, in the modeling assumptions and in the solution algorithms. We can roughly distinguish two types of approaches. The first (see [1], [2], [3], [4], and [5]) consists in decomposing the problem and generating for each school simple routes from children's home. Those routes are then merged in order to optimize the objective function. In this approach no mixed loads are allowed, i.e. children from different schools are never carried in the same bus at the same time. The second one [6] solves the whole problem in one stage allowing mixed loads. From the modeling viewpoint, most authors aim at minimizing the costs, that is the number of buses or a combination of the number of buses and the total travel time. In this paper, we prefer the first approach, but allowing mixed loads, and we explicitly optimize the level of service provided by the bus operator.

3 Problem formulation

The transportation network is given by $G = (\mathcal{V}, \mathcal{E}, \ell)$ where $\mathcal{V}$ is a set of nodes, $\mathcal{E}$ a set of edges and ℓ a weight function associating a non-negative weight to each edge (typically, the travel time necessary to traverse the edge). We denote by $\mathcal{H} = \{h_1, \ldots, h_H\} \subseteq \mathcal{V}$ the set of nodes, called homes, where children get on the bus. Similarly, we denote by $\mathcal{S} = \{s_1, \ldots, s_S\} \subseteq \mathcal{V}$ the set of nodes where schools are located. For each school $s_i \in \mathcal{S}$, we denote by $\mathcal{H}(s_i) \subset \mathcal{H}$ the set of home nodes of children attending school s_i. In the model, a child c may represent in reality a group of several children with the same characteristics, that is the same home location and the same school. This simplifies the formulation with a moderate loss of flexibility. There are B buses to transport the children. Bus b has a given capacity, measured as the same unit as the children's space requirement, and is typically a number of seats. Note that some buses may load more children than the actual number of seats, for example when the bus has benches rather than individual seats and young children are carried.

A solution to the school bus routing and scheduling problem consists in specifying for each bus a tour, that is a list of stops, and for each stop, a list of children to pick-up and/or drop. Consequently, for each bus b and each stop, we impose that children get on bus b only at home, and get off only at school. The bus schedule depend on school starting time.

The number of buses being limited, each child may experience a delay due to the fact that the bus must pick-up other children on its way to the school. For each child, we define the *delay* as the difference between its actual journey time and the shortest possible time between home and school. Also, as buses may have to perform several tours, some children will be dropped at school too early. The time spent by the child waiting at school for the class to begin is referred to as the *waiting time*. The sum of the delay and the waiting time is called the *time loss* of the child c denoted by $f(c)$. We will consider here two related performance measures for the bus tour schedules, leading to two different objective functions to be minimized. The first one is *total* time loss summed over all children or equivalently, the *mean* time loss over all children, to be minimized over all tour schedules

$$\mathcal{F}_1 = \sum_{c \in C} f(c).$$

The second measure is the *maximum* time loss over all children, to be minimized over all tour schedules. It aims at balancing time losses.

$$\mathcal{F}_2 = \max_{c \in C} f(c).$$

The following constraints must be verified in the problem.

- Each child should arrive at school on time.
- Each child is carried by exactly one bus and steps out of the same bus it boarded.
- No bus can carry more children than its capacity allows for.
- For each tour, the schedule must be such that the bus can complete the trip within time, that is the travel time between two stops can not be less than the minimum travel distance between them.

This optimization problem can be formulated as a linear integer program, a program of very large scale, even for small problems.

4 Construction of a feasible solution

The heuristic idea to construct a feasible solution, is to first ignore the number of available buses B, and use as many of them as needed to verify all other constraints. We denote by Q the capacity of these virtual buses such that Q is the minimal capacity of bus fleet. Schools are considered in increasing order of their starting times. For each school s, the associated home nodes in $\mathcal{H}(s)$ are sorted in decreasing order of their distance from s. A first chain is constructed between the two first homes of $\mathcal{H}(s)$ sorted. Then homes $\mathcal{H}(s)$ are successively inserted in that chain, respecting the minimal chain length increase principle and bus capacity Q. Each virtual bus b serves a unique school $\hat{s}(b)$. Finally, for each virtual bus b, we insert the node corresponding to school $\hat{s}(b)$ in the chain. As both chain traversals are possible, we consider inserting its school at either one of its extremities, preferring that resulting in the shortest chain. If needed, the chain is reversed such that the school appears last. Finally, we define the tour associated with the constructed chain, and we impose that the bus is scheduled to arrive exactly when class starts, the bus uses the shortest route, only children going to school $\hat{s}(b)$ step in the bus, there is no pick-up at school, there is no drop off at home, all children that were picked-up are dropped off at school.

If the solution obtained by applying this procedure is composed by more than B buses, tours are merged pairwise the necessary number of times to obtain exactly B buses. Tours are merged in such a way to minimize the modifications of actual schedule. In this operation, we prefer mergers that do not modify the schedule, i.e. that generate a minimal waiting time for the bus driver who has to perform successively two tours. If this is not possible, we prefer mergers that advance the departure time of the first tour the least possible.

5 Algorithmic improvement

In order to improve an initial solution found, e.g. by the above procedure, we developed and compared three heuristic procedures that try to achieve a better value of the objective function ($\mathcal{F}_1$ or $\mathcal{F}_2$) while keeping feasibility. These three local search heuristics are: a variant of simulated annealing (feasible and infeasible version) and tabu search.

We have implemented a variant of simulated annealing (see [7] and [8]), where the temperature reaching a very low value is reheated to its initial value. We consider two neighborhoods: feasible neighborhood $\mathcal{N}_f$ generates only feasible solutions, while infeasible neighborhood $\mathcal{N}_i$ enlarges the set of candidate solutions and may generate solutions violating the bus capacity constraint. In both neighborhoods solutions are modified by removing one child from a bus and trying to place it in an other one. The feasible neighborhood $\mathcal{N}_f$ solution consists to insert child c in bus b' resulting in the largest improvement (or the least deterioration) of the solution, in sense of the objective function chosen $\mathcal{F}_1$ or $\mathcal{F}_2$. The other proposed neighborhood $\mathcal{N}_i$ modifies a solution in a similar way as $\mathcal{N}_f$, except that capacity satisfaction is not enforced when child c is inserted in bus b'. Therefore, $\mathcal{N}_i$ may create infeasible solutions. In order to restore feasibility, we augment the objective function value $\mathcal{F}_1$ or $\mathcal{F}_2$ by a penalty depending on the capacity violation and on the number of consecutive iterations visiting infeasible solutions.

We have also implemented a standard tabu search(see [9] and [10]) considering only the feasible neighborhood $\mathcal{N}_f$. A complete description of these heuristics can be found in [11]. The performance of these methods is analyzed and discussed in Section 6.

6 Case studies

In order to validate the heuristic approaches, these were tested on both real world and synthetic data sets.

The real data comes from two Swiss towns, Savigny and Forel. The network contains 34 nodes, including 12 schools: 4 *kindergartens* starting at 8.35, 6 *primary schools* starting at 8.15 and 2 *high schools* starting at 7.40. For the school year 1997–1998, four buses were used to transport 274 children. The capacity of each bus is 210 units, and each high school child takes up 10 units in the bus, while kindergarten and primary school children take up 7 units.

The synthetic data has been generated in order to analyze the relative efficiency of the methods as a function of the network size. For a given number of nodes randomly distributed in a square, the network is obtained from a Delaunay triangulation, from which links are removed such as to obtain a realistic degree for each node, while maintaining the connectivity. The distributions of homes, schools and trips have been chosen to resemble the proportions in the real data set.

The experimental design is based on four variables: the network, the number of buses, the objective function and the heuristic. In addition to the Savigny and Forel network, we have generated four artificial networks with 10, 20, 30 and 50 nodes, respectively. For each of the five networks, we have run several scenarios for the number of buses (three on average), and the two objective functions $\mathcal{F}_1$ and $\mathcal{F}_2$, to obtain a total of thirty problems. We have considered three types of heuristics, that is tabu search and simulated annealing with feasible neighborhood and infeasible neighborhood. In each case, simulated annealing has been run using four different initial temperatures, and five different seeds for the random number generator. The tabu search algorithm has been run using two different neighborhood sizes, and ten different seeds for the random number generator. Thus, twenty instances of each type of heuristic, totaling 1800 runs were carried out on a Pentium III 499MHz. On this machine an instance with 50 nodes and 9 buses took about 10 hours to complete 50'000 solution evaluations, that is about 0.7 sec. per evaluation. We summarize now the main observations from these runs.

Clearly, the quality of the solution depends on the specific heuristics used and the choice of its parameters. In order to compare the overall quality of the heuristics, we adopt a variant of the performance profiles analysis proposed by Dolan and Moré [12].

If $f_{p,a,i}$ is the performance index of instance i of algorithm a solving problem p, then the *performance ratio* is defined by $r_{p,a,i} = \frac{f_{p,a,i}}{\min_{a,i}\{f_{p,a,i}\}}$. For any given threshold π, the overall performance of algorithm a is given by $\rho_a(\pi) = \frac{1}{n(a)n_p}\Phi_a(\pi)$ where n_p is the number of problems considered, $n(a)$ is the number of instances of algorithm a which have been run, and $\Phi_a(\pi)$ is the number of problems and instances of algorithm a for which $r_{p,a,i} \leq \pi$. We refer the reader to [12] for more details about the benchmarking method.

The performance indices we have selected are the relative improvements of the objective function (that is the ratio between the final and initial values of the objective function) after a given number of candidate evaluations (the *computation budget*). Interestingly, the profiles are qualitatively similar for different values of the computation budget. These results are encouraging, as they indicate that the proposed approach is rather robust and does not depend too heavily on the tuning of the heuristics parameters. We present here the results for an index based on a fixed budget of solution evaluations depending on the problem size ($100|\mathcal{V}|$ evaluations).

We compare tabu search, and the two versions of simulated annealing for the problem with objective function $\mathcal{F}_1$ and $\mathcal{F}_2$. In Figure 1, the plots in the bottom are just zooms of the top plots, in order to emphasize the profile for small values of π. For the sake of clarity, the plots for $\mathcal{F}_1$ and $\mathcal{F}_2$ have been separated, although the performance profile analysis has been performed over all methods.

The most noticeable performance is achieved by the infeasible version of simulated annealing when objective function $\mathcal{F}_1$ is considered (see, Fig-

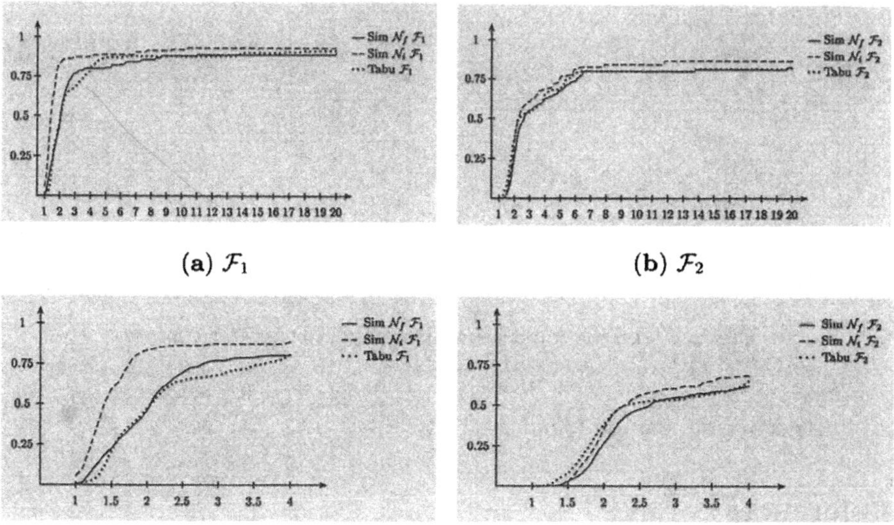

(a) $\mathcal{F}_1$ (b) $\mathcal{F}_2$

(c) $\mathcal{F}_1$ (d) $\mathcal{F}_2$

Fig. 1. Performance profiles for budget=$100|\mathcal{V}|$

ure 1(**c**)). It is interesting to note the slope of the profiles for each objective function (compare, for instance, Figures 1(**c**) and 1(**d**)). The profile is steeper with objective function $\mathcal{F}_1$, illustrating a more robust behavior of the heuristics when $\mathcal{F}_1$ is preferred over $\mathcal{F}_2$. This is due to the fact that the same value of the objective $\mathcal{F}_2$ may be achieved for a large number of different solutions, as only the largest time loss is considered. Therefore, the heuristics are unable to discriminate among a large class of solutions with the same objective, and are more likely to cycle randomly among them.

All heuristics exhibit the same type of behavior: a significant improvement of the objective function in the early iterations, and a slower reduction rate in subsequent iterations. This is illustrated by Figure 2 where the x-axis represents the number of evaluations of candidate solutions, and the y-axis is the ratio of the objective function for the solution computed by each heuristic and the objective function of the initial solution. Results for $\mathcal{F}_1/C$ are represented on the left figure, and results for $\mathcal{F}_2$ on the right.

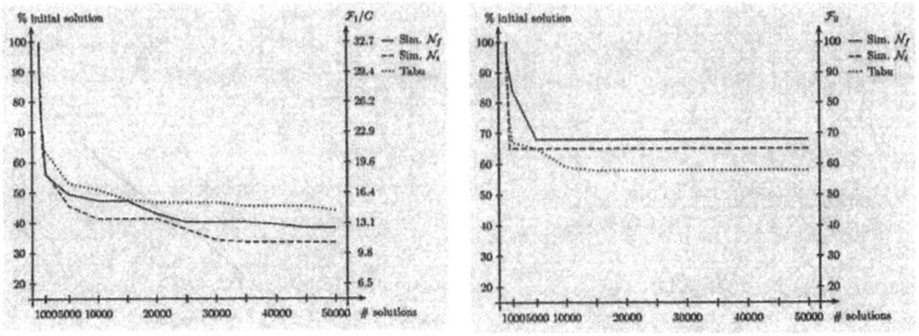

Fig. 2. Heuristic improvements with $|\mathcal{V}| = 50$ and $B = 9$

Further results can be found in [11].

References

1. Bodin, L., Berman, L. (1979) Routing and Scheduling of School Buses by Computer. Transportation Science **13**, 113–129.
2. Angel, R., Caudle, W., Noonan, R., and Whinston, A. (1972) Computer-assisted School Bus Scheduling. Management Science B **18**, 279–288.
3. Bennett, B., Gazis, D. (1972) School Bus Routing by Computer. Transportation Research **6**, 317–326.
4. Desrosiers, J., Ferland, J., Rousseau, J.-M., Lapalme, G., Chapleau, L. (1981) An Overview of a School Busing System. In: N. Jaiswal (Ed.), Scientific Management of Transport Systems, Vol. IX, International Conference on Transportation, New Delhi, November 26-28, 1980, 235–243.
5. Gavish, B., Shlifer, E. (1979) An Approach for Solving a Class of Transportation Scheduling Problems. EJOR **3**(2), 122–134.
6. Braca, J., Bramel, J., Poser, B., Simchi-Levi, D. (1994) A Computerized Approach to the New York City School Bus Routing Problem. Technical report, Graduate School of Business, Columbia University, NY.
7. Kirkpatrick, S., Gelatt, C. D. J., Vecchi, M. P. (1983) Optimization by Simulated Annealing. Science **220**, 671–680.
8. Rossier, Y., Troyon, M., Liebling, T. M. (1986) Probabilistic Exchange Algorithms and Euclidean Traveling Salesman Problems. OR Spektrum **8**(3), 151–164.
9. Glover, F. (1977) Heuristic for Integer Programming Using Surrogate Constraints. Decision Sciences **8**, 156–166.
10. Hansen, P., Jaumard, B. (1987) Algorithms for the Maximum Satisfiability Problem. Technical report, Rutgers University.
11. Spada, M., Bierlaire, M., Liebling, T. M. (2002) Decision-aid Methodology for the School Bus Routing and Scheduling Problem. Report # RO-20020819, École Polytechnique Fédérale de Lausanne.
12. Dolan, E. D., Moré, J. J. (2002) Benchmarking Optimization Software with Performance Profiles. Mathematical Programming, Serie A **91**, 201-213.

Covering Population Areas by Railway Stops

Anita Schöbel[1] and Michael Schröder[2]

[1] Universität Kaiserslautern, Postfach 3049, D-67653 Kaiserslautern, Germany
[2] Fraunhofer Institut für Techno- und Wirtschaftsmathematik, Gottlieb–Daimler–Strasse 49, D-67663 Kaiserslautern, Germany

Abstract. We study the installation of new stops (or stations) along existing links in public transportation, e.g. railway systems. This improves the coverage level of the system, i.e., the number of people living near some stop. On the other hand, additional cost is incurred and travel times tend to increase.
We model this as a network location problem, where the network corresponds to the transportation links. Our main contribution is to model the population distribution by a system of compact subsets in the plane, the *population areas*. The goal is to cover all these areas with as few stops as possible along the railway tracks. We present an efficient algorithm for finding an optimal solution in the case of a single edge and show its applicability to real world problems.

1 Introduction

We define the coverage level of a public transportation system as the percentage of the population that lives within a given distance from any of the stations. A relatively cheap way to improve the coverage level is to establish new stops along existing tracks. The cost incurred are the fixed and variable costs of the stop and the cost of the additional running time of the trains and their crews. Furthermore the travel time of passengers increases if trains stop more often. Therefore we want to achieve a certain coverage level with as few stops as possible.
In the literature, location of stops in transportation systems has been mainly considered within a discrete setting, i.e., given a finite candidate set of possible stops, which of them should be chosen to get a certain coverage level? Such *discrete stop location problems* can be reduced to set covering problems and have been studied in [5,6], and, very recently, in [7], with an application in Brisbane, Australia. Another discrete stop location model has been developed by Laporte et al. [4]. They investigate which candidate stops along one given line in Sevilla should be opened, taking into account constraints on the interstation space. On the other hand, in the *continuous stop location problem* there is no discrete candidate set given, but the stops may be located anywhere along the given track system. This has first been considered heuristically in [2]. Theoretical results about the continuous location of stops can be found in [8], and an algorithm for finding such a set of stops is suggested in [3]. In these papers, the settlements (or *population areas*) are represented by one point each. This assumption is not very practical, but simplifies the

analysis drastically, since only points need to be covered. E.g., defining coverage as the number of customers living closer than 2 km to their closest station, and representing each settlement as a single point yields a coverage level for Germany of 61.3%. In contrast, if we consider the percentage of the covered area, weighted by the population living there, the coverage level is only 52.4%. So it makes sense to take into consideration the whole population sets instead of just one point for each of them.

For our problem, we need the following:

- Let P be a compact set describing the union of the ***population areas*** to be covered.
- Let the transportation network be given as a planar connected graph $G = (V, E)$ with straight-line embedding in the plane. Then

  $$T = \cup_{e \in E} e = \{x \in \mathbb{R}^2 : x \in e \text{ for some } e \in E\} \subseteq \mathbb{R}^2$$

  denotes the set of points along the track system, where we assume that each edge $e \in E$ (representing a single track) is given by a line segment in the plane. We want to locate a set of stops $S \subseteq T$.

We now define what we mean by *covering* P. To this end, let d be a norm-distance, e.g., the Euclidean distance, or the rectangular distance.

Definition 1 $\text{cover}(S) = \{x \in \mathbb{R}^2 : d(x, s) \leq r$ *for some* $s \in S\}$, *where* $r > 0$ *is given, and* $S \subseteq T$. *Moreover, we say that* P is covered *by* S, *if* $P \subseteq \text{cover}(S)$.

Complete Cover (CC) can now be stated as:
Find a minimum-cardinality set $S^* \subseteq T$ *covering* P.

Some properties should be mentioned first.

Lemma 1 *1. (CC) is NP-hard*
2. (CC) is feasible if and only if $P \subseteq \text{cover}(T)$
3. (CC) may be feasible but not finitely feasible.

Proof: Part 1 follows from the NP-hardness of (CSLP), which is a special case of (CC), in which P is a finite set of points, see [8], part 2 is clear, and part 3 is shown by the example of Figure 1. □
For the following we assume that (CC) is feasible, i.e.,

$$P \subseteq \text{cover}(T).$$

In the remainder of this paper we develop an algorithm for solving (CC) on a straight line segment. We then show that this approach can be quite well applied in practice.

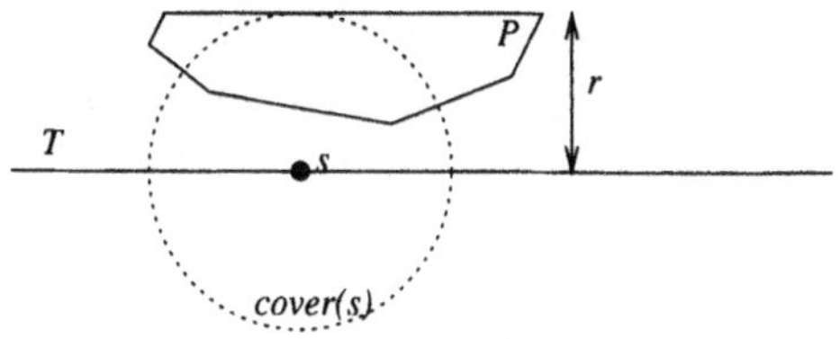

Fig. 1. Example for the Euclidian distance: Infinitely many points of T are necessary to cover P.

2 The Algorithm

We focus on (CC) in the case that T consists of one single edge e with endpoints v_1 and v_2. We will develop an algorithm that constructs a finite optimal solution if it exists.
We first need some notation. Let $\leq$ denote the natural ordering along the edge e induced by $v_1 \leq v_2$. Then, for $x_1 \leq x_2$ let

$$[x_1, x_2] := \{y \in e : x_1 \leq y \leq x_2\}.$$

For each $x \in P$ let us define

$$T(x) := \{y \in e : d(y, x) \leq r\}$$

as the part of edge e that can be used to cover x, i.e., we know that at least one point of $T(x)$ has to be chosen as a new stop. Since $T(x) = e \cap \{y : d(y, x) \leq r\}$ is a convex set contained in e, we can find two points $l(x), r(x)$ such that $T(x) = [l(x), r(x)]$. Note that for $y \in e, x \in P$ we know that

$$x \in \text{cover}(y) \Longleftrightarrow y \in T(x) \Longleftrightarrow l(x) \leq y \leq r(x). \tag{1}$$

Definition 2 *Given some set $Q \subseteq P$, the (left)* fixturing point *of Q is*

$$L(Q) := \inf\{r(x) : x \in Q\} \in e$$

We are now in the position to present our algorithm for solving (CC).

Algorithm for solving (CC) on a single edge

Input: $T = e$, $P \subseteq \text{cover}(T)$
Output: Opt
Step 1. Set $\text{Opt} := \emptyset$, $\bar{P} := P$.
Step 2. $\text{Opt} := \text{Opt} \cup \{L(\bar{P})\}$, $\bar{P} := \bar{P} \setminus \text{cover}(L(\bar{P}))$.
Step 3. If $\bar{P} = \emptyset$, STOP, otherwise goto Step 2.

Note that the algorithm needs not to terminate finitely. But, fortunately, in Theorem 1 we can show that this only happens in the case that no finite optimal solution exists. Moreover, we will show the correctness of the algorithm, i.e., that in case of a finite termination we obtain the optimal solution.

For the following, let y_i denote the fixturing point found in iteration i and let $P^i = P \setminus \text{cover}\{y_1, \ldots, y_{i-1}\}$ be the set $\bar{P}$ at the beginning of iteration i. We collect some simple properties in lemma 2 and prove the correctness of the algorithm in the following theorem.

Lemma 2 *1.* $y_i \le y_{i+1}$ *i=1,2,...*

2. If the algorithm terminates then $y_i < y_{i+1}$ *for i=1,2,...*

3. If the algorithm terminates then Opt *is a feasible solution of (CC).*

Theorem 1 *If there exists a finite solution of (CC) then the algorithm terminates with an optimal solution* Opt.

Proof: Let Opt $= \{y_1, y_2, \ldots\}$ be the solution found by the algorithm. According to part 1 of Lemma 2 we know that $y_1 \le y_2 \le \ldots$. By assumption of the theorem there exists a finite solution $Y^* = \{y_1^*, y_2^*, \ldots, y_S^*\}$ with $y_1^* < \ldots < y_S^*$. Since for all finite Opt we already know the feasibility (part 3 of Lemma 2) our goal is to show that $|\text{Opt}| \le S$. To this end, we first prove that

$$\text{cover}\{y_k^* : k < i\} \cap P \subseteq \text{cover}\{y_k : k < i\} \cap P.$$

$i = 1$: $\emptyset = \emptyset$.

$i \to i + 1$: From the induction hypothesis we know that

$$\text{cover}\{y_k^* : k < i\} \cap P \subseteq \text{cover}\{y_k : k < i\} \cap P,$$

yielding $P^i \subseteq P \setminus \text{cover}\{y_k^* : k < i\}$.

Claim 1: $y_i \ge y_i^*$.

Suppose $y_i^* > y_i$. Since $y_i = L(P^i) = \inf\{r(x) : x \in P^i\}$ and $y_i^* > y_i$ there exists some $x \in P^i$ with $r(x) < y_i^*$. According to (1) this means, x is **not** covered by y_k^* if $k \ge i$. Since $x \in P^i$ yields that $x \notin \text{cover}\{y_k^* : k < i\}$ it is also **not** covered by y_k^* if $k < i$, a contradiction.

Claim 2: $\text{cover}(y_i^*) \cap P^i \subseteq \text{cover}(y_i)$.

Let $x \in \text{cover}(y_i^*) \cap P^i$. Since $x \in \text{cover}(y_i^*)$ we get $l(x) \le y_i^*$ and using Claim 1, $l(x) \le y_i$. On the other hand, $x \in P^i$ and the definition of $y_i = L(P^i)$ yields $y_i \le r(x)$. Together, $l(x) \le y_i \le r(x)$ such that $y_i \in T(x)$ meaning that $x \in \text{cover}(y_i)$ (see (1)).

The induction hypothesis together with Claim 2 shows the result.

Finally, since Y^* is a finite solution we get that $P \subseteq \text{cover}\{y_i^* : i \le S\}$ and hence $P \subseteq \text{cover}\{y_i : i \le S'\}$, where $S' = \min\{|\text{Opt}|, S\} \le S$. This means that at the end of iteration S' the set

$$\bar{P} = P^{S'} = P^{S'-1} \setminus \text{cover}\{y_{S'}\} = P \setminus \text{cover}\{y_1, \ldots, y_{S'}\} = \emptyset$$

and the algorithm terminates with a finite solution $|\text{Opt}| = S' \le S$. □

The algorithm can be transferred to arbitrary graphs if we know in advance that no point $x \in P$ can be covered by two stations belonging to different edges of G. This is formalized below.

Definition 3 *For* $e_1, e_2 \in E$ *let* $C(e_1, e_2) = P \cap \text{cover}(e_1) \cap \text{cover}(e_2)$ *be the conflict zone of edge* e_1 *and edge* e_2*.*

If the conflict zones for all pairs of edges are empty, then (CC) for the whole graph G decomposes into $|E|$ independent subproblems, one for each edge, and can be solved by our algorithm. In the next section we justify that the total population living in these conflict zones is rather small. A weaker definition of conflict zones which assures that the algorithm can be applied on trees, even if (CC) is not decomposable, is currently under research.

3 Application to the German Railway Network

To study the practical relevance of the algorithm we use data from Germany. The towns and cities are given as a set of 30494 population areas of which we know the outlines (given as polygons) and the number of inhabitants. The polygons are not identical with the borders of communities and also do not form a partition of Germany. They represent the population distribution better than community borders since open space is excluded. We assume uniform population distribution inside each polygon.

The network of the German railway DB currently has 6828 stops. For the Euclidian distance with $r = 2\,\text{km}$ their coverage level is 52.4%. However 65.2% of the population live within 2 km distance from the network (nodes and edges), i.e. the coverage level could be increased by almost 13%.

We define P to be the populated area within $(2 - \varepsilon)\,\text{km}$ Euclidian distance from the network (to guarantee a finite solution, cf. Figure 1) that is not already covered by existing stops. Finding a minimal number of new stops to cover P with our algorithm requires that we can apply it to every edge separately. As explained above, this would be possible if all conflict zones were empty. For the data of Germany this requirement is clearly not satisfied (Figure 2).

This does not mean that the algorithm is useless for practical purposes. If we exclude the population in the conflict zones from the covering problem we can still apply the algorithm. To find out how much this would affect the problem we computed all conflict zones. We found that their union contains 1.5% of the total population. Consequently for more than 80% of the possible increase in the coverage level a minimal set of new stops can be calculated with our algorithm.

If the conflict zones are not excluded from the covering problem, our algorithm, applied to every edge, can be used as a heuristic to solve problem (CC). We expect it to produce near-optimal solutions. This will be studied in further research.

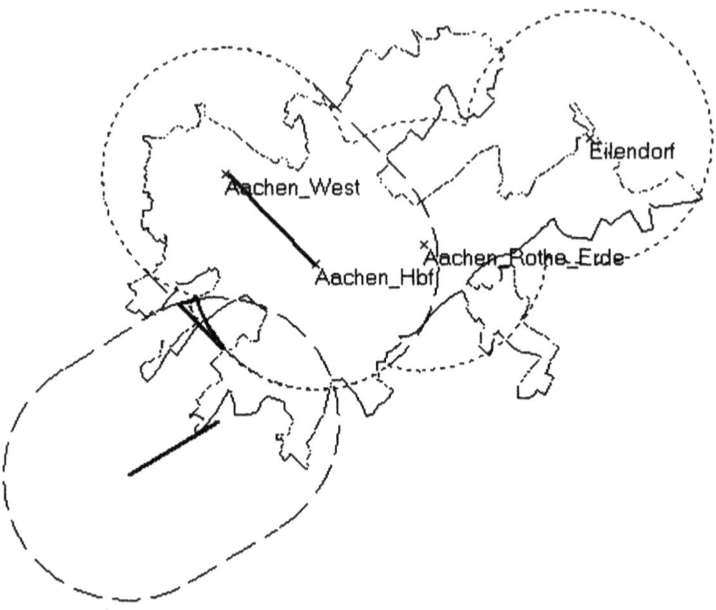

Fig. 2. Two edges with nonempty conflict zone at the city of Aachen. Shown are the union of circles with radius 2 km around stations (dotted), the area covered by the edges (dashed) and the small area containing the conflict zone (bold).

References

1. M. J. Demetsky, M. Asce, and B. B.-M. Lin (1982), "Bus stop location and design", *Transportation Engineering Journal*, 108, 313-327.
2. H.W. Hamacher, A. Liebers, A. Schöbel, D. Wagner, and F. Wagner (2001), "Locating new stops in a railway network", *Electronic Notes in Theoretical Computer Science* 50:1
3. E. Kranakis, P. Penna, K. Schlude, D.S. Taylor, and P. Widmayer (2002), "Improving Customer Proximity to Railway Stations" *Technical report*, ETH Zürich.
4. G. Laporte, J.A. Mesa, and F.A. Ortega (2002), "Locating stations on rapid transit lines", *Computers and Operations Research* 29, 741-759.
5. A. Murray, R. Davis, R.J. Stimson, and L. Ferreira (1998), "Public Transportation Access", *Transportation Research D*, 3:5, 319-328.
6. A. Murray (2001), "Strategic analysis of public transport coverage", *Socio-Economic Planning Sciences*, 35, 175-188.
7. A. Murray (2002), "A Coverage Models for Improving Public Transit System Accessibility and Expanding Access", *Technical Report* Ohio State University.
8. A. Schöbel, H.W. Hamacher, A. Liebers, and D. Wagner (2002), "The continuous stop location problem in public transportation", *Technical Report Wirtschaftsmathematik* 81, University of Kaiserslautern.

Innovative Lösungen im bimodalen Transport Straße / Binnenwasserstraße

Joachim R. Daduna; Johannes Schröter

FHW Berlin; Badensche Straße 50 – 51, D – 10825 Berlin
FW Grundt Logistik & Spedition; Straße 22 Nr. 2 –10, D - 13509 Berlin

1 Güterverkehrsentwicklung und Kombinierter Verkehr

Im Bereich des Güterverkehrs weisen die prognostizierten Entwicklungen für die kommenden Jahre einen deutlichen Anstieg aus, wobei die Zuwächse sich im wesentlichen auf den Straßengüterverkehr beziehen. Die Ursachen hierfür sind sehr vielschichtig, u.a. bedingt durch strukturelle Veränderungen im Güteraufkommen und dem Steigen der Transportnachfrage aufgrund einer zunehmenden Internationalisierung in der Fertigung und im Handel (s. u.a. Daduna 2001). Da ein Anwachsen des Straßengüterverkehrs mit den verkehrspolitischen Zielvorstellungen in keiner Weise zum Einklang zu bringen ist, u.a. aus ökologischen Überlegungen aber auch aufgrund der weitgehend fehlenden Möglichkeiten für Kapazitätserweiterungen im Straßennetz, wird nach Auswegen aus dieser Situation gesucht. Regulierende Maßnahmen staatlicher Institutionen zur Reduzierung der (Straßen-)Güterverkehre bzw. zur Beeinflussung von Verkehrssystementscheidungen erweisen sich in der Regel als wirtschaftspolitisch kontraproduktiv, auch mit Blick auf die Bedeutung der Mobilität als eine der wesentlichen Determinanten für ein nachhaltiges Wirtschaftswachstum. Geeignete Lösungen müssen daher mit Hilfe anderer Ansätze gefunden werden, wobei neben den Informations- und Kommunikations(IuK)technologien (s. u.a. Daduna / Voß 2000 bzw. Daduna 2001) die Bildung *bi-* bzw. *multimodaler Transportketten* im *Kombinierten Ladungsverkehr* (KLV) (s. u.a. Seidelmann 1997) im Vordergrund stehen.

Die IuK-Technologien bilden eine Grundlage, um Verkehrsinfrastrukturen effizienter zu nutzen, d.h., es werden in erster Linie die operativen Abläufe verbessert, z. B. innerhalb eines vorhandenen (Straßen-)Netzes, ohne allerdings mengenmäßige Verlagerungseffekte und damit eine Veränderung des Modal splits im Güterverkehr zu erreichen. Anders sieht es bei bi- bzw. multimodalen Transportketten aus, die explizit auf eine Verkehrsverlagerung ausgerichtet sind. Die (klassische) bimodale Lösung im KLV ist die Kombination Schiene / Straße, die allerdings bisher nicht in dem gewünschten Umfang realisiert werden konnte (s. u.a. von Stackelberg 1998). Ursache hierfür sind die mit den Umschlagvorgängen beim Verkehrsträgerwechsel verbundenen Kosten und Zeitabläufe, sowie die konzeptionellen und organisatorischen Defizite im Bereich der Anbieter von Schienen(güter)verkehrsleistungen. Weniger stark beachtet sind bisher die Kombinationen Binnenwasserstraße / Straße und Binnenwasserstraße / Schiene, u.a. bedingt durch Restriktionen, die in der Binnenschiffahrt auftreten, so auch in den verkehrsträgerbezogenen Schnittstellen. Für diese Form des KLV wird im folgenden ein innovatives (und patentiertes) Konzept vorgestellt, das eine Lösung für die Schnittstellenproblematik anbietet, die auch unter Kostengesichtspunkten als effizient anzusehen ist.

2 Systemkonzept

Das Systemkonzept basiert auf *Ro/Ro-Binnenmotorschiffen* bzw. *Schubleichtern* unterschiedlicher Auslegung mit der Möglichkeit einer Aufnahme von Straßen- und / oder Schienenfahrzeugen. Der aus logistischer Sicht entscheidende Vorteil der Ro/Ro-Technik besteht darin, daß ortsfeste Infrastruktureinrichtungen nur in einem geringem Umfang (u.a. in Form von Rampen) benötigt werden und keine (investitionsintensiven) Krananlagen erforderlich sind. Werden komplette Fahrzeugeinheiten geladen (z.B. Sattelauflieger) sind keine zusätzlichen Ro/Ro-Zugmaschinen notwendig. Sollen allerdings nur Sattelauflieger (oder Eisenbahnwaggons) transportiert werden, müssen die Be- und Entladevorgänge mit Hilfe von (mitgeführten) Zugmaschinen durchgeführt werden. Die Abläufe, die einem solchen Systemkonzept zugrunde liegen, lassen sich nur bedingt in einer einheitlichen Struktur darstellen. Sie werden deshalb im Abschnitt 3 anhand von zwei grundlegenden Ansätzen beschrieben.

2.1 Fahrzeugkonzept

Das Fahrzeugkonzept basiert auf Binnenmotorschiffen und Schubleichtern, wobei in der größenmäßigen Auslegung ein Verband ein Großmotorschiff (GMS) mit einer Abmessung von 110 ö 11.4 m sowie einen Schubleichter mit der Abmessung von 76.5 ö 11.4 m bzw. zwei dieser Schubleichter und eine Schubeinheit umfaßt. Bei dieser Festlegung wird von den in weiten Teilen des Binnenwasserstraßennetzes einsetzbaren Standardfahrzeugen ausgegangen. In Abhängigkeit von den spezifischen Bedingungen von den einzelnen Netzabschnitten können diese Dimensionierungen auch unter- bzw. überschritten werden.

Die wesentliche technische Ergänzung der Fahrzeuge ergibt sich durch den Einbau von hydraulisch betriebenen Stempeln im Bugbereich (s. Abb. 1), mit Hilfe derer sich die Stabilität beim Be- und Entladen an einer Rampe gewährleisten läßt. Außerdem wird hiermit ein Absinken der Ladeöffnung unter die Wasserlinie verhindert, die sich bei einem zunehmenden Ladegewicht ergeben kann. Ergänzt werden kann diese Technik durch an den Hydraulikstempeln angebrachte Lafetten, die ein begrenztes Auffahren an den Rampen ermöglichen. Technische Grundlage können die für das Automatic Loading System (ALS) im KLV entwickelten selbstfahrenden Lafetten sein. Der Laderaum wird mit drei parallel verlaufenden Gleisen ausgerüstet, die für die Aufnahme der Eisenbahnwaggons vorgesehen sind, aber auch der Spurführung für Sattelschlepper bzw. Sattelauflieger dienen, die aufgrund der geringeren Breite in vier Reihen aufgestellt werden können.

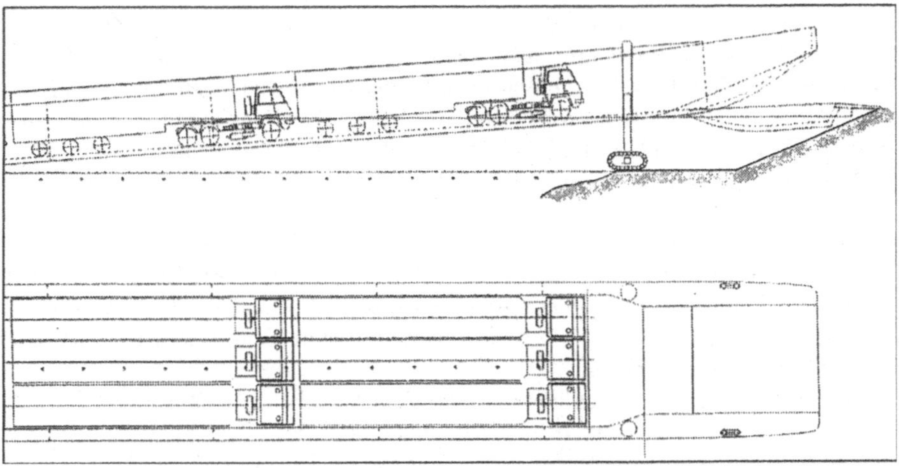

Abb. 1: Schubleichter Seitenquerschnitt und Grundriß (Ausschnitt)

2.2 Infrastrukturanforderungen

Die Infrastruktur beschränkt sich auf Rampen für die Durchführung von Be- und Entladevorgängen. In einer Reihe von Häfen sind diese vorhanden, ohne allerdings eine quantitativ ausreichende Basis für das Systemkonzept zu gewährleisten. Entscheidend ist die (weitgehend) flächendeckende Verfügbarkeit von Rampen durch die aus militärischen Überlegungen eingerichteten Ersatzübergangsstellen im gesamten Binnenwasserstraßensystem der BRD, so z. B. entlang der Elbe mit insgesamt 28 Übergangsstellen. Da nicht alle dieser Rampen unmittelbar an das Straßen- bzw. Schienennetz angeschlossen sind, werden gegebenenfalls für die zu nutzenden Standorte entsprechende Infrastrukturmaßnahmen erforderlich. Erfolgt das Be- und Entladen in Hafenanlagen erfolgt, können auch vorhandene Umschlageinrichtungen genutzt werden, soweit diese die technischen und kapazitiven Anforderungen erfüllen.

Da die Rampen mit der Stromrichtung ausgelegt sind, müssen die Motorschiffe bzw. Schubleichter bei Talfahrt zum Be- und Entladen eine Richtungsänderung vornehmen, da die Ladevorgänge über die geöffnete Bugklappe erfolgen. Es kann hierbei vorausgesetzt werden, daß in der Nähe einer genutzten Rampe geeignete Wendestellen verfügbar sind, soweit die Wasserstraßenbreite die entsprechenden Wendemanöver im Bereich der Rampe nicht zuläßt. Nur im Extremfall ergeben sich hieraus Restriktionen für die Länge der Fahrzeugeinheiten, die u. U. die Wirtschaftlichkeit negativ beeinflussen können.

3 Anwendungen

Anwendungen für das vorgestellte Systemkonzept bestehen sowohl beim Güternah- als auch beim Güterfernverkehr. Die hierbei zugrunde liegenden Zielsetzungen sind sehr vielschichtig, wobei mögliche Kostenreduzierungen im Vordergrund stehen. Diese lassen sich unter den derzeitigen Rahmenbedingungen primär bei den Personal- und Kraftstoffkosten erreichen. Allerdings ergibt sich zukünftig für den Schwerlastverkehr durch die Vermeidung der "streckenbezogenen LKW-Gebühr auf Autobahnen" eine zusätzliche Kosteneinsparung. Daneben bestehen weitere Vorteile, die allerdings nicht unmittelbar quantifizierbar sind, sondern nur als bewerteter Nutzen in die Kalkulationen einbezogen werden können. Vorhandene Verlagerungseffekte spielen bei diesen Überlegungen zur Wirtschaftlichkeit keine wesentliche Rolle, da diese als verkehrspolitische und ökologische Aspekte der makrologistischen Ebenen zu zurechnen sind, und nicht in eine (betriebliche) Kostenbetrachtung eingehen.

Voraussetzung für die Realisierung derartiger bimodaler Konzepte ist die Verfügbarkeit eines geeigneten Binnenwasserstraßennetzes. Dies ist in der BRD in vielen Gebieten gegeben, so daß entsprechende Umsetzungsmöglichkeiten bestehen. Einschränkungen in der Nutzung des Binnenwasserstraßensystems (Eisgang, Wassertiefe) müssen berücksichtigt werden, wobei diese keinen grundsätzlichen Hinderungsgrund für eine Realisierung darstellen. Im Fall einer temporären Nichtverfügbarkeit der Infrastruktur existieren Rückfallebenen, da in einer solchen Situation auf monomodale Abläufe, d.h. die ausschließliche Nutzung des Straßen- bzw. Schienengüterverkehrs zurückgegriffen werden kann.

3.1 Anwendungsbeispiel im Güternahverkehr

Für den Güternahverkehr (Nah- und Regionalbereich) ergibt sich u.a. eine Anwendungsmöglichkeit bei einer Versorgung innerstädtischer Gebiete über peripher gelegene Güterverkehrseinrichtungen (s. Daduna 2000), z. B. einem Güterverkehrszentrum (GVZ). Innerhalb konventioneller Strukturen erfolgt der Einsatz der Lieferfahrzeuge im GVZ normalerweise in den frühen Morgenstunden, wobei häufiger durch die ebenfalls stadteinwärts fließenden (Pendler-)Verkehre temporäre Behinderungen durch Staus auftreten, die zu Störungen in den Abläufen führen können. Vermeiden lassen sich diese durch einen bimodalen Ansatz, bei dem die Lieferfahrzeuge abends bzw. in den Nachtstunden geladen und anschließend per Binnenschiff zu einem innerstädtischen Übergabepunkt transportiert werden. Dort erfolgt die Übernahme der Fahrzeuge durch die Fahrer / Fahrzeugbesatzungen, die dann unmittelbar mit den Liefervorgängen beginnen. Der Rücklauf der Fahrzeuge wird analog abgewickelt. Um effiziente Ablaufstrukturen zu erreichen, ist eine geeignete Strukturierung der Auslieferungszonen vorzunehmen. Hierbei kann bei (zum Beispiel täglich) variierenden Kundenstandorten unter Verwendung (aggregierter) Nachfragepunkte auf *binäre* bzw. *Single Source Warehouse Location-Modelle* (s. u.a. Klose 2001, S. 185ff) zurückgegriffen werden, während bei bekannten (deterministischen) Kundenstandorten *Location-Routing-Modelle* (s. u.a. Klose 2001, S.293ff) zur Anwendung kommen können.

Die Vorteile liegen in den Kosteneinsparungen, die sich durch die Fahrzeugbündelung ergeben, sowie in der Einsparung der anstehenden LKW-Gebühren. Außerdem läßt sich die Bedienungsqualität (u.a. Pünktlichkeit bei den Belieferungen) verbessern, da kritische Streckenabschnitte umgangen werden und die Kundenstandorte näher am aktuellen Einsatzpunkt der Fahrzeuge liegen.

3.2 Anwendungsbeispiel im Güterfernverkehr

Im Güterfernverkehr kann mit diesem Konzept u.a. eine effiziente Verteilung von aus Übersee eingehenden Containern bei stark dislozierten Zielorten durchgeführt werden. Ausgehend von einem Seehafencontainerterminal erfolgt auf der Basis von Sattelaufliegern ein gebündelter Weitertransport zunächst über das Binnenwasserstraßennetz, während die Feinverteilung im Straßengüterverkehrläuft, der an den Landestellen einsetzt. Da ablaufbedingt und auch aus wirtschaftlichen Überlegungen nicht jede der verfügbaren Rampen genutzt werden kann, sind entsprechende Standortplanungen durchzuführen. Diese lassen sich auf *Hub Location-Probleme* (s. u.a. Campbell et al. 2002) zurückführen, wobei die potentiellen Standorte sich in der Regel auf einer Linie liegen. Analog lassen sich auch Sammelabläufe gestalten, z. B. in der Zuführung zu Seehafencontainerterminals. Außerdem besteht die Möglichkeit, Pick-up and Delivery-Verkehre einzurichten, wobei aufgrund der Komplexität der Abläufe ein effizientes Informationsmanagement und der Einsatz geeigneter Planungstools zwingend erforderlich wird.

Aufgrund der geringen Transportgeschwindigkeit eignet sich dieser Systemansatz nur bei Abläufen, die nicht zeitkritisch sind. Somit bilden Massengüterverkehre (u.a. auch in der Entsorgungswirtschaft) sowie Transporte bei ausreichender zeitlicher Planbarkeit die wesentlichen Potentiale. Die ökonomischen Vorteile liegen in den Kosteneinsparungen für Personal und Kraftstoffe sowie in der Vermeidung von Straßennutzungsgebühren. Ein weiterer Vorteil ist in der Erhöhung des zulässigen Fahrzeuggesamtgewichts bei LKW-Transporten im KLV, wodurch sich Kapazitätsausweitungen von ca.15% realisieren lassen. Außerdem ist eine Nutzung bei nicht zeitkritischen Fahrzeugrückläufen möglich, so z. B. für eine wirtschaftliche Lösung bei der Leerwagenbereitstellung im Schienengüterverkehr.

4 Zusammenfassung und Ausblick

Die beiden Anwendungsbeispiele machen deutlich, daß ein derartiges Systemkonzept eine sinnvolle Ergänzung im (bi- bzw. multimodalen) KLV darstellt. Es müssen aber auch die Grenzen gesehen werden, die sich aus der Netzstruktur und deren kapazitiver Verfügbarkeit ergeben sowie auch aus den Anforderungen von Verladern und Kunden. Punkt-zu-Punkt-Verbindungen sind hierbei weniger problematisch, da auf den entsprechenden Relationen lediglich die notwendigen (technischen) Voraussetzungen gegeben sein müssen, die explizit überprüfbar sind. Bei flächenorientierten Problemstellungen werden nicht alle Standorte integrierbar sein, sondern nur solche, die in einem Korridor entlang der zu nutzendenBinnenwasserstraßen liegen. Außerdem lassen sich in dieser Form keine zeitkritischen Transporte abwickeln, da die Transportgeschwindigkeit dieses Verkehrsträgers im

Vergleich zum Straßengütertransport geringer ist. Einschränkungen in der Netzverfügbarkeit aufgrund von Witterungseinflüssen (Eisgang, Hoch- und Niedrigwasser) können auftreten, ohne allerdings ein grundsätzliches Systemproblem darzustellen. Kommt es zu diesen Einschränkungen, ist in soweit eine leistungsfähige Rückfallebene gegeben, da der LKW letztendlich ein geschlossenes monomodales System bildet, das autark operieren kann.

Eine detaillierte Kostenabschätzung für derartige Konzepte kann nur fallbezogen erfolgen, u.a. bedingt durch die Fahrzeugauslegung und den u. U. anfallenden Investitionsaufwand für die Qualifizierung vorgesehener Landestellen. Außerdem besteht bei den Kalkulationen derzeit noch ein Unsicherheitsfaktor, da, wie die derzeitigen Planungen vorsehen, ein Einstieg in diese neue Technologie mit den in den *Richtlinien zur Förderung neuer Kombinierter Verkehre auf Schiene und Wasserstraße* vorgesehen Bezuschussungen investiver Maßnahmen und operativer Kosten durch das Bundesministerium für Verkehr, Bau- und Wohnungswesen gezielt unterstützt werden soll. In der Grundtendenz sind positive Ergebnisse zu erwarten, auch ohne eine Einbeziehung monetarisierter Auswirkungen auf den Modal split im Güterverkehr durch eine Realisierung von Verlagerungseffekten.

Literaturverzeichnis:

Campbell, J.F. / Ernst, A. / Krishnamoorthy, M. (2002): Hub Location Problems. in: Drezner, Z. / Hamacher, H.W. (eds.): Facility Location - Applications and Theory. Berlin et al., 373 - 407

Daduna, J.R. (2000): Logistische Strukturen und Prozesse. in: Ammann, P. / Daduna, J.R. / Schmid, G. / Winkelmann, P.: Distributions- und Verkaufspolitik. Köln, 281 - 310

Daduna, J.R. (2001): Planung und Steuerung im Straßengütertransport unter dem Einfluß von Informationstechnologien. in: Sebastian, H.-J. / Grünert, T. (Hrsg.): Logistik-Mangement. Stuttgart et al., 257 - 269

Daduna, J.R. / Voß, S. (2000): Informationsmanagement im Verkehr. in: Daduna, J.R. / Voß, S. (Hrsg.): Informationsmanagement im Verkehr. Heidelberg, 1 – 21

Klose, A. (2001): Standortplanung in distributiven Systemen. Heidelberg

Seidelmann, C. (1997): Der Kombinierte Verkehr - Ein Überblick. in: Internationales Verkehrswesen 49, 321 - 324

Stackelberg, F. von (1998): Die Binnenschiffahrt im Kombinierten Ladungsverkehr. in: Hartwig, K.-H. (Hrsg.): Kombinierter Verkehr. Göttingen, 93 - 122

Optimal Routing of Snowplows-A Column Generation Approach

Nima Golbaharan, Per Olov Lindberg, and Maud Göthe Lundgren

Linköpings Universitet, Department of Mathematics, SE-581 83 Linköping, Sweden, e-mails: nigol,polin, mabre @mai.liu.se

Abstract. In countries with heavy winters, winter road maintenance, i.e. snow removal, salting etc, is an important problem. In Sweden the government and municipalities together spend close to 0.3 GEUR every year for winter road maintenance. Approximately half of this is snow removal cost, which mainly is the cost for snowplows, which in turn mainly depends on the routing of the snowplows. In this paper we study optimal routing of snowplows after snowfall. One has then to design a set of routes starting and ending at given depots, such that each road segment gets plowed within a prescribed time window, depending on the class of road segment. Our solution approach is based on Dantzig-Wolfe decomposition with column generation. In order to obtain an integer solution to the master problem we have applied two heuristic procedures, one using branch-and-bound on a subset of the columns and the other one a greedy procedure.

1 Introduction

In this paper we consider a special case of the problem of identifying routes with associated schedules for snowplows in order to minimize the total cost of the snow removal operation. The problem can roughly be decomposed into two classes; routing (and scheduling) during snowfall and routing after snowfall. We are interested in modeling the problem for the after snowfall case and finding a solution to this problem that fulfills the constraints defined by the Road Administration Authority. The solution approach is based on Dantzig-Wolf decomposition in which the variables are generated by a column generation scheme. This yields a fractional solution with an objective value that is a lower bound to the total routing cost.

Despite the importance of snow removal, gritting and salt spreading, related work on the subject has been limited. In 1984 Lemieux and Campagna [1] solve a single-depot snow removal problem, which is stated as plowing a set of streets on both sides, once in each direction, while respecting the priorities of the road segments. Cook and Alprin [2] develop a dynamic routing heuristic, Closest Street Heuristic, for snow and ice removal in an urban environment. Eglese [3] presents a solution method for solving a routing problem for winter gritting in a local authority area. Campbell and Langevin [4] describe urban snow removal and disposal operations in detail. They write as conclusion, *"Because of the inherent complexity of the problems, an interactive decision support system would be most useful to allow solutions to be adjusted to incorporate political and difficult to quantify factors. Hence, fast heuristic algorithms that produce good approximate solutions will be preferred over optimal algorithms that require excessive computer time or power".*

2 Problem Description and Mathematical Formulation

2.1 The Snow Removal Routing Problem After Snowfall

The snow removal routing problem after snowfall, SRRPAS, is the mathematical problem of routing a set of snowplows after snowfall, so that every road segment is plowed exactly once and the total cost of plowing is minimized. In practice, there are additional restrictions on the solution that have to be considered. Usually, a time window is associated to each road segment in the road network, which gives rise to a more restricted version of SRRPAS referred to as the *snow removal routing problem with time windows after snowfall*, SRRPASTW. That is, the problem consists of designing a set of routes for snowplows at a minimal cost and such that all road segments that are to be plowed are covered within specified time limits.

2.2 Mathematical Formulation of the SRRPASTW

We formulate SRRPTWAS from an operator's point of view, which means that the goal is to minimize the overall operational cost. Let $\mathcal{G} = (\mathcal{N}, \mathcal{E})$ be a network with node set $\mathcal{N}$ and arc set $\mathcal{E}$. Each edge $e = (i, j) \in \mathcal{E}$, has an associated set of data indicating plowing cost, plowing time, transportation cost and transportation time. A route is a path in the network, starting and ending at a depot, together with an indication of whether the edges in the path are plowed or just used for transport. Let $\mathcal{D}$ be the set of depots in the network, $\mathcal{R}_d$ the set of all possible routes originating at depot $d \in \mathcal{D}$, $\mathcal{R} = \cup \mathcal{R}_d$ and $\mathcal{E}_p \subseteq \mathcal{E}$ the set of road segments that need to be plowed. Let v_d be the number of snowplows available at depot d, w_d a charge for using one extra snowplow in addition to those available at depot d, and c_r the operating cost of route $r \in \mathcal{R}$. Further, let a_{er} be equal to 1 if edge $e \in \mathcal{E}_p$ is plowed in route r, 0 otherwise. The problem is then formulated as follows:

$$[\mathbf{P}] \quad \min_{x,y} \sum_{r \in \mathcal{R}} c_r x_r + \sum_{d \in \mathcal{D}} w_d y_d \qquad (2.1a)$$

s.t.

$$\sum_{r \in \mathcal{R}} a_{er} x_r \geq 1 \qquad \forall e \in \mathcal{E}_p, \quad (2.1b)$$

$$\sum_{r \in \mathcal{R}_d} x_r \leq v_d + y_d \qquad \forall d \in \mathcal{D}, \quad (2.1c)$$

$$x_r \in \{0, 1\} \qquad \forall r \in \mathcal{R}, \quad (2.1d)$$

$$y_d \geq 0, \text{integer}, \forall d \in \mathcal{D}. \quad (2.1e)$$

Here, x_r indicates whether route r is in the solution or not, whereas y_d is the number of extra snowplows used at depot d. The first term of (2.1*a*) is the total cost of routes in the solution and the second one is the charge for extra snowplows. Constraint (2.1*b*) says that every edge in $\mathcal{E}_p$ must be plowed at least once. It is correct to allow overcovering of edges, since if two routes plow the same edge, one of them may be modified to transport on the edge, giving a solution with a lower cost. Constraint (2.1*c*) is the depot fleet size constraint. The problem **[P]** is an extension of a set covering problem, which is a well-known problem in the class of *NP-hard* problems, Papadimitriou et al. [5]. Thus the SRRPTWAS is a hard problem.

3 Application of Dantzig-Wolfe Decomposition to SRRPASTW

3.1 The Master Problem

Relaxing the integrality requirements we obtain $0 \leq y_d \in R$ and $0 \leq x_r \leq 1$, which can be further relaxed to $x_r \geq 0$. To simplify the discussion of the solution algorithm, we will proceed by ignoring the variables y_d, since these do not affect the covering constraints. Lagrangian relaxation of (2.1*b*) and (2.1*c*) by multipliers $\lambda_e \geq 0$ and $\mu_d \geq 0$ gives the relaxed problem

$$[\mathbf{P}_{\mathbf{R}}^{\lambda,\mu}] \; \min_{x \geq 0} \sum_{r \in \mathcal{R}} c_r x_r + \sum_{e \in \mathcal{E}_p} \lambda_e \left(1 - \sum_{r \in \mathcal{R}} a_{er} x_r\right) + \sum_{d \in \mathcal{D}} \mu_d \left(\sum_{r \in \mathcal{R}_d} x_r - v_d\right).$$

The coefficient c_r consist of two parts, c_r^P and c_r^T, the total cost of plowing and the total transport cost of route r, respectively. The reduced cost of x_r is then $\bar{c}_r(\lambda) = c_r^P + c_r^T - \sum_{e \in \mathcal{E}_p} \lambda_e a_{er} + \mu_d$. The term $c_r^P - \sum_{e \in \mathcal{E}_p} \lambda_e a_{er} = \sum_{e \in \mathcal{E}_P} (c_e^P - \lambda_e) a_{er}$ implicitly defines the reduced plowing cost for an edge e, where c_e^P is the plowing cost of an edge e.

3.2 The Subproblems

The problem of identifying the column (route) with the most negative reduced cost, can be seen as a ***shortest path problem with time windows***, SPPTW. To solve a subproblem, one for each depot, we consider a so-called label-setting algorithm.

Labeling: Consider an arbitrary road segment that is represented by an edge $e = (i, j)$. Assume that plowing the road segment takes t_e^P time units at the cost of c_e^P. Transporting on e causes the cost c_e^T and the transportation time t_e^T with $t_e^P \geq t_e^T$ and $c_{ij}^P \geq c_e^T$. Further, there is a time window $[0, b_e]$ specified for edge e. Every label at a node $i \in \mathcal{N}$ corresponds to a path from the given depot to i and is a triplet (c, t, p), where c is the cost of the corresponding path; t is the time spent along the path and p points at the label that precedes the label in the path. Moving from node i, with label $L = (c_i, t_i, p_i)$, gives rise to two different labels, defined below, at each neighbor j accessible from node i.

$$L^{Plow}: \begin{cases} t'_j &= t_i + t^P_{ij} \\ c'_j &= c_i + c^P_{ij} - \lambda_{ij} \\ p'_j &= L \end{cases} \quad L^{Transport}: \begin{cases} t''_j &= t_i + t^T_{ij} \\ c''_j &= c_i + c^T_{ij} \\ p''_j &= L \end{cases}$$

Creating labels in this manner, we conclude that the total number of labels grows rapidly as the number of treated nodes increases, since a label at a node generates two labels at each of its neighbors.

Cycle Avoidance: If $\bar{c}_e = c^P_e - \lambda_e < 0$, then plowing the edge e is advantageous. During the course of the solution process we may thus create a negative cycle by plowing all or some of the edges in a cycle. We have applied a simple procedure with a constant time complexity, to forbid negative cycles. We define a matrix M, where the columns represent edges and the rows represent labels. An entry of M with value 1 indicates that the corresponding edge has been plowed earlier in the route corresponding to the label. Before starting to plow an edge e, using label ℓ, the question of whether the edge is already plowed in the path or not, is answered by checking the (ℓ, e) entry of the matrix. This is done in $O(1)$ time.

Time Discretization: Since the number of labels increases exponentially, problems involving wide time windows often require some special treatment. To simplify the problem, all time values in the problem have been discretized. By doing so, we reduce the number of labels during the column generation procedure. The discretization is done by multiplying all time values by some scalar and rounding up to the nearest integer. The value of the chosen scalar determines the degree of discretization.

To avoid unnecessary labels, some **label dominance criteria** are utilized. Eliminating dominated labels may affect performance of the application dramatically. A label is dominated and can be deleted from further consideration whenever any other label at the same node has better time and cost.

3.3 Finding an Integer Solution

Branch-and-Bound: One way of finding an integer solution is to use a branch-and-bound procedure. However, since branch-and-bound has exponential worst-case performance, it may take unreasonably long time to find and verify an optimal integer solution. We have applied a variable reduction technique, as described below, to speed up the branch-and-bound process.

Variable Reduction: Let $T \in (0, 1)$ be a choosen reduction threshold. Given a master problem with a large pool of columns, U, we select a variable x_r with current value $\bar{x}_r \leq T$, set $\bar{x}_r = 0$ and resolve the master problem. This procedure is repeated untill there is no variable $k \in U$ with $\bar{x}_k \leq T$. Experience and knowledge about the fractional solution obtained from the column generation procedure can be used to choose a proper value for T. After the variable reduction procedure, we solve the remaining integer problem with branch-and-bound.

A Greedy Approach: We have tried a greedy search procedure similar to those used for set covering problems. We successively select columns according to criteria such as $j^* = argmin_j \frac{c_j + w_d}{\sum_i a_{ij}}$. If j^* is selected, the edges plowed by j^* are changed to transport edges in the unselected routes and their costs are changed accordingly. x_{j^*} is then set to 1 and the column j^* is removed from further consideration. This process is repeated untill all edges are covered .

Post Processing: To improve the integer solution obtained from the greedy procedure we have developed a post processing procedure. Assume that $S = T \cup P$ is the set of road segments between nodes n_1 and n_2 in a given route, with P beeing plowed segments and T those used for transport. Applying the greedy procedure may render $P = \emptyset$, which means that a new transport stretch between node n_1 and n_2 is created. The post processing replaces this stretch by the shortest path between n_1 and n_2. These modifications actually result in generation of new routes, which yield an improved integer solution.

4 A Case Study and Computational Results

In order to study the performance and the quality of the solutions generated by the decomposition scheme presented in Section 3, we have performed a number of tests on data representing the Operation District of Eskilstuna, ODE, which is located in the middle of Sweden. The network representing ODE has 814 edges and 363 nodes of which 7 are depots. According to operative standards set forth by the Swedish National Road Administration, road segments are divided into four *standard classes* with different time windows after snowfall when the road segment should be plowed. In ODE the time windows are $[0, k]$ with $k = 4, 6$ or 8 hours. Tables 1 and 2 give some computational results. In the tests, we first generated a varying number of columns, and then applied the greedy or the branch-and-bound heuristic.

4.1 Computational Results

The first and second columns of both tables give the number of generated columns and the cpu-time (in seconds) for generating them, and the third column contains the corresponding objective value. The forth and fifth columns display the integer objective value and the additional cpu-time to obtain an integer solution. The greedy approach is executed on the complete set of columns obtained from the column generation phase, but branch-and-bound on very limited numbers of columns (indicated by column *#Col*) after the variable reduction procedure. The column Z_{post} displays the improved objective value obtained by the greedy procedure after post processing. Table

Table 1. Results of greedy approach

Problem	CPU CG	Z_{lp}	Z_{greedy}	CPU INT	Z_{post}	#SP
ODE2000	718	51 920	68 237.0	32	59 811.5	24
ODE5000	4 141	48 788.5	66 233.8	79	57 615.8	23
ODE8000	10161	47 648.7	71 290.9	136	61 652.9	25
ODE10000	17 046	47 157.9	64 442.4	156	55 578.4	22
ODE15000	35 936	46 192.9	70 244.5	253	60 652.9	24

Table 2. Results of branch-and-bound

Problem	CPU CG	Z_{lp}	Z_{BB}	CPU INT	#Col	#SP
ODE2000	718	51 920	58 581.8	14 714	143	25
ODE5000	4 141	48 788.5	74 103.5	72 268	259	24
ODE8000	10161	47 648.7	57 312.7	6 262	277	24
ODE10000	17 046	47 157.9	56 846.0	12 032	287	23
ODE15000	35 936	46 192.9	56 833.0	28 605	310	24

1 also shows that the greedy approach is both fast and gives good solutions, in particular when post processing is used. A critical point for the greedy approach is the choice of the rule of selecting the columns and for the branch-and-bound approach the right value of the threshold when reducing the number of variables. The last column (indicated by *#SP*) in both Tables 1 and 2 contains the number of snowplows used in the solution.

References

1. Lemieux, P. F., Campagna L. (1984) The Snow Plowing Problem Solved by a Graph Theory Algorithm. Civ. Eng. Sys. **1**: 337-341
2. Cook, T. M., Alprin B. S. (1976) Snow and Ice Removal in an Urban Environment. Man. Sci. **23**: 227-234
3. Eglese, R. W. (1994) Routing Winter Gritting Vehicles. Dis. App. Math. **48**: 231-244
4. Campbell J. F., Langevin A. (1995) The Snow Disposal Assignment Problem. Jour. of the Oper. Res. Soc. **46**: 919-929
5. Papadimitriou C. H., Steiglitz K. (1998) Combinatorial Optimization, Algorithms and Complexity. Dover Publications Inc., New York

Savings Based Ants for Large-scale Vehicle Routing Problems

Marc Reimann and Karl Doerner

University of Vienna, Department of Management Science, Bruenner Strasse 72, A-1210 Vienna, Austria

Abstract. In this paper we present a modified version of our Savings based Ant System for large-scale instances of the Vehicle Routing Problem (VRP). The main idea is to speed up the search by letting the ants solve only sub-problems rather than the whole problem. This is particularly necessary, when one has to solve large real world instances, for which the computation times of classic meta-heuristics are prohibitive. Our model is also inspired by Taillard's work on the VRP ([1]), where the Tabu Search procedure employed solves only some sub-problems.

1 Introduction

The VRP, being one of the central problems in distribution logistics, consists of finding a set of minimum cost vehicle routes that start and end at a central depot, such that each customer is fully served by exactly one vehicle. The fleet is homogeneous and each vehicle can carry goods up to its capacity. Additionally constraints on the maximum possible duration of each vehicle's tour may exist. This problem is a generalization of the TSP. Thus, besides the routing aspect already existing in the TSP one has to find an assignment (or clustering) of customers to vehicles.

In Doerner et al. ([2]) and Reimann et al. ([3]) we have proposed the Savings based Ant System for the VRP. In these papers we have shown the effectiveness of the approach for the standard benchmark problem instances. However, we observed from our simulations that the efficiency in terms of the scaling behavior of the algorithm was poor. More specifically the Savings based Ant System took around three seconds to solve a problem with 50 customers, whereas the solution of a 199 customer problem required up to 40 minutes. This scaling behavior is prohibitive when the Ant System is to be applied to real world instances (with up to many hundreds of customers). Note however that the same argument applies for basically all other meta-heuristics, such that different authors propose various methods to speed up the search (e.g. [4]). Further, we also observed that the average solution quality obtained by our Ant System increased from less than 0.1% for instances with 50 customers to almost 2% deviation from the best known solution for instances with 199 customers (c.f. [3]).

Apart from these algorithmic observations we learned from our industry partners that human dispatchers tend to cluster customers according to postal codes or other regional characteristics before they solve the much smaller sub-problems for each cluster separately.

These three issues led to the development of our new algorithm. The main idea is to split up large problems into a number of smaller problems that can be solved both more effectively and more efficiently. Note however, that the splitting procedure as well as the recombination of the sub-problems are non-trivial problems themselves and we will focus on these issues in this paper. Note also, that a similar approach was used in combination with a Tabu Search algorithm by Taillard ([1]).

The outline of the remainder of this paper is as follows. In the next section we discuss problem decomposition for VRPs. First we start with a brief review of Taillard's algorithm for the VRP. After that we focus on the description of our new approach. We will present first, very preliminary results in section 3 before we conclude with an outlook on future research.

2 Decomposing the VRP

2.1 Taillard's Decomposition Algorithm

Taillard ([1]) realized that good solutions to most VRP instances of moderate to large size feature some spatial characteristics that allow to exploit problem decomposition. In terms of Tabu Search, or more generally Local Search it is obvious that (local) changes only make sense if the nodes involved are close to each other. Moreover, simple change operators generally involve only two tours. Thus, moves can be evaluated and performed simultaneously.

Taillard proposes two distinct methods for partitioning a problem instance. For uniform problems, a partition into sectors is suggested, while for non-uniform problems Taillard uses a partitioning method based on arborescences and associated shortest path. The core of the algorithm consists of the iteration of the following steps. First, the problem is partitioned with the appropriate method (as mentioned above). Each partition is then solved using a standard Tabu Search approach. After a certain number of iterations the sub-problems are re-joined.

Using these methods, Taillard was able to find all the best known solutions for the classic benchmark instances for the VRP. However, the algorithm's performance depended crucially on the initial decomposition and the partitioning method was problem dependent.

2.2 Problem Decomposition and the Savings Based Ant System

Let us now turn to the approach we propose to efficiently solve large-scale VRPs. Initially an Ant System solves the master problem for a given number of iterations (step I). Given the best found solution so far our algorithm determines for each route of this solution the center of gravity according to the modified Miehle algorithm ([5]) (step II) . We then cluster these route centers using the Sweep algorithm as proposed by Gillett and Miller ([6]) (step III). Each of the resulting clusters is then solved independently by applying our Savings based Ant System for a given number of iterations (step IV). After all sub-problems have been solved we re-assemble the global solution and update the global pheromone information and if applicable the global best solution (step V). The steps I to V described above are repeated until a pre-specified time limit is reached.

Let us now turn to each of the five steps in more detail, where we will also discuss implementation issues.

- Step I: Generation of Initial Solutions
  The initial solutions are generated by applying the Savings based Ant System as described and analyzed in detail in Doerner et al. ([2]) and Reimann et al. ([3]).
- Step II: The Modified Miehle Algorithm
  The modified Miehle algorithm is used to compute the center of gravity for each vehicle route. This algorithm iteratively adjusts the coordinates of the center, until the change in the weighted distance of all customers to the center is minimal. The term *modified* in the name denotes that the algorithm can handle a situation where the center of gravity coincides with the location of a customer.
- Step III: The Sweep Algorithm
  Having computed the centers of gravity for each vehicle route, we then ignore the actual customer locations and apply the Sweep algorithm to cluster the nodes corresponding to the centers of gravity. More specifically, a starting node is chosen randomly and the remaining nodes are sorted according to their polar angle with the depot and the randomly chosen starting node. Using the resulting order, the nodes are then assigned to clusters.
- Step IV: Solution of the Sub-Problems
  Sub-problems are solved using the Savings based Ant System already utilized in Step I. However, in this step each Ant System solves only the partial problem it is assigned in Step III.
- Step V: Communication between Processes
  The main notion with respect to the communication between the different processes is *master pheromone information*. In fact, we use one global memory for our algorithm.
  The communication between the sub-problems and the master process is based on two important components. First, each sub-problem is called with the associated part of the master pheromone information. In other words, each sub-problem receives only those parts of the master pheromone information necessary to solve its part of the problem. The sub-problems then change this pheromone information *locally* as they iteratively solve their instances.

Second, after a sub-problem has been solved, the corresponding process returns the best found solution and compares it with the previous best found solution. If an improvement was achieved, the best found solution is updated and pheromone reinforcement in the master pheromone information occurs. Otherwise, only negative reinforcement in terms of pheromone evaporation is applied to the master pheromone information.

3 Computational Experiments

The problem instances used for our computational study are comprised of the 20 large scale instances proposed by Golden et al. ([7]). All of the instances feature one central depot, the customers are located either in concentric circles or squares around the depot. The first 8 instances are distance and capacity constrained, whereas the last 12 instances are only capacity constrained. Service times are equal for all customers and set to zero. The sizes of the instances vary between 200 and 483 customers.

In order to keep things simple we chose to stick to the parameter values that were found to be favorable by Bullnheimer et al. ([8]) for the rank based Ant System in general, and in one of our earlier works (c.f. [3]) for the Savings based Ant System in particular.

Nonetheless, our new approach features a significant number of parameters that need to be adjusted. These are:

- it_{master}...number of master problem iterations
- it_{sub}...number of subproblem iterations
- t_{tot}...total admissible runtime
- n_s...number of sub-problems

For the results presented below we have used the following settings: $it_{master} = 2$, $it_{sub} = 100$. The runtime was chosen to reflect the differences in the problem sizes and varied between $t_{tot} = 10$ and $t_{tot} = 20$ minutes. The actual numbers are given in Table 3 below. Finally, an important parameter is the number of sub-problems. For the computational study presented below we have partitioned each problem instance into $n_s = 4$ sub-problems.

Table 3 summarizes our computational results. Listed are our Ant System (denoted by AS), the GA of Prins ([9]) and the Granular Tabu Search (GTS) approach of Toth and Vigo ([4]). While the results for the GA and for the GTS are best results given a fixed (the most favourable?) parameter constellation, we report average results over 5 runs for our Ant System. The computers used are a 200 MHz Pentium for the GTS, a 1 GHz Pentium for the GA and a 900 MHz Pentium for our Ant System.

From the table two things become obvious. For these large scale instances the trade-off between solution quality and computational effort is significant. More specifically, this can be seen from the last row where we report the average relative percentage deviation (RPD) from the best known solution

Table 1. Comparison of three meta-heuristics on large scale problem instances

Instance #	size	Best known solutions	AS Sol.qual.	AS min.	GA Sol.qual	GA min.	GTS Sol.qual	GTS min.
1	240	5646.43	5734.02	10.00	5648.04	32.42	5736.15	4,98
2	320	8447.92	8629.60	15.00	8459.73	77.92	8553.03	8.28
3	400	11036.22	11278.88	20.00	11036.22	120.83	11402.75	12.94
4	480	13624.52	14192.10	20.00	13728.8	187.60	14910.62	15.13
5	200	6460.98	6460.98	10.00	6460.98	1.04	6697.53	2.38
6	280	8412.80	8507.82	10.00	8412.9	9.97	8963.32	4.65
7	360	10195.59	10445.15	15.00	10267.5	39.05	10547.44	11.66
8	440	11828.78	12158.22	20.00	11865.4	88.3	12036.24	11.08
9	255	587.09	599.32	10.00	596.89	14.32	593.35	11.67
10	323	746.56	765.32	15.00	751.41	36.58	751.66	15.83
11	399	932.68	960.20	15.00	939.74	78.5	936.04	33.12
12	483	1133.79	1167.86	20.00	1152.88	30.87	1147.14	42.9
13	252	868.80	879.16	10.00	877.71	15.3	868.8	11.43
14	320	1086.24	1111.13	15.00	1089.93	34.07	1096.18	14.51
15	396	1363.34	1386.91	15.00	1371.61	110.48	1369.44	18.45
16	480	1650.42	1683.62	20.00	1650.94	130.97	1652.32	23.07
17	240	709.9	712.50	10.00	717.09	5.86	711.07	14.29
18	300	1014.80	1015.66	10.00	1018.74	39.33	1016.83	21.45
19	360	1376.49	1381.92	15.00	1385.6	74.25	1400.96	30.06
20	420	1846.55	1853.29	20.00	1846.55	210.42	1915.83	43.05
Averages			RPD (%)	CPU	RPD (%)	CPU	RPD (%)	CPU
			1.23	14.75	0.56	66.6	2.1	17.54

together with the average computation times for all approaches. Our Ant System consumes approximately a fifth of the computation times reported by Prins, while we still take four times as much runtime as Toth and Vigo (given the differences in the machines used). On the other hand, the GA on average performs best, followed by our approach. With respect to solution quality Toth and Vigo's approach comes last. Thus, none of the three approaches dominates or is dominated by any other approach when assessed by these two goals. However, our algorithm seems to find a nice balance between the two goals, whereas the competing approaches can somehow be viewed as the polar cases.

Analyzing the worst case behavior of our algorithm we observed that on average the solutions found are within 2.3 % of the best known solutions. This worst case deviation is very promising if we consider that the implementation of our approach so far resembles merely a proof of concept.

Finally, let us note that during the course of our experiments we were able to improve 2 of the best known solutions. For problem instance #18 we found 1012.35 and for problem instance #20 we found 1843.62.

4 Conclusions

In this paper we have presented an approach to enhance the efficiency of an Ant System for the VRP. The main idea is to partition a starting solution to the problem into several disjoint sub-problems, each of which is then solved using an Ant System process. We have shown that the algorithm is competitive when compared with other state-of-the-art techniques.

Further analysis should deal with a comparison of the results of our new approach with those of the Savings based Ant System as described in our earlier work. Apart from that more research is required to better understand the effects of communication between the master process and the sub-problems.

The main advantage of our approach is that it can be used in the same way for any kind of vehicle routing problem. Furthermore, the same idea can be used in dynamic, real-time vehicle routing problems to determine whether a new request can be accustomed or not, by optimizing only the subset of tours serving the area where the new request comes from.

Acknowledgments: Financial support from the Oesterreichische Nationalbank (OeNB) under grant #8630 is gratefully acknowledged.

References

1. Taillard, E. D. (1993): Parallel iterative search methods for vehicle routing problems. Networks **23** 661–673.
2. Doerner, K., Gronalt, M., Hartl, R. F., Reimann, M., Strauss, C. and Stummer, M. (2002): SavingsAnts for the Vehicle Routing Problem. in: Cagnoni, S. et al. (eds.): Applications of Evolutionary Computing, Springer, Berlin, 11–20.
3. Reimann, M., Stummer, M. and Doerner, K. (2002): A Savings based Ant System for the Vehicle Routing Problem. in: Langdon, W. B. et al. (eds.): Proceedings of the Genetic and Evolutionary Computation Conference (GECCO 2002), Morgan Kaufmann, San Francisco, 1317–1325.
4. Toth, P. and Vigo, D. (2002): The Granular Tabu Search and its Application to the Vehicle Routing Problem. to appear in: INFORMS Journal on Computing.
5. Spaeth, H. (1977): Cluster-Analyse-Algorithmen zur Objektklassifizierung und Datenreduktion. Oldenbourg, Muenchen.
6. Gillett, B. and Miller, L. (1974): A heuristic algorithm for the vehicle dispatch problem. Operations Research **22** 340–349.
7. Golden, B. L., Wasil, E. A., Kelly, J. P. and Chao, I. M. (1998): The impact of metaheuristics on solving the vehicle routing problem: algorithms, problem sets, and computational results. in: Crainic, T. G. and Laporte, G. (eds.), Fleet management and logistics, Kluwer, Norwell, 33–56.
8. Bullnheimer, B., Hartl, R. F. and Strauss, Ch. (1999): An improved ant system algorithm for the vehicle routing problem. Annals of Operations Research **89** 319–328.
9. Prins, C. (2001): A Simple and Effective Evolutionary Algorithm for the Vehicle Routing Problem. Reseach report, University of Technology of Troyes, France.

Single Machine Scheduling Problems with Exponentially Start Time Dependent Job Processing Times

Aleksander Bachman[1], Adam Janiak[1], and Mikhail Y. Kovalyov[2]

[1] Institute of Engineering Cybernetics, Wroclaw University of Technology, Janiszewskiego 11/17, 50-372 Wroclaw, Poland
[2] Institute of Engineering Cybernetics, National Academy of Sciences of Belarus, Surganova 6, 220012 Minsk, Belarus

Abstract. The paper deals with single machine scheduling problems, where job processing times are given by exponentially start time dependent functions. We consider decreasing and increasing functions of job processing times. For the increasing functions, we proved ordinary and strong NP-hardness of the makespan minimization problem under ready times restrictions. We obtained the same results, i.e., ordinary and strong NP-hardness, for the maximum lateness minimization problem, but for the decreasing functions of job processing times. We proved also ordinary NP-hardness for the total weighted completion time minimization, under the assumption that job processing times are characterized by decreasing functions.

1 Introduction

Scheduling problems investigated in this paper deal with jobs, whose processing times are dependent on the starting moment of their execution. In the scientific literature, such a dependency is called a deterioration of a job processing time. In the past few years, several new models describing the job processing time deterioration have been introduced. Almost all of them are based on the linear functions, which are dependent on the starting moment of job's execution. The recent literature survey for the problems concerning this subject can be found in [1]. The variety of the models, considered in the scientific literature, comes from the applications, where the process can be described by this kind of processing time. The large number of different models describing the job processing time deterioration complicates the indication of the computational complexity for the same optimization criteria. Therefore, we tried to generalize some results concerning the computational complexity of the problems with similar functions characterizing the job processing time for different optimization criteria. The remaining part of the paper is organized as follows. In the next section we formulate the problems considered in the paper. Section 3 deals with the computational complexity results obtained for the makespan, the maximum lateness and the total weighted completion time minimization problems. Section 4 contains some concluding remarks.

2 Formulation of the problems

Each problem studied in this paper can be formulated as follows. There are given a single machine and n independent and non-preemptive jobs available for processing at their ready times $r_i \geq 0$ which should be completed by their due dates d_j. The processing time $p_i(t)$ of job i is a function dependent on the starting moment t of its execution:

- an increasing model

$$p_i(t) = a_i e^{b_i(t-r_i)}, \tag{1}$$

- a decreasing model

$$p_i(t) = a_i e^{-b_i(t-r_i)}, \tag{2}$$

where a_i and b_i denote a normal processing time and an increasing/decreasing rate, respectively. Note that for each job i, the change of its processing time p_i starts at its ready time r_i. For each job i, a weight w_i indicating its relative importance might be given. The objective is to find a job sequence, for which the value of the considered criterion is minimized. In this paper, we investigate the following criteria: the makespan $C_{\max}$, the maximum lateness $L_{\max}$ and the total weighted completion time $\sum w_i C_i$. Adapting the notation $\alpha \mid \beta \mid \gamma$ [3], we denote the above mentioned problems as $1 \mid r_i, p_i(t) = a_i e^{b_i(t-r_i)} \mid C_{\max}$, $1 \mid p_i(t) = a_i e^{-b_i t} \mid L_{\max}$ and $1 \mid p_i(t) = a_i e^{-b_i t} \mid \sum w_i C_i$.

3 NP-hardness results

In this section, we establish the computational complexity results for the following two problems $1 \mid r_i, p_i(t) = a_i e^{b_i(t-r_i)} \mid C_{\max}$ and $1 \mid p_i(t) = a_i e^{-b_i t} \mid L_{\max}$. We prove also that the problem $1 \mid p_i(t) = a_i e^{-b_i t} \mid \sum w_i C_i$ is at least ordinary NP-hard. In order to prove NP-hardness of the problems mentioned above, we use PARTITION and 3-PARTITION [2], respectively. The decision versions of PARTITION and 3-PARTITION, denoted, respectively, by PP and 3PP, are formulated as follows.

PP: Given $m+1$ positive integers $g_1, \ldots, g_m$ and G such that $\sum_{i=1}^{m} g_i = 2G$; is there a partition of the set $\{1, \ldots, m\}$ into 2 disjoint subsets G_1 and G_2 such that $\sum_{i \in G_j} g_i = G$ for $j = 1, 2$?

3PP: Given $3m+1$ positive integers $h_1, \ldots, h_{3m}$ and H such that $\sum_{i=1}^{3m} h_i = mH$ and $H/4 < h_i < H/2$ for $i = 1, \ldots, 3m$; is there a partition of the set $\{1, \ldots, 3m\}$ into m disjoint subsets $H_1, \ldots, H_m$ such that $\sum_{i \in H_j} h_i = H$ for $j = 1, \ldots, m$?

3.1 Makespan

In this subsection, we investigate the problems under the assumption that $w_j = w$ and $d_j = d$ for $j = 1, \ldots, n$.

Theorem 1. *The problem* $1 \mid r_i \in \{0, R\}, p_i(t) = a_i e^{b_i(t-r_i)} \mid C_{\max}$ *is at least NP-hard in the ordinary sense.*

Proof. Given an instance of PP, construct the following instance of our problem. There are $n = m+1$ jobs. Among them, there are m *partition* jobs $1, \ldots, m$, and a single *enforcer* job v with the following parameters:

$$r_i = 0; \qquad a_i = g_i;\ b_i = 0 \quad \text{for } i = 1, \ldots, m,$$
$$r_v = R = G;\ a_v = 1;\ b_v = m.$$

We show now, that PP has a solution **if and only if** there exists a solution for the constructed instance of $1 \mid r_i \in \{0, R\}, p_i(t) = a_i e^{b_i(t-r_i)} \mid C_{\max}$ with the value $C_{\max} \leq y = 2G + 1$.

"Only if". Assume that the subsets G_1 and G_2 give a solution for PP. Let J_1 and J_2 denote the subsets containing the jobs constructed on the basis of the elements from the subsets G_1 and G_2, respectively. Construct a job sequence S such that the machine executes at first the jobs from the subset J_1 in an arbitrary order, then the enforcer job v and finally the jobs from the subset J_2 in an arbitrary order, as well. The completion time $C_{\max}$ of the last job in S is equal to

$$C_{max} = \max\left\{\sum_{i\in G_1} g_i, r_v\right\} + a_v e^{b_v\left(\max\left\{\sum_{i\in G_1} g_i, r_v\right\} - r_v\right)} + \sum_{i \in G_2} g_i. \qquad (3)$$

Note that equation (3) is true for any solution of PP. Since $\sum_{i\in G_1} g_i = \sum_{i \in G_2} g_i = G$, then based on (3) we obtain $C_{max} = 2G + 1$, as it is required.

"If". Assume now that any solution of PP will give $\sum_{i\in G_1} g_i \neq \sum_{i\in G_2} g_i$. Since $\sum_{i=1}^{m} g_i = 2G$, then there are two cases to be considered, namely (a) $\sum_{i\in G_1} g_i = G + \lambda$ and $\sum_{i\in G_2} g_i = G - \lambda$, and (b) $\sum_{i\in G_1} g_i = G - \lambda$ and $\sum_{i\in G_2} g_i = G + \lambda$, where λ is a non-zero integer. Based on (3), we obtain in case (a) $C_{\max} = 2G + e^{m\lambda} > y$ and in case (b) $C_{\max} = 2G + 1 + \lambda > y$, which ends the proof. ∎

Theorem 2. *The problem* $1 \mid r_i, p_i(t) = a_i e^{b_i(t-r_i)} \mid C_{\max}$ *is NP-hard in the strong sense.*

Proof. Given an instance of 3PP, construct the following instance of our scheduling problem. There are $n = 4m$ jobs. Among them there are $3m$ *partition* jobs $1, \ldots, 3m$, and m *enforcer* jobs v_i for $i = 1, \ldots, m$, with the following parameters:

$$r_i = 0; \qquad a_i = h_i;\ b_i = 0; \quad \text{for } i = 1, \ldots, 3m,$$
$$r_{v_i} = iH + (i-1);\ a_{v_i} = 1;\ b_{v_i} = H; \text{ for } i = 1, \ldots, m.$$

It can be shown that 3PP has a solution **if and only if** there exists a solution for the constructed instance of $1 \mid r_i, p_i(t) = a_i e^{b_i(t-r_i)} \mid C_{\max}$ with the value $C_{\max} \leq y = mH + m$. Because of the space limitations, the detailed description of this proof will be omitted. ∎

3.2 Maximum lateness

In this subsection, we investigate the problems under the assumption that $w_i = w$ and $r_i = r = 0$.

Theorem 3. *The problem* $1 \mid p_i(t) = a_i e^{-b_i t}, d_i \in \{d, D\} \mid L_{\max}$ *is at least NP-hard in the ordinary sense.*

Proof. Let's start with the reduction from PP. There are $n = m + 1$ jobs. Among them there are m *partition* jobs $1, \ldots, m$, and a single *enforcer* job v with the following parameters:

$$\begin{array}{llll} d_i = d = 2G + (2e)^{-1}; & a_i = g_i; & b_i = 0; & \text{for } i = 1, \ldots, m, \\ d_v = G + (2e)^{-1}; & a_v = 1/2; & b_v = 1/G. & \end{array}$$

We show now, that PP has a solution **if and only if** there exists a solution for the constructed instance of $1 \mid p_i(t) = a_i e^{-b_i t}, d_i \in \{d, D\} \mid L_{\max}$ with the value $L_{\max} \leq y = 0$.

"Only if". Assume that the subsets G_1 and G_2 give a solution for PP. Let J_1 and J_2 denote the subsets containing the jobs constructed on the basis of the elements from the subsets G_1 and G_2, respectively. A job sequence S is constructed as follows. The machine executes at first the jobs from the subset J_1 in an arbitrary order, then the enforcer job and finally the jobs from the subset J_2 in an arbitrary order, as well. The maximum lateness $L_{\max}$ for the sequence S is equal to

$$L_{\max} = \max \left\{ \sum_{i \in G_1} g_i + a_v e^{-b_v \sum_{i \in G_1} g_i} - d_v, \sum_{i=1}^{m} g_i + a_v e^{-b_v \sum_{i \in G_1} g_i} - d \right\}. \tag{4}$$

Since $\sum_{i \in G_1} g_i = \sum_{i \in G_2} g_i = G$, then based on (4) we have $L_{max} = 0 = y$, as it is required.

"If". Assume now that any solution of PP gives $\sum_{i \in G_1} g_i \neq \sum_{i \in G_2} g_i$. Since $\sum_{i=1}^{m} g_i = 2G$, then there are two cases to be considered, namely (a) $\sum_{i \in G_1} g_i = G - \lambda$ and $\sum_{i \in G_2} g_i = G + \lambda$, and (b) $\sum_{i \in G_1} g_i = G + \lambda$ and $\sum_{i \in G_2} g_i = G - \lambda$, where λ is a non-zero integer. By (4) we have in case (a) $L_{\max} = (2e)^{-1} (e^{\lambda} - 1) > 0 = y$. The result obtained in case (b) is equal to $L_{\max} = 1/2e^{-1-\lambda} - (2e)^{-1} + \lambda > 0$. It means that PP has no solution, which ends the proof. ∎

Theorem 4. *The problem* $1 \mid p_i(t) = a_i e^{-b_i t} \mid L_{\max}$ *is strongly NP-hard.*

Proof. Given an instance of 3PP, construct the following instance of the scheduling problem. There are $n = 4m$ jobs. Among them there are $3m$ *partition* jobs $1, \ldots, 3m$, and m *enforcer* jobs v_i for $i = 1, \ldots, m$, with the following parameters:

$$d_i = d = mH + \tfrac{m-1}{me};\ a_i = h_i; \quad b_i = 0; \qquad \text{for } i = 1, \ldots, 3m,$$
$$d_{v_i} = iH + \tfrac{i}{me}; \qquad a_{v_i} = 1/m;\ b_{v_i} = \left[iH + \tfrac{i-1}{me}\right]^{-1};\ \text{for } i = 1, \ldots, m.$$

We can show that 3PP has a solution **if and only if** there exists a solution for the constructed instance of $1 \mid p_i(t) = a_i e^{-b_i t} \mid L_{\max}$ with the value $L_{\max} \le y = 0$. However, because of space limitation, the detailed proof will be omitted. ■

3.3 Total weighted completion time

We assume that $r_j = r = 0$ and $d_j = d$ are the common values characterizing the jobs in the problem investigated in this subsection.

Theorem 5. *The problem* $1 \mid p_i(t) = a_i e^{-b_i t} \mid \sum w_i C_i$ *is at least NP-hard in the ordinary sense.*

Proof. First a reduction from PP to the scheduling problem is given. Construct an instance of scheduling problem with $n = m+1$ jobs. The parameters of the first m jobs, called the *partition* jobs and the parameters of a single *enforcer* job v are given as follows:

$$w_i = g_i;\ a_i = g_i; \qquad b_i = 0;\ \text{for } i = 1, \ldots, m,$$
$$w_v = G;\ a_v = (2mG)^{-1};\ b_v = 1.$$

We show now, that PP has a solution **if and only if** there exists a solution for the constructed instance of $1 \mid p_i(t) = a_i e^{-b_i t} \mid \sum w_i C_i$ with the value $\sum w_i C_i \le y = \sum_{i=1}^{m} g_i^2 + \sum_{1 \le i < j \le m} g_i g_j + (me^G)^{-1} - G^2$.

"Only if". Assume that the solution of PP is given by two disjoint subsets G_1 and G_2 for which $\sum_{i \in G_1} g_i = \sum_{i \in G_2} g_i = G$. Let J_1 and J_2 denote the subsets containing the jobs constructed on the basis of the elements from the subsets G_1 and G_2, respectively. Let the subsets J_1 and J_2 contain k and $m-k$ jobs $(0 < k < m)$, respectively. If $k = 0$ or $k = m$, then there is no solution for PP. The solution for the considered scheduling problem is given as follows. The machine executes at first the jobs from the subset J_1 in an arbitrary order, then the enforcer job and finally the jobs from the subset J_2 in an arbitrary order, as well. The criterion value for any schedule π described above is given by

$$\sum w_i C_i = \sum_{i=1}^{k} a_{\pi(i)} \sum_{j=i}^{k} w_{\pi(j)} + \sum_{i=k+2}^{n} a_{\pi(i)} \sum_{j=i}^{n} w_{\pi(j)}$$
$$+ a_v e^{-b_v \left(\sum_{i=1}^{k} a_{\pi(i)}\right)} \left[G + \sum_{i=k+2}^{n} w_{\pi(i)}\right], \qquad (5)$$

where the description $\pi(i)$ denotes the job executed in the ith position in the schedule π. Since $a_i = g_i$ and $w_i = g_i$, then expression (5) is equal to:

$$\sum w_i C_i = \sum_{i=1}^{m} g_i^2 + \sum_{1\le i<j\le m} g_i g_j$$

$$+ a_v e^{-b_v\left(\sum_{i=1}^{k} g_i\right)} \left[G + \sum_{i=k+1}^{m} g_i\right] - \left(\sum_{i=1}^{k} g_i\right)\left(\sum_{i=k+1}^{m} g_i\right). \quad (6)$$

Since $\sum_{i\in G_1} g_i = \sum_{i=1}^{k} g_i = G$ and $\sum_{i\in G_2} g_i = \sum_{i=k+1}^{m} g_i = G$, then expression (6) is equal to the required value $y = \sum_{i=1}^{m} g_i^2 + \sum_{1\le i<j\le m} g_i g_j + (me^G)^{-1} - G^2$.

"If". Assume now that any solution of PP represented by two disjoint subsets G_1 and G_2 will give $\sum_{i\in G_1} g_i \neq \sum_{i\in G_2} g_i$. There are two cases to be considered, namely case (a) $\sum_{i\in G_1} g_i = G - \lambda$ and $\sum_{i\in G_2} g_i = G + \lambda$ and case (b) $\sum_{i\in G_1} g_i = G + \lambda$ and $\sum_{i\in G_2} g_i = G - \lambda$, where λ is some positive integer. By (6), in case (a) we have $\sum w_i C_i = y + (me^G)^{-1}\left[e^{\lambda} - 1\right] + \lambda\left[\lambda + (2mG)^{-1}e^{-(G-\lambda)}\right] > y$. The result obtained in case (b) is equal to $\sum w_i C_i = y + (me^G)^{-1}\left[e^{-\lambda} - 1\right] + \lambda\left[\lambda - (2mG)^{-1}e^{-G-\lambda}\right] > y$. Since both results obtained above are greater than y, the theorem is proved. ∎

4 Conclusions

In this paper, we established the computational complexity status of three scheduling problems with exponentially dependent job processing times. We considered the following optimization criteria: the makespan, the maximum lateness and the total weighted completion time. In the NP-hardness proofs, we used the well known PARTITION and 3-PARTITION problems. Since the complexity status is clear now, in the future research we will focus on the construction of the efficient algorithms solving the problems considered in this paper.

References

1. B. Alidaee, N.K. Womer, "Scheduling with time dependent processing times: Review and extensions", *Journal of the Operational Research Society*, 50, 711–720 (1999).
2. M.R. Garey, D.S. Johnson, "Computers and Intractability: A Guide to the Theory of NP-completeness", Freeman, San Francisco, 1979.
3. R.L. Graham, E.L. Lawler, J.K. Lenstra and A.H.G. Rinnooy Kan, "Optimization and approximation in deterministic sequencing and scheduling: a survey", *Annals of Discrete Mathematics*, 5, 287–326 (1979).

Scheduling Problems with Optimal Due Interval Assignment Subject to Some Generalized Criteria

Adam Janiak[1] and Marcin Marek[1]

Institute of Engineering Cybernetics, Wroclaw University of Technology
Janiszewskiego 11/17, 50-372 Wrocaw, Poland

Abstract. The paper deals with two problems of scheduling jobs on identical parallel machines, in which a due interval should be assigned to each job. Due interval is a generalization of well known classical due date and describes a time interval, in which a job should be finished. In the first problem, we have to find a schedule of jobs and a common due interval such that the sum of the total tardiness, the total earliness and due interval parameters is minimized. The second problem is to find a schedule of jobs and an assignment of due interval to all jobs, which minimize the maximum of the following three parts: the maximum tardiness, the maximum earliness and the due interval parameters. We proved that the considered problems are NP-hard and outlined some methods how to solve them approximately as well as optimally.

1 Introduction

In the scientific literature of scheduling theory, an assignment of due dates was considered (see [1], [2], [3]). However group of problems in which to each job a due interval should be assigned are also of permanent interest (see [6], [7], [8]). A due interval is a generalization of the classical due date and describes a time interval, in which a job should be finished. For the considered problems, we should find a schedule of jobs and a common due interval such that the criterion value is minimized. In the first problem we minimize the sum of total tardiness, total earliness and due interval size. In the second problem we minimize the maximum of the following three parts: maximum tardiness, maximum earliness and due interval size. The application of the considered problems can be justified by many manufacturing systems in which the negotiation between the producer and the customer occurs. The negotiation concerns the delivery of the final products. The producer objective is to have the latest time of delivering products, while the customer tries to have them as soon as possible. The compromise of this negotiation is a due interval, i.e., the time period in which the products are completed by producer and are available to be taken by customer.

The reminder of the paper is organized as follows. In next section we formulate problems more precisely. Sections 3 and 4 are, respectively, devoted to properties and the computational complexity of the first problem. In Section

5 we show that both of considered problems are equivalent to each other in terms of optimal solution. We conclude the paper in Section 6.

2 Problems Formulation

There is given a set of n independent and non-preemptive jobs to be scheduled on m identical parallel machines. Each machine can process only one job at a time. We assume that machines execute the jobs without idle times, however, we allow a time gap between the beginning of the optimization process and the start of the execution of the first job on any machine. For each job its processing time p_j and a due interval $\langle d'_j = k_1; d''_j = k_2 \rangle$ are given, where k_1 and k_2 $(k_1 \leq k_2)$ denote common due interval parameters. The first problem (P1) is to find a schedule π (sequence of jobs on each machine) and values of the parameters k_1 and k_2, which minimize the following criterion:

$$f(\pi, k_1, k_2) = \sum_j E_j + (k_2 - k_1) + \sum_j T_j, \tag{1}$$

where $E_j = \max\{d'_j - C_j, 0\}$ - the earliness of job j, $T_j = \max\{0, d''_j - C_j\}$ - the tardiness of job j and C_j is the completion moment of job j.

The second problem (P2) is to find a schedule π and values of the parameters k_1 and k_2, which minimize the following criterion:

$$g(\pi, k_1, k_2) = \max\left\{P_1 \max_j E_j, P_2(k_2 - k_1), P_3 \max_j T_j\right\}, \tag{2}$$

where $P_1 > 0$, $P_2 > 0$ and $P_3 > 0$ are constant weights.

3 Properties of Problem P1

In this section we show some properties of Problem P1. In the following considerations we assume that $C_{min} = \min_j C_j$ and $C_{max} = \max_j C_j$.

Property 1. In Problem P1, for any schedule π, the optimal values of the parameters k_1 and k_2 are equal to $k_1^* = C_{min}$ and $k_2^* = C_{max}$, respectively.

Proof. We consider the following two cases: (1) for any schedule π and any value k_2, the optimal value of k_1 is equal to $k_1 = k_1^* + \varepsilon$; (2) for any schedule π and any value of k_1, the optimal value of k_2 is equal to $k_2 = k_2^* + \varepsilon$, where ε is some positive or negative value. We show how to prove case (1) with $\varepsilon > 0$. The remaining cases are very similar and will be omitted.

Let $f(\pi, k'_1, k_2)$ and $f(\pi, k''_1, k_2)$ denote values of the criterion (1) obtained for the parameters $k'_1 = k_1^* + \varepsilon$ and $k''_1 = k_1^*$, respectively. We have:

$$f(\pi, k'_1, k_2) = \textstyle\sum_j \max\{d'_j - C_j, 0\} + (k_2 - k'_1) + \sum_j T_j \geq$$
$$\max\{k'_1 - C_{\min}, 0\} + (k_2 - k'_1) + \textstyle\sum_j T_j = \max\{k_1^* + \varepsilon - C_{\min}, 0\} +$$
$$(k_2 - k_1^* - \varepsilon) + \textstyle\sum_j T_j = (k_2 - k_1^*) + \sum_j T_j = f(\pi, k''_1, k_2).$$

This result contradicts the assumption that k_1' is optimal. For the remaining cases we obtain similar results, which ends the proof. □

From Property 1 follows that in any schedule π, for the optimal values of k_1^* and k_2^*, there are no early and no tardy jobs. Thus, we have:

$$f(\pi, k_1^*, k_2^*) = \sum_j E_j + (k_2^* - k_1^*) + \sum_j T_j = C_{max} - C_{min}. \tag{3}$$

Property 2. In optimal solution to Problem P1, the jobs executed as first on each processor complete at the same moment.

Proof. Assume that π is an optimal solution to Problem P1 and C_{min} is the completion moment of the jobs executed as first on each processor except for processor $r \in \{1, ..., m\}$. The completion moment of the first job executed on processor r is equal to $C_{min} - \varepsilon$, where $\varepsilon > 0$. Assume also that in schedule π' the first job executed on processor r completes at C_{min}. Now we show, that the criterion (1) for schedule π' is not greater than the criterion for the schedule π. We will use the following notations: $C_{\pi_r \min}$ - the completion moment of the first job on the rth processor in schedule π, $C_{\pi_r \max}$ - the completion moment of the last job on the rth processor in schedule π, $C_{\pi \max}$ - the maximum completion moment in schedule π.

There are only the following two exhaustive cases: (1) the jobs executed on the rth processor complete ε time units later and it doesn't cause the change of maximum completion moment in schedule π, i.e., $C_{\pi \max} \geq C_{\pi_r \max} + \varepsilon$; (2) the jobs executed on the rth processor complete ε time units later and it causes the change of maximum completion moment in schedule π, i.e., $C_{\pi_r \max} \leq C_{\pi \max} < C_{\pi_r \max} + \varepsilon$.

Case (1): It follows from expression (3) that the values of criterion (1) for schedules π and π' are, respectively, as follows: $f(\pi, k_1, k_2) = C_{\pi \max} - C_{\pi_r \min}$, $f(\pi', k_1, k_2) = C_{\pi \max} - (C_{\pi_r \min} + \varepsilon)$. Thus, $f(\pi, k_1, k_2) - f(\pi', k_1, k_2) = \varepsilon$, which contradicts the optimality of the schedule π.

Case (2): If $C_{\pi_r \max} \leq C_{\pi \max} < C_{\pi_r \max} + \varepsilon$ then criterion values for schedules π and π' can be estimated as follows: $f(\pi', k_1, k_2) = C_{\pi_r \max} + \varepsilon - C_{\pi_r \min} - varepsilon = C_{\pi_r \max} - C_{\pi_r \min} \leq C_{\pi \max} - C_{\pi_r \min} = f(\pi, k_1, k_2)$, which contradicts the assumption that schedule π is optimal.

Notice that the cases considered above include also the case where the first job executed on one processor completes ε time units later than the first jobs on remaining processors. □

Property 3. In the optimal solution to Problem P1, jobs completed as first on the processors have the largest processing times.

Proof. Assume that in the optimal solution π, job s with the processing time p_s is scheduled as first on one of the processors. From Property 2 follows that the completion moment of job s is equal to C_{min}. Without loosing generality we assume that $C_{min} = p_{max}$, where $p_{max} = \max_j p_j$. Assume also that job t is scheduled on any position except the first one on any processor and $p_t = p_s + \varepsilon$. Denote by π' the schedule in which jobs s and t are exchanged. In the schedule π' job t has the completion moment C_{min} (according to Property 2). It follows from the expression (3), that the criterion value (1) of the schedules π and π' can be estimated as follows: $f(\pi', k_1, k_2) = C_{\pi' \max} - C_{\min} \leq C_{\pi \max} - C_{\min} = f(\pi, k_1, k_2)$, since $C_{\pi' \max} \leq C_{\pi \max}$. This result contradicts the assumption that schedule π is optimal. □

4 Computational Complexity and Solution Approaches for Problem P1

Theorem 1. *Problem P1 is strongly NP-hard.*

Proof. Form Property 3 follows that in optimal solution to Problem P1, the first m jobs with the largest processing times are scheduled as first on particular processors and have common completion moment $C_{min} = p_{max}$. The remaining jobs should be scheduled so that the maximum completion moment C_{max} is minimized (see expression (3)). Thus, Problem P1 is equivalent to the well known $P||C_{max}$ scheduling problem (for $n-m$ jobs), which is known to be strongly NP-hard in general case [4]. □

In the literature there are many approximation as well as optimal algorithms solving the problem $P||C_{max}$. For example in paper [5] authors propose the LPT (longest processing time) rule with the worst case ratio $R_{LPT} \leq \frac{4}{3} - \frac{1}{3m}$. In the paper [4] the pseudopolynomial time optimal algorithm is proposed and authors of paper [9] propose the fully polynomial time approximation scheme for problem $P2||C_{max}$. After some modifications all mentioned above methods can be used to solve the considered Problem P1 with the same quality of solutions. Modifications concern optimal scheduling of m jobs with the largest processing times, according to Property 2 and Property 3, and determining optimal values of parameters k_1 and k_2 according to Property 1.

5 Equivalence of Problems P1 and P2

In this section we show that the optimal schedule of jobs for the Problem P1 is equivalent to the optimal schedule of jobs for the Problem P2, however, at first we present the following lemmas.

Lemma 1. *In any schedule of jobs for Problem P2, the optimal values of the parameters k_1 and k_2 fulfill, respectively, the following inequalities: $k_1^{\#} \geq C_{min}$, $k_2^{\#} \leq C_{max}$.*

Proof. The proof is very similar to the proof of Property 1 and therefore will be omitted. □

It follows from Lemma 1, that the criterion value (2) is equal to:

$$g(\pi, k_1, k_2) = \max\{P_1(k_1 - C_{min}), P_2(k_2 - k_1), P_3(C_{max} - k_2)\}. \quad (4)$$

The following two lemmas hold for the criterion (4), for fixed schedule π. Those lemmas also hold for a general function:

$$h(u, v, w) = \max\{A_1 u, A_2 v, A_3 w\} \quad (5)$$

subject to:

$$u + v + w = A, \quad (6)$$

where u, v and w are some nonnegative variables, A is given nonnegative constant and A_1, A_2 and A_3 are given nonnegative weights.

Lemma 2. *The value of the expression (5) is minimal if $A_1 u = A_2 v = A_3 w$.*

Proof. The proof is trivial and will be omitted.

Lemma 3. *The optimal values of the variables u, v and w, which minimize the function (5) are as follows: $u^* = \frac{A_2 A_3 A}{A_1 A_2 + A_1 A_3 + A_2 A_3}$, $v^* = \frac{A_1 A_3 A}{A_1 A_2 + A_1 A_3 + A_2 A_3}$, $w^* = \frac{A_1 A_2 A}{A_1 A_2 + A_1 A_3 + A_2 A_3}$.*

Proof. The proof can be easily derived form Lemma 2 and constraint (6) and will be omitted. □

It follows from Lemma 3 and the expression (4), that the criterion value (2) can be described as follows:

$$g(\pi, k_1, k_2) = \frac{P_1 P_2 P_3}{P_1 P_2 + P_1 P_3 + P_2 P_3}(C_{max} - C_{min}). \quad (7)$$

Form Lemma 3 we obtain the following property.

Property 4. In Problem P2, for any schedule π, the optimal value of the parameter k_1 is equal $k_1^{\#} = \frac{P_2 P_3}{P_1 P_2 + P_1 P_3 + P_2 P_3}(C_{max} - C_{min}) + C_{min}$ and the optimal value of the parameter k_2 is equal $k_2^{\#} = \frac{P_1 P_3 + P_2 P_3}{P_1 P_2 + P_1 P_3 + P_2 P_3}(C_{max} - C_{min}) + C_{min}$.

Proof. The optimal values of the parameters k_1 and k_2 can be easily derived from Lemma 3. □

Theorem 2. *The optimal schedule of jobs in Problem P2 is equivalent to the optimal schedule of jobs in Problem P1.*

Proof. From the expressions (3) and (7) follow that for any schedule π and the optimal values of parameters k_1 and k_2, the criterion value (2) is proportional to the criterion value (1): $g(\pi, k_1^{\#}, k_2^{\#}) = \frac{P_1 P_2 P_3}{P_1 P_2 + P_1 P_3 + P_2 P_3} f(\pi, k_1^*, k_2^*)$. □

It follows from Theorem 2 that the considerations concerned the properties of optimal schedule, the computational complexity, and methodology of solution which ware presented in Sections 3 and 4 for Problem P1 are also true for Problem P2. The only exception is the method of determining the optimal values of parameters k_1 and $k2$ given by Property 1 for Problem P1 and by Property 4 for Problem P2.

6 Conclusions

In the paper, we considered two problems of scheduling jobs on identical parallel machines with a common due interval assignment. Some properties of optimal solution were provided for these problems. Based on these properties we proved that solutions of our problems are equivalent to solution of the well known $P||C_{max}$ scheduling problem, which is known to be strongly NP-hard. Therefore we adopted some known methods of solution to solve our problems. The results obtained are very promising and motivating to focus on other scheduling problems with due interval assignment.

References

1. Cheng T.C.E., Gupta M.C. (1989). Survey of scheduling research involving due date determination decision. European Journal of Operational Research. **38**, 156–166.
2. Chengbin C., Gordon V., Proth J.-M. (2002). A survey of the state-of-the-art of common due date assignment and scheduling research. European Journal of Operational Research, **139**, 1–25.
3. Chengbin C., Gordon V., Proth J.-M. (2002). Due date assignment and scheduling: SLK, TWK and other due date assignment models. Production Planning & Control. **13**, 117–132.
4. French S. (1982). Sequencing and Scheduling: An Introduction to the Mathematics of the Job-Shop. Horwood, Chichester.
5. Graham R.L. (1969). Bounds on multiprocessing timing anomalies. SIAM J. Appl. Math. **17** 263–369.
6. Janiak A., Marek M. (2001). Multi-machine scheduling problem with optimal due interval assignment subject to generalized sum type criterion. Operations Research Procedings 2001. Springer, 207–212.
7. Liman S.D., Panwalkar S.S., Thongmee S. (1996). Determination of common due window location in a single machine scheduling problem. European Journal of Operational Research. **93**, 68–74.
8. Liman S.D., Panwalkar S.S., Thongmee S. (1998). Common due window size and location determination in a single machine scheduling problem. Journal of the Operational Research Society. **49**, 1007–1010.
9. Sahni S. (1976). Algorithm for scheduling indenpendent tasks. J. Assoc. Comput. Mach. **23**, 116–127.

A New Exact Resource Allocation Model with Hard and Soft Resource Constraints

Ferenc Kruzslicz

University of Pécs, Department of Business Informatics,
Rákóczi út 80, H-7623 Pécs, Hungary

Abstract. In this paper we present a new bilevel resource allocation model with hard and soft resource constraints for projects. By definition, a hard resource constraint is not resolvable within the given planing horizon, but a soft resource conflict may be managed by a flexible "hiring-firing" strategy.

According to the definition, firstly we solve a "makespan-minimization" problem for the hard resources. After that, we solve a "resource balancing" problem on the set of "hard resource feasible" schedules. In the "well-balanced" schedule searching phase we characterize the soft resource requirements with a new bicriteria measure, namely with the "peak resource requirement" and the "idle time", simultaneously. In the proposed approach, in the efficient solution searching phase we applied a new a "trees in tree" like implicit enumeration method with effective pruning rules.

The practical interpretation of the proposed model is demonstrated in an analysis of a simplified small-scale business software development environment.

1 Introduction

The model introduced in this article is designed to schedule complex multi-resource, multi-project problems with wide range of practical applications. The word multi stays here in its wider sense since both resources and projects may differ substantially from each other [2]. Classification of resource types is based on the flexibility level of resource availability. According to the NP-hard nature of the problem [1], the model provides exact solutions for small to medium size problems.

2 The model

Let $P = \{P_1, P_2, ...P_n\}$ denote the set of projects to be scheduled. Every P_i project consists of n_i different tasks $P_i = \{T_{i,j} : j = 1 \ldots n_i\}$. Tasks are atomic activities with known resource and time requirements. Additionally there are $H_i = \{T_{i,j_1} \to T_{i,j_2}\}$ immediate precedence relations for every P_i project. If activity T_{i,j_1} must be finished, before activity T_{i,j_2} is started, it is written as $T_{i,j_1} \to T_{i,j_2}$.

We have k different classes of resources with hard constraints, denoted by $RH = \{RH^1, RH^2, \ldots RH^k\}$. In similar manner we can define the same set for resources with soft constraints $RS = \{RS^1, RS^2, \ldots RS^l\}$. For every

$R \in RH \bigcup RS$ resource a $R(t)$ resource limit function is given that represents the available amount of resource R at time t.

Definition: Resource R is referred as *hard constraint* if any time t no more than $R(t)$ amount of this resource can be allocated.

In case of hard constraints no auxiliary resources can be applied, not even in short term base. When an allocation conflict is caused by a hard constraint, it can be resolved only with rescheduling the activities, or in the worst case extending the earliest finishing time of the project.

Definition: Resource R is referred as *soft constraint* if any time t more than $R(t)$ amount of it can be allocated by applying extra auxiliary resources.

With these notations every activity can be described with a $T_{i,j} = (D_{i,j}, RH^1_{i,j}, RH^2_{i,j}, \ldots RH^k_{i,j}, RS^1_{i,j}, RS^2_{i,j}, \ldots RS^l_{i,j})$ vector, where $D_{i,j}$ denotes the amount of time, and $R_{i,j}$ the amount of resource units required by activity j of project i for all $R \in RH \bigcup RS$.

A schedule $S = \{S_{i,j} : i = 1 \ldots n, j = 1 \ldots n_i\}$ of projects is given with start times of $T_{i,j}$ activities. For convenience let us introduce two more T_0 and T_1 dummy activities which represent the beginning and the completition of all projects. T_0 becomes the first and T_1 the last activity of all schedules without any time or resource requirement. The following extra immediate precedence relations are added to these dummy tasks: $H = \{T_0 \to T_{i,j}, T_{i,j} \to T_1 : i = 1 \ldots n, j = 1 \ldots n_i\}$. With this notations the S_0 start time of T_0 is the start time of the projects and similarly S_1 the start time of activity T_1 is the earliest finishing time of P.

We can assume that every schedule starts at time $S_0 = 0$. With this assumption $E(S) = S_1$ is the total implementation time of all the projects in P by the schedule S. Throughout in this model only those schedules are considered that preserve the immediate precedence relations.

Definition: Schedule S is said to be *feasible* if inequality $S_{i,j_1} + D_{i,j_1} \leq S_{i,j_2}$ holds for every $T_{i,j_1} \to T_{i,j_2}$ immediate precedence rule. The set of all feasible schedules of projects P is denoted by $\Re = \Re(H \bigcup H_1 \bigcup \ldots \bigcup H_n)$.

If resource constraints are taken into consideration the set of feasible schedules has to be restricted. Only those schedules are acceptable that do not overrun the resource limit of hard constraints. Let us denote $U^R_S(t)$ the utilization of resource R at time t by the schedule S. The function $U^R_S(t)$ can be expressed by the following formula:

$$U^R_S(t) = \sum_{i=1}^{n} \sum_{j=1}^{n_i} S_{i,j}(t) \cdot R_{i,j}\,, where\ S_{i,j}(t) = \begin{cases} 1, if\ S_{i,j} \leq t < S_{i,j} + D_{i,j} \\ 0, \quad otherwise \end{cases}$$

Definition: Schedule S is said to be *resource-feasible* for a C set of resources if $S \in \Re$ and $U^R_S(t) \leq R(t)$ holds for all $t \in [0; E(S)]$ and $R \in C$, moreover it has the earliest finishing time $E(S) = \min\{E(S') : S' \in \Re\}$. The set of resource-feasible schedules for a C set of resources is denoted by $\Re^C$.

It is clear that set $\Re^{RH}$ contains only schedules which are resource-feasible for all hard constraints. This set can be determined by solutions of a resource allocation problem (RAP). Henceforward we are interested in schedules in $\Re^{RH}$ that are efficient for soft constraints as well. There are several well known scalar indicators of efficiency of a schedule. In this model we use a new global bicriteria measure which tracks the "peak resource requirement" and the "idle time" at the same time. This means locating Pareto optimal solutions.

On the one hand the utilization of a resource $R \in RS$ is optimal during a schedule S if minimal extra resource has to be applied. In other words we have to minimize the peak resource utilization level. Our first type of objective functions can be formulated as

$$MU^R = \min_S \{MU^R(S)\}, where\ MU^R(S) = \max_t \{U_S^R(t)\}$$

On the other hand the utilization of a resource $R \in RS$ is optimal during a schedule S if its fluctuation is minimal. The concept of idle times is based on a global concavity measure on histograms introduced in [5]. Let the overline operator $\overline{U}$ denote the quasi-concave hull of utilization histogram U. With this notation our second type of objective functions can be expressed as

$$IT^R = \min_S \{IT^R(S)\}, where\ IT^R(S) = \sum_{t=0}^{E(S)} \left(\overline{U_S^R}(t) - U_S^R(t)\right)$$

This kind of histogram concavity measure perfectly describes the requirement of smooth resource utilization.

Definition: A hard constraint feasible $S \in \Re^{RH}$ schedule is Pareto optimal on MU^R and IT^R objective functions of $R \in RS$ resources if there is no other $S' \in \Re^{RH}$ schedule where both $MU^R(S) \geq MU^R(S')$ and $IT^R(S) \geq IT^R(S')$ hold for all $R \in RS$ resources and at least one of the inequalities is strict.

3 The algorithm

The algorithm almost trivially follows from the model. This algorithm consists of two phases, first a "makespan-minimization" problem is solved for the hard constraints, then a "resource balancing" problem on the set of hard constraints feasible solutions of of the first phase [8].

Phase 1: In this phase the set $\Re^{RH}$ is generated while the $R \in RS$ soft constraints are neglected. Using the Critical Path Method (CPM) we may get schedules $S \in \Re$ that cause resource allocation conflicts. Active tasks at time t may require more hard resources to be allocated than available. First these kind of hard resource violating schedules are repaired in this phase. Problematic schedules are characterized by their *minimal violating sets*. A violating set of activities of S is called minimal if all the activities together cause resource allocation problem but any real subset of them can be realized without conflict. Violations are resolved by delaying activities with not enough resource. The makespan-minimization calculates new schedules from the violating ones by adding new repairing immediate precedence rules between the critical activities. When no hard resource feasible solution meets the deadline determined by the CPM process, the earliest finishing time is increased until a hard resource feasible solution is found.

Phase 2: In the second phase each schedule of $\Re^{RH}$ are taken in turn. It is clear that only the leaf nodes of the search tree used in phase 1 should be analyzed further. These leaf nodes are splitted into subnodes by adding extra resource conflict resolver precedence rules to the original set H. After splitting nodes several new resource leveling problems (RLP) arise where hard constraints can be safely ignored since any feasible schedule of problems in phase 2 is a hard resource feasible schedule of the corresponding problem of phase 1. Efficient schedules of the new RLPs may be determined with the same implicit enumeration method that was used in phase 1 as well. Since there are no conflicts on soft constrained resources the size of the search trees are notable larger. The exact solutions of the whole problem are those Pareto optimal schedules that minimize all the MU^R and IT^R functions of $R \in RS$ soft constraints.

Selecting a suitable implicit enumeration method is the key point in our algorithm. From the wide range of available methods [3] the algorithm published in [6] was found to be a good selection, since it is based on effective pruning rules that makes our algorithm capable to solve larger problems.

4 Application

In software development the life cycle defines the phases and tasks that are essential to systems development, no matter what type or size system is developed. These techniques and methodologies have a common "plan, execute, and test" sequence of tasks to go through. In the problem presented here we assumed that the number of software designers is a hard constraint within the given planning horizon, but the availability of programmers and testers are soft constraints, which can be resolved by "hiring-firing" in short term base. In this example we exploit the fact that in a small-scale business software development process the usual iterative "waterfall" structure might be replaced by a serial "designer-programmer-tester" chain.

The development team consist of five designers RH^1, five programmers RS^1 and five testers RS^2. The level of available resources are constant $R(t) \equiv 5$ during the development process and every activity requires only one type of resource allocated to. In Figure 1 the project structure is represented with a simplified notation.

Projects	D / RH^1	D / RS^1	D / RS^2
New feature development			
P1	2 / 2	3 / 2	4 / 2
P2	5 / 3	3 / 3	4 / 4
P3	3 / 2	3 / 2	4 / 1
Bug fixing			
P4	0 / 0	2 / 1	2 / 2
P5	0 / 0	4 / 3	2 / 3
Extensive system testing			
P6	0 / 0	0 / 0	2 / 3
P7	0 / 0	0 / 0	3 / 3
Available resources			
Person	5	5	5

Corresponding search tree

0. {} {1,{$T_{1,1}$,$T_{2,1}$,$T_{3,1}$},1} 1

1. {$T_{1,1} \rightarrow T_{3,1}$} 2

2. {$T_{3,1} \rightarrow T_{1,1}$} 2

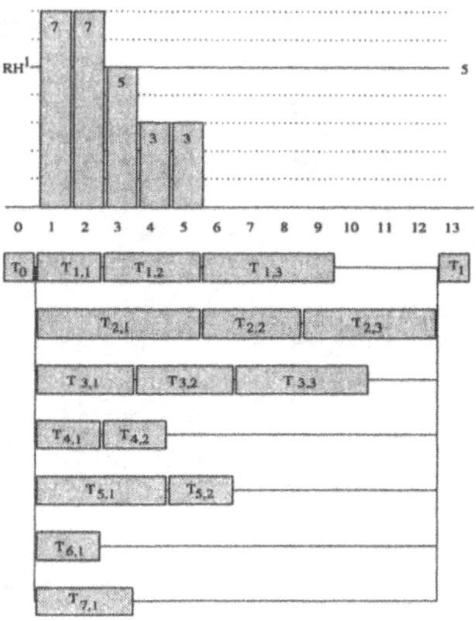

Fig. 1. Time and resource requirements of projects in a software development process and extended Gantt chart of activities in the initial CPM schedule.

The utilization histogram of resource RH^1 in Figure 1 traces out the conflicts of the initial schedule. Since there is no conflict free solution regarding hard resources in 12 days the earliest finishing time has to be delayed. With 1 day enlargement of time horizon in phase 1 the number of feasible schedules grows from 27 720 000 to 731 808 000 which illustrates the complexity of this small example. In 13 days two basically different hard resource feasible solutions can be found. New immediate precedence rules correspond to the branches of the search tree, see Figure 1. Only case of adding the rule $T_{1,1} \rightarrow T_{3,1}$ is investigated hereinafter. The other case needs similar calculations therefore it is omitted here.

The Pareto optimal schedule shown in Figure 2 dominates the ones got from the other case, therefore this Pareto optimal solution is the unique optimum. This optimum makes a smooth workload of all resources, except two testers should be hired for the 9th day.

Computational work was done on a PC with P166 Mhz processor. The program [5] has been written in Visual Basic Version 6.0, but critical parts were compiled as C++ DLL files.

Table 1. Program execution details

Phase	Number of feasible schedules	Number of nodes in the search tree	Execution time in seconds
Phase 1. Node 0.	27 720 000	3	0.150
Phase 2. Node 1.	1 386 000	282	1.382
Phase 2. Node 2.	831 600	150	0.621

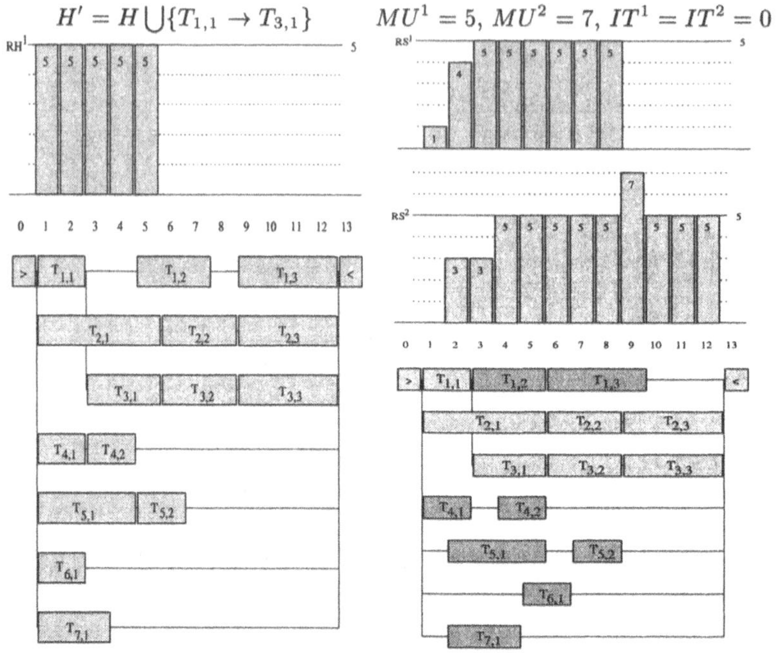

Fig. 2. Hard resource feasible schedule on node 1 of the search tree after the first phase and its Pareto optimal solution after the second phase.

References

1. S. E. Elmaghraby (1995): Activity nets: A guided tour through some recent developments. Eur. J. of Operational Research **82**, 383–408
2. V. Shankar, R. Nagi (1996) A flexible optimization approach to multi-resource, multi-project planning and scheduling. 5th Industrial Engineering Research Conference, Minneapolis **MN**, 263–267
3. M. R. Zamani (2001): A high-performance exact method for the resource-constrained project scheduling problem. Computer & Operations Research **28**, 1387–1401
4. B. D. Reyck, W. Herroelen (1999): The multi-mode resource-constrained project scheduling problem with generalized precedence relations. Eur. J. of Operational Research **119**, 538–556
5. G. Csébfalvi (2001): Optimal resource leveling models for activity networks. Habilitation dissertation, PTE **Pécs** 120p
6. G. Csébfalvi (1998): A fast exact solution procedure for the multiple resource-constrained project scheduling problem. Proc. APMOD '98 Extended Abstracts, Limasol, **Cyprus**, 11–13
7. G. Csébfalvi, P. Konstantinidis (1999): A new resource leveling procedure for the multiple resource-constrained project scheduling problem. Decision Sciences Institute 5th International Conference, Athens, **Greece**, 1723–1725
8. G. Csébfalvi (2000): A new multi-criteria resource leveling procedure for the multiple resource-constrained project scheduling problem. Proc. EURO XVII.

Minimizing Total Weighted Tardiness on Parallel Batch Process Machines Using Genetic Algorithms

Lars Mönch[1*], Hari Balasubramanian[2], John W. Fowler[2], Michele E. Pfund[2]

[1]Institut für Wirtschaftsinformatik, Technische Universität Ilmenau, Germany

[2]Department of Industrial Engineering, Arizona State University, USA

1 Introduction

Recently, the electronics industry has become the largest industry in the world. A key aspect of this industry is the manufacturing of integrated circuits. In the past, sources of reducing costs were decreasing the size of the chips, increasing the wafer sizes and improving the yield, simultaneously with efforts to improve operational processes inside the wafer fabrication facilities (wafer fab). Currently, it seems that the improvement of operational processes creates the best opportunity to realize the necessary cost reductions [6].

This research is motivated by a scheduling problem found in the area of diffusion in a wafer fab. In the diffusion area, there are several diffusion furnaces which can be operated in parallel and can process several jobs simultaneously. However, due to the different chemical nature of the processes, only jobs of the same job family can be processed together. Carefully choosing which products to batch together and which order these jobs should be sequenced is of high importance as the processing times in the diffusion furnaces can take up to 10 hours, versus a 1 hour processing time in other wafer fab processes. Therefore, a proper scheduling of the machines dedicated to the diffusion process steps has a great impact on the performance of the entire wafer fab. Given that many wafer fabs are driven by on-time delivery, the performance measure of interest is the total weighted tardiness. This performance measure is given as $\sum w_{ij} T_{ij}$, where we denote by w_{ij} the weight or priority for the job j of family i. The tardiness of job j of family i is given by $T_{ij} := \max(c_{ij} - d_{ij}, 0)$, where we denote by c_{ij} the completion time for job j of family i in the schedule and d_{ij} represents the due date for job j of family i.

In this paper, we extend previous research on single batch process machines [2,3,4] to the more general case of parallel machines. We use a genetic algorithm in order to assign batches to the parallel machines. We reduce the problem to single machine sequencing problems after performing this decomposition step. For

---

* To whom correspondence should be addressed.

the efficient solution of the single machine problems we use the same techniques as suggested in [2,4].

The paper is organized as follows. In the next section, we describe the problem under consideration in detail. Then we describe a four-phase-scheduling algorithm for obtaining the solution of the problem. We present results of computational experiments in the last section of this paper.

2 Problem Description

By Pinedo (2002), scheduling problems can be represented in the form $\alpha \mid \beta \mid \gamma$. The α field describes the machine environment (single machine, parallel machine, job shop, etc.), the β field describes the process characteristics, restrictions, constraints (such as release dates, batch, set up dependent operations), and the γ field contains the information on the performance measure being considered. For the problem being researched in this paper the notation is:

$$Pm \mid \text{batch, incompatible} \mid \sum w_j T_j \,.$$

There are f families of jobs and m parallel machines. Family i has n_i jobs. The jobs that are waiting to be processed must be batched by family, depending on the maximum batch size B. Each single job has a weight w_{ij} and a due date d_{ij}. Our objective is to minimize the total weighted tardiness of the jobs.

3 Four-Phase-Scheduling-Algorithm

Since the problem is NP-hard (cf. [1] for details) we come up with a heuristic approach in order to solve it. We suggest a four-phase scheduling algorithm:

- formation of the initial batches in the first phase,
- assignment of the batches to the parallel machines in the second phase,
- solution of the resulting single machine sequencing problems of the second phase in the third phase,
- improvement of the objective function value using a swap procedure in the optional fourth phase. The objective is a modification of the batches after they have been scheduled.

3.1 Forming Initial Batches

We use two different heuristics to form the batches. The first one is based on the EDD (Earliest Due Date) dispatching rule. It can be described as follows. For batching the EDD heuristic orders the jobs in descending order of the due dates for each family and then form full batches (except for possibly the last one of each family) starting with the smallest due date. The Apparent Tardiness Cost (ATC) dispatching rule is alternative to EDD for minimizing total weighted tardiness [7,5]. We use the ATC rule in the following way:

1. We calculate for each job j of family i the ATC index given by $I_{ij}(t) := \frac{w_{ij}}{p_{ij}} \exp\left(-\frac{\max\{d_{ij} - p_{ij} - t, 0\}}{k\bar{p}} \right)$, where we denote by k a look-ahead parameter that is varied between 0.5 and 5.0. We use the notation $\bar{p}$ for average processing time of the remaining jobs.
2. The time t is the time when any of the available machines becomes free. At this time t, all available jobs in each family are ordered in descending order of their ATC index. The first B jobs are batched together in each family. Among the different batches the batch with the highest sum of ATC indices (called BATC index) is chosen.

Note, that the ATC dispatching rule in this form (called ATC-BATC rule) can be used to solve the parallel batch process machine problem independently, i.e., without the other phases of the suggested algorithm. We use this heuristic in order to evaluate and compare the performance of our more sophisticated algorithm.

3.2 Assignment of Batches

We use a genetic algorithm (GA) in order to assign the initial formed batches to the parallel machines, evaluate each of the assignments by aggregating the total weighted tardiness of sequences (obtained on each machine by using sequencing heuristics for the single machine case) on all machines. The GA converges towards assignments that give good solutions. The overall framework of the genetic algorithm is shown in Figure 3.1.

We used a batch-based chromosome representation in the GA. A solution of the problem to assign batches to machines is an array, whose length is equal to the number of batches. The s-th element of the array represents the machine that is used to process batch s. Each chromosome is randomly initialized by generating a random number from $\{1,\ldots m\}$ for each single gene. A standard one-point crossover operator is used for performing the crossover operations. According to a predefined probability, a gene of a chromosome is randomly changed to a different machine number for mutation purposes. Each chromosome is evaluated with the total weighted tardiness value. After the batches are assigned to machines, each machine is sequenced and the sum of the total weighted tardiness values of all machines is used as the fitness function of the GA.

A steady state GA with overlapping populations is used (cf. [8]). The GA is controlled by a prescribed number of populations and by using a diversity measure. For parameter setting of the GA we refer to [1].

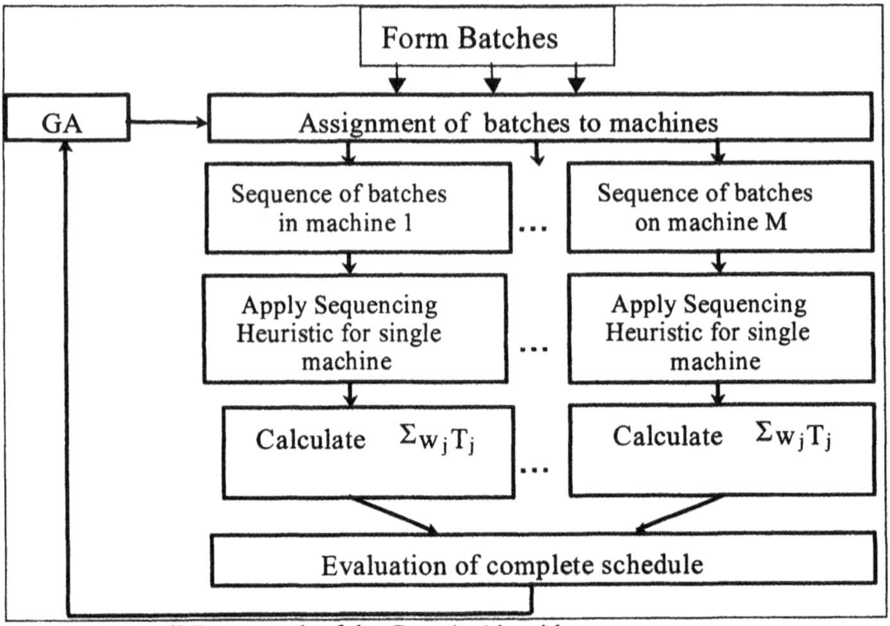

Fig. 3.1. Overall Framework of the Genetic Algorithm

3.3 Solution of the Single Machine Sequencing Problems

For sequencing the batches on each single machine we use the EDD rule, the BATC rule and a sequence decomposition heuristic suggested by Mehta and Uzsoy [3]. From an initial sequence obtained by a heuristic a sub-sequence of size λ is solved into an optimal sequence (with respect to total weighted tardiness) by using total enumeration. The first α batches of this optimal sequence are fixed into the final sequence, and the next λ batches are considered. This process is repeated until all batches are covered and no improvement is made on the performance measure. We refer to [4] for a more detailed description of these heuristics.

3.4 Swapping Procedure

We use the swap procedure suggested in [2] in order to improve the quality of the schedules. The original swap procedure exchanges jobs for all batches of one family across a single machine with the objective to reduce the performance measure. We propose an extension of this technique to the parallel machine case by redistributing jobs not only across compatible batches of the same family but also across the parallel machines.

4 Results

In this section we present results of computational experiments. We used test data generated according to an extension of the test data description given in [4] to the parallel machine case. The test cases depend on the batch size, the number of parallel machines, the number of jobs of each job family, the due date range R and the tightness T of the schedules. In our experiments we considered the case of three different job families.

All experiments were performed on a Pentium III, 800 MHz PC. We used the object-oriented framework GaLib (cf. [8]) in order to implement the suggested GA in an efficient way. The average computing time of the GA was one minute per problem instance. The results of the experiments are given in Table 4.1. We show the ratio for the performance measure value of the suggested variant of the four-phase-scheduling algorithm to the performance measure of the ATC-BATC-heuristic. Clearly, a couple of configurations of our algorithm outperforms the ATC-BATC-heuristic. As can be seen from the results the algorithm is sensitive to the due date range R and the tightness T of the schedules.

5 Conclusions and Outlook to Future Research

In this paper we suggested a four-phase heuristic algorithm for the solution of parallel batch process machine scheduling problems. We do not know how close we are to an optimal solution. Therefore, one direction of future research can focus on the determination of lower bounds for total weighted tardiness. Another direction of future research is to be focus on the incorporation of ready times of the jobs into the algorithm.

Acknowledgement

The authors gratefully acknowledge the support of the Semiconductor Research Corporation (2001-NJ-880). Parts of this research was carried out whilst the first author was visiting the Modeling and Analysis of Semiconductor Manufacturing Laboratory at the Arizona State University, Department of Industrial Engineering, Tempe.

Table 4.1. Results for Different Configurations of the Algorithm

Combinations/ Characteristics	EDD GA EDD	ATC GA BATC	ATC GA BATC swap	ATC GA DH Swap
Machines				
m=4	1.96	0.98	0.92	0.92
m=5	1.86	0.98	0.93	0.89
Batch Sizes				
B=4	2.04	0.99	0.93	0.92
B=8	1.81	0.97	0.92	0.91
Jobs/Family				
80	1.92	1.00	0.91	0.91
100	2.01	0.99	0.92	0.91
Due Date Range				
R=0.5	2.38	0.96	0.87	0.85
R=2.5	1.47	1.00	0.98	0.96
Tightness				
T=0.30	1.96	0.97	0.88	0.86
T=0.60	1.89	1.00	0.97	0.95

References

[1] Balasubramanian, H. 2002. *Minimizing Total Weighted Tardiness on Parallel Batch Processing Machines with Incompatible Job Families.* Master Thesis, Arizona State University, Department of Industrial Engineering.

[2] Devpura, A., J. W. Fowler, M. W. Carlyle, I. Perez. 2000. Minimizing Total Weighted Tardiness on Single Batch Process Machines with Incompatible Job Families. *Proceedings Symposium on Operations Research*, 366-371.

[3] Mehta, S. V., R. Uzsoy. 1998. Minimizing Total Tardiness on a Single Batch Processing Machine with Incompatible Job Families. *IIE Transactions*, 30, 165-178.

[4] Perez, I., J. W. Fowler, M. W. Carlyle. 2002. Minimizing Total Weighted Tardiness on a Single Batch Process Machine with Incompatible Job Families. Submitted to Computer and Operations Research.

[5] Pinedo, M. 2002. *Scheduling: Theory, Algorithms, and Systems.* Prentice Hall, New Jersey, Second Edition.

[6] Schömig, A., J. W. Fowler. 2000. Modelling Semiconductor Manufacturing Operations. *Proceedings of the 9 th ASIM Dedicated Conference Simulation in Production and Logistics*, eds.: K. Mertins and M. Rabe, 55-64.

[7] Vepsalainen, A., T. E. Morton. 1987. Priority Rules and Lead Time Estimates for Job Shop Scheduling with Weighted Tardiness Costs. *Management Science*, Vol. 33, 1036-1047.

[8] Wall, M. 1995. GaLib – A C++ Library of Genetic Algorithm Components, http://lancet.mit.edu/ga/.

Sorting with Line Storage Systems

Thomas Epping[1,2] and Winfried Hochstättler[1,3]

[1] Department of Mathematics, BTU Cottbus, 03013 Cottbus, Germany
[2] epping@math.tu-cottbus.de
[3] hochstaettler@math.tu-cottbus.de

Abstract. We consider a problem that arises in car production. In particular, we focus on the sorting of car bodies with respect to their designated enamel colors for a paint shop. Our objective is to sort a given car body sequence so that the number of color changes within the sorted sequence is minimized. Current technology is to perform the realignment of a sequence by the use of a line storage system.

1 Introduction

Interim storage systems are an essential tool for production control. They are installed in front of production shops to arrange elements of an incoming sequence in a way that increases the efficiency of the particular production shop and thus reduces production costs. Several types of interim storage systems exist (see [1]). Among them, line storage systems are the most popular due to their simplicity and low investment costs.

We deal with line storage systems installed in front of the paint shop of an automobile plant. These systems are used to permute an incoming sequence of car bodies so that the number of color changes arising in the outgoing sequence within the enamel booth of the paint shop is minimized.

Definition 1. Given a color set F and a color sequence $f = (f_1, \dots, f_n)$ with $f_i \in F$ for $i = 1, \dots, n$, we say that we have a color change within f whenever $f_i \neq f_{i+1}$. We denote the number of color changes within f by $\gamma(f)$.

A line storage system (see Figure 1) consists of a lookahead area L of length l that holds the next l elements of the input sequence, a set $Q_1, \dots, Q_r$ of sorting belts, each of length q, and an output belt. We model the lookahead area as a stack, and each sorting belt as a queue. Different colors are represented by different letters. Note, that L and each Q_i is an ordered multiset.

Two problems arise in this context. First suppose that we are given a snapshot of the line storage system. We focus only on the retrieval of already stored colors, ignore the lookahead area and interpret each sorting belt as a stack. Then the problem consists in the retrieval of the colors on the stacks such that the number of color changes within the retrieval sequence is minimized.

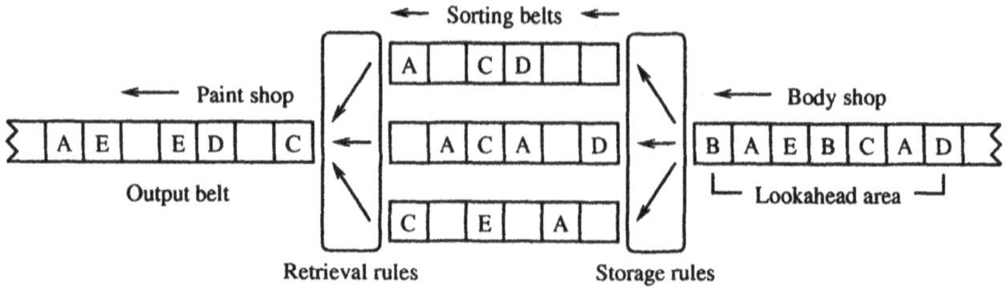

Fig. 1. Example of a line storage system

Problem 1. Color Retrieval Problem (CRP)

Instance A finite set $S_1, \ldots, S_r$ of stacks of length $l_1, \ldots, l_r$ that contain colors of a finite color set F.

Question Retrieve the colors from $S_1, \ldots, S_r$ in a sequence R such that $\gamma(R)$ is minimized.

The second problem is an extension of the first one and includes the distribution of the l colors contained in L on $Q_1, \ldots, Q_r$ together with the retrieval of all colors from $Q_1, \ldots, Q_r$ such that the number of color changes arising on the output belt is minimized. Each Q_i may initially contain colors, and storage and retrieval operations are allowed to take place in any order.

Problem 2. Color Storage and Retrieval Problem (CSRP)

Instance A finite set $Q_1, \ldots, Q_r$ of queues, each having length q, and a stack L of length l that contain colors of a finite color set F.

Question Store the colors in L on $Q_1, \ldots, Q_r$, and retrieve all colors from $Q_1, \ldots, Q_r$ in an order that minimizes the number of color changes within the output sequence.

We describe solution approaches to both problems in the next sections, using fairly standard notation. We consider Problem 1 first.

2 The CRP

We show that the CRP is in a certain sense equivalent to the multiple sequence alignment problem known from molecular biology. This problem can be solved by a fast dynamic programming algorithm (A^* algorithm). We give a short overview of multiple sequence alignment and describe our modifications to the dynamic program (see [2] for more details on this section).

2.1 Multiple Sequence Alignment

Multiple sequence alignment (MSA) is used in molecular biology to detect transformations of DNA or protein sequences. We give a brief sketch of MSA (see [3]). Suppose that we are given r sequences with $a_{ij} \in F$, where F is any

$$a_1 = (a_{11}a_{12}\dots a_{1n_1}) \quad a_2 = (a_{21}a_{22}\dots a_{2n_2}) \quad \vdots \quad a_r = (a_{r1}a_{r2}\dots a_{rn_r})$$

(a)

$$A = \begin{pmatrix} a'_{11} & a'_{12} & \dots & a'_{1c} \\ a'_{21} & a'_{22} & \dots & a'_{2c} \\ \vdots & & \ddots & \vdots \\ a'_{r1} & a'_{r2} & \dots & a'_{rc} \end{pmatrix}$$

(b)

Fig. 2. Illustration of multiple sequence alignment

finite alphabet (see Figure 2(a)). We speak of an *alignment* of $a_1, \dots, a_r$ if we insert elements $- \notin F$ into $a_1, \dots, a_r$ such that we get sequences $a'_1, \dots, a'_r$ that have the same length c with $\max_i\{n_i\} \leq c \leq \sum_i n_i$. In the following, we assume that an alignment of r sequences is given in the form of an $(r \times c)$-alignment matrix A with $a'_{ij} \in F \cup \{-\}$ (see Figure 2(b)).

The MSA problem consists in the computation of an alignment matrix A that has a minimal value $D(A)$ with respect to some scoring function D. It is NP-complete for the *sum of pairs* score (see Section 2.2). Any MSA instance can be solved by dynamic programming with a time complexity of $\mathcal{O}(2^r \Pi_{i=1}^r n_i)$ (see [4]).

2.2 MSA and CRP

We interpret each stack of Problem 1 as an input sequence for the MSA problem and want an optimal alignment matrix to give an optimal retrieval sequence for the colors on the stacks.

First we assume that no stack contains a subsequence of two or more consecutive identical colors by merging any subsequence of that form into a single color element. Correspondingly, we assume that each color on a stack is framed by different colors.

Proposition 1. *The merging of consecutive identical colors does not increase the minimal number of color changes for an instance of the CRP.*

Now the main idea is to calculate an optimal MSA matrix A and to derive an optimal retrieval sequence R for the colors on the stacks by flattening A column by column into the sequence

$$R = (a'_{11}a'_{21}\dots a'_{r1}a'_{12}a'_{22}\dots a'_{r2}\dots a'_{1c}a'_{2c}\dots a'_{rc}),$$

assuming that each stack is open to the left. If we pass through R from the left to the right, we interpret an element $a'_{ij} = f$ as to withdraw color f from stack i and an element $a'_{ij} = -$ as to withdraw nothing from stack i.

For the calculation of an optimal alignment matrix A we use the *sum of pairs* score

$$D(A) := \sum_{i<j}\sum_{j=1}^{c} d(a'_{ij}, a'_{jj}),$$

and define a distance function for two sequence elements $x, y \in F \cup \{-\}$ by

$$d(x,y) := \begin{cases} 0 & \text{, if } x = y \\ 1 & \text{, if } x = - \text{ or } y = - \\ \infty & \text{, if } x \neq y \end{cases}$$

These settings prevent that any column of an optimal MSA matrix A contains more than one color and force A to contain as few columns as possible.

Lemma 1. *Suppose that the solution of a CRP instance results in an optimal $(r \times c)$-MSA matrix and a corresponding withdrawal sequence R. Then R contains $\gamma(R) = c - 1$ color changes. Conversely, we can reconstruct from R an MSA matrix with c columns and r rows, where r is the length of a longest subsequence of consecutive identical colors in R.*

Thus, after applying Proposition 1 and using the distance function defined above, our solution approach solves any instance of the CRP to optimality.

Theorem 1. *The solution of an instance of CRP is equivalent to the solution of an MSA problem as described above. The flattening of the resulting optimal MSA matrix yields an optimal withdrawal sequence R.*

Preliminary computational results on random instances attest the applicability of our solution approach in practice. For example, the solution of a random instance of CRP consisting of $r = 6$ sorting belts, each filled with $q = 20$ colors of an underlying color set of size $|F| = 15$, requires an average running time of 5 seconds on a SUN E450 with 4 SunUltraII 400 MHz processors and 1152 MB memory, using the implementation described in [4].

3 The CSRP

We now describe a dynamic programming approach for the solution of Problem 2. We give a description of the dynamic program and mention some implementation aspects (see [5] for more details on this section).

3.1 A Dynamic Program

We denote a state of our dynamic program by $S = (c; Q_1, \ldots, Q_r; p)$, where c denotes the last color on the output belt, $Q_1, \ldots, Q_r$ denote the content of the queues, and $1 \leq p \leq l$ denotes the current position in the lookahead area. We set $c = \emptyset$ if the output belt contains no colors, and $p = l + 1$ to indicate that L is completely processed. We denote the color at the lookahead position p by L_p.

Given a state S, we assume that new states can be created by applying the following two basic operations to every queue Q_i of S (if possible): Store L_p on Q_i (without moving any queue), or move Q_i by one position. We denote the set of new states obtained after the appliance of the first resp. the second basic operation to every queue of S by $N^+(S)$ resp. $N^{\leftarrow}(S)$. Note, that $|N^+(S)| \leq r$ and $|N^{\leftarrow}(S)| \leq r$ holds for every state S.

Our basic operations admit gaps on the queues and the output belt. We represent gaps by an additional "color" $f_0 \notin F$. Thus a color $f \in F$ or f_0 drops on the output belt whenever we move a queue. For convenience, we ignore the artificial color f_0 on the output belt, as it does not affect the number of color changes.

Proposition 2. *Suppose that we are given a state $S = (c; Q_1, \ldots, Q_r; p)$.*

1. *The set $N^+(S)$ is created as follows: Create a state*

$$S' = (c; Q_1, \ldots, Q_{i-1}, Q_i', Q_{i+1}, \ldots, Q_r; p + 1)$$

with $Q_i' = (f_1, \ldots, f_{q-1}, L_p)$ for every queue $Q_i = (f_1, \ldots, f_{q-1}, f_0)$ of S and add S' to $N^+(S)$.

2. *The set $N^{\leftarrow}(S)$ is created as follows: Create a state*

$$S' = (f_1; Q_1, \ldots, Q_{i-1}, Q_i', Q_{i+1}, \ldots, Q_r; p)$$

with $Q_i' = (f_2, \ldots, f_{q-1}, f_0)$ for every queue $Q_i = (f_1, \ldots, f_{q-1}, f_q)$ of S and add S' to $N^{\leftarrow}(S)$.

We denote the current minimal number of color changes on the output belt of a state S' by $\gamma(S')$ and use the following proposition to compute $\gamma(S')$.

Proposition 3. *Suppose that we are given a state S'. The value $\gamma(S')$ can be computed by*

$$\gamma(S') = \min_{P_1 \in P^+, P_2 \in P^{\leftarrow}} \{\gamma(P_1), \gamma(P_2) + \delta(P_2, S')\}$$

with $P^+ = \{P : S' \in N^+(P)\}$, $P^{\leftarrow} = \{P : S' \in N^{\leftarrow}(P)\}$, and δ defined by

$$\delta(P_2, S') := \begin{cases} 0 \text{ , if } c = c' \text{ or } c = \emptyset \\ 1 \text{ , else} \end{cases}$$

for $P_2 = (c; Q_1, \ldots, Q_r; p)$ and $S' = (c'; Q_1', \ldots, Q_r'; p) \in N^{\leftarrow}(P_2)$.

Thus, to compute $\gamma(S')$ for a state S', we have to consider $\gamma(S)$ of all states S from which S' can be created by the appliance of one of our two basic operations. In addition, if S' was created from S by the movement of a queue, we have to check whether an additional color change occured on the output belt of S' or not.

We use Proposition 3 to formulate our dynamic program (see Figure 3). An initial state $S_0 = (\emptyset; Q_1, \ldots, Q_r; 1)$ has an empty output belt, initially filled queues $Q_1, \ldots, Q_r$, and a lookahead position of $p = 1$. A state $S = (c; \emptyset, \ldots, \emptyset; l + 1)$ is completely processed.

Lemma 2. *The dynamic program depicted in Figure 3 solves an instance of the CSRP requiring a time complexity of $\mathcal{O}((|F| + 1)^{rq} \cdot |F| \cdot l \cdot r)$ and a space complexity of $\mathcal{O}((|F| + 1)^{rq} \cdot |F| \cdot l)$.*

```
1. Create an initial state S_0 with γ(S_0) = 0 and set P ← {S_0}
2. While P ≠ ∅
      Set Q ← ∅
      For all S ∈ P
         For all S' ∈ N^+(S) ∪ N^←(S)
            Compute γ(S') as described in Proposition 3
            If S' is not completely processed, set Q ← Q ∪ {S'}
            Else update best solution
      Set P ← Q
3. Return best solution
```

Fig. 3. The dynamic program for CSRP

3.2 Implementation Details

The huge theoretical complexity of the dynamic program can be drastically reduced in practice. Some non-obvious reasons for that are the admissibility to ignore gaps between colors on the sorting belts and on the lookahead, the avoidance of equivalent states with permuted queue sets by a lexicographical queue ordering, and the use of a suitable hash function for fast access to states.

Nevertheless, it remains a significant running time dependence of the color frequencies and distribution on the queues and the lookahead, the queue and the lookahead size, and the size of the underlying color set. We therefore doubt the practical use of our dynamic program, although it may be applicable for simulation purposes. For example, the solution of random CSRP instances consisting of $r = 3$ sorting belts (each of length $q = 10$ and initially filled with 5 colors), a completely filled lookahead of size $l = 8$, and an underlying color set of size $|F| = 4$, already requires an average time of approximately 400 seconds on a Pentium 3 PC with 900 MHz.

References

1. Wortmann D., Spieckermann S. (1995) Manufacturing Line Simulation of Automotive Industry to Enhance Productivity and Profitability. In: M. R. Heller (Ed.), Automotive Simulation '95, SCS-Society for Computer Simulation International, Erlangen, Germany, 91–106
2. Epping Th., Hochstättler W. (2001) Abuse of Multiple Sequence Alignment in a Paint Shop. Technical report zaik-2001-418, Center of Applied Computer Science, University of Cologne, Germany
3. Waterman M. S. (1995) Introduction to Computational Biology. Chapman & Hall
4. Lermen M., Reinert K. (1997) The Practical Use of the $\mathcal{A}^*$ Algorithm for Exact Multiple Sequence Alignment. Technical Report 97-1-028, Max-Planck-Institut für Informatik, Saarbrücken, Germany
5. Epping Th., Hochstättler W. (2002) Storage and Retrieval of Car Bodies by the Use of Line Storage Systems. Technical report btu-lsgdi-001.02, BTU Cottbus, Germany

On Solvability of the Project Scheduling Problem with Accumulative Resources of an Arbitrary Sign*

Edward Gimadi[1,2] and Sergey Sevastianov[1]

[1] Sobolev Institute of mathematics, 630090, Novosibirsk, Russia
[2] Corresponding author: gimadi@math.nsc.ru

Abstract. A project scheduling problem (PSP) with precedence constraints, deadlines and resource constraints is considered. It was known before that PSP is polynomially solvable if all resource constraints are of so called *accumulative type* (please, don't mix the notion with completely different notions *cumulative* and *cumulatives* lately introduced in the literature and being nothing but a kind of renewable resource constraints!), provided that both resource availabilities and resource requirements are non-negative. Now we show that PSP with accumulative resource constraints remains polynomially solvable if we allow resource availabilities of an arbitrary sign, while all resource requirements remain non-negative. On the other hand, it is shown that in the case of resource requirements of an arbitrary sign the problem becomes NP-hard in strong sense.

1 Introduction

For the resource constrained project scheduling problem (PSP) in parallel with renewable resources [3,4] a notion of *accumulative resources* was known since 1965 [8]. The PSP problem with this type of resource constraints was investigated in a number of papers (see, e.g., [5,7]) mostly written in Russian. It should be noted that the nature of the accumulative resource constraints drastically differs from that of renewable ones (which makes it possible to present polynomial time algorithms for solving quite general cases of the PSP problem under resource constraints of the accumulative type only, while the problem with renewable resource constraints is known to be NP-hard already in very simple cases). Indeed, in the case of *accumulative* resource constraints any quantity of resource allotted for unit time period t ($t = 1, 2, \ldots$) and non-utilized in that period can be consumed at any subsequent period $t' > t$ — unlike it happens with renewable resources, where any part of resource non-consumed at time period t is just wasted. To some extent, money, expendables (with considerably long shelf life), and other materials may be considered as accumulative resources. The term "accumulative" means that at any time period t all resources allotted (and not consumed) for all previous

* Supported by the Russian Foundation for Basic Research (project 02-01-01153), INTAS (project 00-217), and "Russian Universities" (project UR.04.01.012).

time periods are at our disposal: we consume resources **accumulated** for the previous time period.

It should be noted here that the terms *cumulative* and *cumulatives* resource constraints introduced in OR literature quite recently (in 1993 [1] and 2001 [2] respectively), while sound similar, have nothing to do with accumulative resource constraints, and in fact, are kinds of renewable resources. (Cumulative resource is just a special case of the latter, when there is only one renewable resource with constant in time availability, while *cumulatives* constraint extends the classical renewable resource constraints to the case of **negative resource requirements** — which may be treated as resource production — and to the case of resource constraints **from below** — when we need to cover a given amount of job specified in time.) Here the term "cumulative" relates to the **total resource requirement** summed up over all jobs being processed in a given unit time period t, — and one can easily observe that there is nothing new in this type of constraint, for it is typical for renewable resources. So, we find the notion "cumulative" redundant, while the term "cumulatives" is just illiterate, because "cumulative" is not a noun.

It can also be seen that any non-renewable resource may be considered as a special case of an accumulative one, when the whole amount of resource is allotted in the first unit time period, while the resource allotted in subsequent periods is equal to zero.

Known results. Let PSP$^\sigma$ stand for the PSP problem with resource constraints of only the accumulative type. In [7] some basic properties of optimal schedules for the PSP$^\sigma$ problem were esrablished, which allowed the authors to elaborate a pseudopolynomial algorithm.

In [5] for the case of real processing times a polynomial time asymptotically optimal algorithm was suggested with an absolute error tending to zero for increasing number of jobs.

Finally, a fully polynomial time algorithm was presented in [6] for solving the case of PSP$^\sigma$ problem, when both intensities of allotment (further referred to as *availabilities*) of accumulative resources and resource requirements are non-negative.

New results. We prove that PSP$^\sigma$ is still polynomially solvable, if negative availabilities are allowed, while resource requirements should still remain non-negative. On the other hand, it is shown that if negative resource requirements are allowed, the problem becomes NP-hard in strong sense.

2 Problem setting

Let us first formulate a more general problem with resource constraints of two types: accumulative and renewable.

We consider a project with a set of activities (jobs) $J = \{1, \ldots, n\}$. Precedence constraints are given by a directed acyclic graph $G = (V, U)$ (a so-called *reduction graph of a partial order*), where $V = \{v_1, \dots, v_n\}$ is a set of vertices that correspond to jobs, $j \in J$, and U is a set of edges: an edge (v_i, v_j) belongs to U if and only if job i is a direct predecessor of job j.

There are resource constraints of two types: accumulative and renewable. Corresponding sets of resources are denoted by $\mathcal{K}^\sigma$ and $\mathcal{K}^\varrho$ respectively. For each constrained resource $k \in \mathcal{K} \doteq \mathcal{K}^\varrho \cup \mathcal{K}^\sigma$ its available quantity $q_k(t)$ is allotted in the time interval $[t-1, t)$ ($q_k(t)$ may be of an arbitrary sign). For each resource $k \in \mathcal{K}^\sigma$ and each time interval $[t-1, t)$ the whole amount of resource allotted (and not consumed) in the interval $[0, t)$ may be consumed in the time interval $[t-1, t)$.

Processing time p_j and resource requirements $r_{jk}(\tau)$ are given for each activity $j \in J$, resource $k \in \mathcal{K}$, and time interval $[\tau - 1, \tau)$ $(\tau = 1, \dots, p_j)$ calculated from the starting point s_j of the activity.

Deadlines $d_j \in \mathbf{Z}^+$ are specified for all $j \in J_\mathrm{d} \subset J$.

The objective is to compute a schedule $S = \{s_j\}$ that meets all resource and precedence constraints and deadlines and minimizes the makespan.

A formal setting of the PSP problem is presented below.

To minimize the makespan

$$C_{\max}(S) = \max_{j \in J}(s_j + p_j) \tag{1}$$

under the following constraints on variables $\{s_j \mid j \in J\}$:

$$s_j + p_j \le d_j, \quad j \in J_\mathrm{d}; \tag{2}$$

$$s_i + p_i \le s_j, \quad i \in \mathrm{Pred}(j),\ j \in J; \tag{3}$$

$$\sum_{j \in J(t)} r_{jk}(t - s_j) \le q_k(t), \quad k \in \mathcal{K}^\varrho,\ t \in \mathbb{Z}^+ \setminus \{0\}; \tag{4}$$

$$\sum_{t'=1}^{t} \sum_{j \in J(t')} r_{jk}(t' - s_j) \le \sum_{t'=1}^{t} q_k(t'), \quad k \in \mathcal{K}^\sigma,\ t \in \mathbb{Z}^+ \setminus \{0\}; \tag{5}$$

$$s_j \in \mathbb{Z}^+, \quad j \in J, \tag{6}$$

where $J(t) = \{j \mid s_j < t \le s_j + p_j\}$, $\mathrm{Pred}(j)$ is the set of direct predecessors of job j.

The problem (1)–(6) is known to be NP-hard if $\mathcal{K}^\varrho \neq \varnothing$. We will further concentrate on the case $\mathcal{K}^\varrho = \varnothing$ (so, we have $\mathcal{K} = \mathcal{K}^\sigma$). The corresponding PSP problem will be denoted by PSP$^\sigma$.

Our positive results are formulated in the following section.

3 Polynomial solvability of the PSP$^\sigma$ problem under resource availabilities of an arbitrary sign

We assume that intensities of resource requirements and resource availabilities are given by sequences of pairs $\{(r^i_{jk}, p^i_{jk}),\ i = 1, \dots, N_{jk}\}$ and $\{(q^i_k, \alpha^i_k),\ i =$

$1, \ldots, N_k\}$, where i is the index of the ith interval of constancy of functions $r_{jk}(t)$ and $q_k(t)$ respectively, p^i_{jk} and α^i_k are the length and the starting point of the corresponding intervals.

Let T be a positive integer. A schedule $S^T = \{s^T_j \mid j \in J\}$ is called a *T-late schedule* if all s^T_j are maximum possible under constraints (2), (3), (6) and

$$s^T_j + p_j \leq T, \quad j \in J.$$

Theorem 1. *Problem* PSP$^\sigma$ *with nonnegative resource requirements admitting resource availabilities of an arbitrary sign can be solved in polynomial time*

$$O((N^- + \log p_{\max} + \log n)(u + N \log f + N' \log K)),$$

where n is the number of jobs,
u is the number of edges of the reduction graph G,
f is the number of independency of graph G (the width of the partial order on the set of jobs),
K is the number of constrained resources,
N is the total (over all resources and all jobs) number of intervals of constancy of functions $r_{jk}(t)$,
N' is N plus the total (over all resources $k \in \mathcal{K}$) number of intervals of constancy of functions $q_k(t)$,
N^- is the number of points t in which the functions $q_k(t)$ change their sign,
$p_{\max} = \max_{j \in J} p_j$.

Let us briefly outline the polynomial time algorithm for solving the PSP$^\sigma$ problem under assumption that resource requirements are nonnegative.

Algorithm $\mathcal{A}$

Preliminary calculation:

Find the values of each function $Q_k(t) = \sum_{t'=1}^{t} q_k(t')$, $k \in \mathcal{K}$, at nodes $t = \alpha^i_k$, $i = 1, \ldots, N_k$.

Calculate the values $R_k = \sum_{j=1}^{n} \sum_{i=1}^{N_{jk}} r^i_{jk} p^i_{jk}$ of the total need in each resource $k \in \mathcal{K}$.

Find the increasing order of (different) deadlines of jobs.

Compute the length T_{cr} of the *shortest schedule* S', which is equal to the length of the longest (*critical*) path in graph G.

Compute the minimal set of disjoint time intervals $\mathcal{T} = \{[a_l, b_l] \mid l = 1, \ldots, L\}$ containing the set

$$T_{\text{f}} = \{t \in \mathbf{Z}^+ \mid t \geq T_{\text{cr}},\ R_k \leq Q_k(t),\ \forall\, k \in \mathcal{K}\}.$$

(The last interval $[a_L, b_L]$ may be infinite, i.e., $b_L = \infty$ is possible. It is clear that L is no greater than the total number N^- of points t in which the functions $q_k(t)$ change their sign.)

For $l = 1, \ldots, L$ **do begin**
$a := a_l;\ b := b_l;$
if $b = \infty$ **then goto** M;
else if S^b is feasible **then goto** M;
end (for);
output ("There are no feasible solutions"); **STOP**;
M: $\widetilde{T} := \min\{b, a + T_{\rm cr}\}$;
if $S^{\widetilde{T}}$ is infeasible **then**
{ output ("There are no feasible solutions"); **STOP**}
else using dichotomy, find the minimal integer $t \in [a, \widetilde{T}]$ for which the schedule S^t is feasible, and output S^t as the desired optimal schedule.

4 NP-hardness of the PSP$^\sigma$ problem under resource requirements of an arbitrary sign

Theorem 2. *Problem* PSP$^\sigma$ *with resource requirements of an arbitrary sign is NP-hard in strong sense even in the special case, when there are no activities with deadlines;* $p_j = 1,\ \forall\ j \in J$*; graph* G *represents a collection of disjoint chains, and there is only one accumulative resource with a constant availability.*

To prove the above statement, given an instance $\{e_1, \ldots, e_{3k}\}$ of the 3-PARTITION problem ($\sum e_i = kE$; $\frac{1}{4}E < e_i < \frac{1}{2}E$, $\forall\ i$), we define the following instance of the PSP$^\sigma$ problem.

Graph G consists of $(3k+1)$ disjoint chains $C_1, \ldots, C_{3k+1}$. Chain C_i ($i = 1, \ldots, 3k$) contains two vertices: $v_i' \to v_i''$, the resource requirements of jobs corresponding to vertices v_i' and v_i'' being equal to e_i and $-e_i$ respectively. Chain C_{3k+1} consists of k vertices, resource requirements of the corresponding jobs are equal to E. All jobs have unit processing times. Resource availability is constant in time, equal to E in each unit time period.

It can be shown that the above instance of the PSP$^\sigma$ problem has the makespan $C_{\max}(S) = k + 1$, if and only if there exists a required 3-partition for the original 3-PARTITION problem.

It can be easily seen that problem PSP$^\sigma$ remains NP-hard in strong sense if graph G consists of chains having length at most 2, while the resource availability function is not constant.

5 Concluding remarks

Concerning the above positive result mentioned in Theorem 1, it should be noted that in addition to its self-dependent significance (as an important polynomially solvable case of the PSP problem), it can also be utilised in another way: for computing a nontrivial lower bound on the optimum of the PSP in general case, since accumulative resource constraints can be considered as a relaxation of renewable resource constraints.

References

1. Aggoun, A., Beldiceanu, N. (1993) Extending CHIP to solve Complex Scheduling and Packing Problems. *Mathl. Comput. Modelling* **17** (7), 57–73.
2. Beldiceanu, N., Carlsson, M. (2001) A New Multi-Resource cumulatives Constraint with Negative Heights. SICS Technical Report T2001:11, Uppsala, Sweden.
3. Böttcher, J., Drexl, A., Salewski, F. (1999) Project scheduling under partially renewable resource constraints. *Management Science* **45**, 543–559.
4. Brucker, P., Drexl, A., Möhring, R., Neumann, K., Pesch, E. (1999) Resource-constrained project scheduling: Notation, classification, models, and methods. *European J. Oper. Res.* **112**, 3–41.
5. Gimadi, E.Kh., Puzynina, N.M., Sevastianov, S.V. (1979) On some optimisation problems arrising while realization of large-scale projects (such as BAM) (in Russian). *Ekonomika i mat. metody* **5**, 1017–1020.
6. Gimadi, E.Kh., Zaljubovsky, V.V., Sevastianov, S.V. (2000) Polynomial solvability of the project scheduling problem with accumulative resources and deadlines, (in Russian). *Diskret. Analiz i Issled. Oper.* **7** (1), Ser. 2, 9–34.
7. Kozlov, M.K., Shafranskii, V.V. (1977) Project scheduling under given availabilities of acumulative resources (in Russian). *Izvestija AN SSSR. Technicheskaja kibernetika*, **no. 4**, 75–81.
8. Zukhovitskii, S.I., Radchik, I.A. (1965) Mathematical methods in project scheduling (in Russian). "Nauka", Moscow.

Cost Optimized Layout of Fibre Optic Networks in the Access Net Domain

Peter Bachhiesl[1], Gernot Paulus[2], Markus Prossegger[1], Joachim Werner[1], and Herbert Stögner[1]

[1] Carinthian Tech Institute, Department of Telematics and Network Engineering Primoschgasse 8, A-9020 Klagenfurt, email: p.bachhiesl@cti.ac.at

[2] Department of Geoinformatics Europastrae 4, A-9584 Villach

Abstract. During the last two years European network-carriers have invested 7.5 billions Euro in the expansion of the core and the distribution net domain (backbones, city backbones and metropolitan area networks). However, investigations have shown that about 95% of the total costs for the implementation may be expected for the area-wide realization of the last mile (access networks). In order to achieve a return on investment, carriers will be forced to link access net nodes, like corporate clients, private customers or communication nodes for modern mobile services (e.g. UMTS) to their city backbones. We present the methodology behind the planning tool NETQUEST-OPT for the computation of cost optimized and real world laying for fiber optic access networks. Bases on detailed geoinformation data, cluster strategies, exact and approximation algorithms of graph theory, combinatorial optimization and ring closure heuristics form the optimization kernel. Beside the methodology, we present results of a benchmark project which was processed with one of our industrial partners.

1 Introduction

During the last decade the demand on broadband capacities has entailed an increasing changeover from conventional transmission techniques to the fiber optic technology [6]. In order to achieve a return on investment all major network carriers will be forced to push the implementation of the non existing and cost intensive last mile [9].

A variety of strategic planing tools focus on problems of laying optimization, discussion of reliability and traffic optimization in the core net domain [1]. Nowadays carriers hardly use any computational planning tools for cost estimation and optimization in the access net domain. The planning process is performed manually based on expert knowledge and therefore often yield suboptimal decisions. However, since the telecommunication market runs through a period of consolidation the neglection of economization potentials has to be denoted as a critical competitive disadvantage.

In several discussions with our industrial partners it has turned out that the following problem is of substantial interest: minimize the implementation costs for a fiber optic network in the access net domain under consideration of the following constraints: **(Problem Dimension)** Up to 2000 connection objects (access net nodes) in an area up to 4 square kilometers are assumed as appropriate access net dimension. **(Real World Condition)** Computations shall be done with respect to real world conditions including the local spatial topology. The implementation costs are mainly determined by underground work; material costs, e.g. cable costs or pipe installation costs, are taken into account by a fixed penalty per meter. **(Network Topologies and Reliability)** Umbrella networks and considering standards of quality (e.g. required by COLT Telecom) also meshed networks shall be supported such that all access net nodes are supplied in a reliable way under acceptance of branch lines with a predefined maximum length of $K_{\max}$ meters.

2 Umbrella Networks and Meshed Networks in the Access Net Domain

In order to consider real world conditions detailed geoinformation data, usually supplied by the network carriers, are essential. According to the methods in [8] the following information is extracted from the GIS databases: **(Penalty Grid)** A spatially balanced score card combines all relevant land classes with typical implementation costs. The access net domain is regarded as a regular grid $[x_i, x_{i+1}] \times [y_j, y_{j+1}]$, i and j are indices of finite index sets. Each entry $P_{i,j}$ of a penalty matrix P describes the specific implementation costs for the grid pixel $[x_i, x_{i+1}] \times [y_j, y_{j+1}]$. **(Auxiliary geometries)** Plane geometries of all relevant land classes are exported. The shapes of these geometries are determined by a set S of auxiliary nodes. **(Access net nodes)** Finally, the set R of access net nodes is specified by the according coordinate vectors.

Conceptually, we can formulate the route planning problem on an undirected graph $G = (V, E; W)$. V is the disjoint union of R and S, E is the set of possible edges $[i, j]$ for $i, j \in V$. W denotes a symmetric cost matrix measuring the edges of E with respect to the penalty grid P. Let s_k be an element of the finite set of K intersections for the edge $[i, j]$ with the grid lines of P, whereby $s_0 := i$, $s_K := j$ and $COORD\,(s_k)$ are the according coordinates. Furthermore, we denote P_k as the entry of the penalty matrix P corresponding to the predecessor pixels of s_k. Thus the entries of W are defined as:

$$W_{i,j} := \begin{cases} \sum_{k=0}^{K-1} \|COORD\,(s_{k+1}) - COORD\,(s_k)\|_2 \, P_{k+1} F & \text{for } i \neq j \\ \infty & \text{otherwise} \end{cases} \tag{1}$$

F is a fixed penalty including material costs per unit distance and $\|.\|_2$ means the euclidian norm.

Formally, an umbrella network represents a tree T which spans all access net nodes R under the consideration of the auxiliary nodes in S. If $E_T \subseteq E$

is the set of edges of T, we consider the optimization problem:

$$T = \arg\min_{\bar{T}} \left\{ \sum_{[i,j] \in E_{\bar{T}}} W_{i,j} \;\middle|\; \bar{T} \text{ is spanning } R \cup S' \text{ , } S' \in p(S) \right\} \tag{2}$$

$p(S)$ means the power set of S. (2) is known in literature as the steiner network problem in graphs [3]. Upper bounds to the optimal solution (2) are computed by local preprocessing techniques and polynomial time algorithms up to problem dimensions of $|V| \cong 2000$, $|R| \cong |V|/2$ under consideration of sparse graphs [7]. Nevertheless, due to the complexity dealing with real world geometries, the route planning problem lead to dimensions up to $|V| \cong 20000$ nodes. Such problem dimensions prompted us to segment (2) in three hierarchical steps: determination of local clusters, local solution (2) in each cluster and computing cluster shortest spanning trees (cluster SST problem).

Standard procedures based on linkage information and cophenetic correlation are used to cluster groups of access net nodes to local clusters C^k with respect to W [2]. Thus the problem (2) is reduced to the graph $G^k = (V^k = R^k \cup S^k, E^k; W^k)$ for the k-th cluster C^k, preprocessed according to [4] and solved by the algorithm proposed by Hakami and Lawler [3]. Since the algorithm is exponential in the number of auxiliary nodes S^k we restrict the investigations of spanning trees to the vertex sets $R^k \cup S'$, $S' \in p(S^k)$ and $|S'| \leq p_{relax}$. In order to estimate the quality of achieved solutions 100 test problems with OR-Instances B, C, D and E proposed in the SteinLib library [7] were computed and yield the performance ratio estimate $1 \leq \{A(I)/OPT(I) \mid \text{all instances } I\} \leq 1.2$ whereby $A(I)$ and $OPT(I)$ mean the approximation and the optimal solution for the instance I, respectively.

The segmentation to independent clusters allows an efficient distributed processing which is realized by a proprietary development based on the PARMAT toolbox. T^k denotes the optimized tree in the k-th cluster.

In order to achieve a global umbrella network, we consider each cluster as a node of superior complete graph H. The costs of edge $[i,j]$ is given by the cheapest paths between the trees T^i and T^j under consideration of all auxiliary nodes in S. Finally we compute a shortest spanning tree problem on H to obtain the links between the local cluster trees. The union of the found cluster interconnecting links and the local cluster trees yields the cost optimized umbrella network T.

For reliable network structures we have to find a cost optimized redundant network on G under acceptance of branch lines with a maximum distance of $K_{\max}$ meters. In this case we propose an augmentation heuristic: if T is a subgraph of G we search for an cost optimized edge-disjoint subgraph $G^* \subseteq G$ such that $T \cup G^*$ is 2-connected excepting branch lines with a maximum length of $K_{\max}$ meters. The computation of branch lines and their lengths for a given graph is based on depth first search algorithms for the determination of blocks and cut vertices. Furthermore we define $G_r := T \cup G^*$, $G_r = (V_r, E_r)$ as the meshed part and $G_{nr} := G \backslash G_r$ the non meshed part of the considered network, respectively. If B is the set of leaves in T, we consider $P(u,v)$ for $u, v \in B$ as the cost optimal path from the u to v in G under the constraint that $P(u,v) \cap T = 0$. $P(u,v)$ is denoted as a feasible augmentation path with costs $w(P(u,v))$. The united graph $T \cup P(u,v)$ contains a circle $C(u,v)$ depending on $P(u,v)$. The variable $l[C(u,v)]$ represents the sum over all edge lengths $l(e)$ in the circle $C(u,v)$ whereby $l(e) := 0$ for $e \in E_r$ or $e \in P(u,v)$ and $l(e) := \|i-j\|_2$ otherwise. $l[C(u,v)]$ measures the length of that part of T which is made ring closed by including the feasible augmentation path $P(u,v)$. Thus, we define

$$\kappa(u,v) := w[P(u,v)] \ / \ l[C(u,v)] \tag{3}$$

as the ring closure relation for the leaves u and v. $\kappa(u,v)$ estimates the quality of a ring closure between the leaves u and v. An iterative procedure inserts leave connecting paths in G_r with respect to (3) until the maximum branch line length is lower than S_{max}. This greedy typed procedure yields the ring closed network T^R as an upper bound for the searched optimal one. In order to improve T^R we propose an edge deleting stingy postprocessing as described in [5].

3 An Industrial Benchmark Project

We regard an access net domain provided by NetCologne, one of the major private city net carriers in Germany. We were able to benchmark the methods described above under consideration of the following two aspects: validity of the underlying spatially balanced score card and the demonstration of economization with respect to the traditional process of manual planning.

Figure 1 shows the penalty grid resulting from the spatially balanced score card and figure 2 depicts the according auxiliary geometries, the clustering and all auxiliary points in the first cluster. We have recalculated the NetCologne laying of an already existing umbrella network in order to check the validity of our computational approach and the result was confirmed by experienced networks planers within an accuracy of $\pm 5\%$. We have solved the local problems (2) and the cluster spanning tree problem for the umbrella network in distributed mode on three workstations using 223 minutes of CPU-time. The result is depicted in figure 3 and saves 22% distance and 20% costs with respect to the NetCologne laying. We have also computed a meshed network with maximum allowed branch length $K_{max} = 200$ meters as shown in figure 4. Even this meshed network yields a distance and cost reduction of 20% and 11%, respectively.

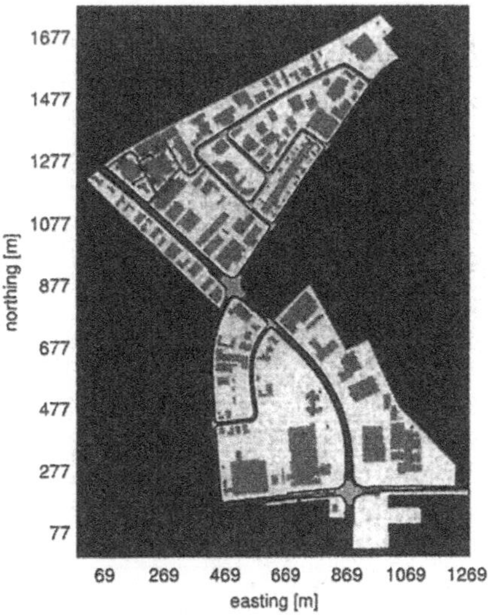

Fig. 1. Penalty grid for the NetCologne project. Gridding is done with a resolution of one meter in square and increasingly colored from blue to red with respect to the average implementation costs [Euro/meter]: sidewalk = 87, stree = 110, unsurfaced area = 54 and crossing = 123. Branch lines for connection objects inside of parcels are weighed by 52 Euro/meter.

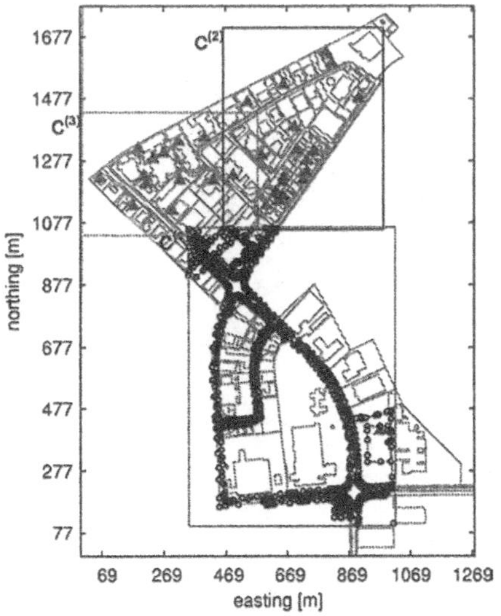

Fig. 2. 37 access net nodes are marked by red triangles. The black circles depict the auxiliary nodes of the first cluster based on the underlying auxiliary geometries. The problem dimension are: for C^1 $|V^1| = 1260, |R^1| = 5$, for C^2 $|V^2| = 1935, |R^2| = 19$ and for C^3 $|V^3| = 2150, |R^3| = 13$, total $|V| = 5345$.

4 Concluding Remarks

The accuracy of the penalty matrix as an output of the spatially balanced score card has to be seen as crucial factor for validity of computational results. In general the specific implementation costs are influenced by a great variety of influencing parameters. Therefore we have to emphasize at this position that the above methodology is dimensioned for a rough planning process of network layings. The details and the according accomplishment on location continue to be task of experienced construction engineers, but may be done in a structured way based on a cost optimized approach.

The consideration of usable infrastructures (subway, pipelines, partial renting existing networks of competitors and others) as well as investigations with respect to improved optimization routines allowing the treatment of enlarged cluster dimensions are topic of current work. It shall be noticed that improvements of the optimization performance have to be challenged

Fig. 3. Computed umbrella network for 37 access net nodes.

Fig. 4. Meshed network. Reliability was exclusively required for the upper part from y-coordinates 877 up to 1677.

under the consideration of the rough planning process and the inaccuracy of the underlying spatially balanced score card.

References

1. Bertsekas, D. (1998) Network Optimization: Continuous and Discrete Models. Athena Scientific, Belmont
2. Blashfield, R. (1984) Cluster Analysis. Sage Publications, Santa Barbara
3. Cieslik D. (1998) Steiner Minimal Trees. Kluwer Academic Publishers, Dordrecht.
4. Duin, C. (2000) Preprocessing the Steiner Problem in Graphs. In Du, D.Z.. et. al. (eds.), Advances in Steiner Trees. Kluwer Acadamic Publishers, Dordrecht
5. Fortz, B. (2000) Design of Survivable Networks with Bounded Rings. Kluwer Academic Publishers, Dordrecht
6. Gilder G. (2002) Telecosm: The World after Bandwith Abundance. Touchstone Books, Charmichael
7. Koch, T., et.al. (2001). SteinLib: An Updated Libary on Steiner Tree Problems in Graphs. In Cheng, X., et.al. (eds.), Steiner Trees in Industry. Kluwer Acadamic Publishers, Dordrecht
8. Paulus, G., et al. (2002) Planning the Next Generation Network Infratructure in Urban Areas. Proc URISA Ann Conf, in print.
9. Stögner, H., et al. (2001) Ein Top-Down Approach für die Planung von photonischen Netzen. Telematik. **4**, 4-6

Ein Algorithmus zur sicheren elektronischen Stimmabgabe über das Internet

Alexander Prosser[1]

Abteilung Produktionsmanagement
Wirtschaftsuniversität Wien, Augasse 2-6, A-1090 Wien
E-mail: alexander.prosser@wu-wien.ac.at

Robert Müller-Török[2]

Beratungsgesellschaft für Beteiligungsverwaltung Leipzig mbH
Ferdinand-Rhode-Straße 16, D-04107 Leipzig
E-mail: mueller-toeroek@BBVL.de

1 Einführung

1.1 Anforderungen

In einem grundlegenden Beitrag identifizieren Nurmi, Salomaa und Santean [1] zwei Dienste im Rahmen eines E-Voting-Systems: den ***Registrator***, bei dem eine elektronische Wahlberechtigung eingeholt wird, und die Stimmabgabestelle (elektronische ***Urne***). Im folgenden wird die Eignung von aus der Literatur bekannten Verfahren diskutiert und ein eigener Vorschlag vorgestellt.[3]

---

1 Die Forschungsarbeit dieses Autors wurde durch den Jubiläumsfonds der Stadt Wien unterstützt.

2 Der Artikel gibt die persönliche Meinung des Autors wieder.

3 Nicht-kryptographische Verfahren, wie etwa die Ausgabe von TANs für Wahl"transaktionen", werden hier nicht berücksichtigt, da sie die Anonymität nicht sichern können. Kolludieren etwa die Stelle zur TAN-Ausgabe und die elektronische „Urne", so sind über die TAN die Stimmen den einzelnen Wahlberechtigten zuordenbar. Die Einhaltung der Wahlgrundsätze (hier Anonymität) hängt daher von der Integrität der Wahlveranstalter ab, garantiert werden kann sie dem Wähler nicht.

1.2 ANDOS-basierte Verfahren

Ein entsprechendes Protokoll wurde von [2] vorgeschlagen und geht von einem Verkäufer aus, der n geheime Bitmuster zu vergeben hat, die von einer Anzahl von Käufern anonym erworben werden (für die Anwendung auf Wahlen s. [3], [4]). Das Protokoll gilt als sicher, gilt aber als schlecht skalierbar. Gemäß Protokoll muß eine zufällig zusammengestellte Gruppe von Wählern so viele Zufallszahlen bilden, wie es Teilnehmer in der Gruppe gibt, und das Protokoll durchführen, außerdem muß eine Gruppe von „Käufern" das Protokoll gleichzeitig durchspielen, was die Praktikabilität weiter einschränkt.

1.3 Zero-knowledge proof (ZKP)-Protokolle

Diese Verfahren wurden für den Dialog einer Chipkarte mit einem Kartenleser entwickelt, wobei die Karte den Kartenleser überzeugt, im Besitz der zur Chipkarte gehörenden geheimen ID zu sein, ohne daß diese bekanntgegeben wird (für konkrete Verfahren von ZKP-Protokollen s. [5] auf RSA-Basis und [6] auf Basis diskreter Logarithmen. Es wurde vorgeschlagen (s. [7] und die darin angeführte Literatur) ZKP-Protokolle für die anonyme Stimmabgabe zu verwenden, wobei aus der Wähler-ID ein nicht-rückrechenbares Token t gebildet würde (bei Verwendung des Fiat-Shamir-Protokolls wäre dies $t = ID^2 \bmod(n)$ mit n als dem Produkt zweier „ausreichend großer" Primzahlen.

Dieses Token wird bei einem Registrator vor der Wahl vom Wählenden beantragt und auf einer Chipkarte zwischengespeichert. Bei der Stimmabgabe identifiziert sich der Wählende mit seinem Token, wobei er seine Berechtigung, das Token einzusetzen, mit der Kenntnis (aber eben nicht Bekanntgabe) der ID nachweist. Wie die Analyse des Einsatzes von ZKP-Protokollen für E-Voting in [8] zeigt, kann die Anonymität dem Wählenden auch hier nicht garantiert werden. Zumindest bei der Registrierung ist der Zusammenhang zwischen der ID des Wählers und dem Token bekannt. Kolludieren daher Registrator und Urne, kann die Anonymität durchbrochen werden.

1.4 Das FOO-Protokoll

Das Verfahren von Fujioka, Okamoto und Ohta [9] basiert auf der blinden Signatur wie sie von David Chaum [10] vorgeschlagen wurde. Es ist die Basis für zahlreiche E-Voting-Implementierungen, beispielsweise das System Sensus von Cranor und Cytron [11][4]. Im FOO-Protokoll erfolgen Registrierung und Stimmabgabe in einem Schritt:

1. Der Wählende bildet ein asymmetrisches Schlüsselpaar (k, k') nach einem beliebigen Verfahren.

---

[4] Das deutsche Produkt i-vote (http://www.e-vote.de, vormals http://www.internetwahlen.de) basiert deklariertermaßen auf der blinden Signatur; eine definitive Aussage bezüglich des verwendeten Protokolls ist aber nicht möglich, da der Algorithmus unseres Kenntnisstandes nach nicht veröffentlicht wurde.

2. Der Registrator bildet für alle Wähler (einer bestimmten Gruppe) ein Schlüsselpaar (e,d) nach dem RSA-Verfahren $\bmod(n)$.
3. Der Wählende W fordert vom Registrator einen Stimmzettel SZ an, füllt diesen aus und sendet $x = r^e k(SZ) \bmod(n)$ mir r als Zufallszahl an den Registrator zurück.
4. Dieser prüft die Wahlberechtigung des Wählenden bzw. ob dieser bereits eine Stimme abgegeben hat und sendet im Gutfall x^d zurück, das gemäß blindem Signaturverfahren zu $\frac{x^d}{r} = k(SZ)^d \bmod(n)$ aufgelöst wird.
5. Der Wählende sendet seine so autorisierte Stimme $\lfloor k(SZ), k(SZ)^d \rfloor$ an die Urne, die die Autorisierung überprüft und die (noch mit k codierte) Stimme abspeichert.
6. Diese codierten Stimmzettel werden veröffentlicht und abschließend sendet der Wählende k' an die Urne, die damit $k(SZ)$ öffnet.

Der letzte Schritt wird im ursprünglichen Protokoll zeitlich getrennt durchgeführt, die den Autoren bekannten Implementierungen lassen diesen Schritt jedoch unmittelbar auf Schritt 5 folgen. Die Darstellung konzentriert sich auf das eigentliche Stimmabgabeprotokoll, selbstverständlich sind alle Kommunikationsvorgänge kryptographisch gesichert sowie durch digitale Signatur authentifiziert (letzteres gilt natürlich nicht für das Einreichen des Stimmzettels und des k' in den Schritten 5 und 6 durch den Wählenden.

1.5 Kritische Würdigung des FOO-Protokolls

Während auf Anwendungsebene die Anonymität des Wählenden durch die blinde Signatur bei Verwendung genügend großer Schlüssel gesichert ist, kann auf Systemebene die Anonymität durch dieses Verfahren nicht sichergestellt werden, da (identifizierte) Registrierung und (anonyme) Stimmabgabe in einem Schritt erfolgen. Eine Rückverfolgung des Wählenden wäre über die IP-Adresse oder über Temporärdateien (z.B. Cookies) möglich. Dieses Problem hat unabhängig vom verwendeten Algorithmus jedes in einem Schritt ablaufende E-Voting-Protokoll.

Weiters haben die Administratoren des Systems die Möglichkeit Stimmen für Nichtwähler einzuschleusen. Sie generieren dazu Schlüsselpaare (k,k'), zu denen die entsprechenden Stimmzettel ins System eingeschleust werden; wobei schließlich die „Wählenden" ihr k' einreichen. Dies kann am Ende der Wahl für Nichtwähler durchgeführt werden, wenn dabei auch die entsprechenden Logfile-Einträge manipuliert werden, kann eine korrekte Historie der Stimmabgabe konstruiert werden.

2 Ein E-Voting-Protokollvorschlag

2.1 Das Protokoll

Der hier gemachte Protokollvorschlag ist eine Weiterentwicklung des Algorithmus von [12]. Wie im FOO-Protokoll existieren ein Registrator und eine elektronische Urne. Das Protokoll läuft aber hier in zwei zeitlich voneinander strikt getrennten Phasen ab: der Registrierung, bei der ein blind signiertes Wahltoken ausgegeben wird, und der Stimmabgabe. Wie in der Darstellung des FOO-Protokolls werden die generelle Verschlüsselung und digitale Signatur der ausgetauschten Nachrichten weggelassen, da es sich um Standardfunktionalität handelt.

Registrierung:

1. Der Registrator hält ein nach dem RSA-Verfahren gebildetes Schlüsselpaar (e,d) pro Wahlsprengel w bereit; jedes für die Wahl verwendbare Trust Center ein Schlüsselpaar (ε,δ).
2. Der Wahlberechtigte sendet sein Trust Center-Zertifikat an den Registrator, der nach Auflösen des Zertifikats gegen das Trust Center und der Prüfung der Wahlberechtigung die Kennung des Sprengels w und den zu w passenden Schlüssel e.
3. Der Wahlberechtigte bildet zwei asymmetrische Schlüsselpaare (u,t) und (ϖ,τ), dies kann, muß aber nicht nach dem RSA-Verfahren geschehen. Es wird $b = r^e t (\mathrm{mod}\, n)$ und $\beta = \rho^\varepsilon \tau (\mathrm{mod}\, n)$ gebildet mit r,ρ als Zufallszahlen. Er signiert nun ein Paket aus b, w und einem standardisierten Antragstext für die Ausstellung eines Wahltokens und sendet dies an den Registrator. Nach der Prüfung der Wahlberechtigung signiert der Registrator b^d und sendet dies an den Wählenden zurück. Dieser dividiert durch r und erhält (t,t^d).
4. Der analoge Prozeß wird mit dem Trust Center durchgeführt, wodurch der Wahlberechtigte den zweiten Teil seines Wahltokens erhält, (τ,τ^δ).

Zwischenspeicherung

1. Der Wahlberechtigte speichert nach Ende der Registrierung seine beiden Tokenteile $\|(t,t^d),(\tau,\tau^\delta)\|$, w und die geheimen u,ω auf einem sicheren Medium ab (zur Implementierung vgl. Abschnitt 2.3).

Stimmabgabe

Bei den folgenden Kommunikationsschritten verwendet der Wählende seine digitale Signatur und auch beim Trust Center gehaltene Public Key Kryptographie nicht, da diese Schritte ja streng anonym erfolgen sollen.

1. Der Wählende generiert ein Paar an Session Keys (k,k') und
2. sendet die beiden Tokenteile, w, die Kennung seines Trust Centers und k' an die Urne.

3. Diese kennt alle relevanten e und ε und überprüft, ob $\left(t^d\right)^e = t$ und $\left(\tau^\delta\right)^\varepsilon = \tau$, ob das Token bereits eingesetzt wurde und ob w und e zusammenpassen. Im Gutfall wird ein laufend numerierter Stimmzettel SZ_m verschlüsselt als $k'(SZ_m)$ an den Wählenden gesandt.
4. Dieser decodiert mit k, füllt SZ_m aus und sendet die beiden Tokenteile, k', die Trust Center Kennung, $SZ_m, \varpi(SZ_m), u(SZ_m)$ an die Urne.
5. Diese prüft nochmals die Berechtigung des Tokens und ob dieses noch nicht eingesetzt wurde und prüft im Gutfall die Authentizität des Tokens wie in Schritt 3.
6. Zusätzlich prüft die Urne, ob $t(u(SZ_m)) = SZ_m$ und $\tau(\varpi(SZ_m)) = SZ_m$. Im Gutfall speichert die Urne die erhaltenen Parameter ab und übergibt SZ_m an die Stimmenauszählung.

2.2 Kritische Würdigung

Die Anonymität des Wählenden wird durch die blinde Signatur auf Anwendungsebene gesichert, da (identifizierte) Registrierung und (anonyme) Stimmabgabe in getrennten Schritten erfolgen, kann eine Rückverfolgung über die IP-Adresse technisch ausgeschlossen werden – nur wenn Registrierung und Stimmabgabe vom selben Endgerät aus bei starrer IP-Adreßvergabe erfolgen, ist eine Rückverfolgung technisch überhaupt machbar.

Anders als beim Algorithmus von Fujioka et al. kann der Administrator keine Stimmzettel für sich der Wahl enthaltende elektronische Wähler einfügen, da nicht nur das vom Registrator signierte Token, sondern auch das eines zertifizierten Trust Center notwendig ist. Fügt der Administrator doch Stimmzettel ein, ist dies erkennbar, die Stimmzettel können nachträglich entfernt und das amtliche Ergebnis entsprechend korrigiert werden. Auch der Wähler kann nicht nachträglich behaupten, eine andere Stimme abgegeben zu haben, da er sich durch die Schlüsselpaare (u,t) und (ω,τ) selbst gebunden hat, wobei t und τ durch den Registrator authentifiziert werden.

2.3 Implementierung

Die Zwischenspeicherung des Wahltokens setzt ein sicherer, an die Person gebundenes Medium voraus. Sinnvoll eignen sich dafür Chip- bzw. Signaturkarten, wobei die Identifizierung des Wählers (ID) erfolgt entweder über die weltweit eindeutige Chipkarten-ID, was aber voraussetzt, daß der Wahlveranstalter die Karte selbst ausgegeben hat und bei der Ausgabe die Verbindung Person-Karte gespeichert hat (wie dies beispielsweise beim Studentenausweis der Wirtschaftsuniversität Wien der Fall ist), oder aber über eine eindeutige Personenkennzahl, die im Zertifikat gespeichert ist, beispielsweise über die Personenzahl des Zentralen Melderegisters in Österreich („Bürgerkarte“).

In Österreich angebotene Signaturkarten allerdings halten das Zertifikat in einem frei auslesbaren Bereich, so daß die Karte für die anonyme Stimmabgabe erst nach einem entsprechenden Redesign geeignet ist; so könnte etwas das Zertifikat PIN-gesichert werden.

Literatur

[1] H. Nurmi, A. Salomaa, L. Santean. Secret ballot elections in computer networks. *Computers and Security* 36 (10):553-560, 1991.

[2] G. Brassard, C. Crepeau, J.-M. Robert. All-or-Nothing Disclosure of Secrets. Proceedings of CRYPTO 86, Springer Verlag, 234-238, 1987.

[3] A. Salomaa. Verifying and recasting secret ballots in computer networks. *New Results and New Trends in Computer Science*. Springer-Verlag, Berlin, 283-289, 1991.

[4] A. Salomaa, L. Santean. Secret Selling of Secrets with Several Buyers, EATCS Bulletin 42: 178-186, 1990.

[5] A. Fiat, A. Shamir. How to prove yourself: Practical Solutions to identifications and signature problems. *Advances in Cryptology – Crypto 86*. Springer-Verlag, Berlin, 1987.

[6] C.P. Schnorr. Efficient identification and signatures for smart cards. *Journal of Cryptology* (4):161-174, 1991.

[7] H.C.A. van Tilborg. *Fundamentals of Cryptology*. Kluwers Academic Publishers, Boston, 316ff, 2000.

[8] A. Prosser, R. Müller-Török. E-Voting – eine neue Qualität im demokratischen Entscheidungsprozeß; akzeptiert durch *Wirtschaftsinformatik*.

[9] A. Fujioka, T. Okamoto, K. Ohta. A Practical Secret Voting Scheme for Large Scale Elections. *Advances in Cryptology – AUSCRYPT92*. Springer-Verlag, Berlin 244-251, 1993.

[10] D. Chaum: Blind Signatures for Untraceable Payments. D. Chaum, R.L. Rivest, A.T. Sherman. *Advances in Cryptology, Proceedings of Crypto 82*, 199-203, 1982.

[11] L.F. Cranor, R.K. Cytron. Sensus: A Security-Conscious Electronic Polling System for the Internet. *Proceedings of the Hawai International Conference on System Sciences (HICSS-97)*. Hawai, 1997 downloaded from http://lorrie.cranor.org/pubs/hicss/hicss.html (4.2.2001).

[12] A. Prosser, R. Müller-Török: Electronic Voting via The Internet. *3rd International Conference on Enterprise Information Systems, ICEIS*. Setubal, 1061-1066, 2001.

Accelerated MILP-Strategies for the Optimal Operation Planning of Energy Supply System

Peter Hackländer and Johannes F. Verstege

University of Wuppertal, Institute for Power System Engineering, Fuhlrottstr. 10, 42097 Wuppertal, Germany, E-Mail: hacklaender@uni-wuppertal.de

Abstract. In this paper a MILP-based optimization algorithm for the optimal operation planning of energy supply systems is presented. For this purpose, the main tasks of optimal operation planning are discussed and the Priority-Based Dynamic Search Strategy (PBDSS) for the solution of the optimization problem is developed taking into account different acceleration strategies. Exemplary results comparing standard branch-and-bound algorithm to the new search strategy demonstrate the significant improvement of the optimization process.

1 Introduction

Since the European directives for the internal markets in electricity and gas have been passed to manage the requirements for competition, the opening of the European energy markets is pushed forward by national laws and agreements. Although new opportunities arise for energy supply systems in liberalized markets, nevertheless, the operators of the systems are forced to produce and distribute energy more efficiently to increase their competitiveness. One substantial chance to reach this aim is the *optimal operation planning* which enables the economical operation of the energy supply system considering the following tasks [2]:

- *Optimal commitment of plants and contracts*: The main problem of the planning process is to determine the optimal commitment of all plants and contracts (i.e. to minimize the operational costs) with respect to all technical, contractual and further restrictions (better known as unit commitment and economic dispatch).
- *Observation and evaluation of energy markets*: Apart from the traditional operation planning, the operator now has to observe and evaluate the short-term electricity and gas markets in order to reduce the operational costs. Thereby, not only the economic benefit has to be taken into account, also the technical feasibility must always be guaranteed.
- *Management of storages*: The economical operation of energy supply systems requires a preferably simultaneous energy generation (or purchase) and consumption. Unfortunately, this criterion can often not be fulfilled due to the varying demand. One way to achieve better exploitation is to use energy storages. Whereas electric power can only be stored with high

costs and only to a small degree, there are various possibilities to store energy (particularly district heating and natural gas) with lower costs and more efficiently [2].

As shown above, the optimal operation planning consists of different operation tasks which have to be considered at once within one closed algorithm to meet the manifold interlinkings and to yield the most synergetic effect. This requirement leads to a complex optimization problem, which can only be solved with computer-aided methods of Operations Research. As shown in many practical applications, the Mixed-Integer-Linear-Programming (MILP) is a useful method to optimize complex power systems because it fulfills exactly every (even time-coupling) restriction introduced, it is able to find the global solution and to calculate the gap between the optimum and suboptimal solutions and it enables the use of continuous as well as discrete variables. Indeed, the MILP requires the mathematical formulation of the entire optimization problem and the optimization algorithm normally needs much computing time to solve complex optimization problems. The MILP-specific mathematical formulation of different energy supply systems has already been presented in several preceding works [3,4]. Therefore, this paper concentrates on the solution process of the formulated optimization problem.

2 Solution Algorithm

In practice, the computing times for the solution of the described optimization problem is limited depending on the considered planning period and the specifications of the operator. Besides, due to the increasing importance of short-term and spot-market contracts, operational decisions have to be made more frequent. However, as mentioned above, the MILP is – compared to other algorithms – rather slow. Thus, a special solution algorithm containing different acceleration strategies has been developed to improve the optimization process:

The standard MILP algorithm first computes the so called primal solution (PS) neglecting the discrete character of the binary variables (relaxed optimization problem) [1]. Thereby, the used simplex algorithm can be significantly accelerated if existing primal results of a basis scenario are used as start values for the solution of similar optimization problems (e.g. scenarios with higher/lower demand or other energy prices). In different case studies it could be shown that the computing time can be even reduced up to 90%.

After calculating the primal solution the subsequent branch-and-bound-algorithm (B&B) is used to find the first feasible integer solution. This is done by branching the binary variables successively to valid values 0 or 1 with the primal objective function value as a natural lower limit (the objective function value of a feasible integer solution can only be either equal or higher than the primal value). The standard B&B (sB&B) calculates the branching order by an algorithm, which first estimates the effect of branching on the objective function for each binary variable and then chooses the variable with the greatest effect [1]. Unfortunately, this method holds the risk that variables, which only have a small effect on the objective function but a high influence on the feasibility of the optimization problem are branched very late resulting in an unsatisfactory convergence behaviour. Figure 1 illustrates the problem on the basis of two binary optimization variables: Whereas the solution algorithm has to examine six branches to notice the infeasibility of the optimization problem (left case), the algorithm can yet be finished after two steps if another branching order is used (right case).

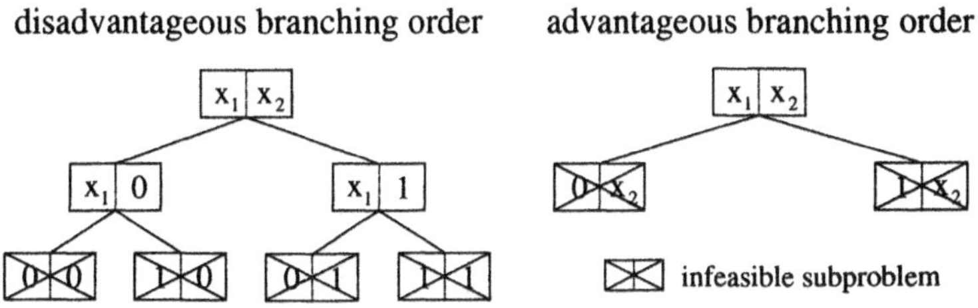

Fig. 1. Comparison of different branching orders

The described lack of the sB&B can be reduced if the branching order is not calculated by the mentioned estimation algorithm, but by a priority-based strategy: This uses the results of the PS – where all binary variables are relaxed to real numbers between 0 and 1 – to control the selection of branching variables more sophisticatedly taking advantage of the fact, that variables with a relaxed value near 0.5 have more influence on the feasibility if they are branched to 0 or 1 than variables nearby 0 or 1. Therefore, any discrete variable gets a priority between 1 (branched soon) and 1000 (branched late) which is calculated on the basis of the primal results as follows

$$\varphi_i = 1 + 1998 \cdot |x_i - 0.5| \tag{1}$$

where φ_i is the priority of variable i at the beginning of the B&B and x_i the relaxed value of variable i after PS.

After finding a first feasible integer solution, further solutions are searched by the Dynamic Search Strategy (DSS), which removes two further drawbacks of the sB&B:

- The objective function values of several feasible solutions often differ only slightly from each other, so valuable computing time is wasted without finding a significant better solution.
- The last integer solution found cannot be treated as the global solution before the whole branching tree has been analyzed. Thus, the time between finding the last solution and the proof that there are no better solutions can be unacceptable long.

The basic idea of the DSS (illustrated in Fig. 2) is not only to search integer solutions (depth-first search), but also to increase the lower limit (breadth-first search). This is used to decrease the accuracy of the last integer solution found, which becomes then

$$\sigma = (v_{Sol} - v_{LL}) / v_{Sol} \qquad (2)$$

where σ is the accuracy, v_{Sol} the objective function value of the last integer solution found and v_{LL} the value of the highest lower limit. While optimization the DSS changes alternately between the depth-first search to find better solutions and the breadth-first search until the accuracy remains under a given value. The changes are controlled by setting a computing time, which is heuristically determined.

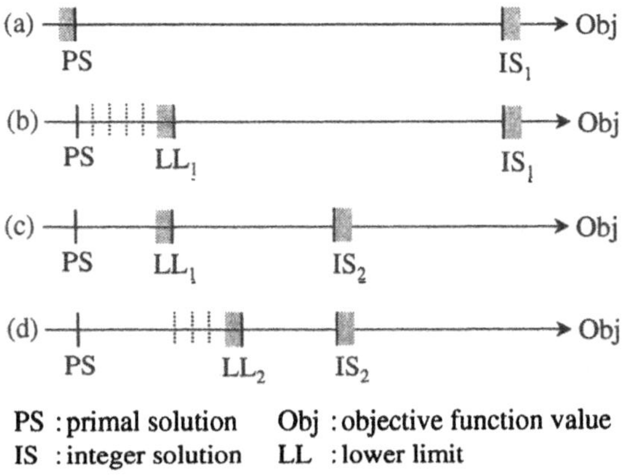

Fig. 2. Exemplary sequence of the DSS

As a result, several possibilities to accelerate the optimization process could be successfully realized and have been merged to the Priority-Based Dynamic Search Strategy (PBDSS).

3 Case Studies

The functionality of the improved MILP-algorithm should be demonstrated on the basis of a complex energy supply system, which consists of 8 plants, 12 purchasing contracts and 20 supplying contracts. The resulting mathematical optimization problem contains 4445 variables (732 of them binary), 5309 restrictions and a coefficient matrix containing 21999 non-zero elements. For the calculation a PC Pentium III 900 MHz was used.

First, the acceleration of the simplex algorithm should be examined: For this purpose, the PS of the considered energy supply system (basis scenario) and three further scenarios (characterized by higher demand, lower demand

and modified energy prices) are calculated. Subsequently, the PS of the three scenarios are calculated again, now using the results of the basis scenario as start solution for the simplex algorithm. In Table 1 the computing times for the PS with start solution are compared to those without start solution. Obviously, the computing times can be significantly reduced for all scenarios.

Table 1. Computing times for the PS (in sec)

scenario	without start solution	with start solution
higher demand	30.18	7.68
lower demand	30.51	6.92
modified prices	24.77	13.56

Following, the first integer solution of the basis scenario is searched once with sB&B (branching order determined by the estimation algorithm) and once with the PBDSS (branching order is affected by priorities calculated on the basis of the PS). The obtained convergence behaviour of the two calculations is presented in Fig. 3. The graph shows the remaining infeasibilities during the B&B process. At the beginning of the B&B there are about 400 infeasibilities, which have to be reduced to zero until a feasible integer solution is found.

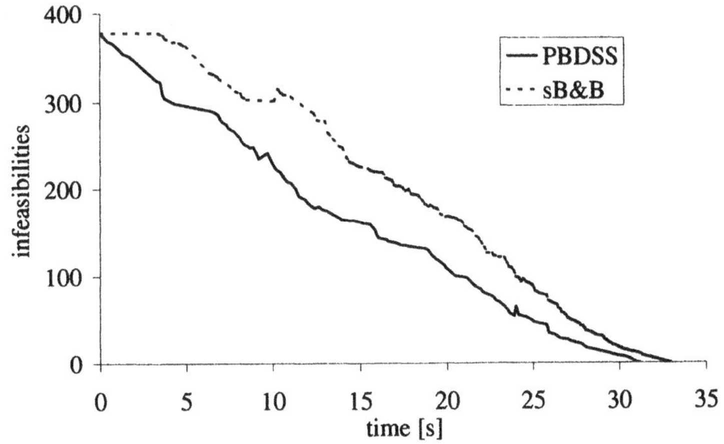

Fig. 3. Convergence behaviour of sB&B and PBDSS

In comparison to the sB&B the priority-based search yields in a more stable and faster convergence. Inspite of less computing time the objective function value of the first PBDSS solution is better than the solution of the sB&B. This shows, that the assignment of priorities is a powerful method to improve the convergence of restrictive optimization problems.

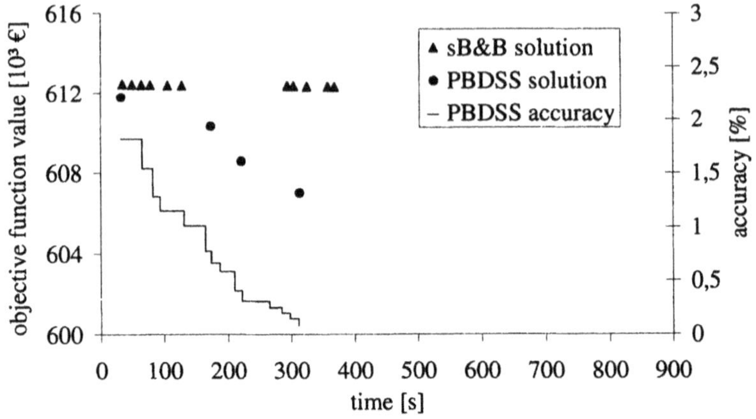

Fig. 4. Comparison of sB&B and PBDSS

Figure 4 finally compares the whole optimization process of the sB&B and PBDSS over the limited time. The sB&B is able to find a multitude of feasible integer solutions whose objective function values differ only slightly from each other, so that a significant improvement of the accuracy can not be achieved. In contrast to the standard algorithm the PBDSS changes alternately between searching an integer solution and improving the accuracy. Therefore, the computing time can be used more reasonable to find better integer solutions in less time. Moreover, the accuracy of the last solution found can be continuously decreased and keeps the system operator – waiting for the optimal schedule – always up to date about the optimization process. Overall, the exemplary results show that a significant acceleration of the MILP-algorithm can be achieved by the PBDSS.

References

1. Beale E. M. L. (1998) Introduction to Optimization. Wiley, Chichester
2. Hackländer P., Verstege J. F. (1999) Public cogeneration in Liberalized Energy Markets: Operation Concepts to Sustain Competitiveness. Euroheat & Power – Fernwärme International, Vol. 29, No. 1/2, pp. 77-85
3. Illerhaus S. W., Verstege J. F. (1999) Optimal Operation of CHP-based Industrial Power Systems in Liberalized Energy Markets. In: IEEE (Ed.) Proceedings of the PowerTech '99 Conference, Budapest, Hungary, BPT99-352-13
4. Maubach K.-D., Verstege J. F. (1993) Long-Term Operation Planning in Combined Heat and Power Suppy Systems. In: PSCC (Ed.) Proceedings of the 11th Power Systems Computation Conference, Avignon, France, pp. 601-607

On On-line Systems for Short-term Forecasting for Energy Systems

Henrik Aalborg Nielsen, Torben Skov Nielsen, and Henrik Madsen

Informatics and Mathematical Modelling, Technical University of Denmark, Richard Petersens Plads, Building 321, DK-2800 Kongens Lyngby, Denmark

Abstract. The paper describes experiences with developing on-line computer systems for short-term forecasting of wind power production and heat consumption in district heating networks. The computer systems are briefly described and some general aspects regarding system modeling with the purpose of forecasting are discussed. One consequence of the approach used is that the stochastic properties of the forecast errors can not be inferred from the models generating the forecasts. With the purpose of using the stochastic properties as input to formal OR-models we discuss how these can be modeled.

1 Introduction

The authors participates in the development of two on-line computer systems, called WPPT and PRESS, for short-term energy forecasts. WPPT (Wind Power Prediction Tool) is a system for forecasting the wind power produced on wind turbines for up to 40-46 hours with a resolution of 30 minutes. PRESS (In Danish: PRognose og Energi Styrings System) is a system which forecasts the heat consumption in district heating networks with the same horizon as above, but with a resolution of 1 hour. Both computer systems run at a number of locations. The systems use on-line meteorological forecasts and local climate measurements together with on-line measurements of the response in order to continuously update the underlying models.

Section 2 briefly describes the computer programs. In Sect. 3 the preconditions and some desirable properties of such computer systems are described, these set the basis for the methods used which are outlined in Sect. 4. Section 5 discuss how the stochastic properties of the forecast errors can be modeled. Furthermore, examples of stochastic decision problems are given. Finally, in Sect. 6 we conclude on the paper.

2 Brief Description of the Computer Systems

The computer systems work on-line. By on-line we understand that the systems continuously receive the most recent information and update the resulting forecasts periodically (every 30 or 60 minutes). Both WPPT and PRESS have been coded in C/C++ and run under Linux / X11. The authors believe that the stability of Linux or an other UNIX-variant is desirable when running an on-line application.

WPPT [14] is a system for forecasting the wind power production in relatively large geographical regions and for individual wind farms. The forecasts for the individual wind farms are upscaled with the purpose of generating regional forecasts, see the end of Sect. 4.

PRESS [6,13] is a system for forecasting the heat consumption in district heating networks and for controlling the supply temperature in order to reduce heat loss. The controller, a Stochastic Generalized Predictive Controller [17], is an essential part of the system. Due to flow restrictions, the forecasts of heat consumption sets a lower limit on the supply temperature during cold periods.

In Fig. 1 an overview of the information flow of the forecasting systems is depicted. Note that measured values of the dependent variable (e.g. wind power production) is used as input to the forecasting system. The reason for using climate measurements is that on the very short term, these are locally more adequate than meteorological (MET) forecasts [8, Sect. 6.5]. Also the climate measurements are needed in models which include low-pass filtered versions of the climate variables [8, Sect. 7.1]. The output of WPPT and PRESS also includes information regrading the uncertainty of the forecasts.

3 Preconditions and General System Considerations

The main information which is supplied to the computer programs is indicated in Fig. 1. Furthermore, information from the physical system, such as the fraction of wind turbines actually running i.e. not being out for maintenance or other reasons, and time/calendar information is supplied to the computer systems.

Except for the meteorological forecasts, typically the information can be sampled with the frequency required for the purpose of being able to update the forecasts with the desired frequency. However, the meteorological forecasts is not updated very frequently, nor is the resolution very high [16] and interpolation is used to circumvent this.

Since the physical systems considered are not stationary it is a precondition for the computer systems to be able to adapt to changes in the physical

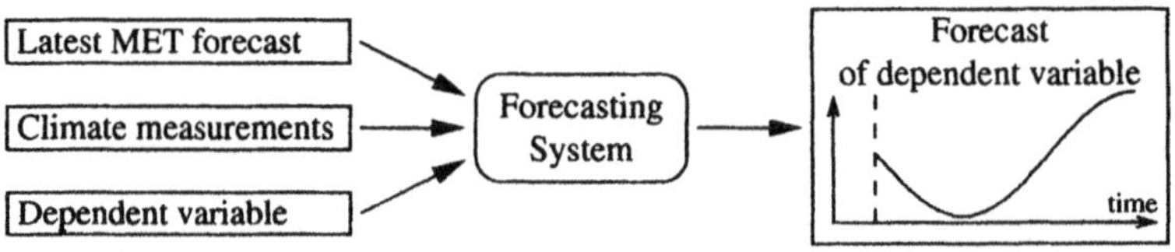

Fig. 1. Overview of the information flow of the on-line forecasting systems. The dashed line on the plot of the forecast indicates the time at which the forecast is generated

systems. A typical example is that additional consumers are added to a district heating network. The computer system should detect this and adapt to the new situation without human intervention.

While both of the physical systems are governed by the climate the heat consumption (PRESS) is also affected by diurnal / weekly variation not related to the climate. Also, while wind turbines react almost instantly on changes in wind speed the time constants in district heating systems are much larger, see [8].

4 Outline of the Methods Used

Let y_t denote the dependent variable at time t. Note that we assume that the time scale is standardized so that the time steps can be indexed as $\ldots, t-1, t, t+1, t+2, \ldots$.

If the possible dynamic dependence on the explanatory variables is neglected the dependence of y_t on a single explanatory variable x_t (e.g the wind speed) can generally be described as

$$y_t = f(x_t) + e_t \,, \tag{1}$$

where f is an unknown function often assumed to be smooth and where e_t denotes the model error and a possible measurement error. The error is assumed to have a mean value of zero.

The k-step forecast of y_{t+k} generated at time t is denoted $\hat{y}_{t+k|t}$. If f in (1) is known and since e_{t+k} is unpredictable then the forecast might be generate as

$$\hat{y}_{t+k|t} = f(\hat{x}_{t+k|t}) \,, \tag{2}$$

where $\hat{x}_{t+k|t}$ is the meteorological forecast, e.g. of wind speed. It is clear that $\hat{x}_{t+k|t}$ will deviate from the actual value x_{t+k} that will eventually act on the physical system. For this reason the use of f, assuming it is known, for the purpose of forecasting is not necessarily an optimal solution. The same holds for estimates of f based on (1). In fact for linear models it has been shown that it is generally better to use estimates based on the forecasts of the explanatory variables rather than on the actual explanatory variables [5]. See also the simulations in [8, Sect. 6.5].

In conclusion the f used for k-step forecasts should be obtained as the estimate $\hat{f}_k$ in the model

$$y_{t+k} = f_k(\hat{x}_{t+k|t}) + e_{k,t+k} \,. \tag{3}$$

Precisely how to estimate f_k depends on which additional assumptions are applied to the problem. For wind power curve estimation local regression proves to be a good solution and especially the adaptive procedure described in [11,16] is well suited for this purpose in that the power curve is only updated and old information weighted down for wind speeds actually occurring.

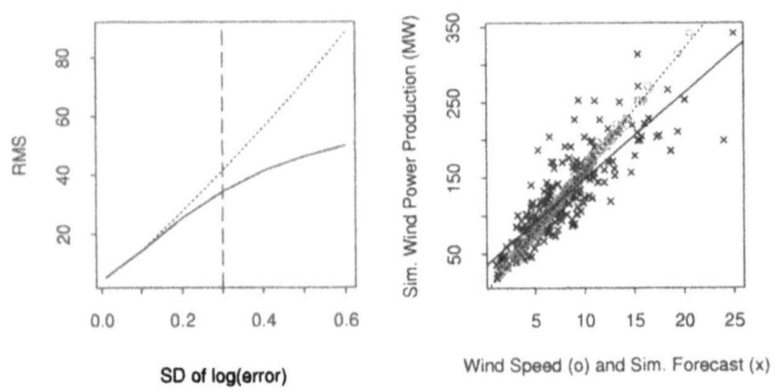

Fig. 2. Forecast results based on (1) (dotted) and (3) (full). Left: Mean over 50 simulations of Root Mean Squared forecast error (RMS) versus the measure of uncertainty of the wind speed forecast. Right: An example of the fits obtained using the uncertainty indicated by the vertical line on the left plot

Figure 2 shows a simulated example where f in (1) is a line trough the origin with a slope of 16 MW/(m/s) and e_t is independent zero-mean Gaussian variables with a standard deviation (SD) of 5 MW. The wind speed consists of 17513 real measurements and the forecasts are simulated by multiplicative log-normal independent errors with a mean chosen to ensure the equality in means of forecasted and actual wind speeds. The fits are obtained using 200 observations, while the remaining data is used for evaluating the forecasts. Note that with the particular error structure it will be advantageous to use a non-linear relationship in (3) although (1) is linear.

Additional reasons exist why model (3) rather than (1) should be used for estimation when adequate forecasts are the main focus. Mainly this is because of the fact that the model, e.g. (1), defines a set of approximations to the physical system. The most appropriate approximation in this set will often depend on the forecast horizon. The reader is referred to [2,19,21] and [3, Sect. 7.3]. See also the results in [8, Sect. 5.7].

In reality model (3) is too simple. Currently the model used in WPPT is given by (18) and (19) in [16]. Besides a forecast of the wind speed the model uses forecasts of the wind direction and also includes a diurnal variation. Furthermore, the most recent observed wind power production is used as an explanatory variable and $\hat{f}_k$ is smoothed over k. For district heating systems the general conclusions outlined above still holds. However, because of the heat dynamics of buildings, in this case the use of low-pass filtered climate variables / forecasts will be appropriate when forecasting the heat consumption, see [8].

The procedure outlined so far assumes that the dependent variable is available on-line. For wind power production the values is available on-line for certain reference wind farms while data for smaller farms and individual turbines is available only trough the total wind power production of sub-areas, which is available at a lower frequency than the real on-line data. The upscaling to regional forecasts are described in [12,15].

5 On the Use of Short-term Forecasts

Generally the forecasts of heat consumption and wind power production is used in the daily decision process of the personal operating the energy supply systems. For district heating systems relevant decision problems are dynamic. For wind power the forecasts is used when trading via the Nordic Power Exchange (http://www.nordpool.com). Unit commitment and load dispatch can most adequately be performed by directly considering the covariance structure of the forecast errors. In reality all the decision problems are of stochastic nature and the covariance must be derived form the actual errors.

5.1 The Covariance Structure of Forecast Errors

Let $e_{t|s}$ denote the forecast error when forecasting at time t given information up to time s. The forecast errors for a finite time span may be collected into a matrix in which the rows $k = 1, \ldots, K$ indicate the forecast horizon and the columns indicate the time at which the forecasts are generated, i.e.

$$\begin{bmatrix} \cdots e_{t|t-1} & e_{t+1|t} & e_{t+2|t+1} & \cdots \\ \vdots & \vdots & \vdots & \\ \cdots e_{t-1+K|t-1} & e_{t+K|t} & e_{t+K+1|t+1} & \cdots \end{bmatrix}, \tag{4}$$

this constitutes a K-dimensional stochastic process. However, the full properties of this process can not be derived from the forecast system itself. By modeling the most important aspects of the process it will be possible to simulate new realizations of the forecast errors, which is a first step towards finding good solutions to the stochastic decision problem. For describing the second-order properties (4) can be modeled as a multivariate ARMA process [4]. However, for wind power forecasts it is important to take into account that the variability depends on the level of forecasted wind power production and to a lesser extend on the time of day [9]. This can not be accomplished by a pure ARMA model, but may be accomplished by first standardizing the forecast errors using an estimate of the variability. This estimate can be obtained by smoothing the squared errors [9], see [20] for a general description.

5.2 Examples of Decision Problems

Control of Supply Temperature: With the purpose of reducing heat loss in district heating networks it is advantageous to keep the supply temperature at a reasonable low level, while water of a certain minimum temperature is supplied to the consumers. Due to restrictions in flow capacity the heat demand sets a lower limit on the supply temperature. This limit is active in periods with a relatively high demand. During other periods the temperature limit at the consumers dictates the supply temperature. The stochastic control problem just outlined is handled by PRESS. For an overview of the principle the reader is referred to [13], whereas [7] gives more details. The stochastic controller is described in [17].

Production Planning for District Heating Plants: At district heating plants with co-generation of heat and power, maybe with time-varying electricity prices, the decision problem may be how to use a heat accumulator for optimizing the operation of the plant [10]. In [10] the criteria function of the dynamic optimization problem includes (i) revenue form selling power at a price varying over time, (ii) cost of purchasing natural gas at a price varying over time, (iii) cost related to starting and stopping the gas turbine, and (iv) maintenance costs depending on the amount of power produced by the plant. The decision problem is complicated by the stochastic nature of the forecast errors of the heat demand. In the future the problem may be further complicated by power and gas prices including a stochastic component.

Optimal Scheduling of CHP Production: The optimal scheduling of CHP production may be performed on a weekly basis and in this case it consists of unit commitment and load dispatch. This problem is solved by SIVAEL [1,18]. One of the major stochastic components of this optimization problem is the deviation of the actual wind power production from the forecasts. Based on the principle described in Sect. 5.1 SIVAEL has recently been updated in order to accomplish this, see also [9].

6 Conclusion

The preconditions and methods for short-term forecasts of heat consumption and wind power are outlined in the paper. Furthermore decision problems related to such systems are briefly described. It is argued that the stochastic properties of the forecast errors must be directly modeled in order to facilitate the use of these properties in a formal decision process.

7 Acknowledgments

The authors wish to acknowledge the cooperation with the partners with whom we have been working on short-term energy forecasts. These include ABB, Elkraft, Elsam, Eltra, KARA, Rambøll, Risø, University of Lund, VEKS, and Vestkraft. Also we wish to acknowledge the financial support which we have received from the Danish Ministry of Energy, the European Commission, and from some of the partners mentioned above.

References

1. P.B. Eriksen. Economic and environmental dispatch of power/CHP production systems. *Electric Power Syst. Res.*, 57:33–39, 2001.
2. M. Gevers. Identification for control. *A. Rev. Control*, 20:95–196, 1996.
3. T. Hastie, R. Tibshirani, and J. Friedman. *The Elements of Statistical Learning*. Springer, 2001.

4. G.M. Jenkins and A.S. Alavi. Some aspects of modelling and forecasting multivariate time series. *J. Time Ser. Anal.*, 2:1–47, 1981.
5. B. Jonsson. Prediction with a linear regression model and errors in a regressor. *Int. J. Forecast.*, 10:549–555, 1994.
6. H. Madsen, T.S. Nielsen, and H.Aa. Nielsen. Intelligent control. *Danish Board of District Heating - News from DBDH*, 3:14–16, 2001.
7. H. Madsen, T.S. Nielsen, and H.T. Søgaard. *Control of Supply Temperature*. Dept. of Mathematical Modelling, Tech. Univ. of Denmark, Lyngby, Sept. 1996.
8. H.Aa. Nielsen and H. Madsen. *Predicting the Heat Consumption in District Heating Systems using Meteorological Forecasts*. Dept. of Mathematical Modelling, Tech. Univ. of Denmark, DK-2800 Lyngby, Denmark, 2000. http://www.imm.dtu.dk/~han/pub/efp98.pdf.
9. H.Aa. Nielsen and H. Madsen. *Analysis and simulation of prediction errors for wind power productions reported to NordPool* (in danish). Informatics and Mathematical Modelling, Tech. Univ. of Denmark, Lyngby, 2002.
10. H.Aa. Nielsen, H. Madsen, and T.S. Nielsen. *Load scheduling for decentralized CHP plants*. Dept. of Mathematical Modelling, Tech. Univ. of Denmark, DK-2800 Lyngby, Denmark, 2000. http://www.imm.dtu.dk/~han/pub/efp98akk.pdf.
11. H.Aa. Nielsen, T.S. Nielsen, A.K. Joensen, H. Madsen, and J. Holst. Tracking time-varying coefficient-functions. *Int. J. of Adapt. Control and Signal Processing*, 14(8):813–828, 2000.
12. T.S. Nielsen, editor. *Using Meteorological Forecasts in On-line Predictions of Wind Power*. Eltra, DK-7000 Fredericia, Denmark, 1999. ISBN: 87-90707-18-4.
13. T.S. Nielsen and H. Madsen. Control of supply temperature in district heating systems. In *Proceedings of the 8th International Symposium on District heating and Cooling*, Trondheim, Norway, 2002.
14. T.S. Nielsen, H. Madsen, and H.S. Christensen. WPPT - a tool for wind power prediction. In *Proceedings of the Wind Power for the 21st Century Conference*, Kassel, 2000.
15. T.S. Nielsen, H. Madsen, H.Aa. Nielsen, G. Giebel, and L. Landberg. Prediction of regional wind power. In *Proceedings of the 2002 Global Windpower Conference*, Paris, 2002.
16. T.S. Nielsen, H.Aa. Nielsen, and H. Madsen. Prediction of wind power using time-varying coefficient-functions. In *Proceedings of the XV IFAC World Congress*, Barcelona, 2002.
17. O.P. Palsson, , H. Madsen, and H.T. Søgaard. Generalized predictive control for non-stationary systems. *Automatica*, 30(12):1991–1997, 1994.
18. J. Pedersen. SIVAEL simulation program for combined heat and power production. In *International Conference on Applications of Power Production Simulation*, Washington (EPRI/SEP), Jun. 11–13 1990.
19. J.A. Rositter and B. Kouvaritakis. Modelling and implicit modelling for predictive control. *Int. J. Control*, 74(11):1085–1095, 2001.
20. D. Ruppert, M.P. Wand, U. Holst, and O. Hössjerin. Local polynomial variance-function estimation. *Technometrics*, 39:262–273, 1997.
21. B. Wahlberg and L. Ljung. Design variables for bias distribution in transfer function estimation. *IEEE Trans. Automat. Control*, 31:134–144, 1986.

Combining Bottom-up and Finance Modelling for Electricity Markets

Christoph Weber

Institute of Energy Economics and the Rational Use of Energy (IER), University of Stuttgart, Germany . E-mail: cw@ier.uni-stuttgart.de

1 Introduction

The high price volatility observed on liberalised electricity spot markets requires detailed modeling to successfully manage price risks. For example, in December 2001 prices at the German power exchange LPX exceeded 1000 €/MWh for certain hours, whereas the yearly average is around 22 €/MWh. In the USA even price spikes with more than 9,000 €/MWh have been observed. Also outside extreme events, price changes of 20 % from one day to another are not unusual. In order to cope with this volatility, price models are needed which reflect the particularities of the electricity market, notably the non-storability of electricity. On the other hand, also the experience from other commodity markets has to be integrated in any model for electricity prices. Therefore, in the following a model is presented which combines several approaches.

2 Methodology

In the past, various models have been used to analyse electricity market prices and predict price levels and price risks (cf. Weber 2002). Each of these approaches has specific strengths and weaknesses, but especially when it comes to risk assessment in the middle and longer term, neither approach is very satisfactory by itself. Finance mathematical models analyse adequately price fluctuations due to market movements but they cannot cope with the impact of changes in the production technology. Furthermore, given the seasonal, climatic and other time-varying exogenous effects, long time-series in stable market environments would be required to estimate stochastic patterns with sufficient accuracy. On the other hand, dealing with stochastic effects in the fundamental models is hardly possible. Therefore, a combination of both approaches seems most appropriate. How this can be achieved, is sketched in Fig. 1.

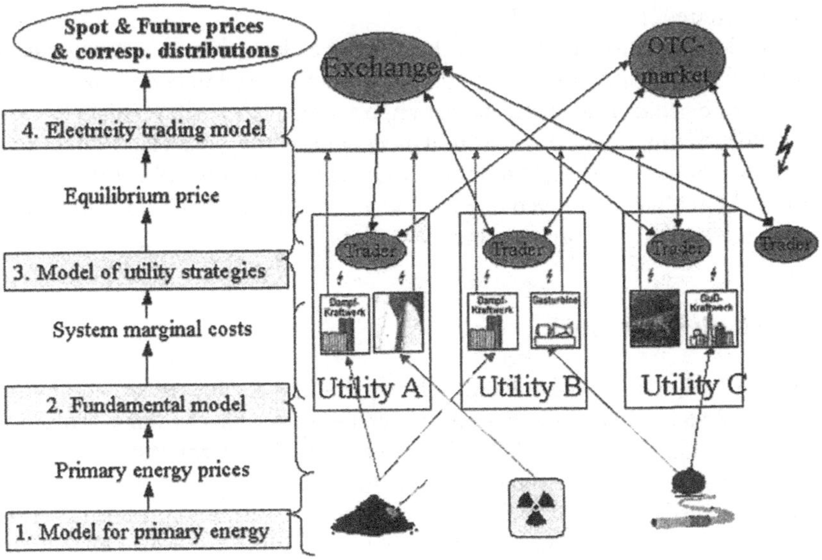

Fig. 1. General approach for the integrated electricity market model

In a first step, the stochastic development of prices on the primary energy market is modelled through a finance-type model (cf. Sect. 2.1). The resulting energy prices are taken as an input to a fundamental model of the European electricity market (cf. Sect. 2.2). This model yields marginal generation costs differentiated by time of the day, type of day and month in the year. These prices could be used as input to a game-theoretic model yielding the prices and mark-up charged by strategic players in the market ("Model of utility strategies"). This part of the model has however so far not been implemented and is therefore not discussed in the following. Rather, the marginal costs are directly used as an input for a stochastic model of the electricity market (cf. Sect. 2.3).

2.1 Model for primary energy prices

An important characteristic of the primary energy markets for oil and gas is that periods of high price volatility are alternating with periods of rather stable prices. This can be dealt with using a model with switching regimes, such as the one proposed by (Erdmann 1995). However, a priori the moments of regime switch can be hardly determined and therefore an interesting alternative is to model the primary energy markets through a mean-reversion process for the derivatives of (logs of) prices. The price changes evolve in this model around an equilibrium price drift, which could be theoretically derived e. g. from Hotelling's rule. The model accounts for the fact that periods with low and with high price drift and volatility are observed. Furthermore, the correlations between price changes for various energy carriers are also accounted for.

Empirical estimation results show that the price changes have a strong tendency to return to their equilibrium growth rate whereas this growth rate is rather close to zero. The stochastic component is most important for heavy fuel oil and lowest for coal. A further important point to be noted is the strong correlation between all price movements in the long run (correlation of price levels). All correlations exceed 0.75. This has to be accounted for when doing Monte-Carlo-simulations of the price movements for primary energy carriers.

2.2 Fundamental model

Starting from given primary energy prices, this model determines the marginal generation costs in the Europe-wide power network. This is done using an LP model describing the minimization of the total variable system costs.

The main restrictions to be included are the demand restrictions for each region, the capacity restrictions for each plant type and the transmission capacity restrictions accounting also for the corresponding availabilities. Further equations describe start-up costs as well as minimum start-up times and minimum down times. Since in the continental European power system hydro power from storage and pump-storage plants is quite of some importance, time coupling constraints for hydro storage have also to be considered. Therefore, one year is taken as optimization period and within that year, producers are assumed to have perfect foresight and static price expectations on fuel prices. In the current version of the model, twelve typical days and six regions are distinguished. Electricity exchanges with regions outside the model scope are modelled exogenously. Data sources for the model include UCTE statistics on electricity load, electricity exchanges and transmission capacities, a database of power plants throughout Europe and fuel price statistics. Power plants are classified according to the main fuel used and the vintage into 21 classes in order to keep the problem size reasonable. Nevertheless, the optimisation problem has about 61,000 variables, 83,000 constraints and 220,000 non-zero elements. Implemented in GAMS and using a CPLEX solver, computation time on a pentium III PC is about 3 minutes for a solve from scratch and about 10 s if an initial solution from a previous run is provided.

The model is operated in two different ways: Firstly, it is used with historical fuel prices and demand to determine the equilibrium prices for past time periods which are then used as an input for estimating the stochastic electricity price model described below. Thereby, simulations are carried out for each historical month using the fuel price information available at that time.

Secondly, the model is operated using Monte-Carlo-simulations of future fuel prices as input, in order to derive stochastic equilibrium prices. Due to the possibilities of fuel switches, variations in hydro-use or changes in import-export flows, the relationship between fuel prices and electricity prices is then far from being linear.

2.3 Electricity trading model

Besides deterministic impacts from marginal generation costs, electricity prices are also subject to stochastic changes – due e. g. to load fluctuations, unforeseen outages or trader speculation. As others have shown (e. g. Gibson and Schwartz 1990; Brennan 1991), electricity prices exhibit mean reversion. However the prices probably do not tend to return to a time-invariant mean, but rather to one which is determined by the fundamentals governing the electricity market. Therefore, the following model has been specified:

$$d \ln p_{E,t} = a_{E,h}\left(b_h + \ln p_{G,t} - \ln p_{E,t}\right) + \varepsilon_{E,t} \tag{1}$$

that is the price change is dependent on the difference between the equilibrium price $p_{G,t}$ and the actual price $p_{E,t}$ adjusted possibly by a constant b_h and superposed by a stochastic term $\varepsilon_{h,t}$. Thereby, similarly to the approach adopted by (Ramanathan et al 1997), each hour is modeled by a separate stochastic process.

It should be noted that part of the stochastic price fluctuations are due to variations in the load or available capacity. Yet, the impact of these fluctuations depends on the slope of the merit order curve[1] .

Therefore, the change Λ in marginal costs for a change in load by ±15 percent has been included as explanatory variable in a GARCH-type model of volatility:

$$\sigma^2\left(\varepsilon_{E,t}\right) = \alpha_h \varepsilon_{E,t-1}^{\ 2} + \beta_h \sigma^2\left(\varepsilon_{E,t-1}\right) + \left(1 - \alpha_h - \beta_h\right)\overline{\sigma}^2 + \gamma_h\left(\Lambda_{E,t} - \overline{\Lambda}_{E,h}\right) \tag{2}$$

Alternatively, a specification has been tested where the variance is dependent on the price level of the previous day – modelling the fact that volatility tends to increase in electricity markets with price levels.

It turns out that the model using the change in marginal cost variable Λ as explanatory variable outperforms an ordinary GARCH-model significantly in 11 out of 24 hours. In those hours, the model mostly also outperforms the alternative model formulation with $\ln p_E$ as variance explaining variable. But in four other hours, the latter model performs considerably better than the first one.

3 Results

In Fig. 2. examples of price simulations are shown, which have been determined using the integrated model. Thereby the prices observed until end of June 2001 have been used as data base for the model. On that basis, prices for August 2001 have been simulated. The real market prices for the first and the last week in August have been also included in the figure. Shape and level of the simulated prices are similar to the observed ones although of course the exact price pattern is not matched.

---

[1] The merit order curve is obtained through ordering the power plants by increasing marginal generation costs. The merit order curve than gives the marginal generation costs as a function of cumulative capacity.

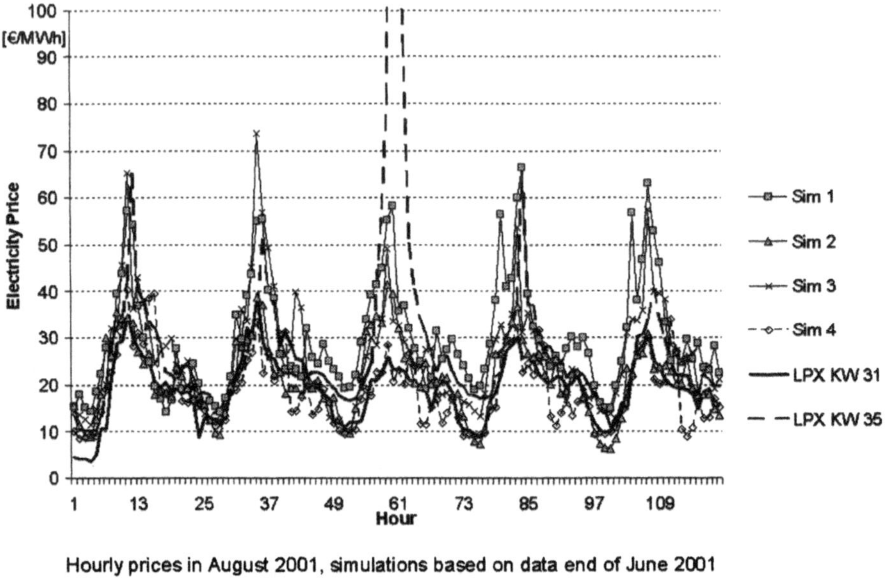

Fig. 2. Comparison of simulated (as from 30 June 2001)and real spot market prices in August 2001

Fig. 4 shows a comparison of price forecasts and uncertainty ranges implied by the integrated model with forward prices and real prices for November 2001. It turns out that the integrated model provides for this month a better forecast then the forward price does. Also the uncertainty range looks plausible. If compared with a pure mean reversion model, the uncertainty range is similar in both cases. But an important difference between the simple mean-reversion model and the integrated model is that the uncertainty range in the integrated model increases with increasing forecasting horizon due to the impact of uncertain primary energy prices. The mean-reversion model on the contrary shows no increase after about one week

4 Final remarks

The combination of fundamental analysis and finance model for analysing electricity prices yields interesting insights and more realistic estimates, both for expected prices and price variations. The model developed so far clearly requires further testing and fine tuning, but it offers a useful basis for estimating prices and volatilities especially in the medium and long run. Furthermore it provides also a basis for estimating the impact of exogenous shocks (e. g. unforeseen outages) on price levels and price volatilities.

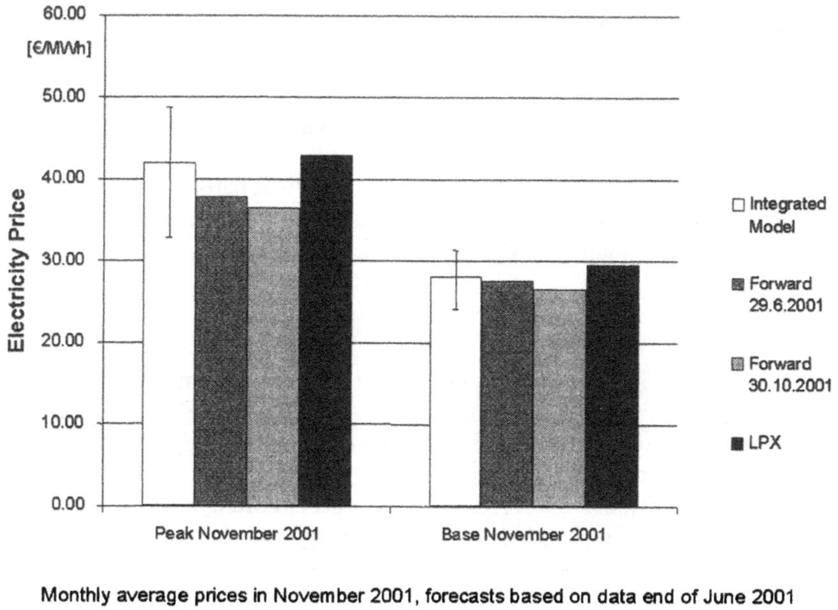

Fig. 3. Comparison of price forecasts of the integrated model (from 30 June 2001), forward prices and actual prices for standard products November 2001

References

Brennan MJ (1991) The price of convenience and valuation of commodity contingent claims. In: Lund D, Oksendal B (eds) Stochastic Models and Option Models. North-Holland, Amsterdam et al, pp 33–72

Erdmann G (1995) An Evolutionary Model for Long Term Oil Price Forecasts. In: Wagner, A Lorenz H (eds) Studien zur Evolutorischen Ökonomik III. Duncker & Humblot, Berlin, München, pp 143–161

Johnson B, Barz G (1999) Selecting Stochastic Processes for Modelling Electricity Prices. In: Jamson R (ed) Energy Modelling and the Management of Uncertainty. Risk Books, London

Gibson R, Schwartz E (1990), Stochastic convenience yield and the pricing of oil contingent claim. Journal of Finance 45:959–976

Ramanathan R, Engle R, Granger CWJ, Vahid-Araghi F, Brace C (1997), Short-run forecasts of electricity loads and peaks. International Journal of Forecasting 13:161–174

Weber C (2002), Electricity markets: Coping with price risk through integrated bottom-up and financial modeling. Working Paper, Stuttgart

Gestaltung von Stoffstrom-Netzwerken zum Produktrecycling[1]

Prof. Dr. Thomas Spengler; Dipl.-Geoökol. Grit Walther

Technische Universität Braunschweig, Institut für Wirtschaftswissenschaften, Abt. BWL, insb. Produktionswirtschaft, Katharinenstr. 3, 38106 Braunschweig, Tel.: +49(531)391 2201, Email: t.spengler@tu-bs.de; g.walther@tu-bs.de

Abstract. Im vorliegenden Beitrag erfolgt die Vorstellung eines Konzepts zur ökonomischen Bewertung von alternativen Stoffstrom-Netzwerken zum Elektro(nik)altgeräterecycling. Dabei finden investitionsabhängige Kosten, variable Stofffluss- sowie Prozesskosten und sonstige Gemeinkosten Beachtung. Für die Ermittlung der Stofffluss- und Prozesskosten wird ein Mengengerüst auf Basis eines linearen Optimierungsmodells erstellt. Es erfolgt eine Anwendung am Beispiel der Unterhaltungselektronik für das Bundesland Niedersachsen.
Keywords. Stoffstrom-Netzwerk, Elektro(nik)altgeräte, Recycling

1 Ausgangslage, Zielsetzung, Vorgehensweise

Vor dem Hintergrund der auf europäischer Ebene vor der Verabschiedung stehenden Richtlinie über Elektro- und Elektronikaltgeräte (WEEE), die die Verpflichtung zur Rücknahme und zum Recycling von Elektro(nik)altgeräten enthält, wird im vorliegenden Beitrag ein Ansatz zur Gestaltung von Stoffstrom-Netzwerken zur Erfassung und Behandlung von Elektro(nik)altgeräten entwickelt.

Dabei wird zunächst auf die aktuellen und zukünftigen Entwicklungen im Bereich der Behandlung von Elektro(nik)altgeräten eingegangen. Anschließend erfolgt der Entwurf eines Konzepts zur Gestaltung und ökonomischen Bewertung von Alternativen des Stoffstrom-Netzwerks. Eine Daten und Ist-Analyse bestehender Stoffstrom-Netzwerke wird im Anschluss an das Optimierungsmodell vorgestellt. Ein Ausblick schließt den vorliegenden Beitrag ab.

2 Aktuelle und zukünftige Entwicklungen im Bereich der Behandlung von Elektro(nik)altgeräten

Im Marktsegment der Elektro(nik)geräte führen immer kürzere Innovationszyklen sowie steigende Ausstattungsgrade der Haushalte zu wachsenden Altproduktmen-

[1] Das Vorhaben „Gestaltung und Lenkung von Stoffstrom-Netzwerken zum Recycling komplexer Verbundprodukte – am Beispiel Brauner Ware“ wird mit Mitteln der Deutschen Forschungsgemeinschaft (Förderkennzeichen SP491/1) gefördert

gen. Das Aufkommen an Elektro(nik)altgeräten in Deutschland wird auf ca. 1,8 Mio. t/a geschätzt [2], davon werden derzeit allerdings nur etwa 0,46 Mio. t/a erfasst und ordnungsgemäß behandelt [4]. Problematisch ist dies vor allem vor dem Hintergrund des Schadstoffgehaltes der Altgeräte, aber auch aufgrund der Wiederverwertbarkeit der in den Geräten enthaltenen Wertstoffe. Eine auf europäischer Ebene diskutierte Richtlinie über Elektro- und Elektronikaltgeräte (WEEE) [3] beinhaltet die Verpflichtung zum Aufbau von Rücknahmesystemen, zur Gewährleistung einer kostenlosen Rücknahme der Altgeräte sowie zur Einhaltung von Sammel-, Recycling- und Verwertungsquoten.

Als Auswirkungen der Umsetzung der EU-Richtlinie WEEE sind zukünftig vor allem für die kleinen und mittelständischen Demontageunternehmen strukturelle Veränderungen zu erwarten. Es stehen die in Tabelle 1 dargestellten Systemalternativen zur Auswahl, wobei die jeweils zuerst genannte Ausprägung der aktuellen Situation entspricht.

Tabelle 1. Alternativen für Stoffstrom-Netzwerke beim Elektro(nik)schrottrecycling

Einflussgröße	Ausprägungsmöglichkeiten der Alternativen
Behandlung der Elektro(nik)altgeräte	– dezentrales System (KMU mit Kapazität von 2500 t/a) – fortgeschrittene Zentralisierung – vollständige Zentralisierung
Innerbetriebliche Ausstattung der Demontageunternehmen	– unverkettete Einzelarbeitsplätze – Linienverkettung – Hochautomatisierte Demontagesysteme
Bereitstellung der Elektro(nik)altgeräte	– Dezentrale Bereitstellung (bei Kommunen) – Zentrale Bereitstellung (Sortierzentrum)
Vorsortierung der Produkte	– unsortiert – wenig sortiert (Großgeräte/ Kleingeräte) – sortiert (große/ kleine Weiße Ware/TV/ Braune Ware/ ITK)
Zusammenarbeit der Demontageunternehmen	– Einzelakteur – regionales Netzwerk – strategisches Netzwerk

Im vorliegenden Beitrag wird auf die folgenden Planungsaufgaben fokussiert:

- (Gesamt-)Kapazität der am Stoffstrom-Netzwerk zu beteiligenden Akteure,
- Anzahl der am Stoffstrom-Netzwerk zu beteiligenden Akteure,
- Art der am Stoffstrom-Netzwerk zu beteiligenden Akteure,
- Technikwahl innerbetrieblicher Demontagesysteme.

Eine Gegenüberstellung und Bewertung zukünftiger Systemalternativen für Demontage- und Recyclingunternehmen fehlt derzeit. Über die ökonomische Effizienz der einzelnen Alternativen liegen keine Informationen vor. Aus diesem Grund wird im Folgenden ein Konzept zur Modellierung und Bewertung alternativer Stoffstrom-Netzwerke zum Elektronikschrottrecycling vorgestellt.

3 Modellierungs- und Bewertungskonzept

Das Ziel des Konzepts besteht in der Gestaltung von Stoffstromnetzwerken mit dem Ziel der ökonomisch effizienten Demontage von Elektro(nik)altgeräten. Dabei werden Spezialisierungen von Demontageunternehmen und somit die Realisierung von Lernkurveneffekten [5] vorausgesetzt.

Der zu erwartende jährliche Gewinn aller Akteure des Gesamtnetzwerks stellt ein allgemein anerkanntes und leicht interpretierbares Kriterium für die ökonomische Vorteilhaftigkeit der einzelnen Systemalternativen dar. Daher wird das folgende Modell [10] als Basis für die ökonomische Bewertung gewählt:

$$K^{KrW} = K_I^{KrW} + K_{Stofffluss}^{KrW} + K_{Prozess}^{KrW} + K_{sonst}^{KrW} \qquad (1)$$

Variablen: K^{KrW} -Entscheidungsrelevante Gesamtkosten, K_I^{KrW} -investitionsabhängige Kosten, $K_{Stofffluss}^{KrW}$ -Stoffflusskosten, $K_{Prozess}^{KrW}$ -Prozesskosten, K_{sonst}^{KrW} -sonst. Gemeinkostenanteile

Zur Ableitung der investitionsabhängigen Kosten werden in einem ersten Schritt jeweils die zum Aufbau der Infrastruktur erforderlichen spezifischen Investitionen ermittelt [6,1] Diese werden modular nach Logistik, Gebäuden sowie Demontagesystemen für unterschiedliche Kapazitäten bestimmt, um die Ableitung der investitionsabhängigen Kosten für variable Abschreibungsdauern und Zinssätze zu ermöglichen (siehe Abb. 1).

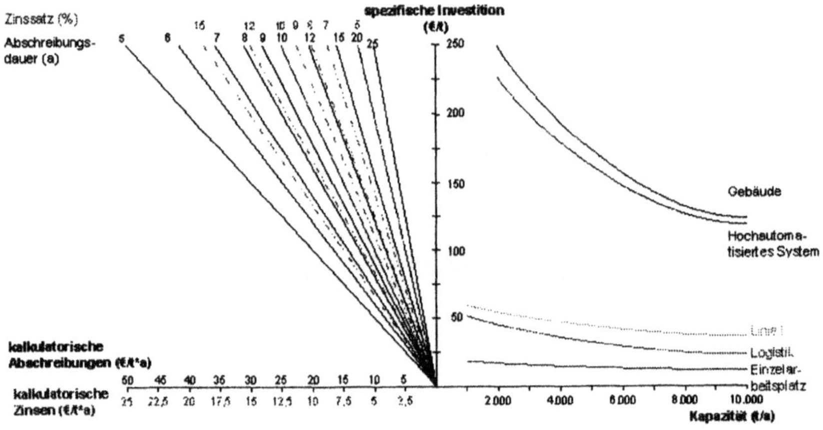

Abb. 1. Ableitung der investitionsabhängigen Kosten aus den spezifischen Investitionen

Sonstige Gemeinkosten wie Energiebedarf, Versicherungen oder Instandhaltungskosten können für die verschiedenen o.g. Module prozentual aus der Investition abgeleitet werden [1].

Die Basis zur Ermittlung der stofffluss- und prozessabhängigen Erlöse und Kosten bildet ein auf der linearen Aktivitätsanalyse [7] beruhendes Stoffstrommodell. Es wird eine ökonomisch effiziente Allokation der Stoffströme auf die Akteure des Netzwerks vorausgesetzt, die mit Hilfe eines linearen Optimierungsmodells erfolgt. Die Systemgrenzen des Stoffstromnetzwerks beinhalten den Transport der Geräte zu den Demontageunternehmen, die Demontage sowie den Verkauf der Demontagefraktionen. Auch Transporte zwischen den Demontageunternehmen sind möglich. Die Modellierung der Demontageoptionen erfolgt in Anlehnung an das gemischt-ganzzahlige Optimierungsmodell zur integrierten Demontage- und Recyclingplanung nach [8]. Aufgrund der hohen Stückzahl an Altgeräten kann bei einer regionalen Betrachtung auf die Ganzzahligkeitsbedingungen verzichtet werden. In Abbildung 2 sind die Stoffströme für das Demontageunternehmen u dargestellt.

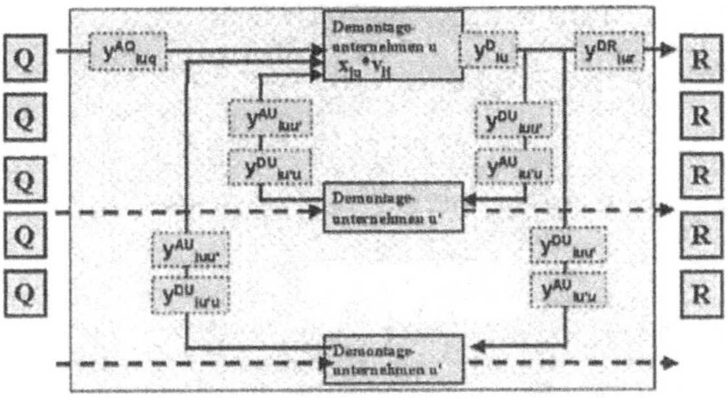

Abb. 2. Stoffstrommodell für das Demontageunternehmen u

Indizes: I-Produkte (i = 1..P) u. Zerlegefraktionen (i = P+1...I), J-Demontageoperationen, Q-Quellen, U-Unternehmen, R-Senken
Variablen: Y-Massen, X-Anzahl der Demontageaktivitäten

Für die ökonomische Bewertung des Stoffstrommodells finden ausschließlich variable Kosten in Form von Annahmeerlösen, Transportkosten, Sortierkosten, Zerlegekosten und Verwertungserlösen der Demontagefraktionen Beachtung. Das Ziel besteht in der deckungsbeitragsmaximalen Allokation der Produkte auf die Demontageunternehmen (2).

$$\textbf{MAX} \quad y^{AQ}_{iuq}, y^{AU}_{iuu}, y^{DR}_{iur}, x_{ju} \tag{2}$$

$$\sum_{u=1}^{U}\left(\sum_{i=1}^{I}\left(\sum_{q=1}^{Q}(e^{A}{}_{i} - c^{TQ}{}_{qu}) \cdot y^{AQ}{}_{iuq} + \sum_{\substack{u'=1 \\ u'\neq u}}^{U}(-c^{TU}{}_{uu'}) \cdot y^{AU}{}_{iuu'} + \sum_{r=1}^{R}(e^{V}{}_{ir} - c^{TR}{}_{ur}) \cdot y^{DR}{}_{iur}\right) - \sum_{j=1}^{J} x_{ju} \cdot c^{Z}{}_{ju}\right)$$

Der Output eines Demontageunternehmens $y^{D}{}_{iu}$ berechnet sich nach Bilanzgleichung (3) aus dem Input an Geräten aller Quellen des Netzwerks ($y^{AQ}{}_{iuq}$) und aus dem Input an Geräten und Bauteilen anderer Demontageunternehmen ($y^{AU}{}_{iuu'}$), sowie aus der Zerlegung dieser Geräte und Bauteile, ausgedrückt durch die Demontageaktivität v_{ij} multipliziert mit der Anzahl der Ausführungen dieser Demontageaktivität (x_{ju}).

$$\sum_{j}(x_{ju} \cdot v_{ij}) + \sum_{q=1}^{Q} y^{AQ}{}_{iuq} + \sum_{\substack{u'=1 \\ u'\neq u}}^{U} y^{AU}{}_{iuu'} = y^{D}{}_{iu} \quad , \quad i = 1..I; u = 1..U \tag{3}$$

Der so berechnete Output des Demontageunternehmens wird nach Bilanzgleichung (4) zum einen an die Senken des Netzwerkes ($y^{DR}{}_{iur}$) sowie zur weiteren Zerlegung an andere Demontageunternehmen ($y^{DU}{}_{iuu'}$) geliefert.

$$y^{D}_{iu} = \sum_{\substack{u'=1 \\ u\neq u'}}^{U} y^{DU}_{iuu'} + \sum_{r}^{R} y^{DR}_{iur} \quad , i = 1..I; u = 1..U \tag{4}$$

Um zu gewährleisten, dass die Massenflüsse zwischen den Demontageunternehmen abgebildet werden, sind diese nach Massenbilanzgleichung (5) so zu bilanzieren, dass Lieferungen des Unternehmens u an das Unternehmen u' ($y^{DU}_{iuu'}$) den Annahmen durch das Unternehmen u' vom Unternehmen u ($y^{AU}_{iu'u}$) entsprechen.

$$y^{DU}_{iuu'} = y^{AU}_{iu'u} \quad , \quad i=1..I; u=1..U, u'=1..U \tag{5}$$

Die Kapazitätsrestriktionen bestehen in der zur Verfügung stehenden Altgerätemenge (6) und den Demontage- (7) sowie Recyclingkapazitäten (8).

$$\sum_{u=1}^{U} y^{AQ}_{iuq} \leq y^{QMAX}_{iq} \quad , \quad i=1..I; q=1..Q \tag{6}$$

$$\sum_{j=1}^{J} c_{ju} \cdot x_{ju} \leq c^{ZMax}_{u} \quad , \quad u=1..U \tag{7}$$

$$\sum_{u=1}^{U} y^{DR}_{iur} \leq y^{RMAX}_{ir} \quad , \quad i=1..I; r=1..R \tag{8}$$

Des Weiteren besteht nach dem derzeitigen Entwurf der EU-Richtlinie über Elektro- und Elektronikaltgeräte der Zwang zur Schadstoffentfrachtung bestimmter Produkte (9). Zudem bestehen Nichtnegativitätsbedingungen (10).

$$y^{RMAX}_{ir} \begin{cases} = 0 & falls\ i\ zerlegepflichtig \\ \geq 0 & sonst. \end{cases} \quad , \quad I=1..P; r=1..R \tag{9}$$

$$y^{AQ}_{iuq}, y^{AU}_{iuu'}, y^{DU}_{iuu'}, x_{ju}, y^{DR}_{iur}, y^{D}_{iu} \geq 0 . \tag{10}$$

Die Zusammenführung der investitionsabhängigen und betriebsbedingten Fixkosten des ersten sowie der variablen Erlöse bei deckungsbeitragsmaximaler Allokation der Stoffströme des zweiten Planungsschrittes nach Gleichung (1) erlaubt schließlich eine ökonomische Bewertung aller Alternativen.

4 Daten- und Ist-Analyse

Das vorgestellte Konzept wird derzeit exemplarisch für das Flächenland Niedersachsen für die Unterhaltungselektronik angewendet. Im Rahmen der Anwendung erfolgte eine Ist-Analyse bestehender Stoffstrom-Netzwerke. Eine Vielzahl von Altgeräten wurde untersucht und insgesamt 7 Referenzgeräten zugewiesen, 39 Demontagefraktionen und 31 Demontageoperationen sowie 13 Demontageunternehmen mit dem Tätigkeitsbereich der Zerlegung Brauner Ware wurden bestimmt. Es existieren 47 Öffentlich-Rechtliche Entsorgungsträger (Quellen) sowie 56 Recycling-/Entsorgungsstandorte (Senken). Eine detailliertere Darstellung der Datenerhebung findet sich in [9].

Das resultierende lineare Optimierungsproblem besteht aus ca. 71.000 nichtganzzahligen Variablen sowie ca. 11.000 Nebenbedingungen und kann durch übliche Lösungsprozeduren für lineare Optimierungsprobleme unter Benutzung des Lingo-Solvers und Zugriff auf Excel-Datenblätter in Sekunden gelöst werden.

5 Ausblick und weitere Vorgehensweise

Zukünftig erscheinen der Einbezug der Sammlung und mechanischen Aufbereitung, eine detailliertere Betrachtung der Transportlogistik sowie die Beachtung aller Produktkategorien aus privaten Haushalten und eine Ausweitung der geografischen Grenzen auf ganz Deutschland interessant.

6 Literatur

1. BFUB (Hrsg) (2001) Kosten der EU-weit vorgesehenen Regelungen zur Behandlung von Elektro- und Elektronikschrott. Projektnummer B-I-01, Abschlussbericht
2. Bvse (Hrsg) (1998) Elektronikschrottrecycling – Fakten, Zahlen und Verfahren
3. European Commission (2001) Proposal for a Directive on waste electrical and electronic equipment. Document 500PC0347(01)
4. Halstrick-Schwenk M (2001) Umfang und Struktur der Entsorgungswirtschaft im Bereich Elektroaltgeräte/ Elektronikschrott in Deutschland. RWI-Papiere Nr 65
5. Hesselbach J v. Westernhagen K (2001) An Integrated Approach for the Planning and Control of flexible Retro-Production Systems. Environmental Information Systems in Industry and Public Administration. IDEA Group Publishing, Hershey, USA
6. Hetzel Elektronik-Recycling GmbH (2001) Konzeption und Realisierung eines industriell geprägten Verwertungsbetriebs für Elektronik-Altgeräte. Teilvorhaben 2 im Verbundprojekt IREAK, Schlussbericht
7. Koopmans T C (1951) Efficient Allocation of Resources. Econometrica 19: 455-465
8. Spengler T (1994) Industrielle Demontage- und Recyclingkonzepte – Betriebswirtschaftliche Planungsmodelle zur ökonomisch effizienten Umsetzung abfallrechtlicher Rücknahme- und Verwertungspflichten. Erich Schmidt Verlag, Berlin
9. Spengler T Walther G Hesselbach J Ohlendorf M (2002) Product Assessment and Recycling Data Analysis as Precondition for Efficient WEEE Recycling. Going Green – Care Innovation 2002. Konferenz-CD-Rom
10. VDI 3800 (2000) Ermittlung der Aufwendungen für Maßnahmen zum betrieblichen Umweltschutz

Ein Ansatz zur Bewertung von Remanufacturingstrategien

Axel Tuma and Baptiste Lebreton

Universität Augsburg, Lehrstuhl für Umweltmanagement,
D-86135 Augsburg, Deutschland

Zusammenfassung Die Verschärfung der europäischen Umweltgesetzgebung wie sie sich etwa in der Richtlinie des europäischen Parlaments über Elektro- und Elektronikaltgeräte zeigt - und hierbei insbesondere die Erweiterung der Produktverantwortung auf den gesamten Produktlebenszyklus - stellt für Hersteller neue Herausforderungen dar. Dabei kommt dem Schließen der Kreisläufe sowohl auf der Material- (Recycling) als auch auf der Komponentenebene (Remanufacturing) eine zentrale Bedeutung zu. In diesem Kontext ist eine Weiterentwicklung der Instrumente des Supply Chain Management im Hinblick auf die Integration von Rückflüssen erforderlich.

Analysiert man bisherige Forschungsarbeiten auf strategischer Ebene, ist eine Fokussierung auf WLP-basierte Modelle zu erkennen [3] [4], deren Aufgabe in erster Linie in der räumlichen Gestaltung von Kreislaufwirtschaftssystemen besteht. Vor diesem Hintergrund wird in der vorliegenden Arbeit ein Ansatz zur Bewertung von Remanufacturingstrategien vorgestellt, der die drei Schlüsselfaktoren *Marktsegmentierung*, *technologische Entwicklung* und *Rückflussverteilung* hervorhebt. Dieser Ansatz wird anschließnd anhand eines Fallbeispiels aus der Reifenindustrie veranschaulicht.

1 Diskussion von Einflußgrößen auf das Remanufacturing-Potenzial

Vergleicht man die Remanufacturingquote unterschiedlicher Produkte (wie etwa Kopiergeräte, Druckerpatronen, Lkw-Reifen, mobile Endgeräte u. Pkws) lassen sich folgende wesentliche Einflussgrößen auf das Remanufacturingpotenzial identifizieren:

- **Marktsegmentierung:** Prinzipiell lässt sich feststellen, dass gewisse Käufergruppen Vorbehalte gegen den Erwerb von Remanufacturingprodukten haben. Gründe hierfür liegen nach Bellmann [1] u.a. in der psychologischen Obsolesenz, die wiederwendeten Produkten unterliegt. Damit ergeben sich unterschiedliche Marktsegmente für Neuprodukte und Remanufacturinggüter. Grundsätzlich ist zu bemerken, dass Güter mit primär funktionalen Eigenschaften wie z.B. Druckpatronen ein höheres Remanufacturingpotenzial im Verhältnis zu modeorientierten Produkten (z.B. mobile Endgeräte) aufweisen.

- **Technologische Entwicklung auf Komponentenebene:** Eine wesentliche Vorausetzung für den Wiedereinsatz einzelner Komponenten (u.a. optische Module, Druckerkartuschen, Karkassen) liegt im Ausbleiben technologischer Sprünge (Beispielsweise beim Übergang von Analog- zur Digitaltechnologie), so dass eine Reintegration der *alten* Komponenten technisch möglich und ökonomisch vertretbar bleibt. Kritisch in diesem Zusammenhang sind einerseits der starke Innovationsdruck andererseits der Trend zur Miniaturisierung, die etwa bei mobilen Endgeräten beobachtet werden können. Im Modell wird die technologische Entwicklung durch den Parameter δ ausgedrückt, der angibt mit welcher Wahrscheinlichkeit eine Komponente, die zum Zeitpunkt t' gefertigt wurde zum Zeitpunkt t noch einsetzbar ist.

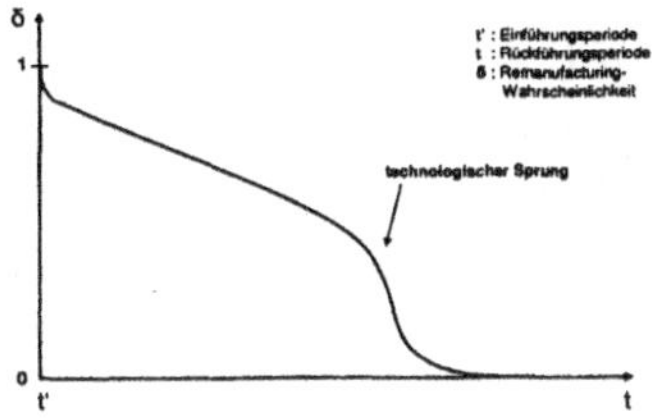

Abbildung1. Technologisch bedingte Wiedereinsetzbarkeit in Abhängigkeit des Komponentenalters

- **Rückflussverteilung:** Neben den technologischen Restriktionen spielt die Verteilung der Rückflüsse zur Planung der Remanufacturingaktivitäten eine zentrale Rolle. In diesem Zusammenhang können dedizierte Rücknahmesysteme oder Produktnutzungskonzepte zu einer Erhöhung der Rückflussquote sowie zur Gewährleistung eines ökonomisch optimalen Rückführungszeitpunktes eingesetzt werden. Die Hersteller von Kopiergeräten setzen hierbei Leasingverträge ein. Pkws hingegen werden nach der Nutzungsphase entweder exportiert (Systemverlust) oder weisen aufgrund ihres Alters eine Abnutzung auf, die ein Remanufacturing kaum konkretisieren lässt.

Aufbauend auf diesen Parametern lässt sich folgendes Modell zur Bewertung von Remanufacturingstrategien formulieren.

2 Modell

Indizes

c : Komponente bzw. Bauteil
p : Produkt
q : Qualitätsniveau
t, t' : Periode

Variablen

$X_{p,t}$: Produktionsmenge von Produkt p in der Periode t.
$\beta_{p,q}$: Marktanteil von Produkt p im Marktsegment q [0,1].
$N_{c,p,t}$: Beschaffungsmenge von neuen Komponenten c für Produkt p in Periode t.
$RF_{c,t',t}$: Rückwärtsfluss von Komponenten c in Periode t, die in Periode t' eingeführt wurden.
$A_{c,p,t',t}$: Anzahl der Komponente c hergestellt in Periode t', die im Produkt p während Periode t wiedereingesetzt werden.
$O_{c,t',t}$: Deponierte Menge in Periode t von Komponente c hergestellt in Periode t'.

Parameter

$\pi_{p,t}$: Verkaufspreis von Produkt p in Periode t.
$c^N_{c,t}$: Beschaffungskosten von einer neuen Komponente c in Periode t.
$c^A_{c,t',t}$: Rückführungskosten der Komponente c in Periode t hergestellt in Periode t' (einschließlich Transport und Demontage).
$c^O_{c,t',t}$: Deponiekosten für Komponente c in period t'.
$c^P_{p,t}$: Herstellungskosten von Produkt p in Periode t.
$D_{q,t}$: Nachfrage im Marktsegment q in Periode t.
$\alpha_{c,p}$: Anzahl von Komponenten c im Produkt p.
$\beta^{max}_{p,q}$: Obere Grenze für den Marktanteil eines Produktes p im Segment q.
$\rho_{p,t',t}$: Rückflusswahrscheinlichkeit von Produkt p nach t-t' Perioden.
$\delta_{c,p,t',t}$: technische Reintegrationswahrscheinlichkeit von Komponente c im Produkt p nach t-t' Perioden.

Zielfunktion

$$Max \quad ! \quad \sum_{p,t} \pi_{p,t} \cdot X_{p,t} - \sum_{c,p,t} c^N_{c,t} \cdot N_{c,p,t}$$
$$- \sum_{c,p,t',t} (c^A_{c,t',t} \cdot A_{c,p,t',t} + c^O_{c,t',t} \cdot O_{c,t',t}) - \sum_{p,t} c^P_{p,t} \cdot X_{p,t}$$

Nebenbedingungen

$$X_{p,t} = \sum_q D_{q,t} \cdot \beta_{p,q} \qquad \forall p,t \tag{1}$$
$$\beta_{p,q} \leq \beta^{max}_{p,q} \qquad \forall p,q \tag{2}$$

$$\sum_p \beta_{p,q} = 1 \qquad \forall q \tag{3}$$

$$A_{c,p,t',t} \leq RF_{c,t',t} \cdot \delta_{c,p,t',t} \qquad \forall c,p,t',t \tag{4}$$

$$\alpha_{c,p} \cdot X_{p,t} = N_{c,p,t} + \sum_{t'<t} A_{c,p,t',t} \qquad \forall c,p,t \tag{5}$$

$$RF_{c,t',t} = \sum_p \rho_{p,t',t} \cdot \alpha_{c,p} \cdot X_{p,t'} \qquad \forall c,t',t \tag{6}$$

$$RF_{c,t',t} = \sum_p A_{c,p,t',t} + O_{c,t',t} \qquad \forall c,t',t \tag{7}$$

Die Zielfunktion drückt das Bestreben des Herstellers aus, den Deckungsbeitrag über die gesamte Produktpalette hinweg zu maximieren. Nebenbedingung (1) stellt mittels des *Marktsegmentierung*-Parameters $\beta_{p,q}$ sicher, dass genügend Produkte $X_{p,t}$ hergestellt werden, um die Nachfrage in den einzelnen Marktsegmenten q zu decken. Mit dem Term (2) kann die psychologische Obsoleszenz berücksichtigt werden.

Term (4) definiert die Obergrenze für die Menge der wiedereinsetzbaren Komponenten. Gleichung (5) gewährleistet, dass ein Produkt p entweder aus neuen Bauteilen $N_{c,p,t}$ oder gebrauchten Bauteilen $A_{c,p,t',t}$ zusammengesetzt wird. Gleichung (6) bildet die Rückwärtsflüsse anhand der Rückflussverteilung $\rho_{p,t',t}$ ab, eine Auflösung auf Komponentenebene erfogt mittels $\alpha_{c,p}$. Auf Basis von Gleichung (7) kann die zu deponierende Mengen der Komponenten c in Periode t bestimmt werden.

3 Fallbeispiel: Remanufacturing in der Pkw-Reifen Industrie

3.1 Bestandsaufnahme

Jährlich fallen in Deutschland 600 000 Tonnen Altreifen an [5], die nach dem 1996 in Kraft getretenen KrW/AbfG einer Verwertung zugeführt werden sollten. Potentielle Verwertungswege sind [2]:

- energetische Verwertung (z.B. als Brennstoff in der Zementindustrie),
- rohstoffliche Verwertung (Pyrolyse zur Rückgewinnung von Rohöl),
- werkstoffliche Verwertung (z.B. Einsatzstoff in der Bitumenherstellung),
- Wiederverwendung der Karkasse zur Herstellung runderneuerter Reifen.

Im Weiteren wird untersucht inwiefern der vierte Verwertungsweg, der nach ökologischen Produktbewertungen [5] [2] vorteilhaft erscheint, auch ökonomisch zu präferieren ist. In diesem Kontext ist anzumerken, dass Flugzeugreifen bis zu zehn Zyklen erreichen, Lkw-Reifen eine Wiederverwendungsquote von 50% aufweisen während Karkassen von Pkw-Reifen nur zu etwa 12% wiederverwendet werden. Hierbei wird zuerst auf die bereits beschriebenen Schlüsselparameter eingegangen.

Marktsegmentierung:

Grundsätzlich lässt sich der Markt für Pkw-Reifen grob in ein so genanntes *Premium* Segment und in ein *Economy* Segment unterteilen. Während im *Premium* Segment neue Markenreifen gehandelt werden, teilen sich runderneuerte Reifen mit *Low-End* Reifen das *Economy* Segment.

Technologische Entwicklung:

Ein Reifen besteht aus einer Karkasse und einem Laufstreifen. Aus technologischer Sicht lassen sich Karkassen mit einer ausreichenden Wandstärke die weniger als sechs Jahre alt sind und keine signifikante Beschädigung aufweisen wiedereinsetzen. Dies ist lediglich bei Karkassen aus dem *Premium* Segment der Fall. Eine Erhöhung der Wandstärke ist allerdings mit zusätzlichen Materialkosten in der Herstellungsphase verbunden und erhöht die Lebensdauer des Reifens in der ersten Nutzungsphase nicht notwendigerweise. Prinzipiell spielt die technologische Weiterentwicklung bei Karkassen eine untergeordnete Rolle und ermöglicht somit einen Wiedereinsatz nach mehreren Jahren.

Rückflussverteilung:

Die Rückflussverteilung der verkauften Reifen hängt im wesentlichen von der Laufleistung sowie von klimatischen Bedingungen ab. So beträgt der Erwartungswert der Nutzungsdauer in Gebieten, in denen Reifen halbjährlich gefahren werden können, ca. sechs Jahre.

3.2 Modellergebnisse

Ausgangszenario

Der Reifenhersteller bietet ursprünglich zwei Produkte an, *Premium* und *Economy*, die sich in den gleichnamigen Segmenten eindeutig zuweisen lassen. Die Segmentnachfrage für das *Premium* Segment wird doppelt so hoch als im *Economy* Segment angenommen.

Einführung eines runderneuerten Reifens

Anschließend wird die Produktpalette um runderneuerte Reifen erweitert, welche sich im *Economy* Segment befinden. Eine Verbesserung der Zielfunktion um 22% liegt darin begründet, dass der *Economy* Segment aufgrund der Verwendung von alten Reifenkarkassen kostengünstiger gedeckt wird. Die Modellergebnisse, d.h. die ermittelte Remanufacturingquote, stimmen mit den empirisch festgestellten Werten überein. Die Gründe für die niedrige Wiederverwendungsquote von 12% liegen in der späten Rückführung und der geringen Wiedereinsetzbarkeit von Karkassen. Vor diesem Hintergrund stellt sich die Frage nach dem Einfluss einer Verbesserung der Karkasse hinsichtlich der Zielgrößen Deckungsbeitrag und Remanufacturingquote. Problematisch hierbei ist, dass der aus dem erhöhten Materialeinsatz resultierende höhere Preis im Allgemeinen nicht an den Kunden des *Premium* Reifens weitergegeben werden darf, da dieser ein für ihn gleichwertiges Produkt zu einem niedrigeren Preis erwerben kann. Ferner ist der Hersteller des ökologischeren *Premium* Produktes nicht in der Lage sicher zu stellen, dass die Karkasse auch wieder zu ihm kommt, um, wie im Modell unterstellt, an einer zweiten Nutzungsphase im Segment der runderneuerten Reifen zu verdienen. Dabei stellen Pfand- bzw. Produktnutzungssysteme potentielle Auswege aus dieser Problematik dar.

Optimierung des Rückführungszeitpunktes

Ein weiterer Ansatzpunkt zur Erhöhung der Remanufacturingquote liegt, zumindest theoretisch, in der Vorverlegung des Rückführungszeitpunktes. Ein entsprechendes Vorgehen führt im Modell zu einer Steigerung der potentiell wiederverwendbaren Karkassen. Bei den Modellberechnungen ergab sich jedoch ein um 9% erniedrigter Deckungsbeitrag. Die Verschlechterung des Ergebnisses liegt im Karkassenüberschuß, der aufgrund einer begrenzten Kundennachfrage nach runderneuerten Reifen nicht wiederverwendet werden kann. Die sprungartige Entwicklung der Entsorgungskosten im Beobachtungszeitraum (+100%) findet ihre Ursprünge in der gestiegenen Menge an überflüssigen Altreifen, die trotz eines positiven δ einer Entsorgung zugeführt werden müssen.

Literatur

1. Bellmann, K. (1990) Langlebige Gebrauchsgüter. DUV, Wiesbaden
2. Ferrer, J. S. (1997) The economics of tire remanufacturing. Resources, Conservation, Recycling. **19**, 221–255
3. Fleischmann, M. (2001) Quantitative models for reverse logistics. Springer, Berlin Heidelberg
4. Jayaraman, V., Guide Jr, V. D. R, Srivastava, R. (1999) A closed-loop logistics model for remanufacturing. Journal of the Operational Research Society **50**, 497–508
5. Umweltbundesamt (1999) Ökologische Bilanzen in der Abfallwirtschaft. Umweltbundesamt, Berlin

Environmental Coordination of Supply Chain Networks Based on a Multi-Agent System

Axel Tuma, Jürgen Friedl

Department of Economics,
University of Augsburg,
Universitätsstraße 16,
86135 Augsburg, Germany

Abstract. Following the actual discussion concerning modern production concepts, an increasing trend to network organizations can be identified. In interconnected production systems materials and different forms of energy are provided, converted, stored and transported. Environmental impacts can be identified at any production unit. In this context the allocation of workload to the different production units, considering simultaneously both ecological and economical goals, is of special interest. In this way it is possible to use available resources more efficiently and to reduce emissions and by-products caused by the production process. To achieve this goal a model of the production system has to be created. This model has to include the input and output streams. Due to the modelling requirements of supply chain networks (modelling of more or less independent production units, dynamic behavior) an agent-oriented simulation seems to be adequate. Secondly evaluation methods for the modelling of the economic and ecological goals have to be integrated. In order to address the multi-criteria structure of such a decision problem, a goal programming approach is discussed.

1. Agent-Oriented Simulation

The fusion of simulation models and agent-oriented modelling techniques is an important step towards the examination and modelling of highly complex, real world applications. Agent-Oriented Simulation embodies the power of both methods: Firstly simulation is a perfect technology for modelling and analyzing complex phenomena; secondly agent-oriented technologies give support towards decentralization on the one hand and "cognitive support", often missed in former simulation systems, on the other hand.

The advantages of using AOS techniques are:

- **Naturalness**: Modelling many collaborative agents is far more natural and can be better controlled than building one big monolithic system.
- **Robustness**: Faulty parts can be easily identified and exchanged and extension and modification of the system cause little update operations.
- **Distribution**: Since there is a clear logical partition, the physical distribution of the system to different machines is straight-forward. This implies a realization of distributed real-time simulation.

What are the (multi-)agent technologies involved in an agent problem solving process? First of all there is the agent itself. From a general point of view there are two schools: the reactive school which supports an efficient real-time behaviour of the system by a fast acting situation-response procedure, and the representational or deliberative school which supports an explicit knowledge representation and reasoning. The decision on one or the other school, or on a hybrid version influences the agent architecture. No matter which school is supported, the agent must be able to communicate and to act on the basis of his world knowledge. Therefore are modelling communication and the individual knowledge base, including sensoric and actoric functionalities, primary agent technology decisions. Secondary technology decisions concern planning, local conflict management, and basic learning abilities. Tertiary technology decisions cover joint planning, global conflict resolution, partner modelling, and agent centred learning.

The transition from one agent to a society of agents raises specific design problems. The functional interaction (as von Bechtholtsheim [1] called it) among the agents has to be modelled. The functional interaction influences every decision of the single agent model by adapting more complex mechanisms. For example, the communication might need language primitives and protocols for complex negotiations in addition to simple REQUEST-REPLY or SEND-RECEIVE patterns. There must be a qualitative step to extend single agent planning to the generation of joint plans including activities of different agents. For this reason explicit coordination mechanisms have to be used for the design of the agent society.

Instead of deepening the AOS discussion (look up [2] for more details) the basic structures of agents representing the production units of the network will be discussed in the following.

2. Configuration of the Production Agents

According to the requirements of the allocation tasks the agents have to be defined, their knowledge and behaviour have to be characterized and their coordination mechanisms have to be described. The agents will be described in an imitation of the CommonKADS knowledge engineering model [2].

PU-agents: (Production Units)

- have the *task* to increase their own marginal income, respectively to decrease the emissions and by-products caused by the production process;
- perform the *actions*: Produce and Consume;
- *communicate* with their direct neighbourhood by a) sending resource/job-requests (need more material) and capacity-requests (need less material/offer jobs) b) receiving push-requests and pull-requests, and c) negotiating on the amount (and possibly the time restrictions) of material flow;
- have *knowledge* about their a) current satisfactory state b) productivity and c) Producing/Consuming relationships and needs;
- *plan* in order to forecast their local situation and plan the communication. The planning process includes the reasoning capabilities of the agents.

Each agent acts essentially on the basis of its state of satisfaction which reflects the fulfilment degree of its task. If an agent is not satisfied, it communicates its needs to the neighbours in order to increase or decrease its workload. The agent's ability to negotiate enables each agent to partially accept the requests and postpone the necessary activities in order to fit into the currently calculated plans. In the following the satisfaction calculation will be discussed in detail. Negotiation principles are discussed in [3].

3. Calculation of the Satisfaction Level of the PU-Agents

The basic idea for the construction of the satisfaction level for the single PU-Agent is the evaluation of the production quantity taking into account economical and ecological criteria. In this context the PU-Agent will be modelled according to the Gutenberg production function [4]. Based on intensity and production time, unit costs, respectively marginal income, and emissions are calculated (Fig. 1).

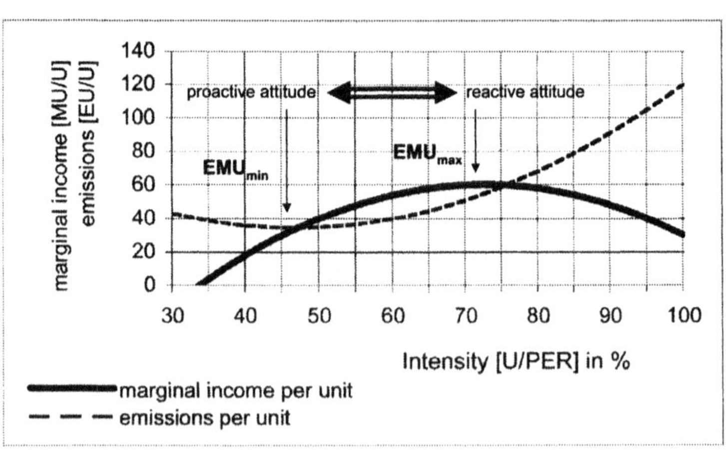

Fig. 1. Marginal Income and Emissions per Units

In order to find an evaluation value for the possible production quantities the multi-criteria character of the problem has to be addressed. In principle this can be done via different methods like scoring-models, goal programming and aspiration level approaches.

In the following a goal programming approach will be explained in more detail (Fig. 2).

$$\begin{aligned}
&Max\ MI(d)\cdot X\\
&Min\ EM(d)\cdot X\\
&(1)\quad MI(d)\cdot X \ge MI_{min}\\
&(2)\quad EM(d)\cdot X \le EM_{max}\\
&(3)\quad d_{min} \le d \le d_{max}\\
&(4)\quad d\cdot t_{min} \le X \le d\cdot t_{max}
\end{aligned}$$

GOAL PROGRAMMING

$$\begin{aligned}
&Max\ a_1 + k\cdot a_2\\
&(1)\quad MI(d)\cdot X - a_1 = MI_{min}\\
&(2)\quad EM(d)\cdot X + a_2 = EM_{max}\\
&(3)\quad d_{min} \le d \le d_{max}\\
&(4)\quad d\cdot t_{min} \le X \le d\cdot t_{max}\\
&(5)\quad a_1, a_2 \ge 0
\end{aligned}$$

d = intensity [U/PER]	k = key parameter [MU/EU]
MI = marginal income function [MU/U]	EM = emission function [EU/U]
EM_{max} = legal emission standard [EU]	X = (fixed) production quantity [U]

MI_{min} = enterprise specific minimal marginal income [MU]
d_{min}/d_{max} = enterprise specific minimal/maximal intensity [U/PER]
t_{min}/t_{max} = enterprise specific minimal/maximal flexible time per period [PER]

Fig. 2. Goal Programming Approach

The key parameter of the evaluation process k has to be calculated first. For k=0 the evaluation process will only focus the economical point of view. The ecological aspects will not be considered. In contrast, for a $k \rightarrow \infty$ the economical evaluation part will be neglected. In this context the model for k=0 represents according to Hansmann [5] a reactive attitude of the production unit (for $k \rightarrow \infty$ a proactive attitude). In order to reflect the behaviour between the two poles proactive and reactive, adequate levels of the parameter k have to be specified. In this context a so-called minimal and maximal emission level will be calculated. The "minimal emission level" EMU_{min} respectively the "maximal emission level" EMU_{max} represents the minimal (maximal) emissions per unit considering the intensity and time constraints (Fig 1.). In the next step the difference between EMU_{max} and EMU_{min} is divided in n equidistant intervals. For each of the intervals i=1,...n the corresponding value k_i (related to the maximal value of the interval) is calculated according to the following equation:

$$EM(d_{k_i}^*) = EMU_{min} + (EMU_{max} - EMU_{min}) * \frac{i}{n} \quad i = 1,\dots,n \tag{3.1}$$

where d_{ki}^* represents the optimal solution of the goal program of Fig 2. Under the assumption that $EM(d)=a_1 \cdot d^2+b_1 \cdot d+c_1$ and $MI(d)= a_2 \cdot d^2+b_2 \cdot d+c_2$, k_i can be calculated (independently of X) by equation 3.2.

$$k_i = -\frac{EM^{-1}[EM_{min} + (EM_{max} - EM_{min}) \cdot \frac{i}{n}] \cdot 2a_1 + b_1}{EM^{-1}[EM_{min} + (EM_{max} - EM_{min}) \cdot \frac{i}{n}] \cdot 2a_2 + b_2} \quad (3.2)$$

Fig. 3 shows an evaluation function for production quantities of a production unit for parameter k_8=0,1 and k_4=0,5 (n=10)[1].

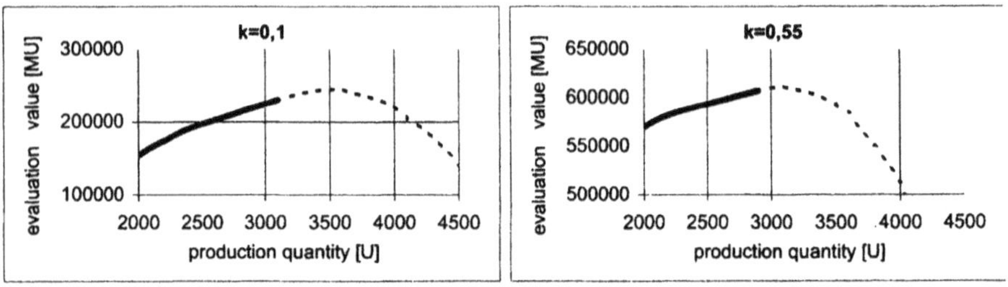

Fig. 3. Evaluation Values for Different Key Parameters

4. Exemplary Negotiation Process

To illustrate the negotiation process a small example consisting of two production units (PU1 and PU2) will be considered. Fig. 4a-b shows the evaluation functions of the two production units (with identical key parameter k) including the actual state of the production (underlined in Fig. 4a-b). Assuming that the production state of PU1 is below a given satisfaction level α of e.g. 85% (of the difference between maximal and minimal evaluation value), a negotiation process will be initiated by this production unit. This means a job-request will be send from PU1 to PU2. In the following the evaluation function of PU2 will be calculated as a function of potential production quantities of PU1 (Fig. 4c). To determine the new allocation of workload between PU1 and PU2 the sum of the evaluation functions of

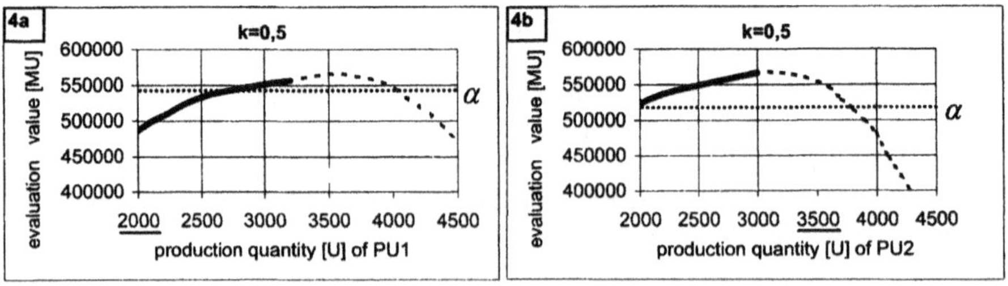

Fig. 4a-4b. Evaluation Values and Satisfaction Level

Fig. 4c is calculated (Fig. 4d). The result (production quantity of PU1) of the negotiating process is determined by the maximal value of the aggregation function in Fig. 4d. The production quantity of PU2 can be computed accordingly to this.

---

[1] Within the dashed part the emission per unit due to the equation 3.2 can not be guaranteed.

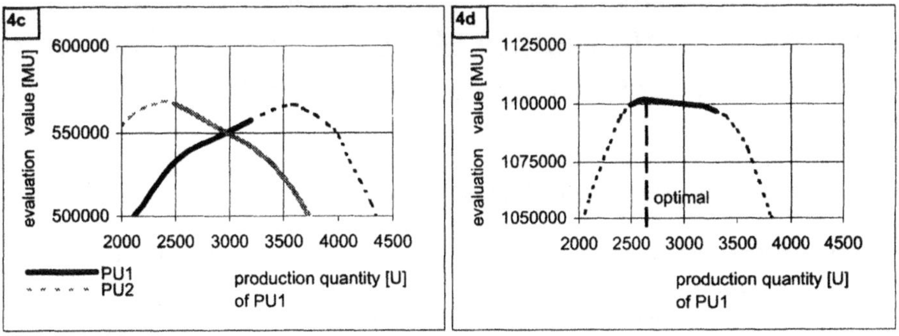

Fig. 4c-4d. Negotiation Process

5. Conclusion

The described approach shows in principle that the allocation of workload in a production network can be based on an agent-oriented simulation system. To describe the single agents the CommonKADS knowledge engineering model seems to be well suited. In order to model the multi-criteria goal functions of the agents a goal programming approach can be applied. In addition the different behaviour between the poles, proactive and reactive, can be modelled. Further research work will focus on different cooperation relations in supply chains as well as on the evaluation of different strategies for the simultaneous addressing of economical and ecological goals. Furthermore the investigation of the dynamic behaviour of a whole production network seems to be interesting. In this context the potential learning capabilities of the single agents seem to be crucial.

REFERENCES

[1] V. Bechtoltsheim, M.: Agentensysteme; Vieweg Verlag, 1993

[2] Müller, H.J.: Towards Agent Systems Engineering, Data & Knowledge Engineering, Vol. 23, pp. 217-245, 1997

[3] Müller, H.J.: Negotiation Principles, IN: Foundations of Distributed Artificial Intelligence, O'Hare, G.M.P. and Jennings, N.R. (Eds), pp. 211-230, Wiley Intersience Pub, 1996

[4] Gutenberg, E.: Grundlagen der Betriebswirtschaftslehre, 1. Band, Die Produktion, 16. Auflage, pp. 314-325, Springer-Verlag, 1969

[5] Hansmann, K.W.: Umweltorientierte Betriebswirtschaftslehre, pp. 10-12, Gabler, 1998

Decision Support for the National Implementation of Emission Reduction Measures by the Dynamic Mass Flow Optimisation Model ARGUS

Jutta Geldermann, Nurten Avci, Stefan Wenzel, Otto Rentz

French-German Institute for Environmental Research (DFIU / IFAREIIP)
University of Karlsruhe (TH), Hertzstr. 16, D-76187 Karlsruhe, Germany
E-mail: {first name.last name}@wiwi.uni-karlsruhe.de;
http://www-iip.wiwi.uni-karlsruhe.de/~voc/

Abstract. A large number of industrial sectors is affected by enforced environmental requirements, and therefore a cost-optimal transformation is essential for the economic and environmental efficiency both of companies and regions. Techno-economic dynamic optimisation models – like ARGUS - have been developed and implemented for various European Countries. The resulting cost functions for various scenarios are essential for the multi-national allocation of emission reductions. Cost-discounting effects and the temporal pathway of the implementation of emission reduction options within a given planning horizon (up to 2020) are considered. The results can be used to analyse the cost-effectiveness of environmental legislation and for emission projections and inventories.

1 Introduction

The availability of reliable cost functions for the considered pollutants (SO_2, NO_x, VOC, NH_3) and countries (UN/ECE) is an important prerequisite for the implementation of the critical load/level approach based on integrated assessment modelling [1; 10]. Beside critical loads/levels and parameters describing the atmospheric transport and transformation of pollutants, cost functions have an important influence on the national emission ceilings, which have been set recently by the European NEC-Directive (81/2001/EC). Thus, the EU Member States have the obligation to draw up programmes to ensure the compliance with these limits by 2010 at the latest, in order to adopt the emission limits negotiated in the Gothenburg Protocol of the UN/ECE to abate acidification, eutrophication and tropospheric ozone (also called "multi-pollutant and multi-effect Protocol"). Other approaches to reach the environmental targets set by the Gothenburg Protocol are the EU-Solvent Directive (13/99/EC) and the IPPC-Directive (taking into account all environmental effects for integrated pollution prevention and control; 61/96/EEC). The 'best available techniques' (BAT) play an essential role in the actual realisation of these regulations.

A large number of industrial sectors is affected by these enforced environmental requirements, and therefore a cost-optimal transformation is essential for the economic and environmental efficiency both of companies and regions. There fore, cost functions have to be as accurate as necessary. This means in particular that all relevant sources and all available emission reduction options have to be taken into account, including technical pollution abatement measures but also structural options (like autonomous technology change during the planning period), which often show an important cost saving potential, especially in countries with economies in transition.

Dynamic energy and mass flow optimisation models have been developed to assess emission reduction strategies on the national level. They are based on a detailed representation of all relevant emission processes and corresponding emission reduction options, including structural options. They provide the "cost optimal" evolution of the production system (production technologies and abatement options in place) over a given planning horizon (up to 2020) which allows for the achievement of emission reduction ceilings and the fulfilment of the demand of products or services specified exogenously on the sectoral level. This contribution explains the underlying principle of dynamic energy and mass flow model ARGUS (Allocation module for a computer aided generation of environmental strategies for emissions), which has been developed for the elaboration of emission reduction strategies for VOC from stationary sources on a national level [7, 6; 11].

2 Methodological Background of the Dynamic Mass Flow Optimisation Model ARGUS

The (quasi-) dynamic mass-flow optimisation model ARGUS is based on a detailed representation of all relevant stationary VOC emission sources and the corresponding applicable emission reduction options. The optimisation criterion is the minimisation of the sum of the discounted costs over the considered planning horizon. Emission sources and abatement options are described in terms of reference installations defined by the UN/ECE Task Forces on abatement options/techniques for VOC [4; 9]. About 2000 emission relevant processes within about 40 industrial sectors (like vehicle coating or production of paints and adhesives) are modelled. The input data are structured as follows:

- Technological data sheets specify the emission factor $e_{s,i,j}$, the investment $I_{s,i,j}$ and the operating costs of an emission reduction option *(s,i,j)* applicable to a reference installation *i* of sector *s*.
- In contrast, the country data sheets are used to characterise the time dependent and country specific structure of emission sources in terms of sectoral activities $A_{s,t}$ the shares of the different reference installation $x_{s,i,t}$ to the sectoral activities $A_{s,t}$ (market shares) and the implementation rate $y_{s,i,j,t}$ of the emission reduction options (share of the option *(s,i,j)* to the sectoral activity of reference installation *(s,i)* in year *t*.

This distinction of the data allows for an wide applicability in different countries, since the technological data can be used as default values and can be adjusted to the current needs, with most necessary modifications regarding the country specific data.

The model determines the evolution of the structure of emission sources and abatement options for planning horizon from 1995 (base year) to 2015. The optimisation variables are the implantation shares $y_{s,i,j,t}$ of the different emission reduction options. The evolution of sectoral activities and market shares of reference installations are specified exogenously and are taken from the country data sheets. The target function is given by

$$G = \sum_{s}\sum_{t}\sum_{i}\sum_{j} \alpha_t \cdot \left(\widetilde{I}_{s,i,j} \cdot \Delta Q_{s,i,j,t} + \widetilde{O}C_{s,i,j} \cdot P_{s,i,j,t}\right) \qquad \text{Eq. (1)}$$

The first term placed within brackets denotes the investment related expenditures due an increase $\Delta Q_{s,i,j,t}$ of the capacity of emission reduction option *(s,i,j)* in the year *t*. $\widetilde{I}_{s,i,j}$ is the annualised specific investment per unit of capacity of option *(s,i,j)*. The second term represents the operating costs of option *(s,i,j)* which is obtained by multiplying the annual activity $P_{s,i,j,t}$ of installations equipped with this option and the specific annual costs per unit of activity $\widetilde{O}C_{s,i,j}$. All these parameters can be expressed as function of the optimisation variables and the parameters specified in the technological and country data sheets [7; 6; 2].

Thus, the decision (optimisation) variables are the capacities of the implemented production or emission reduction technologies for each year. The optimisation is performed by taking into account restrictions concerning inter alia the fulfilment of products or services demand and the limitation of emissions. The total VOC emissions to be considered in constraint on an emission ceiling $E_t^{\max}$ for each period *t* are given by

$$E_t = \sum_{s}\sum_{i} A_{s,t} \cdot \hat{e}^{o}_{s,i} \cdot (1 - \sum_{j} y_{s,i,j,t} \cdot \eta_{s,i,j}) \leq E_t^{\max} \quad \forall\, t = 1, 2, \ldots\ldots T_p \qquad \text{Eq. (2)}$$

where $\eta_{s,i,j}$ is the efficiency of reduction option *(s,i,j)*. In practice, as the time delay for the implementation of the Protocol obligation is relatively long, measures requiring important transformations of existing installations will only be implemented for new installations. The implementation potential of such options is controlled by the changes in sectoral activities (e.g. increase of production capacity) and the capacities of installations reaching the end of their lifetime in a given time period (renewal of installations). The remaining capacities of existing installations are calculated from the age distribution and lifetime of installation using a lifetime model included into ARGUS.

A cost function for the target year T_t is calculated in the following way: In a first step the target function is minimised without any emission ceiling. Let denote $E^0_{T_t}$ the corresponding emission value for the target year. In a second step optimisation runs are performed for a set of decreasing values of emission ceilings $E^{\max}_{T_t}$ (constraints according to Eq. 2) in the range from $E^o_{T_t}$ up to the minimum feasible value. Moreover, intermediate targets can be defined in scenarios, in order to analyse the influence of the implementation delay of emission reduction obligations on cost functions. Figure 1 exemplarily shows the cost functions for Germany [7] corresponding to the different implementation scenarios. Similar results have been obtained for France [6; 11].

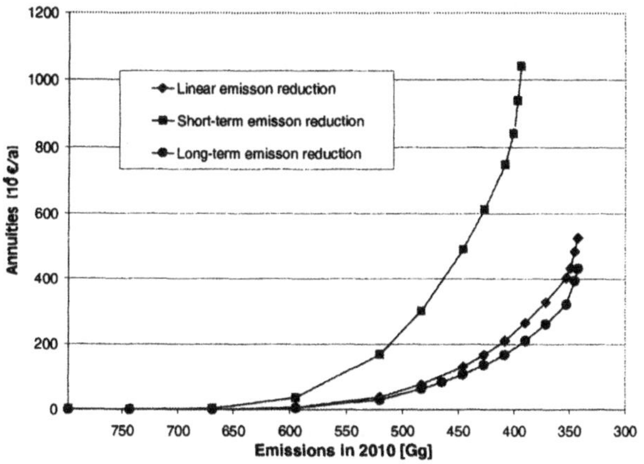

Figure 1: Costs functions for VOC emission sources from stationary sources in Germany for the different implementation scenarios (planning horizon $T_p = 2015$, target year T_t =2010, interest rate i = 0%

3 Uncertainties and variability in Dynamic Mass Flow Optimisation Models

Estimates and long-term forecasts obviously are, by their very nature, uncertain: A strategy considered optimal on the basis of particular assumptions made today is highly unlikely to turn out optimal in the actual situation of 2010 [13]. In this context, the following types of uncertainty and variability can be distinguished [12]:

- ***Parameter uncertainty***, such as a lack of knowledge about emission estimates per unit or environmental degradation rates;
- ***Model uncertainty***, like the assumption of linear relationships between the mass and energy flows, or the assumption that substances are physically and chemically homogeneous;
- ***Uncertainty due to choices***, such as the choice between several allocation methods, if delimitation problems occur with regard to the assignment of the costs for emission abatement and those incurred by other purposes (e.g. industrial health and safety standards); or the choice of weighting factors, if several harmful substances have to be considered simultaneously;
- ***Spatial variability***, like regional differences in emission estimates, or varying sensitivity of the local environment;
- ***Temporal variability***, e.g. differences in yearly emissions or variability caused by seasonal changes;
- ***Variability between sources***, e.g. different techno-economic characteristics of the relevant production plants.

So far, most applied Operations Research approaches concentrate on the treatment of parameter uncertainty, starting from basics like the request for data formats allowing for the characterisation of the data quality. Imprecise data might be modelled by use of Interval Arithmetic or Fuzzy Logic [3]. More advanced approaches cover the treatment of subjective uncertainty estimates, e.g. by Bayesian Statistics, while stochastic modelling describes parameters as uncertainty distributions, including Monte Carlo or Latin Hypercube simulation [14]. Given the size of the dynamic mass flow optimisation models, however, it seems questionable, if a comprehensive analysis of the model's sensitivity can be reached. Here, the reduction of the model size might open the paths for further investigations.

4 Conclusions

Techno-economic dynamic energy and mass flow optimisation models for the determination of national cost functions take into account full sets of emission reduction options including structural options related to changes in sectoral activities and production technologies. While ARGUS has been developed for the analysis of VOC-emission reduction strategies from a large number of stationary sources, e.g. PERSEUS has been widely used to optimise energy systems including the aspects of resources and emissions (SO_2, NO_x, and CO2). Further research is now necessary on the integration of these models for the analysis of combined emission reduction measures. For this cross-media assessment, only the analysis of a discrete set of scenarios seems to be feasible up to now, applying MADM methods (cf. [3]). If the discussion of Life Cycle Assessment (LCA) or sustainability indicators can provide useful input for dynamic mass flow optimisation models, is currently an open issue.

A main factor limiting the accuracy of the required data, however, is the availability and quality of input data, in particular with regard to the structure of emission sources (statistics on activities, information on plants and applied processes, e.g. size and age distribution of installations, control options already in place). Especially the modelling of smaller regions with few installations of similar type requires modified approaches [8]. Further research in the field of Operations Research seems to be necessary for more advanced sensitivity analyses within complex techno-economic dynamic energy and mass flow optimisation models.

5 Literature

[1] Alcamo J., Shaw R., Hordijk L. (eds): The RAINS Model of Acidification. Science and Strategies in Europe. Kluwer, Dordrecht, 1990

[2] Brooke, A.; Kendrick, D.; Meeraus, A.: GAMS, The Scientific Press, 1988

[3] Geldermann, J., Rentz, O.: Entwicklung eines multikriteriellen Entscheidungsunterstützungssystems zur integrierten Technikbewertung. In: Operations Research Proceedings 2000, S.445-451, Springer, Berlin (2001)

[4] Nunge, S.; Geldermann, J.; Rentz, O.: Der Einsatz der Clusteranalyse zur Definition von Referenzanlagen. In: Operations Research Proceedings 2001, SOR 2001, Duisburg, Springer-Verlag, Berlin (2002)

[5] Rentz, O.; S. Nunge, U. Karl, T. Holtmann, T. Zundel: Feasibility Study on the Development of a Design for an Emission Projection Model Based on the CORINAIR-Approach, Final report, December 1999

[6] Rentz, O.; Avci N.; et al.: Elaboration de fonctions de coût pour la réduction des émissions de composés organiques volatils pour la France, final report, research project supported by ADEME, Karlsruhe, 1999

[7] Rentz, O.; Laforsch, M.; Holtmann, T.; Nunge, S.; Zundel, T.; Avci, N.: Technologies, measures and costs for the avoidance of VOC emissions as a basis for integrated assessment models in the framework of the UN/ECE, final report on-behalf of UBA, Berlin, 1998

[8] Rentz, O.; Wenzel, S.; Avci, N.; Geldermann, J.; Joas, R.; Schott, R.: Proposal for a Scheme of Action for the Reduction of the Ozone-Precursors NO_x and VOC in Austria until 2010. On behalf of the Ministries of the Environment and of Economics, Vienna, 2002

[9] Rentz, O.; Nunge, S.; Laforsch, M.; Holtmann, T.: BAT Background Document of the Task Force on the Assessment of Abatement Options/techniques for Volatile Organic Compounds, September 1999

[10] Wietschel M.; Dreher M.; Fichtner W.: Goebelt M.; Rentz. O., PERSEUS: State of Model Development, Applications and Perspectives, in: Hake, J.-Fr., Markewitz, P. (ed..): Modellinstrumente für CO_2-Minderungsstrategien, Jülich, 1997, p. 223-240

[11] Zundel, T.; Fichtner, W.; Avci, N.; Frank, M.; Rentz, O.: Dynamic energy and mass flow optimisation models for the elaboration of cost functions for VOC and NO_x emission reduction: methodological issues, in: Pollution Atmosphérique, Numéro spécial "Atelier d'Angers", 2000

[12] Huijbregts, M.A.J.: Uncertainty and variability in environmental life-cycle assessment, Amsterdam, 2001

[13] Landrieu, G.; Mudgal; S.: Commentaries on the results of studies requested from IIASA by the French Ministry for Land-Use Planning and the Environment. Task Force on Integrated Assessment Modelling, 25th session, Stockholm, 2000.

[14] Insua, D.R,; Martin, J; Proll, L; French, S; Salhi, A.: Sensitivity Analysis in Statistical Decision Theory: A Decision Analytic View. In: Journal of Statistical Computation and Simulation, 1997, Vol. 57, 1-4, p. 197-218

Fuzzy Scheduling for the Dismantling of Complex Products

Frank Schultmann and Otto Rentz

University of Karlsruhe, Institute for Industrial Production, Hertzstrasse 16, D-76187 Karlsruhe
E-mail: {frank.schultmann; otto.rentz}@wiwi.uni-karlsruhe.de

Abstract. In this paper, an approach is presented for the solution of decision-making problems in short-term dismantling planning. Weak data resulting from uncertain dismantling time parameters are modeled as fuzzy sets and integrated into a scheduling model that supports deriving optimal work plans. Case studies from the construction industry and the electronic scrap recycling show the application of the approach.

1 Introduction

Due to extended producer responsibility and stricter environmental regulations an increasing amount of spent products, e.g. electronic equipment, vehicles or buildings, will have to be dismantled in the near future. Consequently, the management of dismantling and the subsequent recycling cannot not only focus on the genuine aim of profitability, but will also have to meet criteria of sustainability, e.g. limitating the discharge of pollutants to the environment. To anticipate an integrated dismantling and recycling procedure, numerous different objectives in environmental, technical and economic means have to be taken into account. Thus, alternative scenarios for the end-of-life treatment need to be considered on strategic, tactical and operational level. While objectives in short-term planning are usually modeled using extended time-based objective functions, different alternatives to meet certain targets in the field of sustainability can be modeled using multiple modes. Mathematically, both aspects can be realized by using project-scheduling models. However, even when using multiple modes, one major problem in modeling dismantling processes is the fact that data is not always available to put mathematical models at work. In particular, the duration of dismantling tasks is seldom precisely known due to uncertainties in the composition of returned materials. The use of fuzzy sets proves to be a powerful tool for modeling this weak data. Fuzzy techniques have been well established in mathematical theory; however, fuzzy programming is still hardly used for dismantling tasks.

2 A fuzzy-scheduling model for dismantling

The well-known multi-mode resource-constrained project scheduling problem (MRCPSP) is used as a basis for the following model formulations. For a survey on modeling concepts as well as scheduling algorithms for resource-constrained project scheduling see [1], for modeling dismantling problems we refer to [2].

Operations Research Proceedings

2.1 Planning of dismantling structures

Dismantling planning aims first at setting up a technology oriented order of the dismantling activities to be carried out. The technological precedence relations of the dismantling process can be illustrated by an topologically ordered activity-on-node network, where the nodes represent the dismantling activities j ($j=1,...,J$) and the arcs indicate the precedence relations between these activities. Regarding the model that will be formulated later, the network contains one unique source ($j=1$) and one unique sink ($j=J$). This can always be guaranteed by introducing a dummy source and a dummy sink, respectively. For examples, we refer to [2]. Usually each activity can be processed in different ways e.g. using different techniques that can be expressed by different resources. Moreover, each different technique may result in different processing times. Several alternatives for carrying out a job can be modeled by introducing different modes m ($m=1,...,M_j$).

The scheduling model contains resource categories. Activity j in mode m is associated with the usage of renewable resources and the consumption of nonrenewable resources. While renewable resources (e.g. machines, workers) are only constrained on a periodical basis (possibly varying from period to period), i.e. after an activity j is accomplished, the renewable resources used by j are available to process another activity. Nonrenewable resources (e.g. financial budget) are limited on the basis of the project's entire duration. Consequently, the consumption of a nonrenewable resource by activity j reduces its availability for the rest of the project. For simplification purposes we will use renewable resources only. Variables for resource usage as well as for constraints are introduced as follows:

q_{jmr}: capacity of renewable resource r, used by dismantling activity j being performed in mode m for each period the activity is in process and

Q_{rt}: capacity of renewable resource r, $r \in R$, available in period t.

2.2 Modeling of uncertain time parameters

Performing activity j in mode m has a nonpreemtable processing time, also referred to as duration, of d_{jm} periods. Due to the uncertainties of these processing times, resulting e.g. from different conditions of the discarded products, time parameters are modeled as fuzzy sets.

The expectations of a decision-maker concerning fuzzy time parameters like the processing time or the start and finish times of an activity j in the mode m ($m=1,...,M_j$) can be modeled as a fuzzy set, where set A in a base set X can be described by a membership function $\mu_{\tilde{A}} : X \rightarrow [0,1]$ with $\mu_{\tilde{A}}(x)=1$ if $x \in A$ and $\mu_{\tilde{A}}(x)=0$ if $x \notin A$. If it is uncertain, whether element x belongs to set A, the model has to be extended in a way that $\mu_{\tilde{A}}$ maps into interval $[0,1]$ with its value representing a membership possibility between poor (low value) and likely (high value). This leads to the definition of a *fuzzy set* [3]:

$$\tilde{A} = \{(x, \mu_{\tilde{A}}(x)) \mid x \in X\}, \text{ with the membership function } \mu_{\tilde{A}} : X \rightarrow [0,1] \quad (1)$$

Other definitions needed are the *fuzzy number* which is a normalized convex fuzzy subset of IR with exactly one $x_0 \in IR$ such that $\mu_{\tilde{A}}(x_0)=1$; and a *fuzzy interval* which, contrary to a fuzzy number, can comprise more than one $x \in IR$ such that $\mu_{\tilde{A}}(x)=1$. Since the precise form of a fuzzy number can rarely be described by an expert in reality, a practicable way of getting suitable member functions of fuzzy data can be obtained if the expert expresses optimistic and pessimistic information about parameter uncertainty on some prominent membership levels α for each mode m in which activity j could be processed. The α-level set of the fuzzy interval (duration) $\tilde{D}_{jm}$ is defined as:

$$(D_{jm})^{\alpha} = \{d_{jm} \in IR_0^+ \mid \mu_{jm}(d_{jm}) \geq \alpha\}\text{, with } \alpha \in [0,1] \tag{2}$$

In this case, the optimistic (pessimistic) value for the duration of a dismantling acitvity j performed in mode m on level α then represents the lower (upper) bound of the corresponding α-level set of the fuzzy interval $\tilde{D}_{jm}$. If L levels $\alpha = \sigma_1, \ldots, \sigma_L \in]0,1[$, $L \in IN$ are introduced, the duration can be expressed as:

$$[{}_o d_{jm};\ {}_p d_{jm}] \qquad \text{if } \alpha = 1, \tag{3}$$

$$[{}_o d_{jm} - {}_o\delta_{jm}^{\alpha};\ {}_p d_{jm} + {}_p\delta_{jm}^{\alpha}] \text{ if } \alpha = \sigma_1, \ldots, \sigma_L \in]0,1[;\ L \in IN \tag{4}$$

The variables ${}_o\delta_{jm}^{\alpha}$ (${}_p\delta_{jm}^{\alpha}$) denote the expert's assessment for the acceleration (prolongation) of the optimistic duration ${}_o d_{jm}$ (pessimistic duration ${}_p d_{jm}$) for activity j in mode m on level α, with ${}_o d_{jm}$ and ${}_p d_{jm}$ fixed on level $\alpha=1$. Apart from this particular case ($\alpha_1=1$), two additional levels ($\alpha_2=\varepsilon=0,1$; $\alpha_2=\lambda=0,6$) are considered in the following, according to [3], with:

$\mu_{\tilde{D}_{jm}}(d_{jm})=1$ indicates that value d_{jm} certainly belongs to the set of possible values,

$\mu_{\tilde{D}_{jm}}(d_{jm}) \geq \lambda$ indicates that value d_{jm} with $\mu_{jm}(d_{jm}) \geq \lambda$ has a good chance of belonging to the set of possible values,

$\mu_{\tilde{D}_{jm}}(d_{jm}) < \varepsilon$ indicates that the expert estimastes that value d_{jm} with $\mu_{jm}(d_{jm}) < \varepsilon$ has only a very poor chance of belonging to the set of possible values.

With ${}_o d_{jm}^{\alpha} = {}_o d_{jm} - {}_o\delta_{jm}^{\alpha}$; ${}_p d_{jm}^{\alpha} = {}_p d_{jm} + {}_p\delta_{jm}^{\alpha}$; $(\alpha \in \{1, \lambda, \varepsilon\})$, $\mu_{\tilde{D}_{jm}}(d_{jm})$ can be approximied by combining linear functions section-wise (cf. Fig. 1).

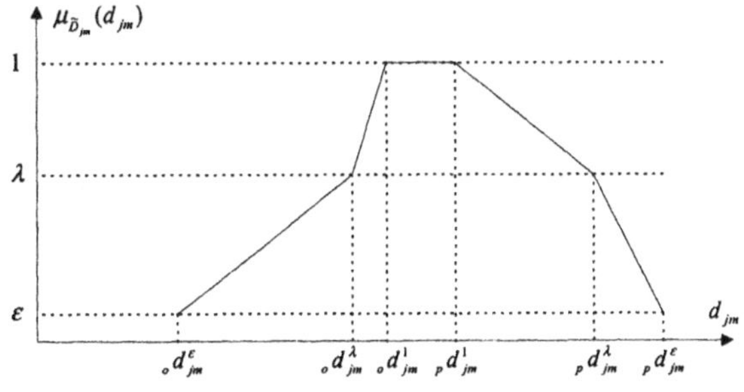

Fig. 1. Membership function of the duration $\widetilde{D}_{jm}$

A fuzzy interval that can be depicted by a membership function as shown in Figure 1 can be transformed in a six-point convention such that it is represented by six real numbers (cf. [4]):

$$\widetilde{D}_{jm} = ({}_o d^{\varepsilon}_{jm},\ {}_o d^{\lambda}_{jm},\ {}_o d^{1}_{jm},\ {}_p d^{1}_{jm},\ {}_p d^{\lambda}_{jm},\ {}_p d^{\varepsilon}_{jm}), \qquad j=1,\ldots J,\ m=1,\ldots,M_j \tag{5}$$

The fuzzy arithmetics that can be applied for such six-point representations are shown in [5].

3 Model formulation

With reference to the resource-constrained project scheduling problem introduced by Pritsker et al. [6], scheduling for dismantling purposes can be formulated as a binary linear program with the decision variables x_{jmt} (dismantling activity j is performed in mode m and completed in period t). As proposed by Hapke et al. [7] this problem can be used for fuzzy scheduling as well. In the following, we assume that the earliest and latest finishing times, the earliest and latest starting times as well as the durations of the dismantling activities are given as fuzzy intervals in six-point representations. Using the expert's estimates for optimistic (o) and pessimistic (p) durations on the three levels $\alpha=1$, λ, ε, the fuzzy-model can be transformed into six crisp resource-constrained project scheduling problems.
In order to reduce the number of variables in the programming formulation, fuzzy time windows can be calculated using a modified critical path analysis. Critical path analysis requires an upper bound $\overline{T}$ for the makespan of the project. It should also be noted that the unique source ($j=1$) and the unique sink ($j=J$) have comprise zero duration, zero resource usage and consumption, respectively. The earliest and lastest starting and finishing times ${}_k ES^{\alpha}_j$, ${}_k EF^{\alpha}_j$, ${}_k LS^{\alpha}_j$, ${}_k LF^{\alpha}_j$, $k \in \{o,p\}$, $\alpha \in \{1, \lambda, \varepsilon\}$ for the dismantling activities $j=1,\ldots,J$ can be calculated as:

$$ {}_k ES^{\alpha}_1 = 0 \qquad k \in \{o,p\},\ \alpha \in \{1,\lambda,\varepsilon\}, \tag{6}$$

$$ {}_k EF^{\alpha}_1 = {}_k d^{\alpha}_{11} \qquad k \in \{o,p\},\ \alpha \in \{1,\lambda,\varepsilon\}, \tag{7}$$

$$ {}_k ES^{\alpha}_j = \max\{{}_k ES^{\alpha}_i + {}_k d^{\alpha}_{i1} \mid i \in P_j\} \qquad j = 2,\ldots,J,\ k \in \{o,p\},\ \alpha \in \{1,\lambda,\varepsilon\}, \tag{8}$$

$$_kEF_j^\alpha = {}_kES_j^\alpha + {}_kd_{j1}^\alpha \qquad j = 2, \ldots, J\,,\; k \in \{o,p\}\,,\; \alpha \in \{1,\lambda,\varepsilon\}\,, \tag{9}$$

$$_kLF_J^\alpha = {}_k\overline{T}^\alpha = \sum_{j=1}^{J} \max_{m=1}^{M_j} \left\{ {}_kd_{jm}^\alpha \right\} \qquad k \in \{o,p\}\,,\; \alpha \in \{1,\lambda,\varepsilon\}\,, \tag{10}$$

$$_kLS_J^\alpha = {}_k\overline{T}^\alpha - {}_kd_{J1}^\alpha \qquad k \in \{o,p\}\,,\; \alpha \in \{1,\lambda,\varepsilon\}\,, \tag{11}$$

$$_kLF_i^\alpha = \min\{ {}_kLF_j^\alpha - {}_kd_{j1}^\alpha \mid j \in S_i \} \qquad i = 1, \ldots, J-1\,,\; k \in \{o,p\}\,,\; \alpha \in \{1,\lambda,\varepsilon\}\,, \tag{12}$$

$$_kLS_i^\alpha = {}_kLF_i^\alpha - {}_kd_{i1}^\alpha \qquad i = 1, \ldots, J-1\,,\; k \in \{o,p\}\,, \alpha \in \{1,\lambda,\varepsilon\}\,, \tag{13}$$

where the modes of the activities are labeled with respect to non-decreasing duration for the levels $\alpha = 1,\ \lambda,\ \varepsilon$ and for the optimistic and pessimistic expert's assessments. Based on these time windows calculated according to (6)...(13), six crisp scheduling problems can be formulated as follows:

Minimize

$$_k\Phi^\alpha(x) = \sum_{m=1}^{M_J} \sum_{t={}_kEF_J^\alpha}^{{}_kLF_J^\alpha} t \cdot x_{Jmt} \qquad k \in \{o,p\}, \alpha \in \{1,\lambda,\varepsilon\} \tag{14}$$

Subject to

$$\sum_{m=1}^{M_j} \sum_{t={}_kEF_j^\alpha}^{{}_kLF_j^\alpha} x_{jmt} = 1 \qquad j = 1, \ldots, J\,,\; k \in \{o,p\}, \alpha \in \{1,\lambda,\varepsilon\} \tag{15}$$

$$\sum_{m=1}^{M_i} \sum_{t={}_kEF_i^\alpha}^{{}_kLF_i^\alpha} t \cdot x_{imt} \le \sum_{m=1}^{M_j} \sum_{t={}_kEF_j^\alpha}^{{}_kLF_j^\alpha} (t - {}_kd_{jm}^\alpha) \cdot x_{jmt} \qquad j = 2, \ldots, J\,,\; i \in P_j\,,\; k \in \{o,p\}, \alpha \in \{1,\lambda,\varepsilon\} \tag{16}$$

$$\sum_{j=1}^{J} \sum_{m=1}^{M_j} q_{jmr} \sum_{\tau=t}^{t+{}_kd_{jm}^\alpha-1} x_{jm\tau} \le Q_{rt} \qquad r \in R,\; t = 1, \ldots, {}_k\overline{T}^\alpha,\; k \in \{o,p\}, \alpha \in \{1,\lambda,\varepsilon\} \tag{17}$$

$$x_{jmt} \in \{0, 1\} \qquad j = 1, \ldots, J,\; m = 1, \ldots, M_j,\; t = {}_kEF_j^\alpha, \ldots, {}_kLF_j^\alpha,\; k \in \{o,p\}, \alpha \in \{1,\lambda,\varepsilon\} \tag{18}$$

The problem (14)...(18) can be solved by using exact or heuristic approaches. Here, the model is solved by a branch and bound algorithm, the basic ideas of which were originally introduced by Talbot [8], [9]. Patterson et al. [10] proposed an enumeration procedure of the branch and bound type and Sprecher [11] accelerated and generalized the algorithm. A slightly modified version of this algorithm has been proposed by Schultmann [12]. The algorithm generates exact solutions as far as regular measures of performance (cf. [11]) are considered. An example of a regular measure of performance is the objective function (15).

As a result of the solution procedure, schedules are generated which represent optimistic and pessimistic time assessment on the given α-levels. Thus, the result is a minimal project duration for each of the six crisp problems (14)...(18) which represent the fuzzy interval of the minimal project duration (Tab. 1).

Table 1. Solutions for the crisp scheduling problems (14)...(18)

problem no.	1	2	3	4	5	6
k	o	o	o	p	p	p
α	ε	λ	1	1	λ	ε
makespan	${}_{o}\Phi^{\varepsilon^*}$	${}_{o}\Phi^{\lambda^*}$	${}_{o}\Phi^{1^*}$	${}_{p}\Phi^{1^*}$	${}_{p}\Phi^{\lambda^*}$	${}_{p}\Phi^{\varepsilon^*}$

4 Conclusion

The model presented has already been successfully applied to the dismantling of buildings and discarded electronic products. Nevertheless, concentrating on processing times which are modeled as fuzzy intervals the model only covers a selection of some uncertainties decision makers are confronted with in practice. Accordingly, further uncertainties like fuzzy due dates, fuzzy capacity constraints, uncertain composition of discarded products or fuzzy precedence relations have to be considered and incorporated in the model. Despite of all dimensions still to be considered, the most important obstacle fuzzy dismantling planning in practice is the generation of adequate data for fuzzy dismantling planning in practice.

References

[1] Brucker, P., Drexl, A., Möhring, R., Neumann, K. and Pesch, E. (1999) Resource-constrained project scheduling: Notation, classification, models, and methods. European Journal of Operational Research 112(1), 3-41

[2] Schultmann, F., Rentz, O (2001) Environment-oriented project scheduling for the dismantling of buildings. OR Spektrum 23 (1), 51-78

[3] Rommelfanger, H. (1994) Fuzzy Decision Support Systeme. Entscheiden bei Unschärfe, 2. Auflage. Springer, Berlin, Heidelberg, New York (in German)

[4] Hapke, M., Slowinski, R. (1996) Fuzzy priority heuristics for project scheduling. Fuzzy Sets and Systems 83, 291-299

[5] Hapke, M., Slowinski, R. (2000) Fuzzy Set Approach to Multi-Objective and Multi-Mode Project Scheduling under Uncertainty. Scheduling under Fuzziness. Slowinski, R.; Hapke, M. (eds.). Springer, Berlin, Heidelberg, 197-221

[6] Pritsker, A.A.B., Watters, L.J., Wolfe, P.M. (1969) Multiproject scheduling with limited resources: A zero-one programming approach. Management Science 16, 93-108

[7] Hapke, M., Jaszkiewicz, A., Slowinski, R. (1994) Fuzzy project scheduling system for software development. Fuzzy Sets and Systems 21, 101-117

[8] Talbot, F. B., Patterson, J. H. (1978) An efficient integer programming algorithm with network cuts for solving resource-constrained scheduling problems. Management Science 24, 1163-1174

[9] Talbot, F. B. (1982) Resource-constrained project scheduling with time-resource tradeoffs: The nonpreemptive case. Management Science 28, 1197-1210

[10] Patterson, J. H., Slowinski, R., Talbot, F. B., Weglarz, J. (1989) An algorithm for a general class of precedence and resource constrained scheduling problems. Advances in Project Scheduling. Slowinski, R. and Weglarz, J. (eds.). Elsevier, Amsterdam, 3-28

[11] Sprecher, A. (1994) Resource-Constrained Project Scheduling - Exact Methods for the Multi-Mode Case. Springer, Berlin, Heidelberg, New York

[12] Schultmann, F. (1998) Kreislaufführung von Baustoffen - Stofffflussbasiertes Projektmanagement für die operative Demontage- und Recyclingplanung von Gebäuden. Erich Schmidt, Berlin (in German)

Group Decision Making Versus Expert Opinion in the Multi-Objective Analysis of Ecosystem Management[1]

Lidija Zadnik Stirn

University of Ljubljana, Biotechnical Faculty, Večna pot 83, 1000 Ljubljana, Slovenia, e-mail: lidija.zadnik@uni-lj.si

1 INTRODUCTION

Ecological issues with demands for preserving the nature and multipurpose use of land have, along with existing economic criteria, become a key part of the modern concept of ecosystem management. Therefore, on one hand, the land owners and experts are faced with the land use decisions which maximize the profit and refer to ecological objectives, while on the other hand, the public, who benefits from the amenity value of the ecosystem, specifically derives its own scenario of decisions. As such, an ecosystem management problem is a satisfactory attainment of multiple, but conflicting, objectives (Zadnik 2001, 2002).

This problem has precipitated a decision support model to determine the multi-objective ecosystem management decisions under achieving the goals from both, owners/experts and public aspects. In the model, the management process is defined in terms of decisions, constraints and objectives. The state of treated ecosystem is described in terms of parameters s_1, s_2, ... s_s, such as area of arable land, forest land, products, labor force, machinery, financial resources, ecological conditions, etc. The parameters s_1, s_2, ... s_s form a state vector $\underline{x}(j)= x(s_1,...,s_s)\in X$, where X is the set of all possible state vectors of the treated ecosystem at the considered time. We assume that there is a finite number of such vectors $\underline{x}(j)$ at each treated time. Furthermore, a considerable attention is paid to the determination of the goal state of the ecosystem $\underline{x}(j^*)=x^*(s_1^*, s_2^*, \ ... \ , s_s^*)$ which is designed to meet demands for sustainable, pro-natural, environment-friendly and socio-economically sound decisions.

The decision maker can influence the virtual ecosystem by invoking management decisions, expressed by decision variables $d(m,\underline{x}(j)) \in D(\underline{x}(j))$, $m=1,2,...,M$. For each state vector $\underline{x}(j)$ there exists a finite discrete set of decisions $D(\underline{x}(j))$. The decisions are mutually exclusive at the treated time and its accompanying state. The set of feasible decisions is constrained by area limitations, budget constraints,

[1] The present paper is part of an EU financed research on 'Tools for evaluating investment in the Mediterranean mountain areas – An integrated framework for sustainable development – MEDMONT (QLK5 – CT-2000-01031)'

available machinery and labor hours, environmental constraints, etc. The decision variables $d(m,\underline{x}(j))$ move the ecosystem from a state $\underline{x}(j)$ to another state $\underline{x}'(j')$ at the next time moment (Eq. 1.1) which should be closer to the goal state $\underline{x}(j^*)$.

$$\underline{x}'(j')=h(\underline{x}(j), d(m,\underline{x}(j))) \tag{1.1}$$

The transformation function h is defined empirically for each ecosystem.

The objective function $Z(\underline{x}(j),d(m,\underline{x}(j)))$ with components $z(k,\underline{x}(j),d(m,\underline{x}(j)))$, $k=1,2,...,K$, such as net revenue from production, recreational opportunities, sustainability, etc., which reflect managerial objectives and link state and decision variables is to be optimized:

$$\max_{d(m,\underline{x}(j))} \text{ and / or } \min \sum_{k=1}^{K} z(k, \underline{x}(j), d(m\, \underline{x}(j))) \tag{1.2}$$

subject to Eq. (1.1), for m=1,2...,M.

The objective functions are first evaluated from a single decision maker (owner/expert) point of view. For each decision, the objectives are given through the multidimensional objective function, viewed as a separable and additive decomposition of the linear utility function which is expressed by indicators (Winston 1994). As the objectives are of different importance to decision makers, weights of objectives are generated (Robinson et al. 1991). Further, the public participation in decision making process is expressed through a group decision making under multiple objectives. An ordinal agreed criteria approach (Hwang and Lin 1987) and conjoint analysis (Hair et al. 1998) are used.

Finally, knowing the effects of ecosystem management decisions on both, owners/experts and public, the analyst incorporates their opinion into the decision model to determine the best compromise decision (Zeleny 1984; Ballestero and Romero 1998). A numerical example is presented to illustrate the problem, model and methods.

2 THE WEIGHTS OF OBJECTIVES AND THE UTILITY FUNCTION

Because the objectives $z(k,\underline{x}(j),d(m,\underline{x}(j)))$ are multiple, conflictive and of different importance (value) to the decision makers, the weights w_k for each objective have to be determined. The method used to determine the preferences of objectives to the decision makers is Saaty's AHP method (Golden et al. 1989; Winston 1994). For the pairwise comparisons of one objective over another a scale with values 1 to 5 is used. The relative comparisons r_i/r_j of objective i over objective j are gathered in a matrix W. The weights w_k are to be obtained by an iterative procedure (Winston 1994) of calculating the sequence of matrices: W^2, $(W^2)^2$, $((W^2)^2)^2$,...., which is followed by summing the rows. Finally, the columns are normalized. Let us illustrate the calculation of weights with an example of 3 decisions, and 3 objectives: investment (I), sustainability (S) and profit (P). Supposing that experts expressed the relative importance of one objective over another by the matrix W, the most important is sustainability (w_2=0.56), the second investment (w_1=0.32) and the least important is profit (w_3=0.12).

$$W = \begin{array}{c} \\ I \\ S \\ P \end{array} \begin{array}{c} \begin{array}{ccc} I & S & P \end{array} \\ \begin{bmatrix} 1 & 3 & 4 \\ 1/3 & 1 & 3 \\ 1/4 & 1/3 & 1 \end{bmatrix} \end{array} \rightarrow W^2,\ (W^2)^2,\ \ldots\ldots \rightarrow \begin{bmatrix} w_1 = 0.32 \\ w_2 = 0.56 \\ w_3 = 0.12 \end{bmatrix}$$

Furthermore, it is assumed that the objective functions $z(k,\underline{x}(j),d(m,\underline{x}(j)))$ for $k=1,2,\ldots,K$ are determined by attributes $a_k(\underline{x}(j),d(m,\underline{x}(j)))$, each having a finite number of possible values called levels. Let $a_{k,i}$ $(\underline{x}(j),d(m,\underline{x}(j)))$ be the i-th level ($i=1,2,\ldots,I$) of the k-th attribute ($k=1,2,\ldots,K$) associated with the state $\underline{x}(j)$ and decision $d(m,\underline{x}(j))$. The vector objective function $Z(\underline{x}(j),d(m,\underline{x}(j)))$ is expressed as the decision maker's multiattribute utility (value) function:

$$v(a_{1,i},\ldots,a_{k,i},\ldots,a_{K,i}) = v(a_{1,i}(\underline{x}(j),d(m,\underline{x}(j))),\ \ldots,\ a_{k,i}(\underline{x}(j),d(m,\underline{x}(j))),\ \ldots\ ,a_{K,i}(\underline{x}(j),d(m,\underline{x}(j)))).$$

Under certain conditions, however, the assessment of $v(a_{1,i},\ldots,a_{k,i},\ldots,a_{K,i})$ can be greatly simplified (Winston 1994). This is the case, if the operator of multicriteria control is additively separable with respect to the space of inputs and control space. The utility function $v(a_{1,i},\ldots,a_{k,i},\ldots,a_{K,i})$ is an additive function if there exist K functions $v_1(a_{1,i}), \ldots, v_k(a_{k,i}), \ldots\ldots, v_K(a_{K,i})$ satisfying Eq. (2.1):

$$v(a_{1,i},\ldots,a_{k,i},\ldots,a_{K,i}) = \sum_{k=1}^{K} w_k v_k(a_{k,i}) \tag{2.1}$$

where w_k present the weights of attributes a_k; in our case determined through the matrix W. If the attribute a_k is a numerical variable (known with certainty), and the decision maker is risk neutral, and the utility function for a single attribute is linear, the utility function $v(a_{k,i})$ is given as Eq. (2.2):

$$v(a_{k,i})=(a_{k,i}-a_{i,worst})/(a_{i,best}-a_{i,worst}) \tag{2.2}$$

where $v(a_{k,i}=a_{i,worst})=0$ and $v(a_{1,i}=a_{i,best})=1$, as illustrated in the second and the third line of Table 1 which considers three decisions d_1, d_2, d_3 and three objectives (sustainability measured by attribute a_1, investment measured by attribute a_2 and profit measured by attribute a_3) and shows the utility functions $v(a_{k,i})$ calculated by the use of Eq. (2.2). If we calculate $v(a_{k,i})$ for each decision and use the normalization for the purpose of commensurability, we get the normalized values in Table 1. Taking into account the weights w_k, calculated for 3 objectives i.e., w_1=0.32, w_2=0.56 and w_3=0.12, and using Eq. (2.1), we obtain for the utility functions of decisions d_1, d_2, d_3 the following values: d_1:0.042; d_2:0.720; d_3:0.544. The normalized values are: d_1:0.03; d_2:0.55; d_3:0.42.

The results show that regarding the experts /owners preferences and the chosen utility functions the most preferred is decision 2, the second is decision 3 and the least preferred is decision 1.

Table 1. The utility function for the attributes for 3 decisions and 3 objectives

Att.	d_1	d_2	d_3	$v(a_{k,i})$	normalized $v(a_{k,i})$ for d_1	normalized $v(a_{k,i})$ for d_2	normalized $v(a_{k,i})$ for d_3
a_1	a	b	c	$(a_{1,i}-1)/2$	0	0.33	0.67
a_2	0	5	2	$a_{2,i}/5$	0	0.72	0.28
a_3	46	30	76	$(a_{3,i}-30)/46$	0.26	0	0.74

3 AGREED ORDINAL APPROACH AND CONJOINT ANALYSIS

Let us assume that in Eq. (1.2) subject to Eq. (1.1) we deal with m decisions d_i, (i=1,...m) which should be evaluated by n decision makers, members of public, residents, or groups of residents A_k (k=1,2,...,n) who are using for the evaluation p attributes a_j (j=1,2,...,p). For each decision maker A_k who evaluates the decisions d_i by ranking, rating or scoring a matrix A_k is generated. In the continuation we will consider only ranking, i.e., the ordinal approach will be used (Hwang and Lin 1987). Furthermore, the agreed criteria approach, which involves each decision maker using the same criteria, is considered, to find the matrices A_k. Starting from the matrices A_k, we calculate for each criterion a matrix B_j (j=1,...,p). Its elements are the ranks of decisions d_i given by the decision makers A_k. By summing up the elements in each row of matrices B_j, a matrix A' is generated. A' provides the ranks for all decisions d_i under all criteria a_j given by decision makers A_k. However, it is possible that some criteria are more important than the others. Therefore, decision makers would want to place more weight on that criterion. To accomplish this, we will use a vector of weights $W=(w_1,\ldots,w_p)$ obtained by the AHP method. Then we formulate an agreement matrix G which entries are g_{ir}, given by Eq. (3.1):

$$g_{ir} = \sum_{j=1}^{p} a_{irj} w_j \qquad (3.1)$$

where $a_{irj} = 1$ if the decision d_i is for the criterion (attribute) a_j placed to the r-th position, otherwise $a_{irj} = 0$.

Furthermore, we want to match the decision d_i with the rank r so that the sum of the corresponding assigned weight value is the largest possible. This task can be achieved by solving assignment problem of linear programming given by Eq. (3.2):

$$\max \sum_{i=1}^{m} \sum_{r=1}^{m} g_{ir} x_{ir} \qquad (3.2)$$

subject to $\sum_{i=1}^{m} x_{ir} = 1, r = 1,\ldots,m$ *and* $\sum_{r=1}^{m} x_{ir} = 1, i = 1,\ldots,m,$ *where* $x_{ir} = 1,$ *if r is assigned to* d_i*, otherwise it is* 0.

With the example for 3 decisions (d_1, d_2, d_3), three objectives (a_1, a_2, a_3) and five decision makers $A_1,\ldots,A_5$, given the preferences by matrices $A_1,\ldots,A_5$, the following results are calculated: d_1 as the most preferred, d_2 as the second preferred and d_3 as the least preferred, which ranks on the last place. For the compromised solution we would need the normalized values given by decision makers. Taking into account the normalized values, d_1 is the most preferred. Let us assign to it 3 points, to d_2 2 points and to d_3 1 point. Normalizing the points, we obtain for decisions the following normalized values: d_1:0.50; d_2:0.34; d_3:0.16.

$$A_1 = \begin{matrix} & a_1 & a_2 & a_3 \\ d_1 & 3 & 3 & 1 \\ d_2 & 2 & 2 & 3 \\ d_3 & 1 & 1 & 2 \end{matrix}, \quad A_2 = \begin{matrix} & a_1 & a_2 & a_3 \\ d_1 & 3 & 1 & 2 \\ d_2 & 1 & 3 & 1 \\ d_3 & 2 & 2 & 3 \end{matrix}, \quad A_3 = \begin{matrix} & a_1 & a_2 & a_3 \\ d_1 & 3 & 3 & 3 \\ d_2 & 1 & 1 & 1 \\ d_3 & 2 & 2 & 2 \end{matrix} \quad A_4 = \begin{matrix} & a_1 & a_2 & a_3 \\ d_1 & 2 & 1 & 1 \\ d_2 & 1 & 3 & 3 \\ d_3 & 3 & 2 & 2 \end{matrix}, \quad A_5 = \begin{matrix} & a_1 & a_2 & a_3 \\ d_1 & 3 & 3 & 3 \\ d_2 & 2 & 2 & 1 \\ d_3 & 1 & 1 & 2 \end{matrix}$$

Furthermore, to assess the public preferences of the decisions the attributes and their levels have to be chosen. We assume three attributes (factors), each at two levels, low (L) and high (H): ecological factor, factor of investment and factor of profit. Using the full profile method and assuming that an additive rule is appropriate, we were able to use factorial design which enabled the evaluation of all 8 stimuli: LLL (low ecology, low investment and low profit), LLH, LHL, LHH, HLL, HHL, HHH, HHL. The conjoint analysis experiment was administered during the personal interview with 24 respondents ranging stimuli with 1 as being the best, while 8 assigned as the worst. The survey results are as follows: respondent 1: 6 5 3 2 4 1 8 7; ....; respondent 24: 6 3 4 2 5 1 8 7.

Using the SPSS subprogram utility (Norusis 1997) the estimation of part-worth (utility (s.e.)) and the relative importance (importance in %) of each attribute are calculated for each respondent separately, and then aggregated to obtain the total worth (TW), using the additive rule (Backhaus et al. 1994) in Eq. (3.3):

TW=const. part-worth of 1. fac.+part-worth of 2. fac.+part worth of 3.fac. (3.3)

The overall utilities (preferences) for all 24 respondents are:

STIMULI	1(LLL)	2(LLH)	3(LHL)	4(LHH)	5(HLL)	6(HHL)	7(HHH)	8(HLH)
UTILITY	5.7917	5.8125	5.7917	5.8125	5.7917	3.1875	3.1875	3.1875

We are interested only in the stimuli HLL, LHL and LLH which represent the decisions d_1, d_2, d_3 respectively. The results reveal that the respondents preference is the highest for d_3, while d_1 and d_2 are for them of equal importance.

4 DETERMINATION OF AN IDEAL COMPROMISED DECISION BY A COMPROMISE PROGRAMMING METHOD

Compromise programming is explained with Eq. (4.1), (Steuer 1986, Tecle et al. 1994, Taha 1997, Ballestero and Romero 1998):

$$L_{si} = \left(\sum_{i=1}^{n} w_j^s \left| \frac{x_{jbest} - x_{ji}}{x_{jbest} - x_{jworst}} \right|^s \right)^{1/s} \tag{4.1}$$

where: L_{si} is a measure of distance from the ideal decision d_i, w_j is the weight for objective a_j, x_{ji} is the attribute level for objective a_j and decision $d_{i,}$, x_{jbest} is the best value for objective a_j, i.e. maximum for benefit criteria and minimum for cost criteria, and x_{jworst} respectively.

Let us assume in our numerical case 3 decisions d_1, d_2, d_3, evaluated by 3 criteria: a_1 by public by agreed criteria, a_2 by experts by utility function, and a_3 by public, owners, experts by conjoint analysis. If we further ascribe to each criterion a_j in dependence of decision d_i a linear function $f(a_{ji})$ in the sense of function given by Eq. (2.2), and use Eq. (4.1), we get Table 2:

Table 2. The evaluation of 3 decisions by experts, owners and public

Criterion	d_1	d_2	d_3	normalized $v(a_{k,i})$ for d_1	normalized $v(a_{k,i})$ for d_2	normalized $v(a_{k,i})$ for d_3
Agreed (a_1)	0.50	0.34	0.16	0	0.32	0.67
Utility (a_2)	0.03	0.55	0.42	0.80	0	0.20
Conjoint (a_3)	0	0	0	0.50	0.50	0

Taking into account the equal weights $w_1=1$, $w_2=1$ and $w_3=1$, and minimizing the expression in Eq (4.1), we can obtain the results: d_1:1.3, d_2:0.82, d_3:0.88. The results show that the most preferred is d_1, the second is d_3 and the least preferred is d_2.

REFERENCES

Backhaus K, Erichson B, Plinke W, Weiber R, (1994) Multivariate Analysemethoden. Springer Verlag, Berlin

Ballestero E, Romero C (1998) Multiple criteria decision making and its applications to economic problems. Kluwer, Boston

Golden B, Wasil E, Harkey P (1989) Analytic hierarchy process. Springer, Heidelberg

Hair JF, Anderson RE, Tatham RL, Black WC (1998) Multivariate data analysis. MacMillan Pub. Co., New York

Hwang CL, Lin MJ (1987) Group decision making under multiple criteria. Springer, Berlin

Norusis MJ (1997) SPSS for Windows: Advanced Statistics, Ver. 10. SPSS Inc., Chicago.

Robinson JP, Shaver PR, Wrightsman LS (1991) Measures of personality and social psychological attitudes. Academic Press, San Diego

Steuer RE (1986) Multiple criteria optimization. John Wiley, New York

Taha HA (1997) Operations research. Prentice Hall, New Delhi

Tecle A, Duckstein L, Korhonen P (1994) Interactive, multiobjective programming. Applied mathematics and computation 63: 75-93

Winston WL (1994) Operations research. Duxbury Press, Bellmont, CA

Zadnik Stirn L (2001) Compromise programming for solving economic and environmental problems. In: Oblak L (eds) Ways for improving woodworking industry for transitional economies. Biotechnical Faculty, Ljubljana, pp 157-162

Zadnik Stirn L (2002) Evaluation of projects regarding the preferences of several decision makers (in Slovene). In: Grad J (eds) Proceedings DSI-SOR, Ljubljana, pp 366-371

Zeleny M (1982) Multicriteria decision making. McGraw-Hill, New York

Capital Market Efficiency – An Empirical Analysis of the Dividend Announcement Effect for the Austrian Stock Market

Roland Mestel[1], Henryk Gurgul[2], Christoph Schleicher[3]

[1] Department of Banking and Finance, University of Graz, Austria; roland.mestel@uni-graz.at
[2] Department of Applied Mathematics, University of Mining and Metallurgy, Krakow, Poland; pgurgul@go2.pl
[3] Department of Economics and Institute of Applied Mathematics, University of British Columbia, Vancouver, Canada; chrschle@interchange.ubc.ca

Summary: This study investigates the effects of dividend announcements using data from the Austrian stock market. Abnormal returns are established as the difference between actual returns and predicted returns generated from time series models. We use the model of expected dividends, where expectations are based on prior dividends. Our results provide evidence that announced dividend changes convey new information to the market as stock prices move in the same direction as dividends. In addition, the speed of stock price reaction to the new information provides a test of the semistrong form of the efficient capital market hypothesis.

1 Introduction

It is widely accepted in financial theory that due to market imperfections a company's dividend policy has an impact on the wealth of its shareholders[1]. At the core of the arguments lies the fact that dividend changes convey valuable information of the better-informed management to the market. Stock price reactions on announced dividend changes therefore reflect information asymmetries between the firm and its investors.

Nevertheless it is still a matter of debate what exactly the information content of changing dividends is. Many authors argue that the firm's decision-makers, via dividends, convey their inside information about current and/or future cash flows to the investors (*cash flow signaling hypothesis*). Following this argument dividend increases (decreases) serve as signals of increased (decreased) current and/or prospective future earnings.

An alternative body of research rests upon *the free cash flow hypothesis*. Free cash flow is the section of cash flow that is left after all investments having posi-

---

[1] The authors would like to thank Christian Gutlederer (Reuters Austria) and Peter Ladreiter (Capital Bank) for supplying data and Peter Steiner (University of Graz) and the anonymous referee for helpful comments.

tive net present value (when cash flows are discounted at the appropriate rate) are realized. If not distributed as dividends, managers tend to invest these free cash flows below the cost of capital, or waste them on inefficient organizational expenditure. From this Lang and Litzenberger (1989) conduct that an increase in dividends is primarily a signal that managers are paying out the free cash flows to the shareholders rather than investing them in projects with negative net present value, thereby reducing agency costs.

Both theories predict a positive correlation between announced dividend changes and stock prices. This effect has been confirmed empirically by several studies, mainly for the US market (e.g. Aharony and Swary (1980) or Best and Best (2001); for the German stock market the study by Amihud and Murgia (1997) has been the most rigorous). For the Austrian stock market our study is the first to test the dividend announcement effect on stock prices.

As dividend announcements have an impact on shareholders' wealth the manner in which this effect is incorporated in stock prices through time also provides a test of the semistrong form of the efficient capital market hypothesis. This hypothesis states that stock prices, at any time, reflect all publicly available information relevant to the valuation of the firm. Therefore, if an announced change in dividends provides new information to the market (i.e. we assume no insider trading), the speed of price reaction supplies information on the degree of market efficiency. The faster the prices adjust to the news the more efficient the capital market.

The remainder of the paper is organized as follows: Section 2 describes our data; in Section 3 we analyze our methodology to determine abnormal (excess) returns; in Section 4 we present our results; Section 5 contains a discussion and conclusion.

2 Data

Our sample contains 22 companies listed on the Austrian stock market. The companies have been quoted on the Austrian Traded Index (ATX) between January 1992 and April 2002, although not all firms have been listed on the stock market for the whole period.

For these firms all daily close-prices are derived from *Datastream* or the *Vienna Stock Exchange*. For the period under consideration we filtered 176 dividend announcements from several thousands included in the Dow Jones and Reuters *Factiva* database. We assume all announced dividend changes to be unexpected. This can be justified by the *reluctance-to-change dividends assertion* which states that managers do not change dividend payments unless they have reasons to expect a significant change in the future prospects of the firm (Aharony and Swary 1980). From the observed announcements 74 were classified as announced dividend increases, 75 as constant dividends and 27 as dividend reductions.

We define the announcement (event) date as the very first day official statements on dividends by the executive board of the analyzed firm can be identified in the *Factiva* database. In many cases these announcements only contain information on the expected direction of dividend changes (increase; constant; decrease), but not on its exact levels. Neither the ex-dividend day nor the day the dividend is paid is considered to be an announcement day.

3 Methodology

To establish the impact of dividend announcements on stock prices for Austrian firms, we conduct an event study analysis based on selected time-series models. This is different from most related studies that use simple linear regressions (based on the *market model* by William Sharpe) to forecast stock returns. As the assumptions of the simple linear regression model are often violated this leads to biased estimators of parameters and biased forecasts of stock returns.

From daily stock prices at close we first calculate log-returns

$$R_{i,t} = ln\left(\frac{P_{i,t}}{P_{i,t-1}}\right), \tag{3.1}$$

where $P_{i,t}$ stands for the stock price of company i on date t, $P_{i,t-1}$ denotes the stock price of company i on date $t-1$, ln denotes the natural logarithm.

For each event (dividend announcement) in the sample we define an event-window and a pre-event window. The event window comprises 5 trading days, the announcement date ($t = 0$) plus the two days before ($t = -2$, $t = -1$) and the two days after the announcement ($t = +1$, $t = +2$). The pre-event window covers the 50 trading days prior to the event window. For each day of the event window we compute the abnormal return AR as the difference between the actual ex-post return and the security's normal return that is predicted in the absence of the event.

Formally, for each announcement of the analyzed companies we computed

$$AR_{i,t} = R_{i,t} - \mathrm{E}[R_{i,t} | X], \tag{3.2}$$

where $R_{i,t}$ stands for the actual return of firm i on date t in the event window and $\mathrm{E}[R_{i,t} | X]$ denotes the predicted return conditional on the information set X, where $X = (R_{i,-52},\ldots,R_{i,-3})$.

To predict returns in the event window we used Box and Jenkins' ARIMA methodology. To illustrate the model identification we consider the general ARIMA(p,d,q) model for a time series z_t

$$\phi_p(B)(1-B)^d z_t = \theta_0 + \theta_q(B)a_t, \tag{3.3}$$

where $\phi_p(B):=1-\phi_1 B-\ldots-\phi_p B^p$, $\theta_q(B):=1-\theta_1 B-\ldots-\theta_q B^q$, $Bz_t:=z_{t-1}$, B is called back-space operator, and a_t is white noise (error term). Model identification refers to the methodology in identifying the required size of sample ($n = 50$ is recommended), proper orders of p, d and q, variance stabilizing transformations, the decision to include a deterministic parameter θ_0 when $d \geq 1$.

We first examined the plots of the individual log-returns in the pre-event windows in order to get a feel for the overall properties (trends, outliers, constancy of variance, stationarity). Our conjecture was that the time series are either integrated of order zero or one.

Next, we identify the orders p and q by matching the patterns in the sample ACF (autocorrelation function) and PACF (partial autocorrelation function) with the theoretical patterns of known models summarized in table 1.

Table 1. Characteristics of theoretical ACF and PACF for stationary time series

Process	ACF	PACF
AR(p)	Tails off as exponential decay or damped sine wave	Cuts off after lag p
MA(q)	Cuts off after lag q	Tails off as exponential decay or damped sine wave
ARMA(p,q)	Tails off after lag (q–p)	Tails off after lag (p–q)

When ACF and PACF are not significant and in the absence of a deterministic trend we model the time series as white noise (i.e. ARIMA(0,0,0)).

In the random walk model, i.e. ARIMA (0,1,0), the value of z at time t is equal to the value at time (t–1) augmented by a random shock. In empirical forecasting it happens very often that the hypothesis of "no change" is the most likely scenario (martingale hypothesis). Therefore the random walk model plays a major role in the empirical analysis of financial time series, also in the modeling of stock prices. If the random walk hypothesis for stock prices is true it can be expected that log-returns would be approximately white noise. Note that the random walk model is the limiting case as $\phi_1 \to 1$ in the AR(1) model.

In addition we also applied the Akaike information criterion in order to compare parsimoniously nested models.

Taking into account these guidelines we estimated four different models for the log-return series in the pre-event window: white noise, AR(1) (with ϕ_1 close to 1), MA(1) (only in 1 case) and linear trend model. All estimated parameters are significant at the 1% level. Most log-return series from our data sample can be modeled approximately as white noise. The goodness of fit was evaluated by inspection of the residuals (tests for mean, autocorrelation and homoscedasticity). We then use the time series models to predict returns (the ARIMA forecasts in the event window) and abnormal returns (the forecast errors).

Next we form three clusters, one for announced dividend increases, one for decreases and one for constant dividends. For each cluster, we compute average abnormal returns $\overline{AR}_t$ across sample members for day t. Finally, we test the null hypothesis that the mean abnormal return on day t of the event window is equal to zero. The test statistic is the ratio of the (cross-sectional) mean abnormal return in the event window and the (cross-sectional – time series) standard deviation of mean abnormal returns in the pre-event window.

Assuming that the $\overline{AR}_t$ are identically, independently and normally distributed, under the null hypothesis t_{stat} has a Student-t distribution with (N–1) degrees of freedom. Although daily excess returns are in general non-normal, by a standard central limit theorem (CLT) the cross sectional mean excess return converges to normality as the number of sample securities increases. Main empirical results are given in the next section.

4 Results

Table 2 summarizes our results for abnormal returns within the event window for each cluster. In order to prove significance by t-Student test we check the hypotheses that the time series of mean abnormal returns in clusters considered are normally distributed. Using the Chi-square goodness-of-fit statistic, the Shapiro-Wilks W statistic and tests based on skewness and kurtosis, we cannot reject the mentioned hypotheses (p-value for all tests is greater than 0.05). Furthermore autocorrelations are not significant at the 5% level.

Table 2. Average daily abnormal returns for the event window in three clusters

	Dividend increases Sample size: 74		Constant dividends Sample size: 75		Dividend decreases Sample size: 27	
Event period day t	$\overline{AR}$ (%)	t_{stat}	$\overline{AR}$ (%)	t_{stat}	$\overline{AR}$ (%)	t_{stat}
−2	+0.454	2.093	−0.196	−1.103	−0.178	-0.390
−1	+0.268	1.237	−0.202	-1.048	−0.276	-0.606
0	+0.617*	2.847	+0.002	0.115	−1.355*	-2.971
+1	−0.030	−0.136	+0.073	0.365	−0.119	-0.262
+2	+0.497	2.292	−0.138	−0.799	−0.089	-0.196
Σ	+1.806	3.540	−0.461	−1.230	−2.017	-2.290

* significant at the 1% level

For the 74 dividend increases the average abnormal return was 0.62% (significant at the 1% level) on the announcement day.

In the case of constant dividends (sample size: 75), the average abnormal returns are not statistically different from 0 on any day within the event window. This supports the hypothesis that companies that leave their dividends unchanged communicate no significant new information to the market.

In our sample (size: 27) of announced dividend decreases we find a statistically significant average abnormal daily return of −1.35% on the announcement day. This result supports the empirical findings for other markets that a cut in dividend payments conveys negative information to the public.

In comparison to the abnormal returns induced by increasing dividends the reported negative return due to an announced contraction in dividends is much higher in absolute terms. This confirms the general observation on financial markets that bad news has a greater impact on stock returns than good news. In the sense of cash-flow signaling theory this is an indication that analysts revise their forecasts for future earnings of companies much more strongly in the case of dividend decreases than increases.

One can see that also CAR (cumulative abnormal return) is in absolute terms highest (2.02%, significant at the 5% level) in the case of dividend decreases. This case is followed by dividend increases (1.81%, significant at the 1% level). Bad news causes greater volatility than good news. Therefore t-values are smaller in the first case than in the second. In the case of constant dividend CAR is negative and not significant.

The fact that significant abnormal returns can be observed only on the immediate day dividend changes are announced implies important findings: firstly, the

Austrian stock market seems to have a high degree of semistrong market efficiency. This means that new information is incorporated into stock prices rather quickly, at least within the same day the news is conveyed to the market. We also find that the reaction of market prices is unbiased. The initial reaction reflects the true implications of the information on the values of the securities as there are no subsequent corrections of the initial reactions on day $t = +1$ or $t = +2$. Finally, as the returns on day $t = -2$ and $t = -1$ are not significant, there is no indication of inside trading prior to the release of new information.

5 Conclusions

The findings of this research show that dividend policy is an important source of information for investors. Announced dividend increases induce a significant positive reaction in stock prices, whereas announced decreases in dividends lead to a significant fall in stock prices. Constant dividends leave stock prices unaltered.

Our results for the Austrian stock market also support the semistrong form of the market efficiency hypothesis. Excess returns can be observed only on the announcement day; there are no subsequent price reactions after that day.

Finally we find little indication of leakage of information prior to the announcement day. This supports the hypothesis of no inside trading on the Austrian stock market prior to dividend announcements.

References

Aharony J, Swary I (1980) Quarterly dividend and earnings announcements and stockholders returns: An empirical analysis. Journal of Finance 35 (1): 1–12

Amihud Y, Murgia M (1997) Dividends, taxes, and signaling: evidence from Germany. Journal of Finance 52 (1): 397–408

Best RJ, Best RW (2001) Prior information and the market reaction to dividend changes. Review of Quantitative Finance and Accounting 17: 361–376

Lang LHP, Litzenberger, RH (1989) Dividend announcements: Cash flow signalling vs. free cash flow hypothesis. Journal of Financial Economics 24: 181–191

On Tail Index Estimation and Financial Risk Management Implications

Niklas Wagner

Department of Business and Economics, Munich University of Technology
D-80290 Munich, Germany
niklas.wagner@wi.tum.de

Summary Estimation bias is a critical issue when drawing inferences about the tails of the return distribution of risky assets. Risk management applications which rely on methods of extreme value theory must consider the statistical properties of the tail index estimator used; see for example recent simulation studies by Gomes and Oliveira (2001), Matthys and Beirlant (2000), and Wagner and Marsh (2000). The present contribution outlines potential effects of bias on quantile estimation thereby considering error sensitivities within the widespread Value-at-Risk-approach. The results show that particularly inference far out in the distribution tails is sensitive to bias. The paper further gives an overview of recent literature documenting small sample bias in tail index estimation and points out some new approaches aiming at its reduction.

1 Introduction

Financial risk management is particularly concerned with the extreme changes in the value of portfolios of risky assets. Depending on the type of risk under consideration, a financial institution may distinguish market, credit and operational risks. Risk management focuses on the distribution of losses where a widespread approach is based on modelling quantiles of the loss distribution also denoted as Value-at-Risk (VaR); for discussions see for example Jorion (1997), Oehler (1998), and Ridder (1998).

Extreme value statistics provide a theory for modelling extremal changes which typically are of main interest in risk management applications. As outlined for example in Beirlant et al. (1996), Coles (2001), Embrechts et al. (1997), and Klüppelberg (2002), extreme value theory characterizes the asymptotic tail behaviour of some unknown, distribution function. It thereby establishes three classes of limiting distributions for normalized sample maxima. An important parameter of these limiting distributions is the so-called "tail index" frequently denoted by α.

Under the quite general assumption that the distribution function is heavy-tailed, i.e. when it belongs to the maximum domain of attraction of the Fréchet limiting distribution (see e.g. Embrechts et al. 1997), a commonly used estimator is based on a proposal by Hill (1975):

$$\hat{\alpha} = \frac{1}{k}\sum_{i=1}^{k} \ln X_{i,T} - \ln X_{k,T} , \tag{1.1}$$

The above estimator is based on k upper order statistics $X_{i,T}$, $i = 1, ..., k$, with $X_{i,T} > X_{i+1,T}$ from T sample observations having a common continuous distribution function. It is constructed as a maximum likelihood estimator *conditional* on some known threshold level $u = X_{k,T}$.

Given that k approaches infinity with the sample size T, but k remains a small portion of the overall sample, one can show that the conditional maximum likelihood estimator is asymptotically normally distributed:

$$\sqrt{k}(\hat{\alpha} - \alpha) \xrightarrow{d} N(b, \alpha^2). \tag{1.2}$$

Note that the distributional limit in (1.2) does not contain an asymptotic bias term as the threshold u is assumed to be known, i.e. one can write $b = 0$.

In applications however, the threshold u is unknown and bias $|b| > 0$ enters. Estimation of the threshold becomes essential to the Hill estimator's bias/variance trade-off: Selecting too many potential tail observations reduces variance but introduces bias and vice versa. In general, the estimator's optimality properties will therefore be weak and estimation performance is measured by mean squared estimation error. Studies which simulate the small sample properties of the Hill estimator and point out estimation bias under various distributional assumptions include Gomes and Oliveira (2001), Matthys and Beirlant (2000), and Wagner and Marsh (2000), for example.

Applications of extreme value statistics in the empirical finance literature typically assume that the returns of some risky asset are drawn from a given common distribution function. Previous studies which apply the Hill estimation approach include for example Danielsson and de Vries (1997), Krämer and Runde (1996), Lux (2001), and Wagner (2002b). A different estimation approach is based on fitting a so-called "Generalized Pareto Distribution" (GPD). Related studies include Bassi et al. (1998), Emmer et al. (1998), and McNeil (1998) as well as Lauridsen (2000) and Frey and McNeil (2000) who, instead of using raw returns which exhibit time-series dependence, model conditional returns and assume that the model innovations are iid draws from some common distribution function.

In any case, tail index estimation is based on maximum likelihood jointly together with asymptotic arguments from extreme value theory. The choice of a threshold which yields a sample fraction of largest (smallest) observation is a potential source of bias in any application irrespective of the estimation methodology chosen. Given the potential impact of bias on practical risk assessment, this contribution provides a sensitivity analysis of percentage biases in risk assessment based on the standard VaR-methodology (Section 2). Section 3 points out new methodological approaches facing the bias problem and gives a brief subject outlook on management implications.

2 Risk Assessment under Biases in Tail Index Estimation

Based on the standard Value-at-Risk approach and qunatile estimation, this section outlines the potential effects of biases in Hill estimates of the extremal index on estimates of VaR. For the methods see also e.g. Embrechts et al (1997), Ridder (1998), and Coles (2001).

2.1 The Model

We start with a simple discrete time model for the prices S_t of some risky asset

$$S_t = S_{t-1}\exp(R_t), \quad 0 \le t \le T, \quad S_0 > 0, \tag{2.1}$$

where the random continuously compounded returns R_t are drawn from a common distribution function F. In the following we are concerned with losses occurring from holding the asset during some given time period. One-period negative-signed changes in asset value, i.e. losses $L_{t+1} = -(S_{t+1} - S_t)$, are given by model (2.1) as

$$L_{t+1} = -S_t[\exp(R_{t+1}) - 1], \tag{2.2}$$

where $S_t > 0$ and hence a loss $L_{t+1} > 0$ occurs if and only if $R_{t+1} < 0$.

Now, given some initial time t asset value $S_t > 0$, the concept of (conditional) Value-at-Risk is concerned with a one-period quantile $\mathrm{VaR}_{p,t,S}$ of the gain/loss distribution such that

$$P(L_{t+1} \ge \mathrm{VaR}_{p,t,S}) = p, \tag{2.3}$$

where p is some small probability. Rearranging (2.3) using (2.2) gives an equivalent condition for the distribution F of the one-period returns R_t:

$$P(R_{t+1} \le \ln[1 - (\mathrm{VaR}_{p,t,S} / S_t)]) = p.$$

Applying the generalized inverse $F^{\leftarrow}$ of the distribution function F and rearranging yields the following expression for the Value-at-Risk

$$\mathrm{VaR}_{p,t,S} = -[\exp(F^{\leftarrow}(p)) - 1]S_t. \tag{2.4}$$

Extreme value statistics can be used to infer the p-quantile $F^{\leftarrow}(p)$ of the lower tail of the distribution function F. In order to derive an estimate of the tail index α, one may choose a sample of T returns, $R_1,\ldots, R_T$, from F and apply the Hill estimator (1.1) to k upper order statistics resulting from negative return sample order statistics: $X_{1,T} = -R_{T,T} > X_{2,T} = -R_{T-1,T} > \ldots > X_{k,T} = -R_{T-k+1,T}$. Selecting the k+1 smallest observation from the return sample, $R_{T-k,T}$, the estimate of the lower tail p-quantile is (see e.g. Embrechts et al. 1997, p 348, for the upper tail quantile):

$$\hat{F}^{\leftarrow}(p) = \left(\frac{T}{k}p\right)^{-1/\hat{\alpha}} R_{T-k,T}\,. \tag{2.5}$$

2.2 Bias and Risk Assessment Implications

As is obvious from equations (2.4) and (2.5) in the preceding section, errors or biases in the tail index α enter nonlinearly in the estimation of conditional Value-at-Risk according to equation (2.4). Conditioning on unit asset value at time t for convenience, the relation is of the form

$$\mathrm{VaR}_{p,t,S=1}(\alpha) = 1 - \exp(d_1 d_2^{-1/\alpha}), \tag{2.6}$$

where d_1 and d_2 are given constants in empirical applications. We can now proceed by calculating bias in VaR-estimates under biases in the tail index estimate. Of course, while we present our results with the background of estimation bias, the analysis is equivalently pointing out the sensitivities of VaR under estimation error in general.

The calculations are done for a set of parameter values which may be considered as typical in financial applications: First, the threshold is set to a fixed level of a 2 percent loss, $d_1 = r_{T-k;T} = -0.02$, a level below which heavy-tailed Pareto-like behaviour can empirically be observed for stock market returns (e.g. Locarek-Junge et al. 2002). Next, the subsample fraction is chosen with a fixed level of 2.5 percent, $k/T = 0.025$. For the other parameters, we allow a range of choices. The probability p is set to typical levels of 0.0001, 0.001, 0.01 and 0.05 which implies analysing VaR-estimation-sensitivity under scaling through the constant $d_2 = pT/k$ in (2.6). Referring to the simulation results in Wagner and Marsh (2000) who find biases of up to roughly 20 percent in the Hill estimator for tail index values of typical financial return models, values of $b = 0.05$, 0.1 and 0.2 are chosen for the percentage estimation biases. Finally, all sensitivity calculations are performed for typical tail index parameter values of 2, 3 and 4.

Based on these settings, Table 2.1 reports relative errors corresponding in VaR-estimates. Using (2.6), relative errors are calculated as

$$\pm\Delta\mathrm{VaR}(\alpha) = \frac{\mathrm{VaR}[\alpha\,(1 \pm b)] - \mathrm{VaR}(\alpha)}{\mathrm{VaR}(\alpha)}$$

and the numbers in the table are given in percentage terms. Hence, the numbers indicate the percentage sensitivities of VaR-estimates under percentage biases in the tail index assessment.

What can we now learn from our numerical results in Table 2.1:

- First of all, a striking observation is that an over-estimation of the tail index α always yields an overestimated VaR for the 5 percent probability level. In other words, assuming a thinner tail yields a larger estimate of the quantile and hence a larger assessment of risk as measured by VaR. This result is clearly counterintuitive and points out one of the well-discussed deficiencies of VaR as a measure of risk. For all the other probability levels, $p \leq 0.01$, consistent with intuition, an over-estimation of α yields under-estimated VaR's as indicated by negative signs for the top $\Delta\mathrm{VaR}(\alpha)$'s under each parameter setting.

Table 2.1. Sensitivity of VaR assessment given by ±ΔVaR(α) in %.

	α:	2	3	4
p	b			
5%	±5%	+1.7%	+1.1%	+0.8%
		−1.8%	−1.2%	−0.9%
	±10%	+3.2%	+2.1%	+1.6%
		−3.8%	−2.5%	−1.9%
	±20%	+5.9%	+3.9%	+2.9%
		−8.3%	−5.6%	−4.2%
1%	±5%	−2.1%	−1.4%	−1.1%
		+2.4%	+1.6%	+1.2%
	±10%	−4.0%	−2.7%	−2.0%
		+5.1%	+3.4%	+2.5%
	±20%	−7.2%	−4.9%	−3.7%
		+12%	+7.8%	+5.8%
0.1%	±5%	−7.0%	−4.8%	−3.7%
		+8.4%	+5.6%	+4.2%
	±10%	−13%	−9.1%	−6.9%
		+18%	+12%	+9.1%
	±20%	−23%	−16%	−12%
		+46%	+30%	+22%
0.01%	±5%	−11%	−7.9%	−6.1%
		+13%	+9.5%	+7.2%
	±10%	−20%	−15%	−11%
		+29%	+21%	+16%
	±20%	−33%	−25%	−20%
		+73%	+53%	+39%

- Due to the nonlinearity of the underlying relation and for the parameters chosen, over-estimation of α gives absolutely smaller percentage deviations in VaR then under-estimation. For $p \leq 0.01$ this implies that over-estimating α results in a smaller lack of VaR-capital than the corresponding capital over-allocation given due to the under-estimation of α by the same magnitude. Hence, while over-estimating α yields failure to protect against losses at the given probability level, under-estimating α is particularly costly in terms of capital requirements.
- For the moderate quantiles based on $p = 0.05$ and $p = 0.01$, percentage errors in α are rather dampened with respect to their impact on VaR. Assume a typical situation where VaR has to be determined at the 1 percent level and the true tail index equals three. Then a positive 10 percent deviation in the tail index value will cause 2.7 percent under-estimation of the true VaR. A negative 10 percent deviation in the tail index value will cause 3.4 percent over-estimation of the true VaR.
- The situation changes dramatically once the probability levels approach the area of extreme quantiles, $p = 0.001$ and $p = 0.0001$. Additionally, under these probability levels, the sensitivities reach enormous values when the true tail index becomes smaller approaching a parameter value of two. In the words of Embrechts (2000): "*produce confidence intervals e.g. beyond 99% VaR; a dangerous job but someone has to do it*".

3 Outlook

The preceding section makes clear that estimation bias is clearly a challenge to applications of extreme value statistics. Also, bias may relate to the magnitude of the tail index then causing systematic bias in capital allocation. When for example, relatively low α's tend to be over- and relatively high α's tend to be under-estimated as indicated e.g. by the results in Wagner and Marsh (2000), even under the moderate probability level of 1 percent, substantial over- and under-allocation with risk capital may arise in management applications.

Recent methodological approaches give promising results in reducing tail index estimation bias. The methods include the exponential regression approach by Feuerverger and Hall (1999) and Beirlant et al. (1999) with a refinement by Matthys and Beirlant (2001). Simulation results of the latter authors indicate that the estimator overcomes typical problems of the Hill plot under various model distributions. However, the authors do not find a dominant quantile estimation technique. Other methodological approaches are based on bias-corrected versions of the Hill estimator. Gomes and Oliveira (2001) analyze bootstrap correction methods. Huisman et al. (2001) use Hill plot regression approach with an application given in Huisman et al. (1998). The analysis in Wagner (2002a) relies on the Hill plot based maximum occupation time estimator of Drees et al. (2000).

Despite the critical view put on VaR estimation in this contribution, one should point out that overall progress in extreme value theory and research in finance during recent years has much improved quantitative risk management methods. Keeping potentials and limitations of those methods in mind will allow further improvements in handling practical risk management tasks.

References

Bassi F, Embrechts P, Kafetzaki M (1998) Risk Management and Quantile Estimation. In: Adler RJ, Feldman RE, Taqqu MS (eds) A Practical Guide to Heavy Tails. Birkhäuser, Boston, pp 111-130

Beirlant J, Dierckx G, Goegebeur Y, Matthys, G (1999) Tail Index Estimation and an Exponential Regression Model. Extremes 2: 177-200

Beirlant J, Vynckier P, Teugels J L (1996) Practical Analysis of Extreme Values. Leuven University Press

Coles SG (2001) An Introduction to Statistical Modeling of Extreme Values. Springer, London

Danielsson J, de Vries CG (1997) Tail Index and Quantile Estimation with Very High Frequency Data. Journal of Empirical Finance 4: 241-257

Drees H, de Haan L, Resnick S (2000) How to Make a Hill Plot. Annals of Statistics 28: 254-274

Embrechts P (2000) Extreme Value Theory: Potential and Limitations as an Integrated Risk Management Tool. Derivatives Use, Trading & Regulation 6: 449-456

Embrechts P, Klüppelberg C, Mikosch T (1997) Modelling Extremal Events for Insurance and Finance. Springer, New York

Emmer S, Klüppelberg C, Trüstedt M (1998) VaR - Ein Mass für das extreme Risiko. Solutions 2: 53-63

Feuerverger A, Hall P (1999) Estimating a Tail Exponent by Modelling Departure from a Pareto Distribution. Annals of Statistics 27: 760-781

Frey R, McNeil AJ (2000) Estimation of Tail-Related Measures for Heteroscedastic Financial Time Series: An Extreme Value Approach. Journal of Empirical Finance 7: 271-300

Gomes MI, Oliveira O (2001) The Bootstrap Methodology in Statistics of Extremes—Choice of the Optimal Sample Fraction.Extremes 4: 331-358

Hill BM (1975) A Simple General Approach to Inference about the Tail of a Distribution. Anals of Statistics 3: 1163-1174

Huisman R, Koedijk KG, Kool CJM, Palm F (2001) Tail-Index Estimates in Small Samples. Journal of Business and Economic Statistics 19: 208-216

Huisman R, Koedijk KG, Pownall R (1998) Fat Tails in Financial Risk Management. Journal of Risk 1: 47-62

Jorion P (1997) Value at Risk: The New Benchmark for Controlling Market Risk. McGraw-Hill, New York

Kearns P, Pagan A (1997) Estimating the Density Tail Index for Financial Time Series. Review of Economics and Statistics 79: 171-175

Klüppelberg C (2002) Risk Management with Extreme Value Theory. Presentation Manuscript, Munich University of Technology

Krämer W, Runde R (1996) Stochastic Properties of German Stock Returns. Empirical Economics 21: 281-306

Lauridsen S (2000) Estimation of Value at Risk by Extreme Value Methods. Extremes 3: 107-144

Locarek-Junge H, Strassberger M, Wagner N (2002) Wann beginnt die Krise?—Ein Blick auf Finanzmarktrenditen. Forthcoming in: Blum U et al. (eds) Krisenkommunikation. Teubner, Stuttgart

Lux T (2001) The Limiting Extremal Behavior of Speculative Returns: An Analysis of Intra-Daily Data from the Frankfurt Stock Exchange. Applied Financial Economics 11: 299-315

Matthys G, Beirlant J (2000) Adaptive Threshold Selection in Tail Index Estimation. In: Embrechts P (ed): Extremes and Integrated Risk Management. Risk Books, London, pp 37-49

Matthys G, Beirlant J (2001) Extreme Quantile Estimation for Heavy-Tailed Distributions, Working Paper, University of Leuven

McNeil A (1998) Calcualting Quantile Risk Measures for Financial Time Series using Extreme Value Theory. Preprint, ETH Zürich

Oehler A (ed) (1998) Credit Risk and Value-at-Risk Alternativen. Schaeffer-Poeschel, Stuttgart

Ridder T. (1998) Basics of Statistical VaR-Estimation. In: Bol G, Nakhaeizadeh G, Vollmer K-H (eds) Risk Measurement, Econometrics, and Neural Networks. Physica, Heidelberg, pp 161-187

Wagner N (2002a) The Hill Estimator in Financial Risk Assessment and an Application to Extremal Exchange Rate Risk. In: Batten JA, Fetherston TA (eds) Financial Risk and Risk Management. Elsevier, Amsterdam, pp 173-187

Wagner N (2002b) Value-at-Risk for Financial Assets Determined by Moment Estimators of the Tail Index. Forthcoming in: Opitz O, Schwaiger M (eds) Classification in Management Science. Springer, Berlin

Wagner N, Marsh TA (2000) On Adaptive Tail Index Estimation for Financial Return Models. Working Paper No. RPF-295, U.C. Berkeley

Project Risk Management by a Probabilistic Expert System

André Ahuja and Wilhelm Rödder

FernUniversität in Hagen, Profilstr. 8, 58084 Hagen, Germany
andre.ahuja@fernuni-hagen.de, wilhelm.roedder@fernuni-hagen.de

Abstract Efficient applications of expert systems to project risk management problems are seldom, if not unusual. In this paper we overcome this lack by using the probabilistic expert system shell SPIRIT. The rule-based shell's power in conditioning, inference and reasoning under incomplete information will work well on risk estimation and classification. A key characteristic of SPIRIT is the possibility to integrate project objectives into the risk management model. So known dependencies between risk variables can be modelled by the user if known beforehand, whereas hidden dependencies might be detected by the proper system. Because of the novelty of projects they suffer from incomplete information and it is this incompleteness which SPIRIT handles at high information fidelity. Furthermore undirected inference is possible, due to the undirected graphical structure in which knowledge is acquired and processed. So, in an early-state risk management situation – where the final model in terms of certain variables and/or their respective dependencies is not yet available – preliminary risk analyses and even recommendations for adequate risk treatment measures are possible, too. A middle size product developement example, including 12 binary variables and 34 rules, shows the inferential power of SPIRIT.

1. Introduction

Novelty, complexity and middle/long size planning periods make a project highly suffer from uncertainty [3], [10]. The novelty implies a lack of experience and concrete numbers so that project risk management (PRM) must be able to identify possible project risks, to evaluate and treat them, if necessary [6], [12]. Because of the complexity of this process the use of information processing computer systems is undisputed [11]. In the present paper we demonstrate how expert systems can support this process. An expert system is a computer program which adapts knowledge about a certain domain and responds to user questions in special situations. Typical examples are related on medical or technical diagnostics [5], [1].

After a short introduction to the probabilistic expert system shell SPIRIT in section 2, we develope the model of a PRM situation in section 3 and then study different scenarios in section 4. The objective is the answer to the question which parts of project risk analysis can be efficiently supported by the expert system. Section 5 is a summary and points out possible future research.

2. The Expert System Shell SPIRIT

Partial knowledge about projects and specially about project risks shall be modelled in SPIRIT under the SPIRIT syntax [8], thus forming a PRM knowledge base. This needs the definition of a finite set of finite valued variables plus a set of probabilistic rules. A probabilistic rule is an expression of the typ "$x\ A \Rightarrow B$" – read A implies B with probability x. Here A and B are propositional sentences built by literals <Variable>=<value> and linked by the operators $\wedge$ (and), $\vee$ (or), $\neg$ (not), $\Rightarrow$ (implies) and respective parentheses. To each such rule there is attached a probability x with which the rule is estimated to be true in the domain under consideration. The set of such probabilistic rules is the rule basis R.

Once all probabilistic rules are provided to the system, it adapts them by generating a probability distribution on the set of all configurations ω, elements of all variables' ranges cross product. This distribution respects the given rules and does not generate any not intented dependency between the variables involved; it obeys the principle of information fidelity [9]. Mathematically speaking we choose the probabilistic extension $W(R)$, which is the set of all probability distributions respecting the linear restrictions given by the set of rules R, solving

$$P^* = \arg\left(\max\left\{H(P) \middle| P \in W(R)\right\}\right). \tag{1}$$

The function H of a distribution P is its entropy

$$H(P) := -\sum_{\omega \in \Omega} p(\omega)\ ld\ p(\omega), \tag{2}$$

here ld denotes the dual logarithm.

Solving (1) yields the distribution P^* which contains exactly the information given by the rule basis and this information measures in [*bit*], c.f. [4]. The system SPIRIT presents the result of (1) in the form of an independency graph including the marginal distributions on the variables, see below. In this graph you might check mutual impacts among variables when focussing different states.

3. Building the Knowledge Base

We consider a product developement project and therefore define the following variables:

- product characteristics: `QUALITY`, `INNOVATION`
- project situation: `ROLL_OUT`, `BUDGET`, `CAPACITY`, `DURATION`, `PROCEED`, `LICENSE`
- valuation/score: `ECONOMY`, `IMPORTANCE`, `SALES`, `CONTINUE`.

Consider the case where the expert estimates a high probability – 85%, say – for a late ROLL_OUT at the project final, due to a risky license situation. This knowledge is informed as the probabilistic rule

$$0.85\ \texttt{LICENSE}=\text{doubtful} \wedge \texttt{PROCEED}=\text{finish} \Rightarrow \texttt{ROLL\_OUT}=\text{delay} \qquad (3)$$

Such a rule represents a linear restriction [4] and thus fixes a hyperplane in the probability space. In general the set of all rules *R* does not determine a unique extension *W*(*R*), but rather a convex polyhedron. Solving (1) we pick out one of these distributions, namely the one with the highest information fidelity. As such rules in their clear syntax are self-explicatory, we give an extract of *R* in figure 1.

Fig. 1: An extract of *R* in SPIRIT's rule-window

Nr.	Pi	Regel
0	0,85	ECONOMY=profitable ⇒ CONTINUE=no_doubt
1	0,95	ROLL_OUT=sure ∧ ECONOMY=profitable ⇒ CONTINUE=no_doubt
2	0,85	LICENSE=doubtful ∧ PROCEED=finish ⇒ ROLL_OUT=delay
3	0,80	LICENSE=doubtful ⇒ PROCEED=start
4	0,95	ECONOMY=not_profitable ∧ SALES=doubtful ⇒ CONTINUE=check
5	0,98	ECONOMY=profitable ∧ SALES=sure ∧ IMPORTANCE=strategic ⇒ CONTINUE=no_doubt
6	0,60	ECONOMY=profitable ∧ SALES=sure ∧ IMPORTANCE=nice2have ⇒ CONTINUE=no_doubt
7	0,75	IMPORTANCE=strategic ⇒ CONTINUE=no_doubt
8	0,95	PROCEED=start ∧ ECONOMY=not_profitable ⇒ CONTINUE=check
9	0,70	INNOVATION=konservativ ∧ BUDGET=slack ⇒ LICENSE=secure
10	0,80	PROCEED=start ∧ ROLL_OUT=delay ⇒ CONTINUE=check
11	0,75	QUALITY=well
12	0,65	SALES=doubtful ∧ DURATION=shortterm ⇒ ECONOMY=not_profitable
13	0,90	PROCEED=start ⇒ CAPACITY=slack

The set Ω of all configuration ω of the given 12 binary variables counts $n = 2^{12} = 4.096$ elements. The uniform distribution on Ω has maximum entropy: $ld\,4.096 = 12$ [*bit*]. Solving (1) and respecting all 34 rules we find a distribution P^* with an entropy of 5.24 [*bit*], significantly inferior than the initial one. The difference 12 - 5.24 = 6.76 [*bit*] is a measure for the knowledge amount adapted in the system [2]. This significant uncertainty reduction we take as a good reason to trust in the system's answers as demonstrated in the next section. Figure 2 shows the corresponding independency graph and the respective marginal probabilities on the variables' values.

Fig. 2: Independence graph for the knowledge base „RISKMGMT.spi“

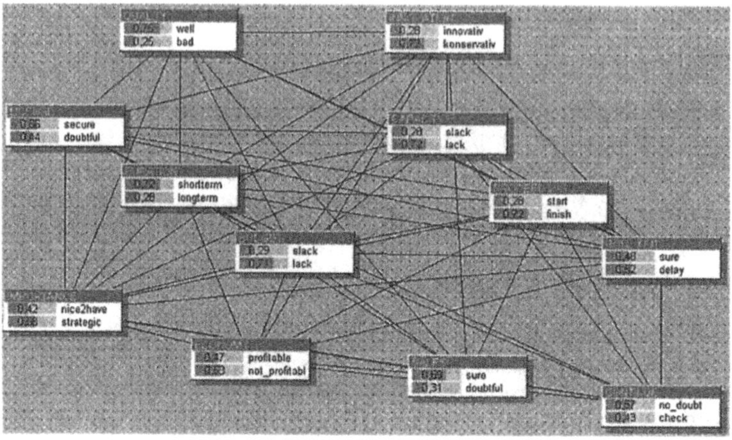

4. Risk Analysis by SPIRIT

After solving (1) of the last section the system disposes of P^*, the knowledge base about the domain. This general knowledge can be modified focussing a certain situation. We use this concept of focussing a situation to study the impact of a probability change of the variable LICENSE upon ROLL_OUT and ECONOMY. More concrete: We are interested in learning about the effect which the risk variable LICENSE exerts on the project's deadline and on its economy. And this effect shall be analysed for the developement of a conservative product of high quality and for a secure sales situation. Thus putting

QUALITY=well, DURATION=shortterm, (4)
INNOVATION=conservative, SALES=secure

yields the desired focus which now permits the analysis, shown in table 1.

Table 1. Objective ROLL_OUT and ECONOMY contingent on LICENSE situation

Variable	LICENSE		ROLL_OUT		ECONOMY	
Value	secure	doubtful	sure	delay	profit.	not_profit.
Probabilities, conditioned by focus (4).	0.79	0.21	0.79	0.21	0.45	0.55
Add. condition by LICENSE≡secure.	1.00	0.00	0.87	0.13	0.46	0.54
Change			+10%	−38%	+2.2%	−1.8%
Add. condition by LICENSE≡doubtf.	0.00	1.00	0.47	0.53	0.38	0.62
Change			−41%	+152%	−38%	+13%

The first line shows the respective marginal probabilities for the focussed situation. The numbers represent average experiences about such projects in general:

There is a 79% probability of a sure ROLL_OUT and a 45% probability of profit. The system is now ready for an analysis with respect to the variable LICENSE. We observe a 10% higher sure ROLL_OUT and a 38% decrease of a delay's probability with a secure license situation. The ECONOMY attributes' probabilities undergo a likewise modest 2.2% and –1.8% change. Things turn out much more significant with a doubtful license situation. Here we have a strong negative effect upon the ROLL_OUT expectation *and* upon the expected economy of the project.

SPIRIT permits the consideration of utilities *u* for each variable thus allowing the calculation of a (numerical) expected value for the model's objectives. This might guarantee a more profound valuation in decision problems than the mere use of probabilities. In the above model for example, attributing *u*(profitable) = 100 and *u*(not_profitable) = –50, the exepted utility of 16.91 for the initial situation raises to 19.72 for a severe and decreases to 6.51 for a doubtful license situation. The project manager should be aware of the fact that, in the loss case – doubtful license situation, which would not be treated – there is a severe 61.5% decrease of profit.

The expert system shell SPIRIT also allows focussing virtual=not certain situations. Going back to table 1 we might be interested in analysing the situation for the case in which the information about a doubtful license situation would be vague. What if the attribute doubtful has a probability of 80% rather than 100%. Such vagueness can be of subjective – linguistic unprecision – or of statistical – 80% of all licenses are there – nature. In either case the system responds like in table 2.

Table 2. (continued, virtual license situation)

LICENSE		ROLL_OUT		ECONOMY	
secure	doubtful	sure	delay	profitable	not_profitable
0.20	0.80	0.55	0.45	0.39	0.61
		–30%	+114%	–13%	+11%

Please notice that virtual focussing is an important instrument in real world situations. On the one hand the effect is not as extreme as the second focussing LICENSE=doubtful shown in table 1. But on the other hand it may be closer to the subjective appraisal of the decision maker, and still the impact on ROLL_OUT and ECONOMY obviously is significant.

5. Summary

The formulation of expert knowledge as a set of probabilistic rules together with the principle of maximizing entropy is a mighty instrument to derive an unbiased knowledge base about a domain such as project risk management, even under partial knowledge, only. The expert system shell SPIRIT permits the focus of special scenarios as well as a sensitivity analysis between the model variables involved. This playing through of different future project situations includes even virtual or vague situations, respectively. Utilities instead of mere probabilities to evaluate such situations also can be handled by the system, if disposable. The reader interested in studying the model RISKMGMT.spi is invited to visit us under www.xspirit.de.

References

1. Breese, J.S., Heckermann, D. (1996): Decision-theoretic Troubleshooting: A Framework for Repair and Experiment, in: Uncertainty in Artificial Intelligence 12, Morgan Kaufman Publishers, San Francisco, California, p. 124-132.
2. Kulmann, F. (2002): Wissen und Information in konditionalen Modellen, Deutscher Universitäts-Verlag, Wiesbaden, p. 56-58.
3. Madauss, B.-J. (1984): Projektmanagement, Stuttgart.
4. Meyer, C.-H. (1998): Korrektes Schließen bei unvollständiger Information, Peter Lang, Frankfurt a.M.
5. Lauritzen, S.L., Thiesson, B., Spiegelhalter, D.J. (1994): Diagnostic systems by model selection: a case study, Lecture Notes in Statistics, 89, Springer, p. 143-152.
6. Raftery, J. (1994): Risk analysis in project management, E.&F.N. Spon, London.
7. Reucher, E., Rödder, W.(2001): Modellierung von Entscheidungsproblemen unter Verwendung von probabilistischen Konditionalen, in Fleischmann, B. et Al.: Operations Research Proceedings 2000, Springer, p. 254-259.
8. Rödder, W. (2000): Conditional Logic and the Principle of Entropy, Artificial Intelligence, 117, p. 83-106.
9. Rödder, W. (2001): Knowledge Processing under Information Fidelity, Proc. IJCAI 2001 - Seventeenth International Joint Conference on Artificial Intelligence, Seattle, Washington, p. 749-754.
10. Rödder W., Ahuja, A. (forthcoming): Projektmanagement: Konzept, Aufgaben Techniken, Kohlhammer, Stuttgart.
11. Schön, D., Diederichs, M., Busch, V. (2001): Chancen- und Risikomanagement im Projektgeschäft, in: Controlling (2001) 7, p. 379-387.
12. Schnorrenberg, U., Goebels G. (1997): Risikomanagement in Projekten, Braunschweig, Wiesbaden.
13. Internet: www.xspirit.de

Regulatory Impacts on Credit Portfolio Management

Ursula Theiler*, Vladimir Bugera**, Alla Revenko**, Stanislav Uryasev**

*Risk Training, Carl-Zeiss-Str. 11, D-83052 Bruckmuehl, Germany,
mailto: theiler@risk-training.org,
**Risk Management and Financial Engineering Lab, University of Florida,
303 Weil Hall, Gainesville, FL 32611-65, USA,
mailto: bugera@ufl.edu, alla@ufl.edu, uryasev@ufl.edu.

Abstract Efficient credit portfolio management is a key success factor of bank management. Discussions of the new capital adequacy proposals by the Basle Committee on Banking Supervision enlighten the necessity to consider the credit risk management both from the internal and the regulatory point of view. We introduce an optimization approach for the credit portfolio that maximizes expected returns subject to internal and regulatory risk constraints. With a simplified bank portfolio we examine the impact of the regulatory risk limitation rules on the optimal solutions.

1 Introduction

Efficient credit portfolio management is a key success factor of bank management. In an adverse market environment and intensifying competition banks are exposed to increasing risks and decreasing return margins of their credit portfolio, while bank shareholders are demanding higher risk premiums for their invested capital. The ability to identify risk-return optimal portfolios becomes a fundamental element of credit portfolio management. The recent discussions of the Basle Committee on Banking Supervision enlighten the necessity to manage credit risk simultaneously from an internal and a regulatory perspective.

In this paper, we give a survey of a new optimization algorithm that determines risk-return efficient credit portfolios under internal and regulatory credit risk constraints. We formulate the optimization problem for the credit portfolio based on the new risk measure, Conditional Value at Risk, and derive risk-return ratios for the optimal portfolios (chapter 2). With an application example, we analyze the risk-return structure of an optimal portfolio. We examine the impact of the regulatory risk limitation rules and visualize how they may lead to inefficiencies in the credit portfolio management (chapter 3).

2 Optimization Approach

2.1 Definition of the CVaR Risk Measure

The risk measure Value at Risk (VaR), commonly applied in finance, lacks the sub-additivity property, when return distributions are not normal. This means that the diversification of the portfolio may increase portfolio VaR. A similar percentile risk measure, *Conditional Value at Risk* (CVaR) does not have this drawback. The term Conditional Value-at-Risk was introduced in [5]. For continuous distributions, CVaR is equal to the conditional expectation beyond VaR, see [5]. However, for general distributions, it is a weighted average of VaR and the conditional expectation beyond VaR, see [6]. CVaR can be applied to measure loss risk from any asymmetric and discontinuous loss distribution with discrete probabilities and it obeys the property of coherence, see [1,4,6], a set of axioms that a risk measure should meet from the point of view of a regulator [2]. CVaR has been proved to be appropriate for credit portfolio risk measurements [4,5,6].

Let $\mathbf{x}=(x_1,\ldots,x_n)'$ be a vector of positions of credit assets of a portfolio, and $\mathbf{y}=(y_1,\ldots,y_n)'$ be a vector of the corresponding market prices. For continuous distributions, we define *CVaR deviation* $\mathrm{CVaR}_\alpha^\Delta(L(\mathbf{x},\mathbf{y}))$ of the portfolio loss risk as

$$\mathrm{CVaR}_\alpha^\Delta(L(\mathbf{x},\mathbf{y})) = E[L(\mathbf{x},\mathbf{y}) \mid L(\mathbf{x},\mathbf{y}) \geq \mathrm{VaR}_\alpha(L(\mathbf{x},\mathbf{y}))], \tag{1}$$

where the loss function $L(\mathbf{x},\mathbf{y})$ is the difference of the uncertain portfolio values and the expected value of the portfolio, i.e. $L(\mathbf{x},\mathbf{y})=E[\mathbf{y}]'\mathbf{x}-\mathbf{y}'\mathbf{x}$, and $\mathrm{VaR}(L(\mathbf{x},\mathbf{y}))$ is the α-quantile of the loss function $L(\mathbf{x},\mathbf{y})$.[1]

2.2 Formulation of the Optimization Model

The optimization problem models the basic goal of the credit portfolio management. We maximize the expected portfolio return $\mu(\mathbf{x})=\boldsymbol{\mu}'\mathbf{x}$ under internal and regulatory loss risk constraints [8], with $\mathbf{x}$ the decision vector and $\boldsymbol{\mu}=(\mu_1,\ldots,\mu_n)'$ the vector of the expected returns of single assets. The internal loss risk is measured by the CVaR deviation of the portfolio loss according to equation (1) and is constrained by the maximal amount of economic capital available, denoted as ec_cap_max. Based on the optimization algorithm of Rockafellar/Uryasev [5], the CVaR constraint is approximated by a set of linear constraints, leading to a linear optimization problem. To implement the algorithm, as input data, we use a sample of market price scenarios $\mathbf{y}_1, \ldots, \mathbf{y}_K$ of the vector $\mathbf{y}$.[2] The regulatory credit risk is measured by the regulatory risk based capital ratios, **reg_cap** = (reg_cap$_1$,..., reg_cap$_n$)' and is limited by the available regulatory core and supplementary capital, denoted by reg_cap_max. The area of the feasible solutions is defined by up-

[1] In the case of nonzero probability atom at the α-quantile, CVaR is defined as the weighted average of VaR and the conditional expectation beyond the VaR [6].

[2] In the application example in the next chapter, these market price scenarios are generated by a Monte Carlo-Simulation according to the CreditMetrics approach of J. P. Morgan.

per and lower position bounds, the vectors **low_bound** and **up_bound**. We solve the following linear optimization model:

$$\text{Objective Function } \mu(\mathbf{x}) = \boldsymbol{\mu}'\mathbf{x} = \sum_{j=1}^{n} \mu_j x_j, \tag{4}$$

Constraint # 1: Internal Risk Constraint

(i) $q + \frac{1}{(1-\alpha)} \cdot \frac{1}{K} \sum_{k=1}^{K} z_k \leq ec\_cap\_max,$

(ii) $L(\mathbf{x}, \mathbf{y}_k) - q \leq z_k, k = 1, \ldots, K,$

(iii) $-z_k \leq 0, k = 1, \ldots, K,$

(iv) $q \in \Re$

"Internal loss risk (CVaR deviation estimate) ≤ Economic capital"

Constraint # 2: Regulatory Risk Constraint

(v) $\mathbf{reg\_cap}'\mathbf{x} \leq reg\_cap\_max.$

Constraint # 3: Boundaries of the Feasible Solutions

(vi) $\mathbf{low\_bound} \leq \mathbf{x} \leq \mathbf{up\_bound}.$

In order to analyze the effects of the regulatory risk constraints on the optimal portfolios, we consider the following optimization models (P') and (P) with and without the regulatory risk constraint, accordingly:

(P): *Maximize* Objective *subject to* Constraints # 1 and 3,
(P'): *Maximize* Objective *subject to* Constraints # 1, 2, 3. (5)

2.3 Risk-Return Analysis of the Portfolio Assets

The contributions of the single assets to the overall portfolio risk and return represent basic information for the risk-return analysis of the optimal portfolios. The return contribution $\mu_j(\mathbf{x})$ of the j-th asset to the portfolio $\mathbf{x}$ is given by the j-th coefficient of the return function, i.e. $\mu_j(\mathbf{x})=\mu_j$, j=1,…,n.

We apply the Euler allocation principle to derive the risk contributions of the single assets [3,7]: The risk contribution $r_j(\mathbf{x})$ of the j-th asset is defined by the partial derivative of the portfolio risk measure with respect to the j-th asset. It corresponds to the conditional expected loss of the j-th component in the tail of the portfolio loss distribution and can be estimated from the given sample of the market prices as the mean of the losses of the j-th asset in the tail of the loss distribution [7]. We achieve the following risk contribution $r_j(\mathbf{x})$ of the j-th asset, j=1,…,n:

$$r_j(\mathbf{x}) = \frac{\partial\, CVaR_\alpha^\Delta(L(\mathbf{x},\mathbf{y}))}{\partial\, x_j} \cdot x_j = E[L_j(\mathbf{x},\mathbf{y}) \mid L(\mathbf{x},\mathbf{y}) \geq VaR_\alpha(L(\mathbf{x},\mathbf{y}))], \tag{6}$$

where $L_j(\mathbf{x},\mathbf{y}) = E[y_j]x_j - E[y_j x_j \mid L(\mathbf{x},\mathbf{y}) \geq VaR_\alpha(L(\mathbf{x},\mathbf{y}))], j = 1, \ldots, n.$

We define the risk-return ratios of single assets, the return on risk adjusted capital $RORAC_j(\mathbf{x})$ of the j-th asset and the return on equity $RoE_j(\mathbf{x})$, i.e. the return on the regulatory capital, of the j-th asset as

$$\text{(i) RORAC}_j(\mathbf{x}) = \frac{\mu_j(\mathbf{x})}{r_j(\mathbf{x})}, \qquad j = 1,\dots,n, \tag{7}$$

$$\text{(ii) RoE}_j(\mathbf{x}) = \frac{\mu_j}{reg\_cap_j}, \qquad j = 1,\dots,n.$$

3 Application Example

An ABC Bank consists of three typical credit assets: asset 1 represents high quality bonds (Rating AA), asset 2 mortgage loans (Rating BB) and asset 3 retail loans (Rating B). 10 units of regulatory capital are available, of which 94% are actually in use. The internal risk (CVaR) level may be varied to some extent according to the risk policy of the bank. The initial portfolio uses 48 units of the economic capital. Our goal is to investigate how the risk-return relations of the initial credit portfolio can be improved and how the regulatory risk constraint effects the optimal portfolios. We applied the optimization models (P) and (P') with different CVaR levels. First, we generated the efficient frontiers and analyzed the overall portfolio risk-return relations. Next, we analyzed the risk-return structures of the single assets of the optimal portfolios.

As shown in the Fig. 1, we observe that the regulatory constraint becomes active at the CVaR-level of 39.9 units. At the given capital levels (ec_cap_max=48, reg_cap_max=10), the expected portfolio returns can be improved by 0.07 units in (P') and by 0.23 units in (P). This means that without the regulatory constraint an additional profit of 0.16 units could be gained. The portfolio RORAC, defined as the expected return $\mu(\mathbf{x})$ divided by the CVaR deviation $\mathrm{CVaR}^{\Delta}_{\alpha}(L(\mathbf{x},\mathbf{y}))$ of the portfolio $\mathbf{x}$, increases from 6.09% to 6.29% in (P') and to 6.63% in (P).

We also observe that the ABC bank can generate higher portfolio RORACs by lowering the level of internal risk. The maximal RORAC of 6.82% can be reached at the interval of ec_cap_max=[34.9,37.7], where the regulatory constraint is not active. However, the implementation of a RORAC optimizing strategy would require reducing the credit volumes and absolute returns. This might be conflicting with other corporate goals and may not be supported by the shareholders.

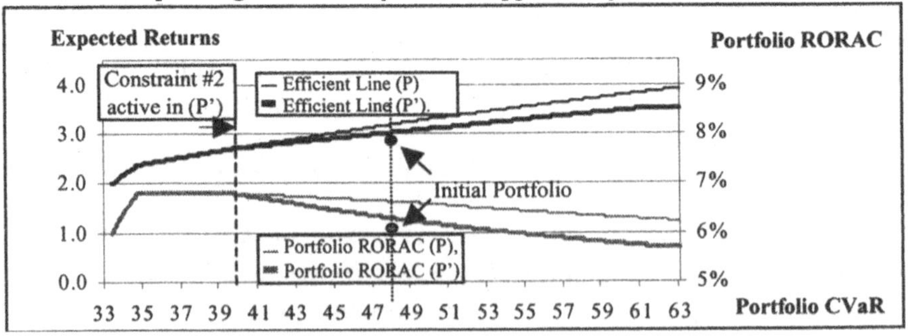

Fig. 1. Efficient Lines and Portfolio RORACs of the Optimization Problems (P) and (P')

In order to analyze the risk-return structure of optimal portfolios we first examine the positions of single assets, which are represented in fig. 2. The narrow and broad lines represent the positions of single assets in the solutions of (P) and (P'), respectively. Starting from the minimal CVaR portfolio, the assets are increased in the optimal solutions in the order of descending RORACs, as defined in the equ. 7 (i). [3] When the regulatory constraint becomes active, we observe the effect of capital arbitrage: assets with higher RoEs are preferred to assets with higher RORACs, and the overall portfolio level of risk is increased.

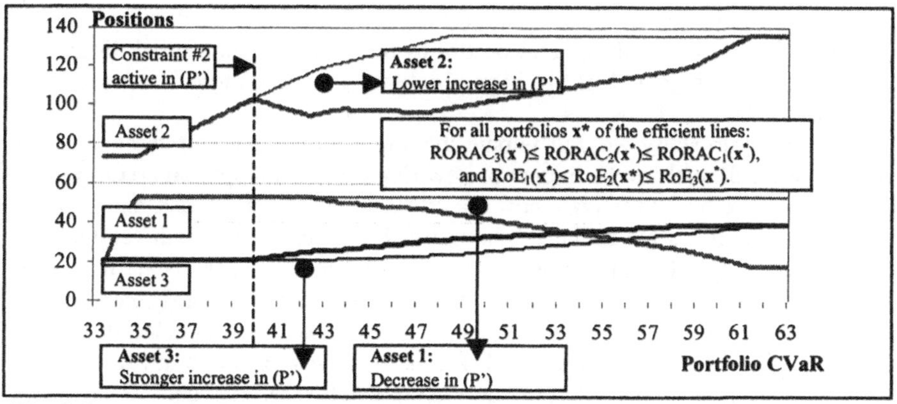

Fig. 2. Impact of the Regulatory Constraint on the Optimal Portfolio Structures

In order to analyze the effect of capital arbitrage more closely, we examine the risk-return structure of the optimal portfolio at the initial CVaR level of 48 units, as described in the fig. 3. Without the regulatory risk constraint, position of asset 1 with highest RORAC is increased by 50%, of asset 2 by 28.3% and position of asset 3 with lowest RORAC is reduced by 21.5%. In (P') asset 1, showing the lowest RoE, is increased less than in (P). Position of asset 3 with the highest RoE is increased, while position of asset 2 with higher RORAC but lower RoE than asset 3 is reduced. The riskier assets are weighted higher in (P'), resulting in lower returns at the given CVaR-level and a sub optimal use of the economic capital, as could be observed in the fig. 1 above.

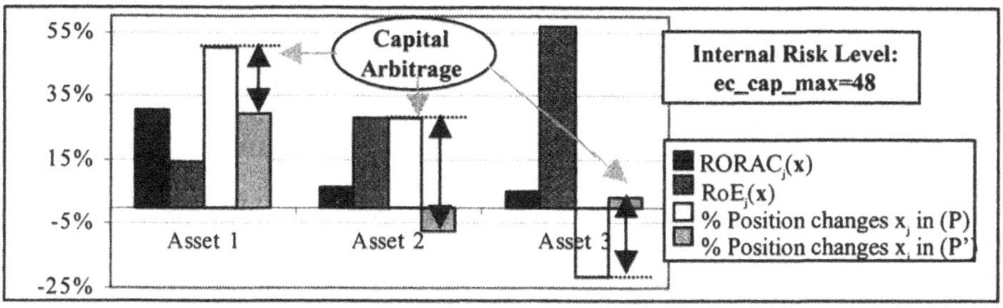

Fig. 3. Portfolio structures of the optimal portfolio in (P) and (P')

[3] Although the RORACs of the single assets of the optimum portfolios **x*** differ slightly along the efficient line, their ranking remains constant, i.e. $RORAC_3(\mathbf{x}^*) \leq RORAC_2(\mathbf{x}^*) \leq RORAC_1(\mathbf{x}^*)$.

4 Conclusion

We have introduced an algorithm that maximizes the expected returns of a credit portfolio subject to the internal and regulatory risk constraints. It is based on the new risk measure, CVaR, which is appropriate for credit portfolio risk measurement and can be solved by linear programming methods. The optimization model allows to spot intervals of efficient use of both capital resources, the available economic and regulatory capital, and of highest portfolio RORACs. It identifies "unrealized" profits due to the regulatory risk constraint. We conducted risk-return analyses of single assets of the optimal portfolios and found evidence of capital arbitrage, that leads to sub optimal portfolios under the regulatory risk limitation rule, as assets of higher RoE but higher risk are weighted higher than assets of lower risk and higher RORACs.

In a follow-up study we will pursue the application example for the Basle II Accord. We will analyze the impact of new risk weights on optimal credit portfolios. Further, we will investigate how the internal and regulatory risk-return structures of the single assets influence the optimal solutions when both capital constraints are active. Also, the intuitive statement that the single assets are increased in the optimum solutions in the order of descending RORACs in (P) can be formally investigated, as the risk contributions are not explicitly modeled in the optimization algorithm. Another point of interest is to develop an optimization algorithm that calculates the RORAC-optimal portfolios of (P) and (P') in one optimization run.

References

[1] Acerbi, C. and Tasche, D., On the coherence of expected shortfall. Working paper (2001), can be downloaded from http://www.gloriamundi.org.

[2] Artzner, Ph., Delbaen, F., Eber, J.-M., Heath, D. (1999): Coherent Measures of Risk, *Mathematical Finance*, Vol. 9, No. 3, pp. 203-228.

[3] Patrik, G., Bernegger, S., Rüegg, M.B. (1999): The use of risk adjusted capital to support business decision making, in: Casualty Actuarial Society (Hrsg.), Casualty Actuarial Society Forum, Spring 1999 Edition, Baltimore.

[4] Pflug, G.Ch. (2000): Some Remarks on the Value-at-Risk and the Conditional-Value-at-Risk, in: Uryasev, S. (Ed.), Probabilistic Constrained Optimization: Methodology and Applications, Kluwer Academic Publishers, pp. 272-281.

[5] Rockafellar, R. T. and Uryasev, S. (2000): Optimization of Conditional Value-At-Risk, *The Journal of Risk*, Vol. 2, No. 4, pp. 21-51.

[6] Rockafellar, R.T. and Uryasev, S. (2002): Conditional Value-at-Risk for General Loss Distributions, Journal *of Banking and Finance*, 27/7.

[7] Tasche, D. (1999): Risk Contributions and Performance Measurement, Working Paper, Technische Universitaet Muenchen.

[8] Theiler, U. (2002): Optimization Approach for the Risk-Return-Management of the Bank Portfolio, Wiesbaden (In German).

Verfahren zur Risikokapitalallokation im Eigenhandel von Banken

Mario Straßberger

Lehrstuhl für Betriebswirtschaftslehre, insbes. Finanzwirtschaft und Finanzdienstleistungen, Technische Universität Dresden, Mommsenstraße 13, D-01062 Dresden, E-Mail: strassberger@finance.wiwi.tu-dresden.de

1 Einführung

Der Value-at-Risk (VaR) als Maß zur Quantifizierung von Marktpreisrisiken bietet neue Möglichkeiten der Erfolgs- und Risikosteuerung, speziell im Eigenhandelsgeschäft der Bank. Die bisherige wissenschaftliche Auseinandersetzung konzentrierte sich v.a. auf Modelle zur Schätzung des VaR. Der nächste Schritt besteht in der Konstruktion eines auf dem VaR-Modell aufbauenden VaR-Limitsystems. Das ist zum einen ein aufsichtsrechtliches Erfordernis und zum anderen eine ökonomische Konsequenz.

Ein VaR-Limitsystem ist ein mehrstufiges, hierarchisch gegliedertes System von VaR-Limiten, dass der simultanen Begrenzung von Marktpreisrisiken aller Portfolios und aggregierten Portfolios des Eigenhandels dient. Es muss so gestaltet sein, dass es gleichzeitig eine optimale Allokation von Risikokapital erzeugt. Als Risikokapital wird diejenige Kapitalreserve der Bank bezeichnet, die mit hoher Wahrscheinlichkeit ausreicht, um unerwartete Verluste des Eigenhandels auszugleichen. Es stellt eine knappe Ressource dar.

Das bekannte analytische Delta-Normal-Modell zur VaR-Schätzung wird in diesem Beitrag auf neue Fragestellungen der Marktpreisrisikosteuerung angewendet. Dabei stehen die Fragen der effizienten Allokation von Risikokapital und der Konstruktion eines konsistenten VaR-Limitsystems für die hierarchische Portfoliostruktur des Eigenhandels von Banken im Mittelpunkt. Das VaR-Modell wird zu diesem Zweck mit einem Optimierungsansatz verknüpft, der die risikoadjustierte Profitabilität des Eigenhandels maximiert.

2 Delta-Normal-Modell zur Value-at-Risk-Schätzung

Der zukünftige potentielle Verlust $L_{t+H} := W_t - W_{t+H}$ eines Portfolios wird als negative Änderung seines Marktwertes W_t definiert. Der VaR als bei gegebener Wahrscheinlichkeit p und Haltedauer H prognostizierte Verlustschranke ist definiert durch das p-Fraktil der Verteilungsfunktion von L_{t+H} [3, 4, 6]

$$\mathrm{VaR}_{t+H} = F^{-1}_{L_{t+H}}(p)\,. \tag{1}$$

Es bezeichne $\boldsymbol{R}_t$ den Vektor marktwertrelevanter Risikofaktoren und $\boldsymbol{Y}_{t+H}$ den Vektor stetiger Renditen $Y_{i;t+H} = \ln R_{i;t+H} - \ln R_{i;t}$, $i = 1,\ldots,M$. Der Portfolioverlust ist dann eine Funktion $L_{t+H} = L(\boldsymbol{Y}_{t+H})$. Das Delta-Normal-Modell arbeitet mit zwei zentralen Approximationen. Es wird erstens angenommen, dass die $\boldsymbol{Y}_{t+H}$ unabhängig und identisch gemeinsam normalverteilt sind mit Erwartungswertvektor μ_Y und Kovarianzmatrix Σ_Y. Zweitens wird $L(\boldsymbol{Y}_{t+H})$ als lineare Funktion approximiert. Mit $\boldsymbol{R}_t^{\,d} = \mathrm{diag}(\boldsymbol{R}_t)$ und δ_t mit $\delta_{i;t} = \partial W / \partial R_{i;t}$, $i = 1,\ldots,M$, gilt die Verlustfunktion

$$L(\boldsymbol{Y}_{t+H}) = -\delta_t{}'\boldsymbol{R}_t^{\,d}\boldsymbol{Y}_{t+H}\,. \tag{2}$$

Aus den Approximationen folgt, dass der Portfolioverlust normalverteilt ist mit Erwartungswert $\mu_{L;t} = -\delta_t{}'\boldsymbol{R}_t^{\,d}\mu_Y$ und Varianz $\sigma^2_{L;t} = \delta_t \boldsymbol{R}_t^{\,d}\Sigma_Y \boldsymbol{R}_t^{\,d}{}'\delta_t{}'$. Für den VaR resultiert die Schätzung

$$\mathrm{VaR}_{t+H}(p) = \mu_{L;t} + \sigma_{L;t} z_p\,. \tag{3}$$

z_p ist das p-Fraktil der Standardnormalverteilung. Über die Korrelationsmatrix $\boldsymbol{P}_t = (\rho_{j,k})$, $j,k = 1,\ldots,K$, der Portfolioverluste lassen sich die im Vektor $\mathbf{VaR}_{t+H}$ zusammengefassten VaR-Schätzer von K Portfolios aggregieren [1, 4]

$$\mathrm{VaR}_{t+H} = \sqrt{\mathbf{VaR}_{t+H}{}'\boldsymbol{P}_t\mathbf{VaR}_{t+H}}\,. \tag{4}$$

3 Risikokapitalallokation

Fragestellungen der Risikokapitalallokation und der Konstruktion eines hierarchischen VaR-Limitsystems wurden bisher behandelt [5, 7, 8, 10], ohne konkrete Lösungsalgorithmen vorzuschlagen. Im folgenden wird eine Portfoliohierarchie betrachtet, wie sie im Eigenhandel einer Bank denkbar und in Abb. 1 vereinfacht dargestellt ist. Innerhalb dieser Portfoliohierarchie mit insgesamt K Portfolios werden $J < K$ Basisportfolios betrachtet, die über $K - J - 1$ Zwischenportfolios zum Gesamtportfolio aggregiert werden.

Für den Eigenhandel wird ein bestimmter Risikokapitalbetrag für ein bestimmtes Zeitintervall als gegeben angesehen. Intertemporale Probleme der Risikokapitalverteilung werden nicht betrachtet. Das bereitgestellte Risikokapital wird als Gesamt-VaR-Limit interpretiert. Die Allokation des Risikokapitals erfolgt mit Hilfe von VaR-Limiten. Ziel ist es, ein VaR-Limitsystem zu konstruieren, dass fol-

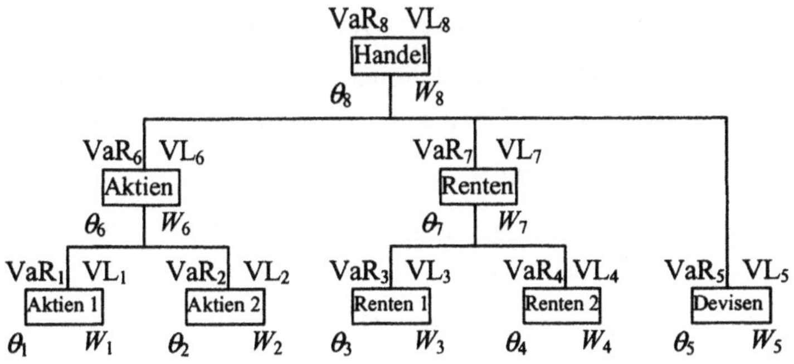

Abb. 1. Beispiel einer Portfolio- und VaR-Limit-Hierarchie

genden Kriterien genügt [2, 9]. Das aggregierte VaR-Limit des Eigenhandels darf das Gesamt-VaR-Limit zu keinem Zeitpunkt überschreiten. Die Einhaltung der VaR-Limite in den dezentralen Basisportfolios muss die Einhaltung des VaR-Limits auf allen Aggregationsebenen sicherstellen. Die VaR-Limite der Basisportfolios sollen so hoch wie möglich gewählt werden. Sie sollen eindeutig und unabhängig von den VaR-Realisationen der anderen Basisportfolios gelten.

Zur Lösung dieser Aufgabe wird der Delta-Normal-Modellrahmen wie folgt modifiziert und erweitert [9]. Über alle Basisportfolios werden N Finanzinstrumente **I** gehandelt. Es sei $\boldsymbol{w}(\boldsymbol{R}_t)$ der Vektor der Marktwerte $w_{i;t}(\boldsymbol{R}_t)$, $i=1,\ldots,N$, aller Finanzinstrumente. Zur Abbildung der Struktur des Basisportfolios $i=1,\ldots,J$ bezüglich der N Finanzinstrumente wird der Strukturvektor θ_i eingeführt. Der Marktwert des Basisportfolios i ergibt sich damit zu

$$W_i(\boldsymbol{R}_t) = \theta_i{}' \boldsymbol{w}(\boldsymbol{R}_t) . \tag{5}$$

Für den Strukturvektor können weiterhin mengenmäßige Begrenzungen des Anlageuniversums $\boldsymbol{u}_i \le \theta_i \le \boldsymbol{o}_i$ mit Untergrenzen $\boldsymbol{u}_i$ und Obergrenzen $\boldsymbol{o}_i$ eingeführt werden. Damit ist es zusätzlich möglich, Volumenlimite in das Modell zu integrieren. Um die Gesamtstruktur zu erfassen, wird aus den Strukturvektoren der Basisportfolios zunächst die Matrix $\Theta_J = (\theta_1,\ldots,\theta_J)$ erzeugt. Zur Abbildung der Portfoliohierarchie wird die Matrix **T** eingeführt, die aus den Elementen 0 und 1 besteht. Die Finanzinstrumente-Struktur der gesamten Portfoliohierarchie ergibt sich damit zu

$$\Theta = \Theta_J \mathbf{T} = (\theta_1,\ldots,\theta_J,\ldots,\theta_K) . \tag{6}$$

Die Struktur der nächst höheren Aggregationsebene ergibt sich jeweils als Summe der Strukturen der darunter liegenden Portfolios; θ_K gibt die Struktur des aggregierten Handelsportfolios wider. Entlang der so erfassten Portfoliohierarchie ist ein System von VaR-Limiten $\mathrm{VL}_i(p)$, $i=1,\ldots,K$, zu konstruieren, dass den oben formulierten Kriterien genügt. Insbesondere muss gelten

$$\mathrm{VaR}(\theta_i, p) \le \mathrm{VL}_i(p) , \; \forall i = 1,\ldots,K . \tag{7}$$

Aus der gegebenen Kovarianzmatrix $\Sigma_{I;t}$ der Finanzinstrumenterenditen kann mit $\boldsymbol{w}_t^d = \mathrm{diag}(\boldsymbol{w}(\boldsymbol{R}_t))$ über die Strukturmatrix Θ die Kovarianzmatrix der Verlustvariablen der gesamten Portfoliohierarchie abgeleitet werden als

$$\Sigma_{L;t} = \Theta' \boldsymbol{w}_t^d \Sigma_{I;t} \boldsymbol{w}_t^d \Theta \,. \tag{8}$$

Dies lässt bei zusätzlicher Approximation von $\mu_L = \mathbf{0}$ die Schätzung des VaR-Vektors für die Portfoliohierarchie zu

$$\mathbf{VaR}_{t+H}(\Theta, p) = \mathrm{diag}(\sqrt{\Sigma_{L;t}}\, z_p)\,. \tag{9}$$

Der VaR-Vektor enthält die VaR-Schätzer aller Basisportfolios und aggregierten Portfolios. Sein letztes Element ist der aggregierte $\mathrm{VaR}_{t+H}(\boldsymbol{\theta}_K, p)$ des Handelsportfolios. Bei exogener Vorgabe von $\mathrm{VaR}_{t+H}(\boldsymbol{\theta}_K, p) = \mathrm{VL}_K(p)$ in Höhe des zu allokierenden Risikokapitals kann durch rekursives Vorgehen und Variation der Basisportfoliostrukturen eine Menge an VaR-Vektoren abgeleitet werden, deren letztes Element jeweils $\mathrm{VL}_K(p)$ entspricht. Diese Menge wird als Menge zulässiger Iso-VaR-Limit-Kombinationen bezeichnet.

Wichtig für das VaR-Limitsystem ist, dass die $\mathrm{VaR}(\boldsymbol{\theta}_i, p)$ der Basisportfolios (und damit auch aller aggregierten Portfolios) durch die Strukturvektoren $\boldsymbol{\theta}_i$ parametrisiert werden. Der VaR hängt von den im Rahmen des definierten Anlageuniversums unabhängigen Entscheidungen über die Struktur der Basisportfolios ab. Für das VaR-Limitsystem hat das weitreichende Konsequenzen. Wenn die Forderung aus Gl. (7) nicht verletzt werden soll, ist bei der Konstruktion des VaR-Limitsystems von denjenigen Basisportfoliostrukturen auszugehen, die zu maximal möglichen Korrelationen der Portfolioverluste führen.

Zur Illustration soll folgendes, einfaches Beispiel dienen. Es seien $J = 2$ Basisportfolios betrachtet, in denen getrennt $N = M = 2$ Finanzinstrumente gehandelt und die zu einem Gesamtportfolio aggregiert werden. Es gilt

$$\boldsymbol{\theta}_1 = (\theta_1 \quad 0)', \; \boldsymbol{\theta}_2 = (0 \quad \theta_2)', \; \mathbf{T} = \begin{pmatrix} 1\,0\,1 \\ 0\,1\,1 \end{pmatrix}. \tag{10}$$

Mittels des dargestellten Vorgehens kann der VaR-Vektor abgeleitet werden als

$$\mathbf{VaR} = (\mathrm{VaR}_1 \quad \mathrm{VaR}_1 \quad \sqrt{\mathrm{VaR}_1^{\,2} + \mathrm{VaR}_2^{\,2} + 2\rho_{L_1,L_2}\mathrm{VaR}_1\mathrm{VaR}_2}\,)'\,. \tag{11}$$

Mit $\mathrm{VaR}_3 = \sqrt{\mathrm{VaR}_1^{\,2} + \mathrm{VaR}_2^{\,2} + 2\rho_{L_1,L_2}\mathrm{VaR}_1\mathrm{VaR}_2} = \mathrm{VL}_3$ resultiert für die Menge an Iso-VaR-Limit-Kombinationen $\sqrt{\mathrm{VL}_1^{\,2} + \mathrm{VL}_2^{\,2} + 2\rho_{L_1,L_2}\mathrm{VL}_1\mathrm{VL}_2}$ die in Abb. 2 dargestellte Ellipse. Dabei wurde $\rho_{L_1,L_2} = -0.5$ und $\mathrm{VL}_3 = 10.0$ unterstellt. Da der VaR als stets positiv definiert ist, kommen nur VaR-Limit-Kombinationen aus dem ersten Quadranten in Frage. Hier liegen in diesem speziellen Fall zwei long-Positionen vor. Das Beispiel verdeutlicht im zweiten und vierten Quadranten für jeweils eine long- und eine short-Position aber auch, dass die VaR-Limite der Basisportfolios mit zunehmender Korrelation der Portfolioverluste um so geringer ausfallen. Damit wird aber auch sofort klar, dass bei unabhängigen Entscheidun-

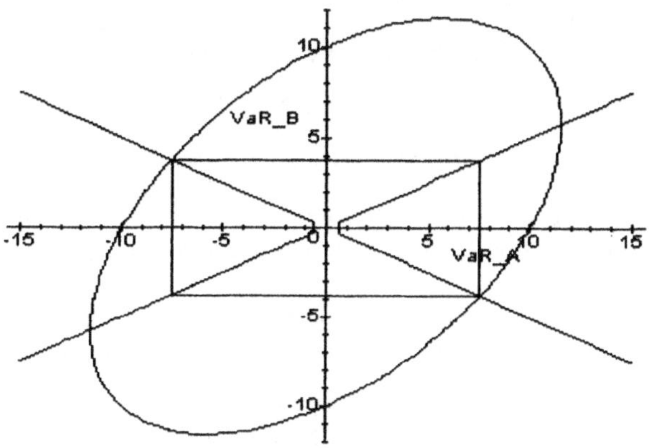

Abb. 2. Zweidimensionales VaR-Limit-Allokationsproblem

gen über die Struktur der Basisportfolios nur VaR-Limite in Höhe der in diesen Quadranten liegenden Kombinationen vergeben werden können. Anderenfalls besteht die Gefahr der Verletzung der Forderung aus Gl. (7).

Zur Selektion einer VaR-Limit-Kombination aus der Menge zulässiger VaR-Limit-Kombinationen wurde die Einführung von Relationen $v_{i,j}$, $i, j = 1,...,J$, zwischen den VaR-Limiten der Basisportfolios vorgeschlagen [9]. Die Relationen können z.B. durch Ergebnis- oder Profitabilitätskennzahlen bestimmt sein. In Abb. 2 wurde $v_{1,2} = \mathrm{VL}_1 / \mathrm{VL}_2 = 2$ unterstellt. Damit ist im Schnittpunkt der Relationsgeraden mit der Ellipse der Iso-VaR-Limit-Kombinationen die Lösung des Allokationsproblems gegeben.

Statt der Einführung von VaR-Limit-Relationen wird zur Selektion einer VaR-Limit-Kombination im folgenden ein Optimierungsansatz aufgesetzt. Ziel des Allokationsprozesses sollte es sein, die knappe Ressource Risikokapital seiner ökonomisch effizientesten Verwendung zuzuführen. Als Optimalitätskriterium wird der sog. Risk Adjusted Return on Risk Adjusted Capital (RARoRAC) benutzt [7]

$$\mathrm{RARoRAC}_i = G_i(\mathrm{VL}_i)/\mathrm{VL}_i \,. \tag{12}$$

Dabei bezeichnet $G_i(\mathrm{VL}_i) = G_i - r_i \mathrm{VL}_i$ den risikoadjustierten Nettogewinn des Portfolios, der als Differenz aus Nettogewinn und Risikokapitalkosten zu verstehen ist. Der RARoRAC stellt nicht nur einen direkten Bezug zum zugeteilten VaR-Limit her, sondern verwendet auch ein risikoangepasstes Ergebnis. Dies erscheint insbes. im Eigenhandel für unterschiedliche Zielverzinsungen r_i des Risikokapitals als geeignet. Die optimale Risikokapitalallokation ist erreicht, wenn der aggregierte RARoRAC für den Eigenhandel bei gegebenem Risikokapital maximal wird. Das führt zu folgendem Optimierungskalkül [1]

$$\mathrm{RARoRAC}_K = \sum G_i(\mathrm{VL}_i) \Big/ \sqrt{\mathbf{VL}' \boldsymbol{P}_t \mathbf{VL}} \to \max! \tag{13}$$

Als Nebenbedingung gilt

$$\sqrt{\mathbf{VL}' \boldsymbol{P}_t \mathbf{VL}} \le \mathrm{VL}_K \,. \tag{14}$$

Bei Verwendung des Lagrange-Ansatzes ergibt sich als Lagrange-Funktion

$$L = \sum G_i(\mathrm{VL}_i) \Big/ \sqrt{\mathbf{VL}' \boldsymbol{P}_t \mathbf{VL}} + \lambda(\mathrm{VL}_K - \sqrt{\mathbf{VL}' \boldsymbol{P}_t \mathbf{VL}}), \tag{15}$$

wobei λ den Lagrange-Multiplikator bezeichnet. Die reelle Lösung des durch

$$\partial L / \partial \lambda = 0, \; \partial L / \partial \mathrm{VL}_i = 0, \; i = 1, \ldots, K \tag{16}$$

bestimmten Gleichungssystems führt eindeutig zu einem globalen Maximum. Zum Aufsuchen der Lösung können bekannte numerische Optimierungsverfahren eingesetzt werden.

Entscheidend für die Lösung des Optimierungsproblems ist zum einen die Formulierung der Funktion $G_i(\mathrm{VL}_i)$ und zum anderen die Korrelationsmatrix $\boldsymbol{P}_t$. Zur Modellierung der risikoadjustierten Nettogewinne sind Polynome höherer als zweiter Ordnung denkbar. Ökonomisch plausibel können s-förmige Funktionen mit zunächst zunehmendem und dann abnehmendem Grenzgewinn sein. Die Korrelationsmatrix der Portfolioverluste ist zeitlich instabil. Außerdem ist sie ex ante unbekannt. Je höhere Korrelationen jedoch eintreten können, desto geringere VaR-Limite dürfen vergeben werden. Diese Sicht erklärt die bekannte Problematik, dass das Gesamt-VaR-Limit auf aggregierter Ebene häufig nicht vollständig ausgenutzt ist.

Literatur

1. Burmester C, Hille CT, Deutsch HP (1999) Risikoadjustierte Kapitalallokation, Beurteilung von Allokationsstrategien über einen Optimierungsansatz. In: Eller R, Gruber W, Reif M (Hrsg) Handbuch Bankenaufsicht und interne Risikomodelle. Stuttgart, S 389–417
2. Delbaen F, Denault M (2000) Coherent allocation of risk capital. E. T. H. Zürich, École des H. E. C. Montréal
3. Duffie D, Pan J (1997) An Overview of Value at Risk. The Journal of Derivatives 4: 7–49
4. Jorion P (2000) Value-at-Risk, The New Benchmark for Managing Financial Risk. New York
5. Kupiec PH (1999) Risk Capital and VaR. The Journal of Derivatives 6: 41–52
6. Linsmeier TJ, Pearson ND (2000) Value at Risk. Financial Analysts Journal 56: 47–67
7. Matten C (1996) Managing Bank Capital, Capital Allocation and Performance Measurement. Chichester
8. Merton RC, Perold AF (1993) Theory of risk capital in financial firms. Journal of Applied Corporate Finance 6: 16–32
9. Ridder T (1999) Konsistente VaR-Limitsysteme. SGZ-Bank, Frankfurt a. M.
10. Saita F (1999) Allocation of Risk Capital in Financial Institutions. Financial Management 28: 95–111

Process Optimization via Conventional Factorial Designs and Simulated Annealing on the Path of Steepest Ascent for a CSTR

Pongchanun Luangpaiboon[1]

Department of Industrial Engineering, Faculty of Engineering, Thammasat University (Rangsit Campus), KlongLuang, Pathumthani, 12121, THAILAND

Abstract This work determines the efficiency of sequential algorithms for automatic optimization of a chemical process. A method of steepest ascent and an integrated approach between the method of steepest ascent and Simulated Annealing, are compared on a simulated continuous stirred tank reactor (CSTR) with various levels of signal noise. The results suggest that the method of steepest ascent seems to be the most efficient on the CSTR surface at the lower levels of noise. However, the integrated approach with the Simulated Annealing element works well when the standard deviation of the noise is at higher levels. Although the average, the standard deviation of the greatest actual concentration of the product and percentage of sequences ended at the optimum from the integrated algorithm are better, it needs more runs, on average, to converge to the optimum when compared.

1 Introduction

The objective of Response Surface Methodology is to describe how the response of a process varies with changes in k predictor variables (Myers and Montgomery 1995). The predictor variables determined will depend on the specific field of the application. Most industrial processes have some predictor variables. These predictor variables can be adjusted by plant operators or by automatic control mechanisms to enhance the efficiency of the machine.

There is much current interest in optimization methods with the stochastic element, such as Genetic Algorithms (GA) and Simulated Annealing (SA). The genetic algorithm is introduced for finding the global maximum on a hypersurface (Jennison et al. 1995). The genetic algorithm is a set of rules for searching large solution spaces in a manner similar to natural selection in biological evolution (Holland 1975; Goldberg 1989). Simulated Annealing has been used in an interesting analogy between problems in statistical mechanics and optimization. Its properties expose useful information and overcome the large and noisy systems. A recent study (Luangpaiboon et al. 2000) compared a modified simplex method and a genetic algorithm for a variety of response surfaces and levels of measurement

[1] lpongch@engr.tu.ac.th

noise. The GA appears to work well in the area of the RSM. However, high variability of the GA when applied to on-line optimization could be a serious disadvantage (Luangpaiboon 2000).

The objective of this study is to compare the efficiency of sequential algorithms for on-line optimization of a chemical process in the presence of noises. The method of steepest ascent and the integrated approach between the method of steepest ascent and Simulated Annealing are selected and implemented on the CSTR. The context is maximizing the concentration of a desired product of a chemical reactor with respect to feed rate, concentration and temperature.

2 Related Methods

2.1 Method of Steepest Ascent

The procedure of steepest ascent is that a hyperplane is fitted to the results from the initial 2^k designs. The direction of steepest ascent on the hyperplane is then determined by using principles of least squares and experimental designs. The next run is carried out at a point, which is some fixed distance in this direction, and further runs are carried out by continuing in this direction until no further increase in yield is noted. When the response first decreases another 2^k design is carried out, centered on the preceding design point. A new direction of steepest ascent is estimated from this latest experiment. Provided at least one of the coefficients of the hyperplane is statistically significantly different from zero, the search continues in this direction. Once the first order model is determined to be inadequate, the area of optimum is identified via a finishing strategy (Luangpaiboon 2001).

2.2 Simulated Annealing

This algorithm is a set of rules for searching large solution spaces in a manner that mimics the annealing process of metals (Kirkpatrick et al. 1983). The algorithm simulates the behavior of an ensemble of atoms in equilibrium at a given finite temperature (Bohachevsky et al. 1986). In case of maximization the procedures of this algorithm start at a corresponding initial value of the objective function. The new objective value will be then determined. The new solution will be unconditionally accepted if its objective value is improved and the process regularly continues. However, the stochastic element occasionally allows the algorithm to accept the new solution to the problems, which deteriorate rather than improve the objective function value. Simulated Annealing also includes a number of parameters and they have been claimed that affect the efficiency of the algorithm (Luangpaiboon 1995).

3 Continuous Stirred Tank Reactor (CSTR)

For the CSTR a stream rich in chemical A of feed concentration $C_{A(in)}$ is flowing into a reactor at a feed flow rate of $F_{(in)}$, and a feed temperature of $T_{(in)}$. The reaction in the CSTR is an irreversible, first order exothermic reaction. The proportion of chemical A is converted to a desired product B, which, in turn, at high temperature undergoes further reaction and is decomposed to form an undesired by-product C (Fig.3.1). The stated objective is to explore the operating conditions corresponding to higher concentration of product.

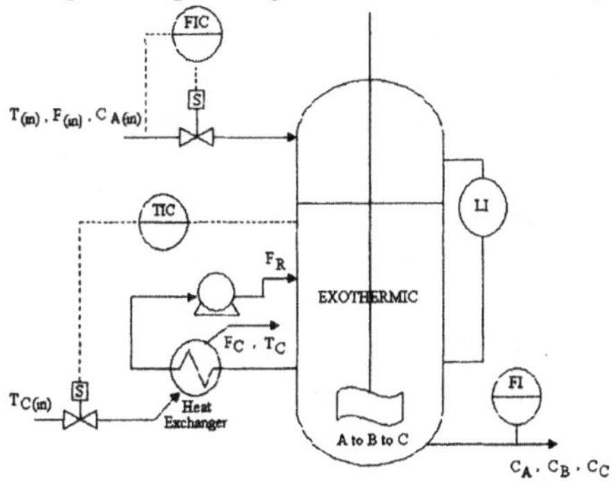

Fig. 3.1. The continuous stirred tank reactor

It is also assumed that level is perfectly controlled, so the volume of material in the tank is constant. This implies that the flow out equals the flow in. The temperature in the reactor may be regulated by manipulating the flow rate of the cooling water (F_C) in the heat exchanger. There are three predictor variables, which can be set to any chosen values within safe limits. These predictor variables relate to the feed flow are shown in Table 3.1. The response variable of the process is defined to be the concentration of the desired product B, C_B.

Table 3.1. Predictor variables of feed flow and their safe limits

Predictor variables	Description	Unit	Feasible Region
$T_{(in)}$	Feed temperature of reactant A	Celsius	60-100
$F_{(in)}$	Feed flow rate of reactant A	Liter/minute	1-10
$C_{A(in)}$	Concentration of reactant A	Mole/liter	1-15

4 Details of the Proposed Method

Parameters: the volume of the factorial design [8]; the step length [1]; the significance level for tests of significance of slopes [10%]; g [−1]; β [6.5].
Step 1: Perform a 2^3 design at a random centre point.
Step 2: Fit a regression plane to the data.
Step 3: Test whether there is evidence that either β_1, β_2 or β_3 is different from zero at the 10% level of significance.
Step 4a: If the result is significant, move one step along the path of steepest ascent and determine the yield and go to *Step 5a*. Otherwise go to *Step 4b*.
Step 4b: Test whether there is evidence that the interaction or curvature check is significant. If the check is significant, go to *Step 6*. Otherwise, replicate the design and return to *Step 2*.
Step 5a: If the yield is greater than the previous yield or the stochastic element meets the requirement of acceptance, continue by moving another step in the same direction.
Step 5b: If the yield is not greater than the previous one (y_0), then calculate the objective increment (Δy) and test the element as follows: Randomly generate a random variable, x, ~ Uniform (0, 1). If $x < P(\Delta y) \equiv EXP(\beta y_0^g \Delta y)$ then go to *Step 5a*. Otherwise return to the preceding point then carry out another 2^3 design and return to *Step 2*. If the first step leads to a yield less than the yields obtained in the preceding 2^3 designs then replicate the design and go to *Step 2*.
Step 6: Implement the finishing strategy. This is a central composite design (CCD) centred on the point ($T_{(in)p}$, $F_{(in)p}$, $C_{A(in)p}$), and fit then a quadratic surface to find the maximum ($T_{(in)p}$, $F_{(in)p}$, $C_{A(in)p}$). If ($T_{(in)p}$, $F_{(in)p}$, $C_{A(in)p}$) is within the volume of the designs, then ($T_{(in)p}$, $F_{(in)p}$, $C_{A(in)p}$) is taken as the optimum operating condition. If ($T_{(in)p}$, $F_{(in)p}$, $C_{A(in)p}$) is not within this volume, another CCD is carried out, centered on the point from the first CCD with greatest yield. A quadratic surface is now fitted to all the data. If the maximum is outside the volume of the union of the two containing cubes, the ridge is searched for the greatest value of the function, using a step length of 0.05 (from additional experiments by using fewer runs).

5 Experimental Results and Discussion

The comparison is made with the measurement noise on the concentration of the desired product B (normal and independent with zero mean and standard deviation of 0.5, 1, 2 and 3). The typical three-dimensional response surfaces, with $C_{A(in)}$ fixed at 1 and 15, are shown in Fig 5.1. There are four performance measures over 100 realizations in this study. The first and second measures are an average and a standard deviation of greatest actual concentration of the desired product B from the finishing strategy respectively. The third is an average number of runs until the algorithms converge. Finally it is the percentage of sequences ended at the optimum.

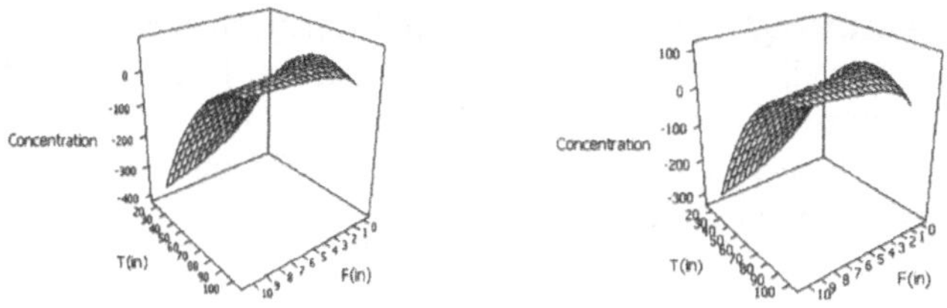

Fig. 5.1. The surface plot with $C_{A(in)}$ fixed at 1 and 15 respectively

The process settings for all the scenarios are given in Table 5.1. The performance of the method of steepest ascent and the integrated approach can be explained by the box plots in Fig. 5.2 when the error standard deviation was 2.0 and 3.0. Note that since the efficiency of these algorithms is related to their initial points, it would be helpful to set random starting points for all algorithms. These results show that the performance of the integrated approach under the stochastic element of Simulated Annealing seems superior to the algorithm based on the method of steepest ascent at the higher levels of error standard deviations.

Table 5.1. Four achievements over 100 realisations

Stdev of Noise	Algorithm	Average Greatest Actual Concentration	Stdev. of Greatest Actual Concentration	Average Number of Runs	Percentage (ending at optimum)
0.5	Steepest Ascent	59.2011	7.6823	33.3	0.85
	Integrated	56.7053	10.2178	38.17	0.85
1	Steepest Ascent	57.4528	7.3680	33.75	0.90
	Integrated	57.0683	6.8899	34.9	0.90
2	Steepest Ascent	59.3067	9.5568	32.6	0.85
	Integrated	61.1069	7.7166	34.75	0.90
3	Steepest Ascent	60.0803	7.7723	31.35	0.80
	Integrated	61.2676	6.3431	33.3	0.95

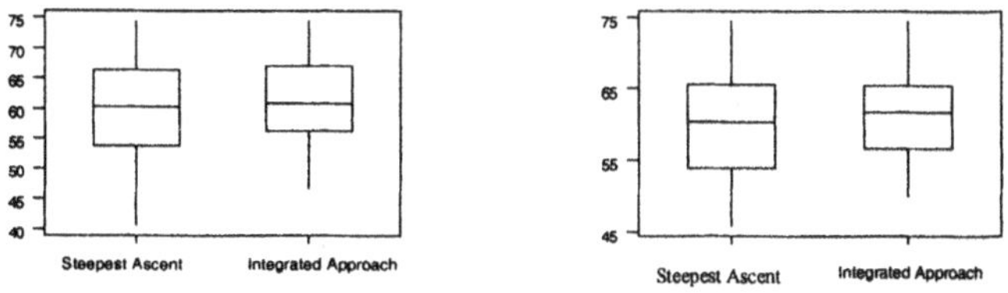

Fig. 5.2. Two independent box plot comparisons showing the performance of the method of steepest ascent and the integrated approach when the error standard deviation was 2.0 and 3.0 respectively.

Moreover, percentage of sequences ended at the optimum or near optimum of radius equalling two from the integrated approach is better at higher levels of error standard deviation although the greater number of runs were required to converge to the optimum. As stated earlier, the function in this research was restricted to three predictor variables. Consequently, comparisons and conclusions between the two algorithms may not be valid for other families of functions Other stochastic approaches could be extended to the method based on conventional factorial designs to increase its performance, especially in terms of speed of convergence, when the error standard deviation is at higher levels.

6 Acknowledgements

The author wishes to thank the Faculty of Engineering, Thammasat University, THAILAND for the financial support. I gratefully acknowledge the computing assistance of Pawabutra A and Kansompod S in the early phase of this research.

References

Bohachevsky IO, Johnson ME, Stein ML (1986) Generalized simulated annealing for function optimization. Technometrics: 209-217

Goldberg DE (1989) Genetic algorithms in search optimization and machine learning. Addison-Wesley

Holland JH (1975) Adaptation in natural and artificial systems. Ann Arbor, The University of Michigan Press

Jennison C, Franconi L, Sheehan N (1995) Stochastic optimization: simulated annealing and the genetic algorithm. Institute of Mathematics and its Applications Conference Series 54: 209-213

Kirkpatrick S, Gelatt CD, Vecchi MP (1983) Optimization by simulated annealing. Science 220: 671-680

Luangpaiboon P (1995) Dynamic process layout planning. Published Master of Engineering Thesis, Kasetsart Uiversity, Thailand

Luangpaiboon P (2000) A comparison of algorithms for automatic process optimization. Published Doctor of Philosophy Dissertation, University of Newcastle upon Tyne, UK

Luangpaiboon P (2001) Proposed finishing strategies based on experimental designs for process optimization. Thammasat International Journal of Science and Technology: 39-45

Luangpaiboon P, Metcalfe AV, Rowlands RJ, Tham MT, Willis MJ (2000) Comparison of a modified simplex and a genetic algorithm for optimizing a chemical process. Proceedings of the 1st International Conference on Industrial Statistics in Action 2000, Newcastle upon Tyne, UK

Myers RH, Montgomery DC (1995) Response surface methodology: process and product optimization using designed experiments. John Wiley & Sons, Inc.

Optimization on Directionally Convex Sets

Vladimir Naidenko

Institute of Mathematics, National Academy of Sciences of Belarus
11 Surganov str., Minsk, 220072, Belarus
E-mail: naidenko@im.bas-net.by

Abstract. Directional convexity generalizes the concept of classical convexity. We investigate OC-convexity generated by the intersections of C-semispaces that efficiently approximates directional convexity. We consider the following optimization problem in case of the direction set of OC-convexity being infinite. Given a compact OC-convex set A, maximize a linear form L subject to A. We prove that there exists an OC-extreme solution of the problem. A Krein-Milman type theorem has been proved for OC-convexity. We show that the OC-convex hull of a finite point set represents the union of a finite set of polytopes in case of the direction set being finite.

1 INTRODUCTION

In the paper, we consider directional convexity that generalizes the concept of classical convexity. It arises in many geometrical problems such as image processing, databases, VLSI design, etc. [1–4].

Let O be a set of vectors in R^n. We assume that the set O is symmetric, i.e., $O = -O$, which is not a restriction in the context of directional convexity. For $a, b \in R^n$, we write $[a, b]$ for the segment with endpoints a,b, i.e., $[a, b] = \{\alpha a + (1 - \alpha)b \mid \alpha \in [0, 1]\}$.

Definition 1. A set $A \subseteq R^n$ is called a *O-convex* if for any two points $x_1, x_2 \in A$ such that the segment $[x_1, x_2]$ is parallel to some nonzero vector of O, we have $[x_1, x_2] \subseteq A$ (O-convexity is also called directional convexity or restricted-orientation convexity in the literature).

The main object of our investigation is a suitable notion of an "O-convex hull" of a set. One can define the O-convex hull of a set A as the intersection of all O-convex sets (according to definition 1) containing A; this O-convex hull will be denoted by $\mathrm{conv}^o[A]$.

We shall concentrate on another kind of O-convex hull, namely one defined by means of C-semispaces. By a convex cone we mean a set C such that for any $x, y \in C$ it contains $\lambda x + \mu y$, $\lambda \geq 0, \mu \geq 0$. A convex cone C is called *acute* if C does not contain any one-dimensional linear subspace. A set $M \subset O$ is called a half-set if $M \cup -M = O$ and $M \cap -M = \emptyset$. By $\mathrm{CH}[X]$ we denote the *convex conic hull* of a set $X \subseteq R^n$, i.e., the smallest convex cone containing the set X.

Definition 2. Let C be an acute cone that is maximal among all acute cones containing O. Then the set $S_a^c = R^n \setminus (a + C)$ is called a *C-semispace* of directional convexity in $a \in R^n$.

Let us introduce the notion of OC-convexity. The class OC of subsets from R^n yields the OC-convexity if OC contains only the following sets: 1) R^n, 2) all C-semispaces, 3) the intersections of arbitrary C-semispaces.

Definition 3. Let $A \subseteq R^n$. The set $\mathrm{conv}^{oc}[A]$, called the *OC-convex hull* of A, is defined as the intersection of all C-semispaces containing A.

It is easy to check that the O-convex hull is always contained in the OC-convex hull; also, if O contains all vectors (i.e., for the usual convexity), both these hulls of A coincide. Hence the OC-convexity is an approximation of the O-convexity. Many problems, difficult in case of the O-convexity, become easy in case of the OC-convexity. A point $a \in A$ is called OC-extreme for A, if $A \setminus \{a\}$ is OC-convex. The set of all OC-extreme points for A is denoted by $\mathrm{ext}^{oc}[A]$. Then the OC-convex hull $\mathrm{conv}^{oc}[A]$ can be determined through the OC-extreme points as follows. For any $x \in R^n \setminus A$: $x \in \mathrm{conv}^{oc}[A]$ iff $x \notin \mathrm{ext}^{oc}[A \cup \{x\}]$. By (x, y) we denote the scalar product of vectors $x, y \in R^n$.

2 MAIN RESULTS

We shall use the following properties of OC-convexity [2–4].

- All classically convex sets are OC-convex.
- Every OC-semispace at an arbitrary point $x \in R^n$ represents a C-semispace O-convexity at the same point x.
- Let $M \subset O$ be a half-set. If $\mathrm{CH}[M]$ is an acute convex cone then $R^n \setminus (\mathrm{CH}[M] + x)$ is an OC-semispace at an arbitrary point x.
- For every OC-semispace $S_x^{oc} = R^n \setminus (C + x)$, there exists a half-set $M \subseteq C$ such that $R^n \setminus (\mathrm{CH}[M] + x) = S_x^{oc}$.

The following theorem holds.

Theorem 1. *Let $R^n \setminus (C + a)$ is an OC-semispace at an arbitrary point a. Then, for any OC-convex compact A, there exists a point $x \in A$ such that the OC-semispace $R^n \setminus (C + a)$ includes the whole set A except the point x.*

Proof. The theorem is proved by induction on the dimension of linear space R^n. The theorem is obvious for R^1. Suppose that it holds for R^{n-1}. Let us prove the theorem for R^n. Let $H = \{x \in R^n \mid a \cdot x = 0\}$ be a hyperplane of support to C, and let the halfspace $H^+ = \{x \in R^n \mid a \cdot x \geq 0\}$ include the whole set C. Let us consider the following optimization problem in the variable $x \in R^n$:

$$(a, x) \to \max \\ x \in A. \tag{1}$$

Since the set A is compact, the function (a, x) of the problem (1) attains the maximum value at some point $x_0 \in A$ by the Weierstrass theorem. We denote the hyperplane $\{x \in R^n \mid a \cdot x = a \cdot x_0\}$ by H_0. If $H \cap O = \emptyset$ then any point $x \in H_0 \cap A$ obviously proves the theorem in the case. Suppose that $H \cap O \neq \emptyset$. The hyperplane H_0 can be described as an affine (n-1)-dimensional subspace. Let us associate a linear space (n-1)-dimensional space with the affine subspace as follows. Fix a point h_0 in H_0 as the origin of coordinates. Then every point $x \in H_0$ corresponds to a vector $v = \overrightarrow{h_0 x}$. Let O' be a set of vectors in H_0, such that each vector of O' is collinear to some vector of $H \cap O$ and vice versa. In the space H_0, the point set $H_0 \setminus (C + h_0)$ represents an OC-semispace at the point h_0 for the direction set O' of directional convexity. Since the space H_0 can be identified with R^{n-1}, there exists a point $a_0 \in H_0 \cap A$ such that $(H_0 \setminus (C + a_0)) \cap (H_0 \cap A) = (H_0 \cap A) \setminus \{a_0\}$ by the induction hypothesis. Hence $(C + a_0) \cap H_0 \cap A = \{a_0\}$. Note that $(C + a_0) \cap A \subseteq H_0$. Therefore $(C + a_0) \cap A = \{a_0\}$, i.e. $(R^n \setminus (C + a_0)) \cap A = A \setminus \{a_0\}$. Thus, the OC-semispace $R^n \setminus (C + a_0)$ in the space R^n proves the theorem in the case. This concludes the proof.

Let A be a compact OC-convex set in R^n, and O be infinite. Let us consider the following optimization problem

$$(a, x) \to \max \\ x \in A, \tag{2}$$

where a is fixed and x is a variable vector.

As shown in [5], in case of O being finite, there exists at least one solution of (2), which is an OC-extreme point of A. We shall show that an analogous statement holds in the infinite case as well.

The following theorem holds.

Theorem 2. *There exists an OC-extreme point of A, which is a solution to problem (2).*

Proof. Let us consider the hyperplane $H = \{x \in R^n \mid a \cdot x = 0\}$ and the halfspace $H^+ = \{x \in R^n \mid a \cdot x \geq 0\}$ bounded by H. Note that there exists a classically convex semispace S_0 at the point $\mathbf{0}$, such that H is a hyperplane of support to S_0 and $R^n \setminus S_0 \subset H^+$, according to the properties of semispaces of classical convexity. Either $v \in R^n \setminus S_0$ or $-v \in R^n \setminus S_0$ for any nonzero vector $v \in R^n$ (but not simultaneously). Then $R^n \setminus S_0$ includes some vector half-set M of O. Since the cone $R^n \setminus S_0$ is acute, the cone CH[M] is acute as well. Hence $R^n \setminus$ CH[M] represents an OC-semispace at the point $\mathbf{0}$. Note that H is a hyperplane of support to CH[M], and the halfspace H^+ includes CH[M]. Thus, using the proof of theorem 1, one can find an OC-extreme point for A, which is a solution to problem (2). This concludes the proof.

For the OC-convex sets we have proved a Krein-Millman type theorem.

Theorem 3. *Let O be finite. If $A \subset R^n$ is a compact set then* $\mathrm{conv}^{oc}[A] = \mathrm{conv}^{oc}[\mathrm{ext}^{oc}[A]]$.

Proof. Suppose there exists a compact set $A \subset R^n$ such that $\mathrm{conv}^{oc}[A] \neq \mathrm{conv}^{oc}[\mathrm{ext}^{oc}[A]]$. Then there exists a point $x \in A$ such that the OC-convex hull $\mathrm{conv}^{oc}[\mathrm{ext}^{oc}[A]]$ does not contain x. Hence $x \in \mathrm{ext}^{oc}[\mathrm{ext}^{oc}[A] \cup \{x\}]$. Moreover, one can find an OC-semispace $S_x^{oc} = R^n \setminus (C + x)$ such that $\mathrm{ext}^{oc}[A] \subseteq S_x^{oc}$, i.e. $(C + x) \cap \mathrm{ext}^{oc}[A] = \emptyset$. Since O is finite, $(C + x)$ is a closed polyhedral set with the corner point x. Hence, for some constants $a \in R^n$ and $\alpha \in R$, there exists a closed halfspace $H = \{v \in R^n \mid (a, v) \geq \alpha\}$ such that $(C + x) \subset H$ and $\{v \in R^n \mid (a, v) \geq \alpha\} \cap (C + x) = \{x\}$. In other words, there exists a hyperplane $\{v \in R^n \mid (a, v) = \alpha\}$ which intersects with $(C + x)$ only at the corner point x. Let us consider the following problem:

$$\begin{array}{l}(a, v) \to \max \\ v \in (C + x) \cap A\end{array} \tag{3}$$

Since $(C+x)$ is closed, the set $(C+x) \cap A$ is compact. Hence the function (a, v) attains the maximum value on $(C+x) \cap A$ by the Weierstrass theorem. Therefore, problem (3) has at least one solution. Fix a point v^* such that (a, v^*) is the maximum of (a, v). Consider the set $(C + v^*)$. Since H contains the whole set $C + x$, the halfspace $H^* = \{v \in R^n \mid (a, v) \geq (a, v^*)\}$ contains the whole set $(C + v^*)$. On the other hand, $\{v \in R^n \mid (a, v) \geq (a, v^*)\} \cap (C + v^*) = \{v^*\}$. Since $v \in (C + x) \cap A$ implies $(a, v) \leq (a, v^*)$, then $((C + x) \cap A) \cap (C + v^*) = \{v^*\}$. The statement $A \cap (C + v^*) = \{v^*\}$ follows from $(C + v^*) \subseteq (C + x)$, i.e. $v^* \in \mathrm{ext}^{oc}[A]$. It contradicts the conjecture $(C + x) \cap \mathrm{ext}^{oc}[A] = \emptyset$. This concludes the proof.

In what follows, we shall show that the OC-convex hull $\mathrm{conv}^{oc}[A]$ represents the union of a finite set of polytopes in case of both the sets A and O being finite. Let us derive a formula describing $\mathrm{conv}^{oc}[A]$ in that case. Let $A = \{a_1, ..., a_k\}$. We set an enumeration of OC-semispaces $R^n \setminus (C_1 + a), ..., R^n \setminus (C_m + a)$ at an arbitrary point $a \in R^n \setminus A$. Note that the point a belongs to $\mathrm{conv}^{oc}[A]$ iff $a \notin \mathrm{ext}^{oc}[\{a_1, ..., a_k\} \cup \{a\}]$. This means that

$$(C_i + a) \cap A \neq \emptyset \quad (\forall i \in \overline{1, m}) \tag{4}$$

Any polyhedral set $(C_i + a)$ can be described in terms of all solutions to the system of linear equations $\{x \in R^n \mid A^{(i)}x \geq A^{(i)}a\}$, where $A^{(i)}$ is a matrix corresponding to C_i. Then, using relation (4), one can derive the following formula:

$$\mathrm{conv}^{oc}[\{a_1, ..., a_k\}] = \bigcap_{i=1}^{m} (\bigcup_{j=1}^{k} \{a \in R^n \mid A^{(i)}a_j \geq A^{(i)}a\}) \tag{5}$$

Since any classically convex set is OC-convex, the classically convex hull $\text{conv}[\{a_1, ..., a_k\}]$ includes $\text{conv}^{oc}[\{a_1, ..., a_k\}]$. Moreover, since the classically convex hull of any finite point set is bounded, $\text{conv}^{oc}[\{a_1, ..., a_k\}]$ is bounded as well. Using formula (5), we obtain that $\text{conv}^{oc}[\{a_1, ..., a_k\}]$ represents the union of a finite set of polytopes.

We shall consider the following problem of finding $\text{ext}^{oc}[A]$. Given a finite A, n and O, does an arbitrary point a of R^n belong to $\text{ext}^{oc}[A]$? The following theorem holds.

Theorem 4. *The decision problem whether or not an arbitrary point belongs to the OC-extreme point set, is NP-complete.*

Proof. We shall show how to reduce SAT to the problem. Let $F = \bigwedge_{i=1}^{p} C_i$ be a CNF where every C_i is a disjunction of literals from $\{x_1, \bar{x}_1, ..., x_n, \bar{x}_n\}$. Suppose that $O = \{e_1, -e_1, ..., e_n, -e_n\}$ where the vectors $e_1, ..., e_n$ yield a basis of R^n. Let A be $\{a_i \mid i \in \overline{0,p}\}$ where a_0 is the origin of coordinates, $a_i = \alpha_1^{(i)} e_1 + ... + \alpha_n^{(i)} e_n$, $\forall i \in \overline{1,p}$. The values $\alpha_1^{(i)}, ..., \alpha_n^{(i)}$ is determined as follows: if x_j lies in C_i then $\alpha_j^{(i)} = -1$; if $-x_j$ lies in C_i then $\alpha_j^{(i)} = 1$; if both the literals x_j and $-x_j$ don't lie in C_i then $\alpha_j^{(i)} = 0$. We shall show that the point a_0 belongs to $\text{ext}^{oc}[A]$ iff the CNF is satisfied. Every valuation of $x_1, ..., x_n$ bijectively corresponds to some set $M(x_1, ..., x_n) = \{e_1(x_1), ..., e_n(x_n)\}$ where $e_j(x_j) = e_j$ if $x_j = 1$, and $e_j(x_j) = -e_j$ if $x_j = -1$ for any $j \in \overline{1,n}$. Note that any C-semispace at a_0 for the set O can be determined as an appropriate set $R^n \setminus \text{CH}[M(x_1, ..., x_n)]$ with the values of $x_1, ..., x_n$ fixed in some way. Suppose we can find values of $x_1, ..., x_n$ such that $a_i \notin \text{CH}[M(x_1, ..., x_n)]$, where $i \geq 1$. Then there exists $\alpha_j^{(i)} \neq 0$ such that $\alpha_j^{(i)} = -1$ if $e_j(x_j) = e_j$, and $\alpha_j^{(i)} = 1$ if $e_j(x_j) = -e_j$. Hence C_i is satisfied with the values of $x_1, ..., x_n$. In other words, if there exists a valuation of $x_1, ..., x_n$ such that $A \setminus \{a_0\} \subseteq R^n \setminus \text{CH}[M(x_1, ..., x_n)]$, then $F = \bigwedge_{i=1}^{p} C_i$ is satisfied with the values of $x_1, ..., x_n$. The relation $A \setminus \{a_0\} \subseteq R^n \setminus \text{CH}[M(x_1, ..., x_n)]$ is equivalent to $a_0 \in \text{ext}^{oc}[A]$. This concludes the polynomial-time reduction of SAT to the problem of finding $a_0 \in \text{ext}^{oc}[A]$. Now we shall show that the problem lies in *NP*. Note that the relation $a_0 \in \text{ext}^{oc}[A]$ is equivalent to $(\exists M \subset O)(\forall y \in A \setminus \{a_0\})[y \in R^n \setminus \text{CH}[M], \quad M \cup -M = O, \ M \cap -M = \emptyset]$. It can be checked in polynomial time for any $M \subset O$. This concludes the proof.

Corollary 1. *It is NP-hard to verify the membership of an arbitrary point in the OC-convex hull of a finite point set.*

References

1. Wood, D. (1985) Computational Geometry /ed. G.T. Toussant. North-Holland, Amsterdam

2. Metelskii, N.N., Martynchik, V.N. (1996) Directional convexity. Matematicheskie zametki. **60**, 406–413 (in Russian)
3. Metelskii, N.N., Naidenko, V.G. (1999) Directionally convex hulls in R^n. Vesti Natsionalnoi Akademii nauk Belarusi. Seriya fiz.-mat. nauk. **4**, 39–42 (in Russian)
4. Metelskii, N.N., Naidenko, V.G. (2000) On a class of directionally convex semispaces. Vesti Natsionalnoi akademii nauk Belarusi. Seriya fiz.-mat. nauk. **1**, 56–59 (in Russian)
5. Naidenko V.G. (2000) Directionally convex set and its application to an optimization problem. Proc. Int. WorkshopDiscrete Optimization Methods in Scheduling and Computer-Aided Design, Minsk, Republic of Belarus, September 5-6. Institute of Engineering Cybernetics, Minsk, 159–161.

Meta-Heuristiken in virtuellen Lernumgebungen

Torsten Reiners, Imke Sassen und Stefan Voß

Technische Universität Braunschweig, Institut für Wirtschaftswissenschaften, Abteilung ABWL, Wirtschaftsinformatik und Informationsmanagement, Abt-Jerusalem-Str. 7, D-38106 Braunschweig, {t.reiners, i.sassen, stefan.voss}@tu-bs.de.

Zusammenfassung. In der Lehre gewinnt der Einsatz von virtuellen Lernumgebungen an Bedeutung, wobei für das Operations Research aufgrund der Anforderungen an eine praxisorientierte Ausbildung ein erhöhter Bedarf an qualitativ hochwertigen Lernangeboten besteht. Anhand von konkreten Lerninhalten soll aufgezeigt werden, auf welche Weise eine Verknüpfung der theoretischen Grundlagen mit praktischen Anwendungen hergestellt werden kann. Hierzu erarbeiten die Lernenden grundlegende Begrifflichkeiten und Zusammenhänge, die in einem weiteren Schritt anhand von realen Problemstellungen aus einer neuen realitätsbezogenen Perspektive betrachtet und dadurch vertieft und gefestigt werden.

Einleitung

Angesichts einer steigenden Anzahl von Studierenden, verringerten finanziellen Ressourcen der Ausbildungsinstitutionen und darüber hinaus der Forderung nach der Bereitstellung von Angeboten zum lebenslangen Lernen, bieten virtuelle Universitäten weitere Möglichkeiten zur Förderung der Qualität in der Lehre und werden aus diesem Grund von zahlreichen Ländern in ihr Bildungskonzept integriert (siehe Hazemi et al. (1998)). In diesem Beitrag soll der Bereich Operations Research in Bezug auf die Darstellung und Präsentation in virtuellen Lernumgebungen (VLU) am Beispiel der Meta-Heuristiken näher betrachtet werden. Da gerade im Operations Research eine praxisorientierte Lehre zur Aneignung von Kenntnissen und Fähigkeiten in der Modellierung und Lösung von (mathematischen) Problemstellungen, Algorithmen, Softwareentwicklung sowie Projektmanagement notwendig ist, sind die Entwicklungen der Forschung hinsichtlich neuer didaktischer Methoden und virtueller Lernumgebungen zu verfolgen und umzusetzen. Insbesondere Meta-Heuristiken, als zeitgemäße und wichtige Werkzeuge zur Optimierung, haben in der Lehre sowohl in Lehrbüchern als auch in praktischen Umsetzungen ein Schattendasein und sind nur mangelhaft vertreten, so dass traditionelle Präsenzveranstaltungen im Bereich OR/MS sich nur unzureichend mit Meta-Heuristiken auseinandersetzen. Daraus und auch aufgrund fehlender guter Beispiele aus der Praxis ergibt sich eine unzureichende Vermittlung der wesentlichen Vorteile von Meta-Heuristiken gegenüber anderen Methoden der Optimierung, wobei hier bekannte Algorithmen wie z.B. die genetischen Algorithmen ausgenommen sind.

Meta-Heuristiken besitzen in vielerlei Hinsicht Analogien zu weiteren Gebieten des Operations Research. Beispielsweise operiert der Simplex-Algorithmus (siehe u.a. Domschke und Drexl (2002)) zur Lösung von linearen Optimierungsproblemen durch das gezielte Traversieren eines den Lösungsraum beschreibenden Polyeders. Analog gehen z.B. Verfahren zur Lösung von Transportproblemen vor, bei denen im Anschluss an eine Eröffnungsheuristik zur Ermittlung einer zulässigen Lösung Verbesserungsverfahren wie die MODI-Methode (siehe z.B. Domschke (1995)) auf stark strukturierten Lösungsräumen zum Einsatz kommen. Für die Studierenden ergeben sich durch die Behandlung von Meta-Heuristiken aufgrund von Analogien der Prinzipien zur Lösung von Optimierungsverfahren aus fachdidaktischer Sicht weitreichende Transfer- und Anwendungsmöglichkeiten. Daher bietet sich eine inhaltliche Behandlung der lokalen Suche sowie Meta-Heuristiken im Rahmen der Lehre an. In diesem Beitrag stellen wir einen einführenden Überblick zu unserem Konzept vor, für eine ausführlichere Darstellung siehe Reiners und Voß (2002).

Neben dem einfachen Angebot von Lehrmaterialien ohne eine weitere mediendidaktische Aufbereitung existieren bereits Lernumgebungen im Internet, die weiterführende Präsentationsformen als Hypertexte verwenden. Ein Beispiel für Offline-Software ist OR Welt (2002); frei nutzbare Lernumgebungen sind tutOR von Sniedovich und Byrne (2002) und darauf basierend tutORial von der IFORS (2002). Kommerzielle Systeme, im Wesentlichen zur Unterstützung der Lehre und der Verwaltung von Kursmaterialien, sind u.a. WebCT (2002). In einigen Ländern wie den USA oder Australien ist es auch üblich, anstelle web-basierter Lernumgebungen urheberrechtlich geschützte Software wie Microsoft Excel im Unterricht zu verwenden; siehe z.B. Bell (2000). Gemein ist den Ansätzen, dass das Angebot im Bereich Meta-Heuristiken zurzeit höchstens auf statische untereinander verlinkte Texte oder auf unstrukturierte Sammlungen interaktiver Applets ohne weitere theoretische Ausführungen begrenzt ist. Weitere Darstellungsformen von Optimierungsergebnissen sind z.B. bei Jones (1995) zu finden.

Darüber hinaus fehlt in aktuellen Ansätzen eine ausreichende Integration von Problemstellungen aus der Praxis unter Einbindung interaktiver Darstellungsmöglichkeiten; siehe Wolsey (1979) oder Reisman (1997): "This profession currently has more algorithms than applications".

Aufbau der virtuellen Lernumgebung

Für die interne Repräsentation eines Kurses über Meta-Heuristiken soll ein weiterführendes Konzept für eine VLU verwendet werden; insbesondere die Unterteilung in semantische Module unterschiedlicher Größe soll eine dynamische Darstellung bezüglich der Konfiguration durch Studierende ermöglichen, um Bedürfnisse der Lernenden zu erfüllen und dadurch insgesamt eine Motivationssteigerung zu erzielen. Der Einsatz weiterer Komponenten soll der Schwierigkeit Rechnung tragen, im Bereich Operations Research dauerhaft Interesse und Motivation der Lernenden zu erzielen und die Teilnahme an web-basierten Kursen zu steigern, wobei keine Unterscheidung zwischen der Art der Lernenden (z.B. „Distance-Lerner“ oder „Lifelong-Lerner“) gemacht werden soll. Hierzu erfolgt die Einbindung einer relationalen Datenbank zur Verlinkung der Lernobjekte unter

Ausnutzung der dort gespeicherten Meta-Informationen. Dies erlaubt eine Adaption an Studierende und ihre Bedürfnisse hinsichtlich Lerngewohnheiten sowie den Einbezug von verschiedenen Kommunikationsformen. Eine detaillierte Beschreibung der konzipierten Architektur ist in Reiners et al. (2002a), ein Aufbau der Lehrmaterialien in Reiners et al. (2002b) gegeben.

Das dargestellte Lehrangebot besitzt den Anspruch, dass Lernende ohne Vorkenntnisse später eigenständig Meta-Heuristiken zum Lösen komplexer praktischer Problemstellungen einsetzen können. Hierzu werden grundlegende Prinzipien in einer vorgelagerten Kurseinheit durch interaktive Beispiele vermittelt. Eine Meta-Heuristik ist definiert als ein iterativer Generationsprozess, der eine untergeordnete Heuristik steuert und eine Methode darstellt, ein Optimierungsproblem in adäquater Zeit zu lösen. Jedes Optimierungsproblem besitzt einen speziellen Lösungsraum, der hinsichtlich der Restriktionen alle zulässigen (und ggf. darüber hinausgehende nicht zulässige Lösungen, wie sie sich z.B. vermöge geeigneter Problemrelaxationen ergeben) enthält. Der Lösungsraum ist jedoch in den meisten Fällen zu groß, um die optimale Lösung durch eine vollständige Enumeration zu finden bzw. eine derartige Vorgehensweise in der vorgegebenen Zeit zu realisieren. Daher grenzen Meta-Heuristiken durch die intelligente Kombination verschiedener Methoden den Lösungsraum auf bestimmte Bereiche ein und verwenden Lernstrategien, um Informationen zu strukturieren und effektiv optimale oder gute Lösungen zu finden, siehe Voß et al. (1999).

Die grundlegenden Komponenten von Meta-Heuristiken werden zur Erklärung von speziellen Algorithmen wie Steepest Descent, Simulated Annealing oder Tabu Search genutzt. Der Lernende wird hierbei durch einfache interaktive Beispiele unterstützt, welche die zugrunde liegenden Prinzipien veranschaulichen und das Verhalten der verschiedenen Methoden verdeutlichen. Abbildung 1 zeigt die Schritte einer interaktiven Animation zur Verdeutlichung einer Nachbarschaft. Die Interaktion für die Studierenden liegt z.B. darin, dass der nächste „beste" Nachbar gewählt werden muss, wobei eine Bewertung durch die VLU erfolgt. Über die Berücksichtigung von Analogien lassen sich hier wiederum Kenntnisse z.B. über den Simplex-Algorithmus einbinden (sowie umgekehrt die Vorgehensweise des Simplex-Algorithmus als „einfache" lokale Suche auf einem gut strukturierten Lösungsraum erläutern). Betrachtet man die Vorgehensweise des Bestimmens einer ersten zulässigen Lösung und die daraufhin folgende Anwendung eines Verbesserungsverfahrens, wie sie im Bereich der Meta-Heuristiken i.d.R. angewendet wird, so besitzt dieses gleichermaßen eine Korrespondenz in der so genannten Zweiphasen- oder M-Methode.

Abb. 1 Darstellung einer Nachbarschaft für eine vorgegebene Lösung

Aufbauend auf den grundlegenden Prinzipien sollten Studierende erworbenes Wissen in einer komplexeren und in weiten Teilen realistischen Problemstellung anwenden, um eine Intensivierung der Materie zu erreichen. Hierzu können Problemstellungen von Forschungsprojekten oder aus der Praxis eingesetzt werden. Unterstützt durch die virtuelle Lernumgebung entwickeln die Studierenden eine Lösungsmöglichkeit unter der Verwendung bestehender und konfigurierbarer Softwarepakete. Diese Software kann entweder ein kommerzielles Produkt mit einer graphisch aufbereiteten Benutzeroberfläche zur Eingabe des Problems und mit Parametern für die Lösungsalgorithmen sein, oder aber eine Softwarebibliothek, die z.T. wieder verwendbare Codes enthält, aber dennoch auch einige Programmierkenntnisse voraussetzt. Um eine weitgehenden Integration in unsere virtuelle Lernumgebung ohne Medien- oder Technologiebrüche zu gewährleisten, verwenden wir HOTFRAME (Heuristic OpTimization FRAMEwork) von Fink und Voß (2002). Ein Benutzerinterface ermöglicht die Konfigurierung der wesentlichen Komponenten der Meta-Heuristiken. Außerdem wird ein automatisch generierter Source-Code ausgegeben, mit Todo-Teilen für die Lernenden, die in ihrem Schwierigkeits- und Komplexitätsgrad variabel sind. Schließlich stellt die VLU ein Experimentierfeld zur Verfügung, in dem das erreichte Ergebnis an diversen Problemstellungen mit verschiedenen Parametereinstellungen der Meta-Heuristiken getestet werden kann. Die Resultate werden gesammelt und können von den Lernenden durch den Einsatz statistischer Methoden und Darstellungsmöglichkeiten eigenständig evaluiert werden (siehe auch Abbildung 2).

Ausblick

Auch hochgradig interaktive Lernumgebungen unterstützen zumeist nicht die Idee, dass Studierende erworbenes Wissen im Rahmen eines bestimmten Forschungsfeldes anwenden können, um Fragestellungen der Praxis oder aktuellen Forschung zu untersuchen. Gerade die Erfahrung von Praxisrelevanz und Anwendungsmöglichkeiten des erlernten Wissens kann jedoch hochgradig zu einer Steigerung der Motivation der Lernenden beitragen. Darüber hinaus kann aufgrund der Möglichkeit der eigenständigen Evaluation des erreichten Ergebnisses bei Studierenden der Anspruch entstehen, ihr Ergebnis noch weiter zu verbessern, was ebenfalls zu einer weiteren Auseinandersetzung mit den Lehrinhalten führt. So beinhaltet das Generieren und Testen eigener Lösungen unter lernpsychologischen Gesichtspunkten lernförderliche Momente, da durch den Wechsel von Hypothesengenerierung und -testen das entdeckende und explorierende Lernen unterstützt wird und Lösungen beliebig oft konstruiert werden können. Im Vergleich zu anderen Methoden werden Selbstlernprozesse besonders gefördert, da auch eigenständige Fehlersuche und deren Beseitigung in schwierigen Programmteilen durch die Möglichkeit der Selbstevaluation motiviert werden, aber optional auch Hilfestellungen von Seiten der Lernumgebung gegeben werden können. Gerade bei den Anfängen des Transfers von erlerntem Wissen auf das Lösen komplexer Problemstellungen sollten die Lernenden sich nicht allein gelassen fühlen, sondern im gewünschten Maße Unterstützung in Anspruch nehmen können. Vor allem die Erfahrung, die bereitgestellte Software in richtiger Weise zum Lösen komplexer Problemstellungen eingesetzt zu haben, gibt den Studierenden auf der einen Seite

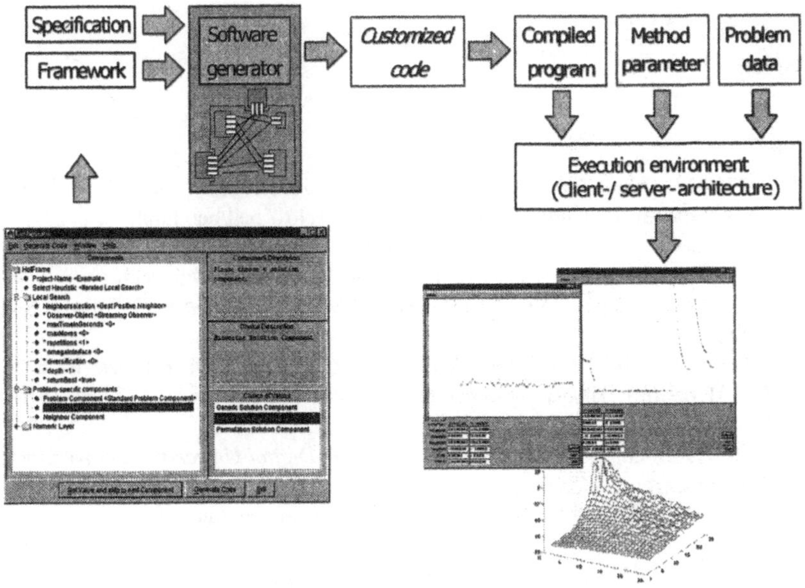

Abb. 2 Anwendungssequenz einer Meta-Heuristik

Aufschluss über ihren Lernerfolg und versichert ihnen auf der anderen Seite auch den Wert des Gelernten, sowie dessen interdisziplinären Verwendungsmöglichkeiten, entgegen dem Vorurteil, dass sich mit Methoden des Operations Research nur in theoretischen Anwendungen gute Ergebnisse erzielen ließen.

Danksagung

Diese Arbeit wurde von dem BMBF (Bundesministerium für Bildung und Forschung) in dem Programm „Neue Medien in der Bildung" (08NM094D) unterstützt.

Literatur

Bell, P.C., 2000. *Teaching Business Statistics with Microsoft Excel.* INFORMS Transactions on Education. 1(1). http://ite.informs.org/Vol1No1/ bell/bell.html.

Björck, U., July 2002. *List of Virtual Universities.* http://www.ped.gu.se/ulric/vus.html.

Domschke, W., 1995. *Logistik: Transport.* 4. Auflage. Oldenbourg, München.

Domschke, W. und A. Drexl, 2002. *Einführung in Operations Research.* 5.Auflage. Springer, Berlin.

Fink, A. und S. Voß, 2002. *HotFrame: A Heuristic Optimization Framework.* In: S. Voß und D.L. Woodruff (Hrsg.). *Optimization Software Class Libraries.* Kluwer, Boston. 81–154.

Hazemi, R., S. Hailes und S. Wilbur (Hrsg.), 1998. *The Digital University: Reinventing the Academy.* Springer, London.

IFORS, July 2002. *tutORial.* http://www.ifors.ms.unimelb.edu.au/tutorial.

Jones, C.V., 1995. *Visualization and Optimization.* Kluwer, Boston.

ORWelt, July 2002. *ORWelt.* http://dsor.uni-paderborn.de/de/forschung/wbs/orwelt.

Reisman, A., 1997. *Flowshop Scheduling/Sequencing Research: A Statistical Review of the Literature, 1952–1994.* IEEE Transactions on Engineering Management. 44, 316–329.

Reiners, T. und S. Voß, 2002. *Teaching Meta-Heuristics.* Working Paper, Technische Universität Braunschweig.

Reiners, T., D. Reiß und S. Voß, 2002a. *Using Hyperbolic Trees and SmartBars within Virtual Learning Environment Concepts.* Proceedings of the World Congress Networked Learning in a Global Environment, Challenges and Solutions for Virtual Education (NL 2002). ICSC-Naiso Academic Press, Millet Alberta [ISBN: 3-906454-31-2], #100029-03-TR-026, 1–7.

Reiners, T., D. Reiß und S. Voß, 2002b. *XML-basierte Kodierung von Lernobjekten: Dokumentation zu VORMS.* Arbeitspapier. Technische Universität Braunschweig, in Vorbereitung.

Sniedovich, M. und A. Byrne, July 2002. *tutOR.* http://www.tutor.ms.unimelb.edu.au/ frame.html.

VORMS, July 2002. *Virtuelles Studienfach Operations Research/Management Science.* http://www.vorms.org.

Voß, S., 2001. *Meta-Heuristics: The State of the Art.* In: A. Nareyek (Hrsg.). *Local Search for Planning and Scheduling.* Lecture Notes in Artificial Intelligence 2148. Springer, Berlin, 1–23.

Voß, S., S. Martello, I.H. Osman und C. Roucairol (Hrsg.), 1999. *Meta-Heuristics: Advances and Trends in Local Search Paradigms for Optimization.* Kluwer, Boston.

WebCT, Inc., July 2002. *WebCT.com.* http://www.webct.com.

Wolsey, R.E.D., 1979. *Pragmatism Triumphant or Past Sophistication and Future Elegance.* In: K.B. Haley (Hrsg.). *Operational Research '78.* North Holland, Amsterdam. 80–86.

An Evolutionary Algorithm for Bayesian Network Triangulation

Tomasz Łukaszewski

Institute of Computing Science, Poznań University of Technology, ul. Piotrowo 3a, 60-965 Poznań, Poland, luki@man.poznan.pl

Abstract: The problem of triangulation (decomposition) of Bayesian networks is considered. Triangularity of a Bayesian network is required in a general evidence propagation scheme on this network. Finding an optimal triangulation is NP-hard. A local search heuristic based on the idea of evolutionary algorithms is presented. The results obtained using existing and proposed approaches are compared on a basis of a computational experiment.
Keywords: Bayesian networks, graph triangulation, evolutionary algorithm

1 Introduction

Some of the earliest Artificial Intelligence approaches to reasoning under uncertainty were based on Bayesian and decision-theoretic schemes. Both the computational and the representational complexity of probabilistic schemes caused a long-lasting departure from these approaches. Only recently, the development of probabilistic graphical models, such as *Bayesian networks*, caused a renewed interest in applying probability theory in intelligent systems. Today, Bayesian networks (directed acyclic graphs) are successfully applied in a variety of problems, including machine diagnosis, user interfaces, natural language interpretation, planning, vision, robotics, data mining and many others. They provide a natural, efficient method for representing probabilistic dependencies among variables. It is enough to consider only the known dependencies among variables in a domain, rather than to assume that all variables are dependent on all the other variables.

The most common task performed on Bayesian networks is a *general evidence propagation* – the calculation of the posterior marginal distributions of all non-evidence variables for a given set of evidence. This kind of reasoning is performed mainly by *message-passing* algorithms in a *join tree*. These messages are passed from leaves towards the root of the join tree (*inward phase*), and then backwards (*outward phase*). The best known algorithms are Lauritzen-Spiegelhalter, Hugin, Shafer-Shenoy [5,1,7]. They differ mainly in the construction of messages. However, all of them are based on a join tree created for a given Bayesian network. Such a join tree can be created by *triangulation* of the (moralised) graph for a given Bayesian network. For a given Bayesian network many different triangulated graphs can be built. Moreover, for each triangulated graph many different join trees can be built. Finding an optimal join tree for a given Bayesian network is NP-hard, however, the second stage can be done in a very efficient way [2]. Different heuristic approaches are used at the first stage – Bayesian network triangulation [3].

The Bayesian network triangulation is also used in our approach to reasoning with knowledge updating in Bayesian networks [6]. Therefore, we developed a local search heuristic based on the idea of evolutionary algorithms to solve this problem. We started with examining our algorithm for a general evidence propagation problem. We compared the efficiency of some heuristic approaches known from the literature with the proposed evolutionary algorithm for different number of graph vertices and different graph densities. The results of the computational experiment are presented in this paper.

2 Problem formulation

Triangularity (chordality) of a Bayesian network is required in the general evidence propagation scheme on this network. Depending on the problem to be solved by graph triangulation, optimality may be defined differently. The typical optimality criteria are the minimum fill criterion, the minimum size criterion, the minimum weight criterion [3]. Since our interest in graph triangulation originates from the general evidence propagation algorithm, we want to obtain a Bayesian network triangulation with small probability tables. Therefore, our objective is to obtain triangulations of *minimum weight.* Let $|V_i| = n_i$. The base 2 logarithm of the number of states (state size) of a vertex V_i is the weight of V_i (denoted $w(V_i)$). Let $C_i = \{V_1,\ldots,V_k\}$ represent a clique of a triangulated graph G_T. The weight of C_i is defined as follows:

$$w(C_i) = \sum_{V_i \in C_i} w(V_i) \tag{2.1}$$

and the weight of G_T is defined as follows:

$$w(G_T) = \log_2 \sum_{C_i} 2^{w(C_i)} = \log_2 \sum_{C_i} \prod_{V_i \in C_i} n_i \tag{2.2}$$

Note that if all vertices have equal weights the minimum size criterion and the minimum weight criterion are identical. The size of C is $s(C) = k$ and the size of G_T is the sum of the sizes of all of its cliques.

The basic technique applied to triangulate a graph G (that represents a given Bayesian network) is to add the extra edges T produced by eliminating the vertices of G one by one. A vertex V_i is eliminated in the following way:

1. Adding edges such that the vertices adjacent to V_i are pairwise adjacent.
2. Deleting V_i and its incident edges.

This technique is not guaranteed to produce optimal triangulations (in terms of the weight of the triangulated graph) when the vertices are selected at random. However, the number of elimination sequences grows exponentially with the number of vertices making the enumerative approach inefficient. Thus it is justified to apply a more sophisticated elimination technique.

Several heuristic algorithms were suggested and compared for establishing elimination orderings in [3]. The tests were based on two "real world" graphs (43 and 56 vertices, state sizes of vertices range from 3 to 21) and two "artificial" ones (50 vertices, state sizes of vertices range from 2 to 5). The goal function was to minimise the weight of the triangulated graph. The best average results were produced by the minimum weight heuristic. A simulated annealing metaheuristic was also applied. For the real and artificial graphs, simulated annealing performed better than the minimum weight heuristic.

3 Evolutionary algorithm

The proposed evolutionary algorithm is based on principles of evolution (selection and recombination of individuals). However, we implemented three different versions of the algorithm:

1. Crossover removes parents from the population when children are generated. Mutation is possible for all the individuals of the population.
2. Crossover adds children to the population – parents are not removed, the population size is extended. Mutation is possible for all the individuals of the extended population. The (population) extended size is reduced to the initial size during the selection phase.
3. Crossover adds children to the population – parents are not removed, the population size is extended. Mutation is possible only for the new generated individuals. The (population) extended size is reduced to the initial size during the selection phase.

We used tournament selection method (constant tournament size equal 3). Different crossover (PMX, CX, OX1a, OX1b, OX1c, OX2, POS) and mutation operators (ISM, EM, DM, IVM) were implemented. These operators are described in [4]. The crossover and mutation rates changed from 10% to 100% with the 10% step. The number of iterations changed from 1000 to 2000 depending on the number of graph vertices.

4 Description of experiments

The experiments were carried out for randomly generated graphs – from 20 to 100 vertices, two states of each vertex and different densities – Table 1. The minimal graph density when a directed acyclic graph G is connected (the minimal relative number of edges) is the following:

$$k = \frac{n-1}{(n^2-n)/2} * 100[\%] = \frac{2}{n} * 100[\%] \tag{4.1}$$

Table 1. Specification of graph parameters used in the experiment

n	k [%]	d [%]
20	10	15, 20, 25, 30
33	6	12, 18, 24, 30
50	4	8, 12, 16, 20, 24
80	2,5	5, 10, 15, 20, 25, 30
100	2	4, 6, 8, 10, 12, 14

n the number of graph vertices, *d* densities of graphs.

The objective function was to minimise the sum of sizes of probability tables of triangulated graph (the modified minimum weight criterion):

$$w'(G_T) = \sum_{C_i} \prod_{V_i \in C_i} n_i \tag{4.2}$$

The following heuristics were implemented: the minimum size heuristic (equal weights of vertices), the lexicographic search, the maximum cardinality search [3].

For each graph configuration of parameters the following steps were carried out:

1. All heuristics were run *n* times (each vertex was eliminated as the first one, the best result for each heuristic was accepted).
2. The evolutionary algorithm was run 2-times for each configuration of parameters (the best result for each configuration was accepted).
3. The best result, obtained either by the heuristics or by any configuration of parameters of the evolutionary algorithm, was treated as a *reference point*. Then, we computed the *relative deviation* (*RD*) between this reference point and results obtained by the heuristics and the evolutionary algorithm in the following way:

$$RD = \frac{v - \min(H, EA)}{\min(H, EA)} * 100[\%] \tag{4.3}$$

where:
- *RD* is the relative deviation,
- min(*H*,*EA*) is the value of the reference point,
- *v* is a result obtained by heuristics or a configuration of parameters of the evolutionary algorithm.

For each graph configuration of parameters about 20 graphs were randomly generated in order to obtain the average values of *RD* for all heuristics and all configurations of parameters of the evolutionary algorithm. For each number of graph vertices the one density was chosen for which average values of *RD* were the largest. For this density *d** about 100 graphs were generated in order to obtain a better assessment of the average values of *RD*.

Experiments were conducted on 7 machines (with the Intel Pentium4–1,5GHz processor, 256 MB RAM memory and Windows 2000 operating system.

5 Results

For each version of the evolutionary algorithm there are 2800 configurations of parameters. In order to reduce the computational time we were rejecting some configurations of parameters after experiments for a given number of graph vertices. Therefore, experiments for the biggest graphs were carried out only for the best crossover operator, mutation operator and recombination rates. Moreover, at the beginning we rejected the third version of the evolutionary algorithm, whose results were worse than those of the other versions. Nevertheless, experiments took 55 days (maximum 7 machines were used). Performing these experiments only on a single machine would take 173 days.

Almost all the best results were obtained for the second version of the evolutionary algorithm. For small graphs (20 and 33 vertices) the best results were obtained for the ISM and EM mutation operators, mutation rate 50–100%; crossover operators and their rates were not so significant. For medium graphs (50 vertices) the best results were obtained for the ISM mutation operator, mutation rate 80–100%, crossover operators CX, OX2, OX1c, POS, PMX, crossover rate 10–50%. For the biggest graphs (80 and 100 vertices) the best results were obtained for the ISM mutation operator, mutation rate 100%, crossover operator OX2, crossover rate 40%. This configuration of parameters was chosen as the best one – EA^*.

Let $\min(RD_H)$ represent the minimal relative deviation obtained by heuristics for a given graph configuration of parameters; RD_{EA^*} represents the relative deviation obtained by the best configuration of parameters of the evolutionary algorithm for a given graph configuration of parameters. Below we present the comparison of average values of these values (for d^* density) – Table 2.

Table 2. Comparison of average values of *RD* (for d^* density)

n	d^* [%]	ave $\min(RD_H)$ [%]	ave RD_{EA^*} [%]	ave t_H [s]	ave t_{EA^*} [s]
20	20 (= 2k)	8	0	<1	2
33	18 (= 3k)	78	1,00	<1	5
50	16 (= 4k)	374	5,14	<1	25
80	10 (= 4k)	2685	41,76	<1	108
100	8 (= 4k)	7768	40,74	<1	240

n the number of graph vertices, t_H the computational time of all heuristics, t_{EA^*} the computational time of the evolutionary algorithm for the best configuration of parameters.

6 Conclusions

As we had expected, results obtained by the evolutionary algorithm are much better than results obtained by the heuristics. However, if an on-line triangulation is required the evolutionary algorithm can be too slow.

Secondly, such an extensive computational experiment showed which configuration of parameters of the evolutionary algorithm are the best ones: parents are not removed by children during the crossover phase and mutation operator is applied to all individuals (second version of the evolutionary algorithm) results are significantly better than in two other approaches; ISM mutation operator with the rate 100%; crossover operator OX2 with the rate 40%.

Analysis of the obtained results was possible to make in a reasonable time using graphical representations of these numbers. Using several PC's reduces the computational time of the experiment. The computational time of such experiments can be very limited performing them in distributed architectures.

The very high rate of the ISM mutation operator was the reason that we tested the evolutionary algorithm for two benchmark graphs [3]. The evolutionary algorithm was able to achieve the optimal results. Concluding, we suggest improving crossover operators.

Results of this experiment are the basis for our further research on the reasoning in Bayesian networks with knowledge updating.

References

[1] Andersen SK, Jensen FV, Olesen KG (1990) An algebra of Bayesian Belief Universes for Knowledge-Based Systems. Networks 20: 637–659

[2] Jensen FV, Jensen F (1994) Optimal Junction Trees. Proceedings of the Tenth Conference on Uncertainty in Artificial Intelligence. Morgan Kaufmann Publishers, San Francisco, pp 360–366

[3] Kjærulff U (1990) Triangulation of Graphs – Algorithms Giving Small Total State Space. Research Report R90-09, Department of Computer Science, Aalborg University, Denmark

[4] Larranaga P, Kuijpers C, Poza M, Murga R (1997) Decomposing Bayesian networks: triangulation of the moral graph with genetic algorithms. Statistics and Computing 7: 19–34

[5] Lauritzen SL, Spiegelhalter DJ (1988) Local computations with probabilities on graphical structure and their application to expert systems (with discussion). Journal of Royal Statistical Society, Series B 50: 157–224

[6] Łukaszewski T (2001) Knowledge Updating in Bayesian Networks. Procedings of the 14th International Conference on Systems Science, Oficyna Wydawnicza Politechniki Wrocławskiej, Wrocław, pp 274–281

[7] Shenoy P, Shafer G (1990) Axioms for probability and belief-function propagation. In Shachter RD, Levitt TS, Lemmer JF, Kanal LN (eds) Uncertainty in Artificial Intelligence 4: 169–198

Approximation Algorithms for the k-center Problem: An Experimental Evaluation

Jurij Mihelič[1] and Borut Robič[1]

Faculty of Computer and Information Science
University of Ljubljana, Tržaška 25, 1000 Ljubljana, Slovenia
{jurij.mihelic, borut.robic}@fri.uni-lj.si

Abstract. In this paper we deal with the vertex k-center problem, a problem which is a part of the discrete location theory. Informally, given a set of cities, with intercity distances specified, one has to pick k cities and build warehouses in them so as to minimize the maximum distance of any city from its closest warehouse. We examine several approximation algorithms that achieve approximation factor of 2 as well as other heuristic algorithms. In particular, we focus on the clustering algorithm by Gonzalez, the parametric pruning algorithm by Hochbaum-Shmoys, and Shmoys' algorithm. We discuss several variants of the pure greedy approach. We also describe a new heuristic algorithm for solving the dominating set problem to which the k-center problem is often reduced. We have implemented all the algorithms, experimentally evaluated their quality on 40 standard test graphs in the OR-Lib library, and compared their results with the results found in the recent literature.

1 Introduction

Problems of finding the best location of facilities in networks or graphs abound in practical situations. One of the well known facility location problems is the *vertex k-center* problem, where given n cities and distances between all pairs of cities, the aim is to choose k cities (called centers) so that the largest distance of a city to its nearest center is minimal. More formally, the vertex k-center can be defined as follows. Let $G = (V, E)$ be a complete undirected graph with edge costs satisfying the triangle inequality, and k be a positive integer not greater than $|V|$. For any set $S \subseteq V$, and vertex $v \in V$, we define $d(v, S)$ to be the length of a shortest edge from v to any vertex in S. The problem is to find such a set $S \subseteq V$, where $|S| \leq k$, which minimizes $\max_{v \in V} d(v, S)$. The vertex k-center problem is NP-hard [5].

A popular way to solve the k-center problem consists of solving a series of *set cover* problems [2,4,9,10]. At each step, a threshold for the cover distance is chosen and it is checked whether all vertices can be covered within this distance using at most k centers; if so, the threshold is decreased, otherwise it is increased. (One can also use the *dominating set* problem instead of the set cover problem [7].) For example, Minieka [10] solved the k-center problem as a series of set cover problems. More elaborate versions of this approach were described by Daskin [2,3], where also the *maximum cover* problem was used, and Ellumni et al. [4] and Ilhan et al. [9], which applied more efficient definition of the problem. Usually, these set cover problems were solved with integer programming. Another way to solve the k-center problem was recently given by Mladenović et al. [11], where the *tabu search*, *variable neighborhood search* and various greedy methods were used. The greedy method was also applied by Gonzalez [6], Hochbaum and Shmoys [8], and Shmoys [12]. The last three describe 2-approximation algorithms which are the *best possible* in the sense that no r-approximation algorithm exists with $r < 2$, unless $P = NP$ [7]. (No approximation algorithm exists in case the triangular inequality does not hold, unless P=NP.)

In the following we briefly describe various heuristics for the k-center problem that are not based on the integer programming (Section 2). We then describe a new heuristic, which combines the greedy approach with solving the dominating set problem, and returns surprisingly good results (Section 3). We have experimentally evaluated all these heuristics as well as the new one. The experimental results are given in Section 4.

2 Heuristics

By using greedy heuristics we often locate centers one by one until there are k centers. For the selection of the first center there may be several possibilities. For example, the center can be located at random, it can be the result of the 1-center problem, or we can apply a heuristic n-times, $n = |V|$, each time with different starting vertex, and then choose the best of the solutions. These approaches will be called *random*, *1-center*, and *plus* version, respectively.

A very simple heuristic is the *pure greedy method*, where centers are located one by one so that the objective function is each time reduced as much as possible. For the selection of the first center we have implemented random, 1-center and plus version. It is easy to see that the time complexity of this pure greedy method is $O(kn^2)$.

Another greedy heuristic for the k-center problem was described by Gonzalez [6], who was able to prove the approximation factor of 2. The algorithm builds final solution in k steps so that, given a partial solution C_{i-1}, it forms a new partial solution C_i by extending C_{i-1} with the vertex v which is the farthest from the C_{i-1}, i.e. the vertex v which maximizes $d(v, C_{i-1})$ at step i. We have implemented random, 1-center, and plus version of this algorithm. The time complexity of Gonzalez's algorithm is $O(kn)$.

Shmoys [12] briefly describes 2-approximation algorithm for the decision version of k-center problem, i.e. where radius r is also given and the aim is to decide if there exist k vertices so that the coverage distance from these vertices is at most r. The algorithm repeatedly chooses one of the remaining vertices v, adds it to the partial solution, and deletes all vertices whose distance to v is at most $2r$. At the end, if the size of the solution exceeds k, the algorithm outputs "no", otherwise "yes". We implemented two versions where either random vertex or vertex with maximum degree can be chosen on each step. The algorithm for the optimization version of problem runs the algorithm for the decision version several times with increasing value of r. Time complexity of this algorithm is $O(kn^3)$.

Hochbaum and Shmoys [8] introduced the algorithmic technique called *parametric pruning* for solving k-center problem. Initially, edge costs are sorted in nondecreasing order. For each edge cost t the graph is pruned by removing edges with cost greater than t. The aim is to find a *minimum dominating set* in the pruned graph, i.e. the smallest set S of vertices such that every vertex not in S is adjacent to one of the vertices in S. If the cardinality of the minimum dominating set of the pruned graph is at most k, then such a dominating set is also the optimal solution for k-center problem.

Unfortunately, to compute the minimum dominating set is NP-hard optimization problem [5]. Consequently, instead of searching for the minimum dominating set we rather search for the *maximal independent set*[1], i.e the subset S of V such that no two vertices of S are connected in G and no vertex can be added to S while S retaining this property.

Define the *square* of the graph G to be the graph G^2 containing an edge (u, v) whenever G has a path of at most two edges between u and v, $u \neq v$. It is well known that every maximal independent set is also dominating set. The fact that the cardinality of the maximal independent set of G^2 is at most the cardinality of the minimum dominating set of G can be used to construct a 2-approximation algorithm for solving the k-center problem. More specifically, instead of searching for the minimum dominating set of G the algorithm constructs the maximal independent set of G^2. The overall time complexity of the algorithm is estimated to be $O(kn^5)$.

3 Elimination Heuristic

We designed a new algorithm for the k-center problem. The algorithm is based on the standard approach that solves a series of dominating set problems. First, edge costs are sorted in a nondecreasing list which is used for getting the threshold values r and for solving the series of dominating set problems. When the cardinality of the dominating set S becomes at most k, the set S is returned as the result of the k-center problem and the algorithm is completed.

For solving the dominating set problem we developed a new heuristic algorithm. Informally, a pair of numbers $(c(v), s(v))$ is initially assigned to each vertex $v \in V$, where $c(v)$ (*cover count*) is the number of vertices that can cover v within distance r, while $s(v)$ (*vertex score*) is used in the following selection process ($s(v)$ is initially set to $c(v)$). At each step of the selection process the vertex v with the smallest $s(v)$ is chosen. If there is a vertex $u \in V$ such that $d(u, v) \leq r \wedge c(u) = 1$, then v is added to the set S of centers;

[1] *Maximum independent set* is NP-hard problem, while *maximal* independent set is one of the suboptimal solutions.

otherwise, $s(u)$ is incremented for all $u \in V$ for which $d(u,v) \leq r$. Next, the cover count $c(u)$ is decremented for all vertices $u \in V$ with $d(u,v) \leq r$. These steps are repeated until all the vertices of the graph have been processed.

Notice, that we use $c(v)$ to ensure that every vertex is covered at least once. In addition, the way $c(v)$ is used makes it possible to easily adapt the algorithm for solving the (*fault-tolerant*) α-neighbor k-center problem where every node must be covered by at least α centers. Moreover, one can adapt the algorithm to solve the minimum set cover problem instead of the minimum dominating set problem. This is useful when solving the so-called k-supplier problem, where k centers must be chosen from a predefined set of vertices.

4 Experimental Results

We tested the described algorithms on 40 OR-Lib test problems, which were originally designed for testing p-median problems [1]. The number of vertices ranges from 100 to 900 while k ranges from 5 to 90. The preprocessing phase runs the all shortest paths algorithm (time complexity $O(n^3)$).

All the algorithms were implemented in Borland Delphi 6.0, and were tested on a computer with Intel processor running at 1.7 GHz with 512MB of system memory. Designations and names of algorithms appear in Table 1. Although our primary aim was to compare the quality of the solutions, let us mention that Gonzalez algorithms were the fastest (running below 1 second). The average time of `HS` was about 100 seconds. The pure greedy methods were quite fast (about 2 seconds on average), but their execution time was very variable and dependent on the parameter k, Shmoys' variants were also fast as well as our `Scr`. (Notice that plus variants run much slower due to the algorithm which tries all vertices for the first center.)

Recall that approximation factor is the ratio between approximated and the optimal objective value. Since sometimes we do not know the optimal solution, we take the best known so far. We call such a ratio an *approximation degree.* Nevertheless, in our case most of the best known objective values were proved to be optimal (see for example [4,9]). Approximation degrees for each algorithm are given in Table 1 below, where we also included the results for Daskin's and tabu search approach.

Objective values for all of the 40 problems are in Table 2. The pure greedy method is the worst, while only slightly better results were with the plus variant. The solution quality strongly depends on the parameter k, and is much better for low values of k. Gonzalez algorithms are very fast and solutions are about 50% worse than best known. The algorithms `HS`, `ShR`, `ShD` exploit very similar problem properties, and consequently return very similar results. Our algorithm `Scr` proved to be quite competitive since it achieved better results than any of the implemented algorithms, with the exception of the integer programming approaches [3,4,9] and tabu search and variable neighborhood search [11].

Algorithm	Level	Deviation	Description
GrR	1,697	0,559	Pure greedy first random
Gr1	1,675	0,570	Pure greedy first 1-center
Gr+	1,512	0,550	Pure greedy plus
GonR	1,495	0,130	Gonzalez first random
Gon1	1,398	0,128	Gonzalez first 1-center
Gon+	1,317	0,139	Gonzalez plus
ShR	1,432	0,112	Shmoys random
ShD	1,343	0,105	Shmoys degree
HS	1,462	0,177	Hochbaum-Shmoys
Scr	1,058	0,043	Elimination heuristic
Das	1,002	0,007	Daskin
TS	1,025	0,045	Tabu search

Table 1. Approximation degrees

References

1. J. E. Beasley. A note on solving large p-median problems. *European J. Oper. Res.*, 21:270–273, 1985.
2. Mark S. Daskin. *Network and Discrete Location: Models Algorithms and Applications.* Wiley, New York, 1995.
3. Mark S. Daskin. A new approach to solving the vertex p-center problem to optimality: Algorithm and computational results. *Communications of the Operations Research Society of Japan*, 45:9:428–436, 2000.
4. Sourour Elloumi, Martine Labbe, and Yves Pochet. New formulation and resolution method for the p-center problem. 2001.
5. M. R. Garey and D. S. Johnson. *Computers and Intractability: A Guide to the Theory of NP-Completeness.* W. H. Freeman and Co., San Francisco, 1979.
6. T. Gonzalez. Clustering to minimize the maximum inercluster distance. *Theoretical Computer Science.*, 38:293–306, 1985.
7. Dorit S. Hochbaum, editor. *Approximation Algorithms for NP-hard Problems.* PWS publishing company, Boston, 1995.
8. Dorit S. Hochbaum and David B. Shmoys. A best possible heuristic for the k-center problem. *Mathematics of Operations Research*, 10:180–184, 1985.
9. Taylan Ilhan and Mustafa Pinar. An efficient exact algorithm for the vertex p-center problem. 2001.
10. E. Minieka. The m-center problem. *SIAM Rev.*, 12:138–139, 1970.
11. N. Mladenović, M. Labbe, and P. Hansen. Solving the p-center problem with tabu search and variable neighborhood search. 2000. http://smg.ulb.ac.be/Preprints/Labbe00_20.html.
12. David B. Shmoys. Computing near-optimal solutions to combinatorial optimization problems. Technical report, Ithaca, NY 14853, 1995. http://citeseer.nj.nec.com/shmoys95computing.html.

#	n	k	Best	Das	TS	GrR	Gr1	GrP	Gon	Gon1	Gon+	HS	ShR	ShD	Scr
1	100	5	127	127	127	143	133	133	186	162	155	184	188	171	133
2	100	10	98	98	98	117	117	110	131	124	117	160	128	135	109
3	100	10	93	93	93	126	116	106	154	133	124	160	140	120	99
4	100	20	74	74	74	127	127	92	114	99	92	124	109	84	83
5	100	33	48	48	48	87	87	78	71	64	62	77	62	59	48
6	200	5	84	84	84	98	94	89	138	99	98	126	138	106	90
7	200	10	64	64	64	78	79	77	96	87	85	90	88	90	70
8	200	20	55	55	55	72	72	72	82	72	71	84	74	68	60
9	200	40	37	37	37	73	73	63	57	51	49	62	50	52	38
10	200	67	20	20	20	44	44	38	31	29	29	32	28	28	20
11	300	5	59	59	59	68	67	61	73	68	68	82	73	74	60
12	300	10	51	51	51	62	72	56	71	70	66	78	74	70	53
13	300	30	35	36	36	64	64	52	59	51	49	60	54	52	38
14	300	60	26	26	26	60	60	46	40	36	36	44	36	34	27
15	300	100	18	18	18	42	42	40	25	25	23	30	22	20	18
16	400	5	47	47	47	52	51	47	84	55	52	64	83	58	48
17	400	10	39	39	39	50	50	43	56	51	48	56	56	52	41
18	400	40	28	28	28	50	50	42	44	41	39	46	40	38	31
19	400	80	18	18	19	40	40	31	28	28	27	30	26	24	20
20	400	133	13	13	14	32	32	32	19	19	17	22	18	16	14
21	500	5	40	40	40	48	48	42	53	51	45	52	53	45	40
22	500	10	38	38	38	48	49	43	56	54	47	54	54	48	41
23	500	50	22	22	23	41	41	35	34	33	32	36	32	30	24
24	500	100	15	15	16	35	35	32	23	23	21	24	22	20	17
25	500	167	11	11	12	27	27	27	15	15	15	18	16	14	11
26	600	5	38	38	38	44	43	39	50	47	43	52	50	52	41
27	600	10	32	32	32	37	39	35	43	42	55	42	44	44	33
28	600	60	18	18	19	33	33	27	28	28	25	28	28	28	20
29	600	120	13	13	13	34	36	34	19	19	18	22	18	18	13
30	600	200	9	9	11	29	29	29	14	14	13	16	12	12	10
31	700	5	30	30	30	35	34	31	42	38	36	40	42	44	30
32	700	10	29	29	29	35	35	32	45	43	37	40	44	40	31
33	700	70	15	15	16	32	26	24	26	25	23	26	24	22	17
34	700	140	11	11	12	30	30	27	17	17	16	18	16	16	11
35	800	5	30	30	30	37	32	31	38	37	34	40	38	38	32
36	800	10	27	27	27	34	34	30	41	41	34	38	42	38	28
37	800	80	15	15	16	26	26	26	25	24	23	24	22	22	16
38	900	5	29	29	29	42	35	31	36	38	31	38	40	38	29
39	900	10	23	23	24	27	28	25	35	35	28	32	36	34	24
40	900	90	13	13	14	25	22	22	21	20	19	22	20	20	14

Table 2. Objective function values

MaxFlow-MinCut Duality for a Paint Shop Problem

Thomas Epping[1], Winfried Hochstättler[1], and Marco E. Lübbecke[2]

[1] Department of Mathematics, BTU Cottbus, 03013 Cottbus, Germany
{epping, hochstaettler}@math.tu-cottbus.de

[2] Department of Mathematical Optimization, TU Braunschweig, Germany
m.luebbecke@tu-bs.de

Abstract. Motivated by an application in car manufacturing we consider the following problem: How can we synthesize a given word from restricted reservoirs of colored letters with a minimal number of color changes between adjacent letters?

We focus on instances in which each letter occurs exactly twice, once in each of two given colors. In this case the problem turns out to be the dual of a MinCut problem for one point extensions of a certain class of regular matroids.

We discuss consequences of the MaxFlow-MinCut duality and describe algorithmic approaches.

1 Introduction

A sequence of car bodies has to pass various shops during the production process, including a press and a body shop, a paint shop, and an assembly shop. The daily sequencing of the car bodies has to be done with respect to the minimization of the specific objective function of each of these shops. We focus on the paint shop, where the objective function consists in the minimization of the number of color changes that occur whenever two consecutive car bodies have to be colored in different colors, giving rise to non-negligible costs and water pollution.

The number of color changes may be reduced by the use of interim storage systems (see [1]) that permute a car body sequence before it enters the paint shop. However, the efficiency of succeeding production shops may also significantly decrease for the permuted sequence. We therefore consider the car body sequence to be a fixed external parameter.

Together with the fact that current technology heads for the detachment of car bodies and their features (what allows us to uncouple car bodies and enamel colors), the minimization of color changes for a car body sequence yields a new type of combinatorial problem.

We first give a formal problem description, a review of complexity results, and previous results and conjectures on the minimal number of color changes for regularly structured instances. Among these, we examine particular instances from a matroid point of view in more detail. Our notation is fairly standard.

2 Problem Formulation and Previous Results

We are given a fixed sequence of car bodies together with a set of orders that specify the demand of each car body type in each color. We assume that these orders are given by an inital coloring of the car body sequence. We associate a letter of an alphabet Σ with each car body type and represent a sequence of n car bodies by a word $w \in \Sigma^n$. A coloring of w is represented by a vector $f \in F^n$ for some color set F, where f_i denotes the color of w_i for all i. We say that we have a color change within f whenever $f_i \neq f_{i+1}$.

Our problem consists in finding a permutation that minimizes the number of color changes within f and leaves the sequence of letters in w unchanged.

Problem 1. Paint Shop Problem for Words (PPW)

Instance A finite alphabet Σ, a word $w = (w_1, \ldots, w_n) \in \Sigma^n$, a finite color set F, and a coloring $f = (f_1, \ldots, f_n) \in F^n$ of w.

Question Find a permutation $\sigma : \{1, \ldots, n\} \to \{1, \ldots, n\}$ such that $w_{\sigma(i)} = w_i$ for $i = 1, \ldots, n$, and the number of color changes within $\sigma(f) = (f_{\sigma(1)}, \ldots, f_{\sigma(n)})$ is minimized.

The following complexity results hold.

Theorem 1 ([2]). *Any instance of PPW can be solved by a dynamic program with a space and time complexity of $\mathcal{O}(|F|n^{(|F|-1)|\Sigma|})$. PPW is NP-complete for either $|F| = 2$ or $|\Sigma| = 2$, that is, if at least one parameter is unbounded.*

2.1 Upper Bounds for Regular Instances

We now turn to structured instances of PPW. In the following, we denote an instance of PPW by $(w; f)$, and the minimal number of color changes for $(w; f)$ by $\gamma(w; f)$. Recall that the coloring f determines reservoirs of colored letters for $(w; f)$. We denote the reservoir of letter i in color j by $R(i, j)$.

Definition 1. Given a fixed integer $k \geq 1$, we call an instance $(w; f)$ of PPW k-regular, if $R(i, j) = k = \frac{n}{|\Sigma||F|}$ for all letters i and colors j.

The following upper bounds on $\gamma(w; f)$ hold for k-regular instances.

Lemma 1 ([2]). *If $|\Sigma| = |F| = 2$, then $\gamma(w; f) \leq 2$. If $|\Sigma| = 2$, then $\gamma(w; f) \leq 2(|F| - 1)$.*

We conjecture the following general upper bound on $\gamma(w; f)$. A linear algebra argument yields the correctness of Conjecture 1 for $k = 1$.

Conjecture 1 ([2]). For a k-regular instance holds $\gamma(w; f) \leq |\Sigma|(|F| - 1)$.

Theorem 2. *For a 1-regular instance holds $\gamma(w; f) \leq |\Sigma|(|F| - 1)$.*

Examples (see [2]) show that the bound given in Conjecture 1 is tight if the conjecture is correct. Natural solution approaches like a greedy coloring algorithm or an improvement algorithm based on color exchanges of letters or letter blocks fail to yield or even approximate an optimal solution in general.

3 1-Regular Instances and Two Colors

This section focuses on 1-regular instances with a color set of size $|F| = 2$, thus every letter is available exactly once in each color. We do not know whether PPW is polynomially solvable when restricted to such instances, but we give some indications that might support such a conjecture. In the following we denote by $\mathbf{0}$ resp. $\mathbf{1}$ the vector (of appropriate dimension) of all zeros resp. ones.

(a)

$$\begin{pmatrix} 1&1&1&0&0&0&0&0&0 \\ 0&1&1&1&1&1&1&0&0 \\ 0&0&1&1&1&1&1&1&1 \\ 0&0&0&0&1&0&0&0&0 \\ 0&0&0&0&0&0&1&1&0 \end{pmatrix}$$

(b)

Fig. 1. An example instance and the associated matrix A

First we associate an interval $I(b)$ to each letter $b \in \Sigma$, where $I(b)$ runs from the first occurence of b in w to the second (see Figure 1(a)). We may consider each $I(b)$ being an interval on the real line. The restriction of PPW to 1-regular instances with $|F| = 2$ is then equivalent to the following problem.

Problem 2. Odd Intersection of Intervals

Instance A set of closed intervals on the real line, where no two intervals share a common endpoint.

Question Find a minimal set of points on the real line that intersects each interval in an odd number of points.

We denote the set of intervals of an instance $(w; f)$ of Problem 2 by $I(w; f)$ and interpret each point that intersects one or more intervals of $I(w; f)$ as a color change or a cut between two adjacent letters of w. In the most easiest case, $I(w; f)$ is a clutter.

Lemma 2. *If $I(w; f)$ is a clutter, then $\gamma(w; f)$ can be computed by a greedy algorithm.*

Otherwise, we consider the $(|\Sigma| \times n - 1)$-matrix A (siehe Figure 1(b)) that is defined by

$$a_{ij} := \begin{cases} 1 \text{ , if a cut between } w_j \text{ and } w_{j+1} \text{ intersects } I(i)\text{, and} \\ 0 \text{ , otherwise.} \end{cases}$$

Obviously A has the consecutive ones (CO) property for rows (see [3]). Thus, A is totally unimodular. Furthermore, A contains at least all maximal cliques of the interval graph of $I(w; f)$. Note that any instance of Problem 2 can be solved by the integer program depicted in Figure 2(a).

$$\begin{aligned} \mathbf{1}^T \boldsymbol{x} &\rightarrow \min! \\ A\boldsymbol{x} - 2I\boldsymbol{y} &= \mathbf{1} \\ x_i &\in \{0,1\} \\ y_i &\in N_{\geq 0} \end{aligned} \qquad\qquad \begin{aligned} \mathbf{1}^T \boldsymbol{x} &\rightarrow \min! \\ A\boldsymbol{x} - 2I\boldsymbol{y} &= \boldsymbol{b} \\ x_i &\in \{0,1\} \\ y_i &\in N_{\geq 0} \end{aligned}$$

(a) (b)

Fig. 2. Integer programs for the solution of Problem 2

3.1 Lower Bounds for the Number of Interval Cuts

The fact that A is a node-clique matrix of an interval graph enables us to compute a lower bound on the number of intersections of each interval in an optimal solution for Problem 2. Therefore we consider the partial ordering on $I(w;f)$ given by proper containment, i. e. $I(b') < I(b) :\Leftrightarrow I(b') \neq I(b)$ and $I(b')$ is properly contained in $I(b)$. We define the set $C(b) := \{I(b') : I(b') < I(b)\}$ for all $b \in \Sigma$ and assign a lower bound of 1 to all $I(b)$ with $C(b) = \emptyset$. Then, we compute a lower bound for an interval $I(b)$ whenever all intervals in $C(b)$ are already assigned a lower bound. Therefore we have to compute the value of a maximum weighted stable set in $C(b)$, where the weight of each interval is given by its lower bound. The result has to be rounded up to the next odd integer. Note that the maximum weighted stable set problem can be solved in polynomial time on interval graphs by computing a maximum weighted clique on its complement graph (see [3]).

Theorem 3. *Let P resp. P' denote the IP shown in Figure 2(a) resp. (b), and let $F(P)$ denote the set of vectors feasible for P. Then the set of vectors feasible for P' is given by $F(P') = \{(\boldsymbol{x}, \boldsymbol{y} - \frac{1}{2}(\boldsymbol{b} - \mathbf{1})) : (\boldsymbol{x}, \boldsymbol{y}) \in F(P)\}$.*

Thus a feasible solution for P' can be derived from a feasible solution for P (and vice versa) by changing only the vector $\boldsymbol{y}$. In the following we denote the vector of lower bounds, computed as described above, by $\boldsymbol{b}$. Note that we get (in addition to Theorem 2) a lower bound on $\gamma(w;f)$ if we enclose w by an extra letter λ and compute the lower bound on the number of intersections of $I(\lambda)$, this time without rounding the result up to the next odd integer.

3.2 A Dual Pair of Linear Programs

In this section we apply the "Big-M"-method and consider the dual pair of LPs depicted in Figure 3. Recall that A (and thus (A, I)) is totally unimodular, so both LPs have integer solutions. We call an optimal solution $(\boldsymbol{x}^*, \boldsymbol{y}^*)$ of the primal LP R-feasible (R-optimal), if $\boldsymbol{x}^*$ is a feasible (optimal) solution for Problem 2. Note that any $(\boldsymbol{x}^*, \boldsymbol{y}^*)$ with $\boldsymbol{y}^* = \mathbf{0}$ is R-feasible, but not necessarily R-optimal.

Theorem 4. *If $(\boldsymbol{x}^*, \boldsymbol{y}^*)$ is an optimal solution of the primal LP with $\boldsymbol{y}^* = \mathbf{0}$, then $x_i^* \in \{0,1\}$ for all i.*

$$\begin{array}{rl} \mathbf{1}^T \boldsymbol{x} + \boldsymbol{M}^T \boldsymbol{y} \to \text{min!} & \quad \boldsymbol{u}^T \boldsymbol{b} \to \text{max!} \\ A\boldsymbol{x} - I\boldsymbol{y} = \boldsymbol{b} & \quad \boldsymbol{u}^T A \leq \mathbf{1} \\ \boldsymbol{x}, \boldsymbol{y} \geq \mathbf{0} & \quad \boldsymbol{u} \geq -\boldsymbol{M} \end{array}$$

Fig. 3. The primal LP and its dual

If $(\boldsymbol{x}^*, \boldsymbol{y}^*)$ is not R-optimal, the lower bound $\boldsymbol{b}$ is not tight. A rough statistic shows that the percentage of such instances increases from less than 5% for $|\Sigma| = 5$ to more than 90% for $|\Sigma| = 25$. For example, both instances shown in Figure 4 have a lower bound of $\boldsymbol{b} = \mathbf{1}$, which is not tight. In such cases we are searching for an adaption of $\boldsymbol{b}$ so that the solution of the primal LP yields an R-optimal $(\boldsymbol{x}^*, \boldsymbol{y}^*)$. Due to recent computational experiments we conjecture the following.

Conjecture 2. Suppose that $\boldsymbol{b}$ is not tight and let $U := \{j : u_j < 0\}$ denote the index set of negative dual variables after the solution of the primal LP. Then there exists $I \subset \Sigma$ so that $I \subset U$, and the primal LP yields an R-optimal solution if we increase b_i by 2 for all $i \in I$.

Figure 1(a) shows an example of an instance for which the primal LP yields an R-feasible solution and an objective value of 4, while we get $\gamma(w; f) = 3$ if we increase the lower bound on the number of intersections of $I(B)$ by 2.

3.3 MaxFlow-MinCut Duality

We get an even stronger duality than the LP duality described in Section 3.2 if we formulate Problem 2 in terms of matroid theory.

Therefore, we identify the set of feasible solutions to Problem 2 with the set of feasible solutions of the equation $A\boldsymbol{x} \equiv \mathbf{1} \bmod 2$ over GF(2). If we replace A by $(A, \mathbf{1})$, this equation is equivalent to $(A, \mathbf{1})\boldsymbol{x} \equiv \mathbf{0} \bmod 2$, if we demand that $x_n = 1$ always. Thus we are seeking a minimal element in the kernel of $(A, \mathbf{1})$, or, in other words, a shortest circuit in the binary vector matroid $(A, \mathbf{1})$, that contains the element $\mathbf{1}$. This yields an equivalent formulation of Problem 2.

Problem 3. Shortest circuit in a clutter

Instance A matrix A with the CO property for rows, where no two sequences of consecutive ones start or end in a common column.

Question Find a shortest circuit in the clutter of all circuits of the binary vector matroid $(A, \mathbf{1})$ that contain the element $\mathbf{1}$.

Note that computing a shortest circuit in a binary matroid is NP-complete in general. Furthermore, recall that A is totally unimodular. Thus $(A, \mathbf{1})$ is a one point extension of a regular matroid. We cite the dual version of Seymour's famous MaxFlow-MinCut theorem.

Theorem 5 (Seymour [4]). *Given a binary matroid $M = (E, \mathcal{I})$ and a specific element $e \in E$, the value of a maximum disjoint packing of cocircuits, each of which contains e, equals the value of a minimum circuit that contains e for all length functions $f : E \to N_{\geq 0}$, if and only if M has no F_7-minor that contains e.*

Now we consider the binary vector matroid $M = (A, \mathbf{1})$ with $e = \mathbf{1}$ and fix the length function to $f \equiv 1$ except for $f(\mathbf{1}) = 0$. Then the length of a shortest circuit that contains **1** corresponds to the minimum number of color changes for any instance of Problem 2. Note that we are allowed not only to pack disjoint rows, but also odd sums of symmetric differences of rows.

Theorem 6. *If $(A, \mathbf{1})$ has no F_7-minor that contains* **1**, *then the maximal value of a disjoint odd row sum packing equals the minimum number of color changes for any instance $(w; f)$ of Problem 3.*

For example, the instance shown in Figure 4(a) has a value of $\gamma(w; f) = 4$, and $(A, \mathbf{1})$ contains no F_7-minor. A maximal odd row sum packing is given by $\{I(A), I(B), I(C)\}$, $\{I(C), I(D), I(E)\}$, $\{I(B)\}$, and $\{I(E)\}$. The instance in Figure 4(b) does contain an F_7-minor (contract the first column, add $I(E)$ to $I(C)$, and contract the eighth column within $(A, \mathbf{1})$). It has a value of $\gamma(w; f) = 3$, whereas a maximal odd row sum packing consists of only two disjoint odd row sums ($\{I(B)\}$ and $\{I(E)\}$).

(a) (b)

Fig. 4. Instances of Problem 3 without (a) and with (b) an F_7-minor in $(A, \mathbf{1})$

References

1. Epping Th., Hochstättler W. (2002) Storage and Retrieval of Car Bodies by the Use of Line Storage Systems. Technical report btu-lsgdi-001.02, BTU Cottbus, Germany
2. Epping Th., Hochstättler W., Oertel P. (2002) Complexity Results on a Paint Shop Problem. Submitted to: Discrete Applied Mathematics
3. Golumbic, M. C. (1980) Algorithmic Graph Theory and Perfect Graphs. Academic Press
4. Seymour P. D. (1977) The Matroids with the Max-Flow Min-Cut Property. Journal of Combinatorial Theory, Series B 23, pp 189–222
5. Spieckermann S., Voß S. (1996) Paint Shop Simulation in the Automotive Industry. ASIM Mitteilungen 54, pp. 367–380
6. Oxley, J. G. (1992) Matroid Theory. Oxford University Press

From Edge Decomposition Formulae to Composition Algorithms

André Pönitz

Hochschule Mittweida

Abstract For some graph invariants simple decomposition formulae are known. Unfortunately, a straight forward implementation of such formulae usually leads to algorithms exponential in the number of edges or nodes of the graph even if the graph has a structure that would allow polynomial algorithms. This paper describes a way to derive polynomial algorithms for graphs of bounded pathwidth from edge decomposition formulae by using a variant of the composition method originally proposed by the author for the computation of graph invariants in graphs of small bandwidth.

1 Introduction

For several graph invariants and other graph-related values decomposition formulae are known. So we know e.g. for the all-terminal reliability $R(G)$ of a graph G that $R(G) = p_e\, R(G_e) + (1 - p_e)\, R(G_{-e})$ holds with G_{-e} denoting the minor derived from G by deleting some edge e, G_e the minor derived by contracting, and p_e the working probability of edge e. Similarly, we have $N(G) = x\, N(G_{-e}) + y\, N(G_e)$ for the Negami polynomial or $\chi(G) = \chi(G_{-e}) - \chi(G_e)$ for the chromatic polynomial. For other invariants like the residual connectedness reliability node decomposition formulae are known.

The existence of polynomial algorithms for the computation of such invariants in graphs of bounded pathwidth (and even for the larger class of graphs of bounded treewidth, see e.g. [And95,Bod97]) is well-known; it is rather the simplicity and performance of algorithms created by the composition method which makes our approach appealing. Starting from a decomposition formula and assuming cost of $O(\omega(n))$ for elementary operations on the values, a working $O((m + n)\, n\, (\omega(n) + \log n))$ (n being the number of nodes, m the number of edges in the graph) implementation can be obtained completely automatically within a few minutes.

After introducing some essential notation in the next section, we will give a description of the algorithm, a sketch for proving its correctness, an evaluation of its complexity, and two examples.

2 Notation

Let $G = (V, E)$ be a finite, undirected graph, $n = |V|$, $m = |E|$ and $t = 2n + m$. A sequence $S = (s_i)_{i=1}^t$, $s_i = (x_i, c_i)$ with an *element* $x_i \in V \cup E$, and a *code* $c_i \in \{A, K, D\}$ is called a *composition sequence* if the following holds:

$$\forall v \in V(G)\, \exists v_A, v_D \in \{1 \dots t\} : v_A \leq i < v_D \;\wedge\; s_{v_A} = (v, A) \;\wedge\; s_{v_D} = (v, D)$$

$$\begin{aligned} \forall E = (uv) \in E(G)\, \exists u_A, v_A, k, u_D, v_D \in \{1 \dots t\} : \\ u_A, v_A < k < u_D, v_D \;\wedge\; s_{v_A} = (v, A) \;\wedge\; s_{u_A} = (u, A) \\ \wedge\; s_k = (e, K) \;\wedge\; s_{v_D} = (v, D) \;\wedge\; s_{u_D} = (u, D)\,. \end{aligned} \tag{1}$$

A composition sequence induces a sequence (A_i) of *active node sets* by

$$A_i = \{v \in V(G) : \exists v_A < v_D : s_{v_A} = (v, A) \;\wedge\; s_{v_D} = (v, D)\}\,.$$

The *width* $\mathrm{wd}(S)$ of a composition sequence S is defined as

$$\mathrm{wd}(S) = \max_i \operatorname{card} A_i\,.$$

The set of all composition sequences for a given graph will be denoted by $\mathcal{S}(G)$. It is easy to see that for a graph of pathwidth p there is a composition sequence of width p as one can be derived from a path decomposition of the graph.

Given some graph invariant R which takes values in some ring $\mathcal{W}$, we call a triplet $T = (a, b, r)$ of functions $a, b : E(G) \to \mathcal{W}$, $r : \mathbb{Z} \mapsto \mathcal{W}$ an *edge decomposition formula for the invariant* R if the following holds:

$$R(G) = a(e)\, R(G_e) + b(e)\, R(G_{-e}) \tag{2}$$

$$R(\overline{K_n}) = r(n), \quad r(0) = 1\,. \tag{3}$$

In this formula, $\overline{K_n}$ denotes an edgeless graph on n nodes.

Let $\mathbb{P}(V)$ denote the set of all set partitions of subsets of the node set V. A pair $[\pi, d] : \pi \in \mathbb{P}(V), d \in \mathbb{Z}$ will be referred to as an *index*, an element w of $\mathcal{W}$ as a *value*. Moreover, a pair $z = (i, w)$ of an index $i = [\pi, d]$ and a value w will be called a *state*. A multiset Z of states will be called a *stateset*, the set $Z_0 = \{([\emptyset, 0], 0)\}$ the *initial stateset*. $\mathcal{Z}$ shall denote the set of all statesets.

Furthermore, let $\pi|v$ denote a partition consisting of the blocks of π and an additional block containing only v, $\pi_{(uv)}$ a partition which results from π by merging the blocks containing u and v, π^u the block of π containing u, and $\pi - v$ the partition π with node v removed from π^v. Finally, for an edge

$e = (uv)$ let $[\pi, d]_{(e)}$ denote $[\pi_{(uv)}, d']$ where $d' = d$ if $\pi^u = \pi^v$ and $d' = d - 1$ otherwise.

Given a graph G and an edge decomposition formula T, a function $f : \mathcal{S}(G) \times \mathcal{Z} \to \mathcal{Z}$, $f(s_i, Z_{i-1}) = Z_i$ is called a *transfer function* if

$$f((v, A), Z) = \cup_{([\pi,d],w)\in Z} \{([\pi|v, d+1], w)\}$$
$$f((e, K), Z) = \cup_{([\pi,d],w)\in Z} \{([\pi, d]_{(e)}, a(e)\, w), ([\pi, d], b(e)\, w)\}$$
$$f((v, D), Z) = \cup_{([\pi,d],w)\in Z} \{([\pi - v, d], w)\}$$

for all $Z \in \mathcal{Z}$ and

$$f\,(s_t, \ldots f\,(s_2, f\,(s_1, Z_0)) \ldots) = f^* = \text{const}$$

for all composition sequences $S = (s_i) \in \mathcal{S}(G)$ of the given graph G.

3 The Algorithm

The following algorithm can be used to compute a graph invariant given by some decomposition formula T for a graph instance G:

input $G \ldots$ a graph, $T \ldots$ a decomposition formula

compute some composition sequence $S(G) = (s_i)$

set $Z_0 = \{([\emptyset, 0], 1)\}$

for i from 1 to t do

  set $Z_i = \text{collect}\,(f(s_i, Z_{i-1}))$

output $\sum_{((\pi,d),w)\in Z_t} r(d)\, w$

In this algorithm, we use a function collect: $\mathcal{Z} \to \mathcal{Z}$, $Z \mapsto Z'$ that replaces states with the same index in Z with a single state of the same index in Z'. The value of the new state equals the sum of the values of the replaced states. This collection step does not actually change the correctness of the algorithm, but it is crucial for the performance as it guarantees that statesets never contain more than one state for a given index at the end of each iteration of the loop.

3.1 Correctness

Let $G = (V, E)$ a graph. If we apply (2) repeatedly, we end up with a sum of the form

$$R(G) = \sum_{X \subseteq E} r\,(k(G{:}X))\; p_E(X) \tag{4}$$

with $k(G{:}X)$ denoting the number of connected components of $G{:}X = (V, X)$ and $p_E(X)$ being an abbreviation for $\prod_{e \in X} a(e) \times \prod_{e \in E \setminus X} b(e)$.

Let us fix some subset A of V. We call the index $[\pi, d]$ *induced by* $X \subseteq E$ *in* A (written as $[\pi, d] = \mathrm{ind}_A X$) if there are d connected components in $G{:}X$ and the blocks of π correspond to the intersection of A with $G{:}X$. By collecting the terms in (4) into blocks with the same induced index we obtain

$$R(G) = \sum_{[\pi,d]} r(d) \sum_{X \subseteq E \,:\, \mathrm{ind}_A X = [\pi,d]} p_E(X) \,. \tag{5}$$

By introducing the set

$$\mathrm{desc}_A(G) := \left\{ \left([\pi, d], \sum_{X \subseteq E \,:\, \mathrm{ind}_A X = [\pi,d]} p_E(X) \right) \right\}$$

(5) can be written as

$$R(G) = \sum_{([\pi,d],w) \in \mathrm{desc}_A(G)} r(d)\, w \,.$$

With $G_i = (V_i, E_i)$ being the subgraph of G which consists of all nodes activated and edges handled in the first i steps of S and A_i the ith active node set we are prepared to state the following lemma:

Lemma: The stateset Z_i after step i equals $\mathrm{desc}_{A_i}(G_i)$.

Proof (Sketch): In the beginning we have $A_0 = \emptyset$ and an empty graph G_0 with zero connected components. Consequently there is only a single index $[\emptyset, 0]$, and the products in (5) both range over empty sets, so their value is 1. This coincides with the initial stateset $Z_0 = \{([\emptyset, 0], 1)\}$ as used by the algorithm. The proof continues by assuming $Z_{i-1} = \mathrm{desc}_{A_{i-1}}(G_{i-1})$ and distinguishes three cases according to the three possible values for the code c_i in step $s_i = (x_i, c_i)$. In each case Z_i is built according to the rule given in the algorithm and by a sequence of simple operations this can be shown to be equal to $Z_i = \mathrm{desc}_{A_i}(G_i)$. These operations are elementary, albeit a bit lengthy, so the details should be skipped here.

3.2 Complexity

Let us assume we are given a composition sequence $S \in \mathcal{S}(G)$ of some graph G of width p, some decomposition formula $T = (a, b, r)$ and assume further that an operation like addition or multiplication of two values as well as the evaluation of a, b, and r can be done in $O(\omega(n))$ time.

For each node of the graph we have an activation and a deactivation step, and for each edge an edge step, resulting in a total of $O(n+m)$ iterations in the outer loop of the algorithm.

In each iteration, some stateset is transformed by looping over all states. The index of such a state consists of a set partition of the current set of active nodes (as in the node deactivation steps nodes are removed from the indices) and an integer between 0 and n. The number of set partitions over p nodes is given by the pth Bell number, which is, given bounded p, bounded by some constant. As the collection step replaces multiple states with the same index by one state, there cannot be more the $B(p)\,(n+1)$ different indices – and as a result of the collect function no more states than that – at the beginning of the transformation of the stateset.

Each individual state transformation consists of a bounded number of simple operations. The most difficult ones include the melting of two blocks of a given set partition of a set of maximum cardinality p, which can be done in unit time for bounded p, and at most four operations on values taking time of at most $O(\omega)$ leading to a maximum complexity $O(n\,\omega)$ for the state transformation.

Each state gets transformed to at most two new states such that the total number of states produced in can't be larger than $2\,B(p)\,(n+1)$ i.e. $O(n)$. In order to perform the collection step after the transformation, we can sort the stateset in $O(n\,\log n)$ and collapse sequences of states with identical index in a total of $O(n\,\omega)$. So the total amount of work for one iteration is $O(n\,(\log n+\omega))$, leading to a total for the whole algorithm of $O(n\,(n+m)\,(\log n+\omega))$.

Depending on the actual nature of value operations or the level of abstraction one could assume $\omega = O(1)$ and $n = O(m)$ which leads to $O(n^2\,\log n)$.

4 Examples

Negami polynomial: The triplet $T = (a, b, r)$ with $a(e) = x$, $b(e) = y$ and $r(n) = t^n$ is a description of the Negami polynomial (see e.g. [Neg87]). Some experimental results for $p \times p$ grids are shown in the following table with "#total" denoting the total number of states, "#max" the maximum size of a stateset, "time" the approximates running time in minutes on a AMD1700+:

p	#total	#max	time	#total	#max	time
6	125.898	4.770	0 : 05	10.087	264	0 : 04
7	790.878	22.760	1 : 03	44.032	858	0 : 40
8	4.078.059	93.252	17 : 04	179.910	2.860	5 : 30

The second block of results show the results of an manually improved algorithm which uses a slightly different transfer function that leads to fewer states during the course of the algorithm (but, of course, identical final results).

As there is a well-known linear transformation between the Negami polynomial and the Tutte polynomial of a graph, we get composition algorithms for the Tutte polynomial as well as for several invariants derived from the Tutte polynomial without much additional work.

All-terminal reliability: The triplet $T = (a, b, r)$ with $a(e) = p_e$, $b(e) = 1 - p_e$, $r(0) = r(1) = 1$ and $r(n) = 0$ for $n > 1$ constitutes a decomposition formula for the all-terminal reliability. For $p \times p$ grids we obtain:

p	# total	# max	time	result
9	745.202	9.724	$0:03$	$1.58e-5$
10	3.092.058	33.592	$0:12$	$2.24e-6$
11	12.734.918	117.572	$0:52$	$2.27e-7$
12	53.276.048	416.024	$4:22$	$2.85e-8$
13	218.954.187	1.485.800	$20:36$	$2.46e-9$
14	917.861.182	5.348.880	$95:50$	$1.97e-10$

The column "result" shows the actual all-terminal reliability for all edge failing independently with probability 0.9. It should be noted that the algorithm does work equally well if these probabilities are not equal for all edges.

5 Conclusion

The proposed algorithm is very easy to implement. If performance is crucial, it gives a decent starting point for further manual optimization. Some special graph structures automatically lead to better performance without any additional work as e.g. for planar graphs, only non-crossing partitions are created. The method seems to be easily generalizable to solve problems with more complex decomposition formulae and/or other kinds of descriptions like splitting formulae.

References

[And95] Artur Andrzejak (1995) A polynomial-time algorithm for computation of the Tutte polynomials of graphs of bounded treewidth. http://citeseer.nj.nec.com/andrzejak95polynomialtime.html

[Bod97] Hans Bodlaender (1995) Treewidth: Algorithmic techniques and results. Proceedings 22nd Intern. Symposium on Math. Foundations of Computer Science, MFCS'97, Lecture Notes in Computer Science, volume 12/95

[Neg87] Seiya Negami (1987) Polynomial invariants of graphs. Transactions of the American Mathematical Society, 299: 601-622

[Poe99] André Pönitz (1999) Computing invariants in graphs of small bandwidth. Mathematics in Computers and Simulation, 49: 179-191

The Complexity of Some Problems on Maximal Independent Sets in Graphs

Igor Zverovich[1] and Yury Orlovich[2]

[1] RUTCOR-Rutgers Center for Operations Research, Rutgers University, 640 Bartholomew Rd, Piscataway, NJ 08854-8003, USA, e-mail: igor@rutcor.rutgers.edu

[2] Institute of Mathematics, National Academy of Sciences of Belarus, 11 Surganov str, 220072 Minsk, Belarus, e-mail: orlovich@im.bas-net.by

Abstract. Let $\text{mi}(G)$ be the number of maximal independent sets in a graph G. A graph G is *mi-minimal* if $\text{mi}(H) < \text{mi}(G)$ for each proper induced subgraph H of G. As it is shown in [6], every graph G without duplicated or isolated vertices has at most $2^{k-1}+k-2$ vertices, where $k = \text{mi}(G) \geq 2$. Hence the extremal problem of calculating $m(k) = \max\{|V(G)| : G \text{ is a mi-minimal graph with } \text{mi}(G) = k\}$ has a solution for any $k \geq 1$. We show that $2(k-1) \leq m(k) \leq k(k-1)$ for any $k \geq 2$ and conjecture that $m(k) = 2(k-1)$. We also prove NP-completeness of some related problems.

1 Introduction

We consider finite undirected graphs without loops or multiple edges. Standard graph-theoretical terminology not presented here can be found in [3]. If G is a graph, $V(G)$ (respectively, $E(G)$) is the vertex set (respectively, the edge set) of G. We denote by $n = |V(G)|$ the order of G. We write K_n for the complete graph of order n. The *open neighborhood* of a vertex $x \in V(G)$ is the set $N_G(x) = N(x) = \{y \in V(G) : xy \in E(G)\}$. The *closed neighborhood* of x, denoted by $N_G[x] = N[x]$, is the set $N_G(x) \cup \{x\}$. In addition, we define $N_G(X) = N(X) = \bigcup_{x \in X} N_G(x)$ and $N_G[X] = N[X] = N_G(X) \cup X$ for a subset X of $V(G)$. The subgraph of G induced by X is denoted by $G(X)$. A *clique* in a graph G is a maximal (by inclusion) vertex set which induces a complete subgraph in G. A subset $I \subseteq V(G)$ is *independent* if $G(I)$ is an edgeless graph. A *maximal* independent set is an independent set that is not a proper subset of any other independent set. Let $\text{Ind}(G)$ be the set of all maximal independent sets of a graph G and $\text{mi}(G) = |\text{Ind}(G)|$.

Erdös and Moser raised the problem of calculating the function

$$f(n) = \max\{\text{mi}(G) : |V(G)| = n\}$$

and determining those graphs G on which this maximum value is achieved. This problem was solved by Erdös, and later by Moon and Moser [8]. It was then extensively studied for various classes of graphs, including trees, forests, graphs with at most one cycle, bipartite graphs, connected graphs, k-connected graphs, triangle-free graphs; for a survey see [5].

The reverse problem of calculating

$$g(k) = \max\{|V(G)| : \text{mi}(G) = k\}$$

is trivial because $|V(G)|$ is not bounded above on $\text{IN}(k) = \{G : \text{mi}(G) = k\}$. However, we can restrict the set $\text{IN}(k)$ to minimal graphs in the following sense. A graph G is *mi-minimal* if $\text{mi}(H) < \text{mi}(G)$ for each proper induced subgraph H of G. In the sequel we will use the proposition below. This proposition is proved in [6].

Proposition 1. *If H is an induced subgraph of G then* $\text{mi}(H) \leq \text{mi}(G)$.

Denote by $\text{MIN}(k)$ the set of all mi-minimal G with $\text{mi}(G) = k$ and put

$$m(k) = \max\{|V(G)| : G \in \text{MIN}(k)\}.$$

Vertices u and v are called *duplicated* if $N(u) = N(v)$. If u and v are duplicated vertices of a graph G then $\text{mi}(G - u) = \text{mi}(G)$. Also, deleting isolated vertices from a graph $G \neq K_1$ produces an induced subgraph H with $\text{mi}(H) = \text{mi}(G)$. The following theorem was proved in [6].

Theorem 1. *Let G be a graph without duplicated or isolated vertices and* $\text{mi}(G) = k \geq 2$. *Then* $|V(G)| \leq 2^{k-1} + k - 2$.

As an easy consequence of Theorem 1 we obtain that the set $\text{MIN}(k)$ is finite for all $k \geq 1$ and $m(k) \leq 2^{k-1} + k - 2$ for all $k \geq 2$.

In section 2, we establish a better bound on $m(k)$. Namely, we show that $2(k-1) \leq m(k) \leq k(k-1)$ for any $k \geq 2$ and conjecture that $m(k) = 2k-2$.

In section 3, we prove NP-completeness of some problems associated with the parameter $\text{mi}(G)$. Next, we discuss the complexity of the problems arising from the theory of general partition graphs [7]. These graphs recently have been characterized in terms of covers by universal cliques. We prove that the problem NON-UNIVERSAL CLIQUE is NP-complete even when the graphs in question are weakly chordal and therefore perfect.

2 Bounds for $m(k)$

Our first result is the following.

Theorem 2. $2(k-1) \leq m(k) \leq k(k-1)$ *for every* $k \geq 2$.

Proof. Let K_{k-1} be the complete graph with vertex set $V(K_{k-1}) = V = \{v_1, v_2, \ldots, v_{k-1}\}$, and let $W \cap V = \emptyset$, where $W = \{w_1, w_2, \ldots, w_{k-1}\}$. We define a graph G_k in the following way: $V(G_k) = W \cup V$, $E(G_k) = E(K_{k-1}) \cup \{v_i w_i : i = 1, 2, \ldots, k-1\}$. It easy to check that $\text{mi}(G_k) = k$ and $\text{mi}(G_k - u) <$

k for any vertex $u \in V(G_k)$. Proposition 1 implies $\mathrm{mi}(H) < \mathrm{mi}(G)$ for every proper induced subgraph H of G. Thus, $G_k \in \mathrm{MIN}(k)$ and $m(k) \geq |V(G_k)| = 2(k-1)$.

Further, let $G \in \mathrm{MIN}(k)$. Then G contains k pairwise distinct maximal independent sets $I_1, I_2, \ldots, I_k$. For all $1 \leq i \neq j \leq k$ there are adjacent vertices $u_{ij} \in I_i$ and $u_{ji} \in I_j$. We denote $U_i = \{u_{ij} : j = 1, 2, \ldots, i-1, i+1, i+2, \ldots, k\}$, $U = \bigcup_{i=1}^{k} U_i$ and $H = G(U)$. The set U_i is independent in H since $U_i \subseteq I_i$. We extend U_i to a maximal independent set J_i of H. We have $J_i \neq J_j$ when $i \neq j$ because there are adjacent vertices $u_{ij} \in J_i$ and $u_{ji} \in J_j$. Hence $\mathrm{mi}(H) \geq k$. By mi-minimality of G, we have $G = H$ and $m(k) \leq |V(H)| = k(k-1)$. The proof of the theorem is complete.

We conjecture that $m(k) = 2k - 2$ for any $k \geq 2$.

3 Complexity results

We shall investigate the complexity of the following problems.

Decision Problem 1 (MI-MINIMAL GRAPH).
Instance: a graph G.
Question: is G a mi-minimal graph?
Decision Problem 2 (MI-CRITICAL VERTEX).
Instance: a graph G and a vertex u of G.
Question: is it true that $\mathrm{mi}(G - u) < \mathrm{mi}(G)$?

A set $X \subseteq V(G)$ *dominates* a set $Y \subseteq V(G)$ if $Y \subseteq N[X]$. The following lemma will be useful for the proofs of Theorem 3 and Theorem 4.

Lemma 1. *Let x be a vertex of a graph G. Then $\mathrm{mi}(G - x) < \mathrm{mi}(G)$ if and only if there exists a maximal independent set J in $G - N[x]$ that does not dominate $N(x)$ in G.*

Proof. We denote

- $A = \{I \in \mathrm{Ind}(G) : x \notin I\}$,
- $B = \{I \in \mathrm{Ind}(G) : x \in I$ and $I \setminus \{x\}$ dominates $N(x)\}$,
- $C = \{I \in \mathrm{Ind}(G) : x \in I$ and $I \setminus \{x\}$ does not dominate $N(x)\}$,
- $A' = \{I \in \mathrm{Ind}(G - x) : I \cap N(x) \neq \emptyset\}$, and
- $B' = \{I \in \mathrm{Ind}(G - x) : I \cap N(x) = \emptyset\}$.

It is clear that $A = A'$ and $I \to I \setminus \{x\}$ is a bijection between B and B'. We have

$$|A \cup B| = |A| + |B| = |A'| + |B'| = |\mathrm{Ind}(G - x)|.$$

Therefore $\mathrm{mi}(G - x) < \mathrm{mi}(G)$ if and only if $C \neq \emptyset$. Finally, there exists I in C if and only if the set $J = I \setminus \{x\}$ is a maximal independent set in $G - N[x]$ that does not dominate $N(x)$ in G. Lemma 1 is proved.

Theorem 3. *MI-MINIMAL GRAPH and MI-CRITICAL VERTEX problems are polynomially equivalent.*

Proof. It follows from Proposition 1 that a graph G is mi-minimal if and only if $\mathrm{mi}(G - u) < \mathrm{mi}(G)$ for each vertex u of G. Hence MI-MINIMAL GRAPH is polynomially reducible to MI-CRITICAL VERTEX.

Conversely, let G and u be an instance to MI-CRITICAL VERTEX. We construct a new graph F by

- taking two disjoint copies G and G' of G,
- fixing an isomorphism $\varphi : V(G) \to V(G')$, $v \mapsto v'$, and
- adding new edges $v'w$ for all $v' \in V(G')$ and $w \in N_G[v]$.

Now we delete the vertex u' from F, and we denote the resulting graph by H. By the construction, $N_H(v) = N_H(v')$ for every vertex $v \neq u$. This fact and Lemma 1 imply that $\mathrm{mi}(H-x) < \mathrm{mi}(H)$ for every vertex $x \in V(H)\setminus\{u\}$.

Thus, H is a mi-minimal graph if and only if $\mathrm{mi}(H - u) < \mathrm{mi}(H)$. It is clear from the definition of H and Lemma 1 that the last inequality holds if and only if $\mathrm{mi}(G - u) < \mathrm{mi}(G)$. In other words, MI-CRITICAL VERTEX problem for G and u is polynomially reducible to MI-MINIMAL GRAPH problem for H. This completes the proof of Theorem 3.

Theorem 4. *Both MI-MINIMAL GRAPH and MI-CRITICAL VERTEX are NP-hard problems.*

Proof. According to Theorem 3, it is sufficient to consider MI-CRITICAL VERTEX problem only.

We shall construct a polynomial-time reduction from SATISFIABILITY which is an NP-complete problem (see [2]). An instance of SATISFIABILITY is a set $C = \{c_1, c_2, \ldots, c_m\}$ of clauses over the set $X = \{x_1, x_2, \ldots, x_n\}$ of Boolean variables (each clause being an elementary conjunctions of some literals over X). The set of literals over X is

$$L_X = \{x_1, x_2, \ldots, x_n\} \cup \{\overline{x}_1, \overline{x}_2, \ldots, \overline{x}_n\}$$

Question: is there a truth assignment to X that satisfies all the clauses in C?

Given an instance X, C to SATISFIABILITY, we construct a graph G with $V(G) = \{u, v\} \cup L_X \cup C$ as follows:

- $N_G(u) = \{v\}$ and $N_G(v) = \{u\} \cup C$,
- the set L_X induces a matching $x_1\overline{x}_1, x_2\overline{x}_2, \ldots, x_n\overline{x}_n$,
- the set C is independent, and
- a vertex $y \in L_X$ is adjacent to a vertex $c_i \in C$ if and only if the clause c_i includes the literal y.

We consider G and u as an instance to MI-CRITICAL VERTEX. By Lemma 1, $\mathrm{mi}(G - u) < \mathrm{mi}(G)$ if and only if there exists a maximal independent set J in $G - N[u] = G(C \cup L_X)$ that does not dominate $N(u) = \{v\}$ in

G. Since each vertex of C is adjacent to v and J does not dominate v, J must be a maximal stable set in $G(L_X)$. Moreover, J dominates C (otherwise J is not a maximal independent set in $G - N[u]$).

An independent subset of L_X that dominates C defines the following truth assignment to X that satisfies all the clauses in C : a literal y is true if and only if $y \in J$. Conversely, given a truth assignment to X that satisfies all the clauses in C, we can define a required maximal independent set J : a vertex y of L_X is in J if and only if the literal y is true.

Thus, SATISFIABILITY is polynomially reducible to MI-CRITICAL VERTEX. The proof of the theorem is complete.

A clique in a graph G is called *universal* if it intersects all maximal independent sets of G. We consider the complexity of the following problem.

Decision Problem 3 (NON-UNIVERSAL CLIQUE).
Instance: A graph G and a clique C of G.
Question: Is C is not a universal clique?

We use a construction as in [1] for the problem NOT-WELL COVERED GRAPH. To show that determining if C is not universal is NP-complete, we will show that there is a polynomial time reduction from 3-SATISFIABILITY.

Decision Problem 4 (3-SATISFIABILITY).
Instance: Collection $C = \{c_1, c_2, \dots, c_m\}$ of clauses on a finite set $U = \{u_1, u_2, \dots, u_n\}$ of variables such that $|c_i| = 3$ for $i = 1, 2, \dots, m$.
Question: Is there a truth assignment for U that satisfies all the clauses in C?

Theorem 5. *NON-UNIVERSAL CLIQUE is an NP-complete problem.*

Proof. One can easily verify the independence and maximality of a proposed vertex set I with $I \cap C = \emptyset$, showing that the problem is in NP.

Let C and U be an instance of 3-SATISFIABILITY. We construct a graph G on a vertex set

$$\{u_1, \overline{u}_1, u_2, \overline{u}_2, \dots, u_n, \overline{u}_n\} \cup V,$$

where $V = \{v_1, v_2, \dots, v_m\}$, as follows. For each $i = 1, 2, \dots, n$ form a K_2 on vertices u_i and $\overline{u}_i$; form K_m on a vertex set V; if $c_j = (x_{j,1}, x_{j,2}, x_{j,3})$ (where each $x_{j,k}$ is some u_i or $\overline{u}_i\}$) make v_j adjacent to $x_{j,1}, x_{j,2}$, and $x_{j,3}$), $j = 1, 2, \dots, m$. Clearly, G can be constructed in time polynomial in the length of C.

If there exists a vertex u ($u = u_i$ or $u = \overline{u}_i$) which is adjacent to all vertices in V, then 3-SATISFIABILITY has a trivial solution (all clauses contain a common literal). Suppose it is not the case. Hence V is a clique in G. We consider G and V as an instance for NON-UNIVERSAL CLIQUE.

Assume that there is a satisfying truth assignment

$$t : \{u_1, u_2, \dots, u_n\} \to \{T, F\}.$$

Put $I = \{u_i : t(u_i) = T\} \cup \{\overline{u}_i : t(u_i) = F\}$. Because every c_i contains a true literal, each v_i is adjacent to a vertex in I. So I is a maximal independent set and $I \cap V = \emptyset$. We obtain that V is not a universal clique.

Conversely, suppose that there is no satisfying truth assignment. Consider a maximal independent set I in G which does not intersect V. Clearly, I contains exactly one of u_i and $\overline{u}_i$. Let us define the truth assignment on

$$\{u_1, \overline{u}_1, u_2, \overline{u}_2, \ldots, u_n, \overline{u}_n\}$$

by letting a literal be true if and only if it is in I. Maximality of I implies that each v_j is adjacent to a vertex in I. So we construct a satisfying truth assignment, a contradiction. Hence V intersects all maximal independent sets in G and V is a universal clique. The proof of the theorem is complete.

A graph is called *weakly chordal* if neither it nor its complement contains a chordless cycle with more than four vertices. It is possible to show that the graph G obtained in the proof of Theorem 5 is a weakly chordal graph. Hence, NON-UNIVERSAL CLIQUE is NP-complete for weakly chordal graph. Since it has been proven by Hayward [4] that weakly chordal graphs are also perfect graphs, then we conclude that the problem NON-UNIVERSAL CLIQUE for perfect graphs is also NP-complete.

4 Acknowledgements

The research of the second author financed by the Institute of Mathematics of the NASB within the framework of the State program "Mathematical Structures", and partly supported by INTAS (Project INTAS 00-217). Also, the authors would like to thank the referees for their useful comments and suggestions.

References

1. Chvátal, V., Slater, P. J. (1993) A note on well-covered graphs. Ann. Discrete Math. **55**, 179–182
2. Garey M. R., Johnson D. S. (1979) Computers and Intractability. W. H. Freeman and Company, San Francisco
3. Harary F. (1969) Graph Theory. Addison-Wesley
4. Hayward, R. B. (1985) Weakly trianqulated graphs. J. Comb. Theory. Ser. B **39**, 200–208
5. Jou, M.-J., Chang, G. J. (1995) Survey on counting maximal independent sets. Proc. Second Asian Math. Conf. World Scientific, Singapore, 265–275
6. Jou, M.-J., Chang, G. J., Lin, C., Ma, T.-H. (1996) A finiteness theorem for maximal independent sets. Graphs and Combin. **12**, 321–326
7. McAvaney, K., Robertson, J., DeTemple, D. (1993) A characterization and hereditary properties for partition graphs. Discrete Math. **113**, 131–142
8. Moon, J. W., Moser, L. (1965) On cliques in graphs. Israel J. Math. **3**, 23–28

Testing Solution Quality in Stochastic Programs

David P. Morton

Graduate Program in Operations Research, The University of Texas at Austin, Austin, TX 78712, USA, morton@mail.utexas.edu

Abstract. We describe a statistical procedure for testing the quality of a feasible candidate solution for an important class of stochastic programs. Quality is defined via the so-called optimality gap and the procedure's output is a confidence interval on this gap. We review a multiple-replications procedure for constructing the confidence interval. Then, we present a result that allows the procedure to be computationally simplified to a single-replication procedure.

1 Introduction

We consider a stochastic optimization problem of the form

$$z^* = \min_{x \in X} Ef(x, \xi), \qquad (SP)$$

where f is a real-valued function measuring the performance of a system of interest, x is a decision vector constrained to obey physical and policy rules represented by the set X, ξ is a random vector, and E is the expectation operator. We denote a solution to (SP) as (x^*, z^*). Throughout we assume f can be evaluated exactly, given specific values for its arguments.

Unless the random vector ξ has a small number of realizations (also called scenarios) or f has a particularly simple structure, it is usually impossible to solve (SP) exactly. For problems in which ξ is of moderate-to-high dimension and is continuous or has a large number of realizations, Monte Carlo simulation is widely regarded as the method of choice for estimating $Ef(x, \xi)$, when x is fixed. So, one approach for approximately optimizing (SP) is to sample n independent and identically distributed (i.i.d.) observations ξ^i, $i = 1, \ldots, n$, from the distribution of ξ (of course, other sampling schemes are also possible) and then solve the approximating problem

$$z_n^* = \min_{x \in X} \frac{1}{n} \sum_{i=1}^{n} f(x, \xi^i). \qquad (SP_n)$$

Solving (SP_n) in place of (SP) is justified by consistency results that establish conditions on f, X, ξ and the sampling procedure under which solutions (x_n^*, z_n^*) to (SP_n) are optimal to (SP) as the sample size n grows to infinity [3,10,15,17]. These consistency results are clearly needed, but they are not enough because they tell us nothing about the quality of a solution,

x_n^*, obtained by solving (SP_n) for finite n. In this paper we first review a multiple-replications methodology for establishing the quality of the solution x_n^* obtained by solving (SP_n) and then prove a result that allows the procedure to be computationally simplified to a single-replication procedure.

The optimal solution value z_n^* to (SP_n) plays an important role in establishing the quality of a candidate solution. So, before proceeding, we give a simple example that illustrates a number of properties of z_n^*.

Example 1. Define (SP) through $X = [-1,1]$, $\xi \sim N(0,1)$, and $f(x,\xi) = x\xi$. Clearly, $z^* = 0$ and every feasible solution is an optimal solution to (SP). Even though this is a trivial problem we can form the approximating problem

$$z_n^* = \min_{-1 \le x \le 1} \left(\frac{1}{n} \sum_{i=1}^{n} \xi^i \right) x$$

and we find $x_n^* = 1$ if $\sum_{i=1}^{n} \xi^i \le 0$ and $x_n^* = -1$ if $\sum_{i=1}^{n} \xi^i > 0$. The objective coefficient is an average of n i.i.d. standard normals and so $z_n^* = -|N(0,1/n)|$. In this example, z_n^* has the following properties:

1. $Ez_n^* \le z^* \forall n$ *negative bias*
2. $Ez_n^* \le Ez_{n+1}^*$ *monotonically shrinking bias*
3. $z_n^* \to z^*$, *wp1*   *strong consistency*
4. $\sqrt{n}(z_n^* - z^*) = -|N(0,1)|$ *non-normal errors*
5. $B_z(n) \equiv Ez_n^* - z^* = a_1/\sqrt{n}$ $O(n^{-1/2})$ *bias.* □

These first three properties hold much more generally, although some mild conditions are needed (as can be seen by replacing $X = [-1,1]$ with $X = \Re$). The fourth condition is not in a form to hold for more general instances of (SP). Instead of "=" we usually have "⇒", i.e., convergence in distribution, and the limiting distribution will differ. That said, the result is representative of the more general case both with respect to the $n^{-1/2}$ rate of convergence and the "folded" or "projected" normal random variables that arise [2,8,17]. The fifth property concerns the rate at which z_n^*'s bias shrinks to zero. When (SP) has multiple optimal solutions (as in the example), $O(n^{-1/2})$ bias arises in a very general setting. However, when (SP) has a unique optimal solution, $Ef(x,\xi)$ is sufficiently smooth, and X satisfies a regularity condition, the bias is of order $O(n^{-1})$ [18].

2 A Multiple Replications Procedure

A key observation from the discussion in Section 1 is that z_n^* gives a lower bound, in expectation, on the optimal solution value z^*. In integer programming, and other areas of optimization, lower bounds (for minimization problems) arise via relaxations of integer constraints and relaxations of other complicating constraints. Such bounds play fundamental roles in proving optimality, or near-optimality, of candidate solutions. Our procedure for stochastic programs is analogous except that our lower bounds are statistical in nature and therefore yield different types of optimality statements compared to deterministic problems.

We will measure the quality of a candidate solution $\hat{x}$, e.g., $\hat{x} = x_n^*$, by the *optimality gap*, $Ef(\hat{x}, \xi) - z^*$. If the gap is sufficiently small then $\hat{x}$ is of high quality. Unfortunately, in our setting this gap cannot be computed exactly. In earlier work [11], we establish the bias result, $Ez_n^* \le z^*$, and circumvent the issue of non-normal errors for z_n^* via a replications procedure to construct a confidence interval (CI) of the form

$$P\{Ef(\hat{x}, \xi) \le z^* + \epsilon\} \approx 1 - \alpha. \tag{1}$$

Here, $\hat{x} \in X$ is a candidate solution, $Ef(\hat{x}, \xi)$ is its "true" and unknown expected performance measure, ϵ is the (random) CI width, and $1 - \alpha$ is the confidence level, e.g., 0.95. We summarize below our procedure for constructing (1) from [11].

Procedure MCSP
Input: Data for (SP). Batch size n, number of batches n_g, and n_x which is the size of the approximating problem used to obtain the candidate solution. Confidence level $1 - \alpha$ and t distribution quantile $t_{n_g-1,\alpha}$.
Output: Candidate solution $\hat{x}$ and approximate $(1 - \alpha)$-level CI $[0, \bar{G}_{n_g} + \epsilon_g]$ on $Ef(\hat{x}, \xi) - z^*$.

1. Sample observations $\xi^1, \ldots, \xi^{n_x}$ and solve (SP_{n_x}) to obtain $\hat{x}$
2. Sample i.i.d. batches $\xi^{i1}, \ldots, \xi^{in}$ for $i = 1, \ldots, n_g$
3. For each $i = 1, \ldots, n_g$ calculate

$$G_n^i = \frac{1}{n} \sum_{j=1}^{n} f(\hat{x}, \xi^{ij}) - \min_{x \in X} \frac{1}{n} \sum_{j=1}^{n} f(x, \xi^{ij}) \tag{2}$$

4. Let $\bar{G}_{n_g} = \frac{1}{n_g} \sum_{i=1}^{n_g} G_n^i$, $s_g^2 = \frac{1}{n_g - 1} \sum_{i=1}^{n_g} (G_n^i - \bar{G}_{n_g})^2$, and $\epsilon_g = t_{n_g-1,\alpha} s_g / \sqrt{n_g}$
5. Print("Candidate solution:",$\hat{x}$, "Confidence interval on optimality gap:", $[0, \bar{G}_{n_g} + \epsilon_g]$)

In Step 1, (SP_{n_x}) is solved to obtain the candidate solution. However, any technique that generates a feasible solution, $\hat{x} \in X$, can be used instead. The CI $[0, \bar{G}_{n_g} + \epsilon_g]$ on $Ef(\hat{x}, \xi) - z^*$ is inferred from the central limit theorem (CLT)

$$\sqrt{n_g}\left[\bar{G}_{n_g} - EG_n^1\right] \Rightarrow N(0, \sigma_g^2) \text{ as } n_g \to \infty \text{ where } \sigma_g^2 = \operatorname{var} G_n^1,$$

and the fact that $EG_n^1 \ge Ef(\hat{x}, \xi) - z^*$.

MCSP has been applied in several settings [1,7,12,14,19–21]. Other related work on establishing solution quality for stochastic programs includes [4,16] and algorithm-specific work [5,6,13].

One of the shortcomings of the MCSP approach is the cost of performing multiple replications, i.e., having to solve (say) $n_g = 20$ instances of (SP_n), can be prohibitive. The technical reason for performing n_g replications is that z_n^* can have non-normal errors. Despite this, we develop an approach in the next section that enables us to assess solution quality of $\hat{x}$ by solving a single instance of (SP_n).

3 A Single Replication Procedure

In this section we show how solving a single instance of (SP_n) yields sufficient information regarding a lower bound on z^* so that we can make a valid statistical inference concerning the quality of a candidate solution $\hat{x}$. We will assume: $\xi^1, \ldots, \xi^n$ are i.i.d. from the distribution of ξ, $Ef^2(x, \xi) < \infty \, \forall x \in X$, X is compact, and $Ef(x, \xi)$ is continuous on X. We will also assume that all limit points of $\{x_n^*\}_{n=1}^{\infty}$ are optimal to (SP). This consistency hypothesis holds provided $f(\cdot, \xi)$ is convex, wp1, and has integrable subgradients [10]. An alternative sufficient condition is that X is finite [9]. [?]. should do so.

Let X^* denote the set of optimal solutions to (SP), $x^* \in X^*$, $\bar{f}_n(x) = \frac{1}{n}\sum_{i=1}^n f(x, \xi^i)$, $\sigma^2(x) = \text{var} f(x, \xi)$, and $s_n^2(x) = \frac{1}{n-1}\sum_{i=1}^n (f(x, \xi^i) - \bar{f}_n(x))^2$. Let z_α denote the normal quantile satisfying $P(N(0,1) \le z_\alpha) = 1 - \alpha$. We assume $\alpha < 1/2$ and typically take $\alpha = 0.05$. Finally, for technical reasons that will become apparent we let x_v^* solve $\min_{x \in X^*} \text{var} f(x, \xi)$, i.e., x_v^* is a solution with minimum objective function variance among all optimal solutions.

Because z_n^* is optimal for (SP_n) we have $z_n^* \le \bar{f}_n(x)$, wp1, $\forall x \in X$. Hence,

$$
\begin{aligned}
P\left[\frac{z_n^* - z^*}{s_n(x_n^*)/\sqrt{n}} \le z_\alpha\right] &\ge P\left[\frac{\bar{f}_n(x_v^*) - z^*}{s_n(x_n^*)/\sqrt{n}} \le z_\alpha\right] \\
&= P\left[\frac{\bar{f}_n(x_v^*) - z^*}{s_n(x_v^*)/\sqrt{n}} \le z_\alpha \left(\frac{s_n(x_n^*)}{s_n(x_v^*)}\right)\right]. \qquad (3)
\end{aligned}
$$

By the standard CLT for i.i.d. random variables

$$\lim_{n\to\infty} P\left[\frac{\bar{f}_n(x_v^*) - z^*}{s_n(x_v^*)/\sqrt{n}} \le z_\alpha\right] = 1 - \alpha. \qquad (4)$$

Now, $s_n^2(x_v^*)$ is a strongly consistent estimator of $\text{var} f(x_v^*, \xi)$, $\text{var} f(x_v^*, \xi) \le \text{var} f(x, \xi) \; \forall x \in X^*$, and $\{x_n^*\}_{n=1}^{\infty}$ has limit points in X^*. Thus,

$$\liminf_{n\to\infty} \frac{s_n(x_n^*)}{s_n(x_v^*)} \ge 1, \text{wp1}. \qquad (5)$$

Combining (3), (4), (5) and a converging-together result yields:

Theorem 1. *Under the hypotheses stated above*

$$\liminf_{n\to\infty} P\left[\frac{z_n^* - z^*}{s_n(x_n^*)/\sqrt{n}} \le z_\alpha\right] \ge \lim_{n\to\infty} P\left[\frac{\bar{f}_n(x_v^*) - z^*}{s_n(x_v^*)/\sqrt{n}} \le z_\alpha\right] = 1 - \alpha. \qquad (6)$$

When X^* is not a singleton we cannot expect $\{x_n^*\}_{n=1}^{\infty}$ to have a unique limit. Hence, $\{s_n(x_n^*)^2\}_{n=1}^{\infty}$ will not, in general, have a unique limit. This is why the "lim inf" appears on the left-hand side of (6) instead of "lim."

As described above, lower bounds on z^* are key to establishing the quality of a candidate solution. The importance of (6) is that from it we infer that for sufficiently large values of n

$$P\left\{z_n^* - z_\alpha s_n(x_n^*)/\sqrt{n} \le z^*\right\} \approx 1 - \alpha. \tag{7}$$

So, even though z_n^* is not asymptotically normal we can construct a valid one-sided CI on z^* using normal quantiles. Intuitively, (7) holds because the lack of normality comes from a projection of normal random variables that occurs in a way to make the interval more conservative (see Example 1).

For many problems the cost of performing multiple replications is the primary computational bottleneck in efficiently establishing solution quality. In such cases, Theorem 1 will yield computational savings in a revised MCSP procedure. In particular, we can use $n_g = 1$ batch instead of (say) $n_g = 20$ batches. Result (6) can to be revised, in straightforward fashion, to include the upper bound estimator so that we obtain a CI on the optimality gap and not just a lower bound on z^*. There are two ways to do this. The MCSP procedure in Section 2 uses common random number streams (CRNs) in the upper bound estimator and the lower bound estimator in (2). Alternatively, we can use independent streams for upper and lower bound estimation. We have previously investigated this issue in [7,11,12]. For problems in which evaluating $f(x, \xi)$ is computationally expensive, e.g., two-stage stochastic linear programs, using CRNs affords significant variance reduction (ranging from a factor of 17 to 4100 in [11]). On the other hand, if $f(x, \xi)$ is relatively inexpensive to evaluate and we can use a dramatically larger sample size for upper bound estimation then using independent streams can be preferable (e.g., [7,12]).

References

1. M. Bertocchi, J. Dupačová, and V. Moriggia. Sensitivity of bond portfolio's behavior with respect to random movements in yield curve: A simulation study. *Annals of Operations Research*, 99:267–286, 2000.
2. J. Dupačová. On non-normal asymptotic behavior of optimal solutions for stochastic programming problems and on related problems of mathematical statistics. *Kybernetika*, 27:38–52, 1991.
3. J. Dupačová and R.J.-B. Wets. Asymptotic behavior of statistical estimators and of optimal solutions of stochastic optimization problems. *The Annals of Statistics*, 16:1517–1549, 1988.
4. J.L. Higle and S. Sen. Statistical verification of optimality conditions for stochastic programs with recourse. *Annals of Operations Research*, 30:215–240, 1991.
5. J.L. Higle and S. Sen. Stochastic decomposition: an algorithm for two-stage linear programs with recourse. *Mathematics of Operations Research*, 16:650–669, 1991.

5. J.L. Higle and S. Sen. Stochastic decomposition: an algorithm for two-stage linear programs with recourse. *Mathematics of Operations Research*, 16:650–669, 1991.
6. J.L. Higle and S. Sen. Statistical approximations for stochastic linear programming problems. *Annals of Operations Research*, 85:173–192, 1999.
7. A.S. Kenyon and D.P. Morton. Stochastic vehicle routing with random travel times. *Transportation Science*, 2001. To appear.
8. A.J. King and R.T. Rockafellar. Asymptotic theory for solutions in statistical estimation and stochastic programming. *Mathematics of Operations Research*, 18:148–162, 1993.
9. A.J. Kleywegt, A. Shapiro, and T. Homem-de-Mello. The sample average approximation method for stochastic discrete optimization. *Stochastic Programming E-Print Series*, 1999. http://dochost.rz.hu-berlin.de/speps/.
10. A.J. King and R.J.-B. Wets. Epi-consistency of convex stochastic programs. *Stochastics*, 34:83–91, 1991.
11. W.K. Mak, D.P. Morton, and R.K. Wood. Monte Carlo bounding techniques for determining solution quality in stochastic programs. *Operations Research Letters*, 24:47–56, 1999.
12. D.P. Morton and R.K. Wood. On a stochastic knapsack problem and generalizations. In D.L. Woodruff, editor, *Advances in Computational and Stochastic Optimization, Logic Programming, and Heuristic Search: Interfaces in Computer Science and Operations Research*, pages 149–168. Kluwer Academic Publishers, Boston, 1998.
13. V.I. Norkin, G.Ch. Pflug, and A. Ruszczyński. A branch and bound method for stochastic global optimization. *Mathematical Programming*, 83:425–450, 1998.
14. E. Popova and D. Morton. Adaptive stochastic manpower scheduling. In *Proceedings of the Winter Simulation Conference*, pages 661–668, 1998.
15. S.M. Robinson. Analysis of sample-path optimization. *Mathematics of Operations Research*, 21:513–528, 1996.
16. A. Shapiro and T. Homem-de-Mello. A simulation-based approach to two-stage stochastic programming with recourse. *Mathematical Programming*, 81:301–325, 1998.
17. A. Shapiro. Asymptotic properties of statistical estimators in stochastic programming. *The Annals of Statistics*, 17:841–858, 1989.
18. A. Shapiro. Stochastic programming by Monte Carlo simulation methods. *Stochastic Programming E-Print Series*, 2001. http://dochost.rz.hu-berlin.de/speps/.
19. B. Verweij, S. Ahmed, A. Kleywegt, G. Nemhauser, and A. Shapiro. The sample average approximation method applied to stochastic vehicle routing problems: a computational study. *(Working paper)*, 2001.
20. D.W. Watkins, Jr., D.P. Morton, and D.C. McKinney. Monte Carlo techniques for estimating solution quality in stochastic groundwater management models. In *Proceedings of the XII International Conference on Computational Methods in Water Resources, Crete*, pages 67–74, 1998.
21. G. Zakeri. Metaneos project: Verify optimization solver. http://www-unix.mcs.anl.gov/metaneos/. Argonne National Laboratory.

Scenario Updating Method for Stochastic Mixed-integer Programming Problems

Guglielmo Lulli[1] and Suvrajeet Sen[2]

[1] Department of Statistic, Probability and Applied Statistics, University of Rome "La Sapienza" - I-00185 Rome, ITALY. e-mail: guglielmo.lulli@@uniroma1.it
[2] Department of Systems and Industial Engineering University of Arizona - Tucson AZ85721, USA. e-mail: sen@@sie.arizona.edu

Abstract. In this paper, we propose an approximation scheme to solve large stochastic mixed-integer programming (SMIP) problems with fixed recourse. We refer to this as the Scenario Updating Method. The algorithm is based on solving instances of the problem, which contain only a subset of the scenarios in the scenario tree. At each iteration, the subset of scenarios is updated by adding only those scenarios which suggest a significant potential for change in the objective function value. The algorithm is terminated when the potential for change is insignificant. Different selection and updating rules are discussed.

We test the effectiveness of our method on a multi-stage stochastic batch-sizing problem, which is solved using a branch-and-price algorithm, at each iteration. The quality of the computational results demonstrates the viability of the proposed method.

1 Introduction

Stochastic mixed-integer programming problems are mixed-integer programming problems in which some problem data are uncertain. More precisely, the problem data are given by a discrete time stochastic process $\{\xi_t\}_{t=1}^{|\mathcal{T}|}$ defined on some probability space $(\Xi, \mathcal{F}, \mathcal{P})$. $\mathcal{T} = \{1, \ldots, |\mathcal{T}|\}$ is the decision horizon. Decisions are based on the information available at that time, i.e. on the set of decisions already made and on the outcome of the random variable in the previous stages. If $\underline{x}_t = (x_1, \ldots, x_t)$ is the vector of all decisions made from stage 1 to stage t and $\underline{\xi}_t = (\xi_1, \ldots, \xi_t)$ is the vector of the random variable outcomes during the same interval then, a prototypical multi-stage stochastic program is given by the following problem:

$$min\{c_1(\xi_1)x_1 + Q_1(x_1) : W_1 x_1 \leq h_1(\xi_1), x_1 \in \mathbf{X_1}\},$$

where

$$Q_t(\underline{x}_t) = E_{\tilde{\xi}_{t+1}|\underline{\xi}_t} min\{c_{t+1}(\tilde{\xi}_{t+1})x_{t+1} + Q_{t+1}(\underline{x}_{t+1}) :$$
$$T_{t+1}(\tilde{\underline{\xi}}_{t+1})\underline{x}_t + W_{t+1}x_{t+1} \leq h_{t+1}(\underline{\xi}_{t+1}), \;\; x_{t+1} \in \mathbf{X_{t+1}}\}$$

for $t = 1, \ldots, |\mathcal{T}| - 1$ with $Q_{|\mathcal{T}|} \equiv 0$. Here we assume that ξ_1 is known at time $t = 1$ and $E_{\tilde{\xi}_{t+1}|\underline{\xi}_t}$ denotes expectation with respect to the distribution of $\tilde{\xi}_{t+1}$ conditioned on the observation $\underline{\xi}_t$. For all realizations of ξ and time stages, we suppose that $T_t(\underline{\xi}_t), W_t, c_t(\xi_t), h_t(\underline{\xi}_t)$ are matrices and vectors of conformable dimensions. The set $\mathbf{X_t}$ denotes restrictions that require some or all the decision variables to be integer.

In keeping as much with much of the literature, we assume that the random vector ξ has a finite support; that is $\Xi = (\xi^1, \dots, \xi^r)$ with probabilities $p^1, \dots, p^r$. This hypothesis allows us to represent uncertainty by means of scenarios. In most of the real applications, building a representative scenario tree is a crucial task. It calls for compromises between a manageable problem size and the desired precision of results. In this area of research, two main techniques are used according to the data available. Cluster analysis is used when the main random factors have been detected and enough data paths can be generated in accordance with a stochastic model. A sampling procedure is adopted whenever a well-calibrated stochastic model is available. In Dupačová, Consigli and Wallace [4], several methods for generating representative scenario trees for multi-stage stochastic programs of a general structure are given. Høyland and Wallace [5], Klassen [6], Kouwenberg [7] and Pflug [10], to mention a few, provide several insights in the generation of representative scenario trees in the field of financial applications, one of the most prolific area of research.

At the same time, one of the challenges in stochastic programming is the development of efficient algorithms which are able to solve problem with a larger and larger number of scenarios. Løkketangen and Woodruff [8] combine the progressive hedging algorithm with a tabu search to solve multi-stage SMIP with binary variables. A Lagrangian relaxation for use within a branch and bound algorithm for multi-stage SMIP has been proposed by Carøe and Schultz [2]. In our paper [9], we propose a branch-and-price algorithm to solve special structured multi-stage SMIP problems.

In this paper, we propose a heuristic which approximately solves large SMIP problem with complete fixed recourse. We refer to this method as the Scenario Updating Method. The algorithm is based on solving instances of the problem, which contain only a subset of the scenarios in the scenario tree. At each iteration, the subset of scenario is updated adding those scenarios, which imply either a certain degradation or an improvement of the objective function. The algorithm is terminated, when no more of such scenarios are available to enter in the current scenarios subtree. Different selection and updating rules are discussed. We validate our procedure on a multi-stage stochastic batch-sizing problem.

The paper is organized as follows. In § 2, we describe the algorithm, while the computational analysis is given in § 3. Finally, § 4 contains conclusions and future research.

2 A Heuristic Method for Complete Fixed-Recourse SMIP Problems

In this section we propose an approximation method to solve the SMIP problem with complete fixed-recourse. The complete recourse assumption allows us to compute a feasible solution which accommodates all the scenarios of the scenario tree once the heuristic one is computed. Such a solution provides an upper bound on the optimal solution. The purpose of this procedure is to solve models arising in realistic applications.

The procedure, we present in this paper is motivated by the contamination method, first proposed in Dupačová [3]. The idea behind the contamination method consists of estimating lower and upper bounds for the value function of a multi-stage stochastic program when a new scenario is added to the current scenario tree. The term "contamination," refers to the fact that the probability distribution P, is contaminated by the probability distribution associated with an additional scenario. While the motivation for the method lies in post-optimality analysis of stochastic programs, the application to a sequential tree generation process is one of its intended purposes. This paper extends the ideas of the contamination method in two ways: first, we deal with SMIP problems, rather than MSLP problems, and second, we discuss and compare explicit rules for algorithm development. In contrast, previously reported work (e.g. Dupačová, Consigli and Wallace [4]) has not discussed issues related to algorithmic realization of the method in any detail.

We start the procedure with a subset of scenarios $\bar{\mathcal{S}}_0 \subset \mathcal{S}$, where $\mathcal{S}$ is the set of scenarios. At each iteration k, the algorithm solves an instance of the problem, which is composed by the subset of scenarios $\bar{\mathcal{S}}_k$. The solution of the instance, is referred to as the current solution. Then, the subset of scenarios is updated by adding those scenarios, which imply a significant change in the objective function. The algorithm is terminated, when no more of such scenarios are available to enter in the current scenario subtree. To verify if a scenario can be elected or not to enter in the current scenario subtree, we compute an upper bound of the current objective function, resorting on the complete recourse hypothesis. Under this hypothesis, any scenario solution is feasible for all the scenarios, thus allowing to compute a feasible solution for the original scenario tree.

The Scenario Updating procedure is summarized below.

Initialization *Start selecting a subset of scenarios of the scenario tree.*

Step 1 *Solve the "reduced" problem with an appropriate algorithm.*

Step 2 *Compute an upper and a lower bound on the value function induced by adding a scenario not in the current scenario subtree. Those scenarios, which if added to the current scenario tree imply a certain change in the objective function, are candidates to enter in the next subtree.*

Step 3 *If there are no scenario candidates then stop, otherwise add those scenarios or some of them to the current scenario subtree, and go to Step 1.*

When implementing the proposed algorithm, several issues have to be addressed. Most of these have an impact on the trade-off between solution time for each iteration and number of iterations. The first issue pertains to the number of scenarios to use in the initial subtree, and which ones. At the beginning of the procedure, it is desirable to have a subtree whose scenarios are most "representative." A "representative scenario" is one whose solution can be used for other scenarios, without incurring very high penalty costs. Effective scenario selection assures a fastest convergence of the procedure. Several decision rules may be developed for this initial task. However, it is not possible, a priori, to decide which one is most representative. Analogous considerations arise when deciding which of the candidate scenarios should be used to update the scenario tree. In the following section we will discuss different selection rules.

Finally, for Step 1 we need to specify the algorithm to be adopted for the solution of the restricted problem. As in many successive approximation methods, there is a fair amount of art in deciding the appropriate quality of solution to require in Step 1. Obtaining an optimal solution (for Step 1) in each iteration may involve greater computations per iteration, than using approximate solutions. In this case, we may seek progressively better approximations, which ultimately provide an optimal solution to the entire (unreduced) problem.

3 Computational Analysis

Both a tuning analysis and performance evaluation of the algorithm has been executed on the stochastic batch-sizing problem. The stochastic batch-sizing problem belongs to the class of economic lot-sizing (ELS) models, in which the demand, production, inventory and set-up costs are uncertain problem parameters evolving as discrete random variables. Furthermore, production takes place in a batch mode, i.e. production level is a multiple of the batch size and consequently the corresponding decision variables are discrete. In [9], a detailed description of the problem is given.

In our computational analysis we use randomly generated instances. We generate different sets of problems by varying the number of time periods (stages) of the system. As for scenario trees, we generate binary trees with conditional probability for any branch being p_n, and the other branch having conditional probability $1 - p_n$ where n denotes a node of the tree. Here p_n is chosen from the uniform [0,1] distribution. By choosing alternative values of p_n, as well as cost and demands we can generate different problem instances corresponding to any given planning horizon.

We first evaluate different rules for selecting the initial subset of scenarios $\bar{S}_0$. We analyze the dissimilarity, the highest probability, the random and the mixed rules. The mixed rule selects scenarios using all the others. The computational results, for a seven-stage class of instances, are given in the following table. For each scenarios selection rule, the following statistics are reported: the computational time in seconds (Time), the objective function value (O.F. value), the number of iterations of the algorithm (It.s) and the number of scenarios accommodated in the approximated solution (SG). The best, worst and average (avg) instances are those instances with reference to the optimal solutions computed using a branch-and price algorithm (B&P) proposed in [9]. The computational time and the objective function value of such optimal solutions, have the following values: 3227, 16505 (avg) - 1406, 96431 (worst) and 3890, 8221 (best) respectively. In the Table 1, the first set

Table 1. Scenario selection rules

Mixed				Highest Probability				Dissimilarity			
Time (secs.)	O.F. value	It.s	SG	Time (secs.)	O.F. value	It.s	SG	Time (secs.)	O.F. value	It.s	SG
210	8173	8	30	607	8247	9	33	686	8318	7	30
1461	95097	21	45	1450	94595	19	44	2465	94064	19	42
791	16383	11	36	872	16304	10	34	1066	16365	9	31
124	8151	5	34	230	8255	4	35	544	8305	5	32
782	95133	7	48	501	94563	6	44	1369	93926	5	43
482	16384	5	37	379	16300	4	35	650	16364	4	31

of three rows refers to using a single-scenario updating rule and the second set of three rows refers to the multiple-scenario updating rule. The three rows of both the sets, refer to the best, the worst and in the average instance respectively.

According to these results, the multiple-scenario selection with a mixed selection rule provides better results for the problem tested. To solve instances of larger dimensions, in term of number of stages of the problem, we combine cplex branch-and-bound (B&B) method and the B&P algorithm. In fact, B&B provides good feasible solutions in a reasonable amount of time, which are then used to generate the initial columns in the B&P algorithm.

Table 2 reports the computational results for eleven-stage instances. In particular the following statistics are given: solution time in seconds, objective function value, number of iterations executed (of B&B and of B&P respectively) and the number of scenarios accommodated in the solutions. Even though the computational time is high, it is relatively small considering the dimension of the instances solved. It is important to note, that the number of scenarios accommodated in the solutions is much smaller than the number of scenarios in the scenario tree.

Table 2. Instances with 1024 scenarios

Instance	Time (secs.)	O.F. value	It.s	SG
I	14659	15341	4/2	68
II	7673	13517	2/1	107
III	7895	14687	2/1	130
IV	8245	13976	2/1	145
V	8733	14363	4/2	120

4 Conclusions

In this paper, we propose a heuristic, which solves instances with a large number of scenarios of SMIP problems with complete recourse. The quality of our results suggests a deeper investigation on the scenario selection rules and on stopping criteria. On this subject, the computation of stronger bounds would allow the solution of larger instances.

References

1. Birge J.R., Louveaux F. (1997) Introduction to Stochastic Programming.Springer, Berlin Heidelberg.
2. Carøe C.C., Schultz R. (1999)Dual Decomposition in Stochastic Integer Programming.Operations Research Letters **24** 37–45.
3. Dupačová J. (1995)Postoptimality for Multistage stochastic linear programs. Annals of Operations Research **56**, pp. 65-78.
4. Dupačová J., Consigli G. and Wallace S.W. (2000)Scenarios for Multistage Stochastic Programming. Annals of Operations Research **100** 25–53.
5. Høyland K., Wallace S.W. (2001)Generating scenario trees for multistage decision problems.Management Science **47** 295–307.
6. Klaassen P. (1998)Financial Asset-Pricing Theory and Stochastic Programming Models for Asset/Liability Management: A Synthesis.Management Science **44** 31–48.
7. Kouwenberg R.R.P. (2001)Scenario generation and stochastic programming models for asset liability management.European Journal of Operational Research **134** 51–64.
8. Løkketangen A., Woodruff D.L. (1996)Progressive Hedging and Tabu Search Applied to Mixed Integer (0,1) Multi-stage Stochastic Programming.Journal of Heuristics **2** 111–128.
9. Lulli G., Sen S. (2002) A Branch-and-Price Algorithm for Multi-stage Stochastic Integer Programming with Application to Stochastic Batch-Sizing Problems. Submitted for pubblication.
10. Pflug G.Ch. (2001)Scenario tree generation for multiperiod financial optimization by optimal discretization.Mathematical Programming **89** 251–271.

Pricing of Multidimensional Resources in Revenue Management (Multidimensional Dynamic Stochastic Knapsack Problem)

Jens Feller

University of Dortmund, Fakultaet WiSo, Fachgebiet Operations Research und Wirtschaftsinformatik, j.feller@wiso.uni-dortmund.de

Abstract. Revenue management deals with selling a limited amount of a perishable resource. This resource becomes valueless at a known point of time (deadline). There are many papers dealing with one-dimensional resources (e. g. seats or hotel rooms) using different approaches as linear programming or Markov decision processes which consider the problem as a dynamic stochastic knapsack problem (DSKP).

In this article we address the question of choosing an optimal price of these resources in consideration of the remaining time, the resources left and the expected demands and focus on the characteristics of multidimensional resources which often have to be taken into account, e. g. when transporting freight with limitations to volume and weight or when pricing package holidays as a simultaneous selling of seats and rooms.

We present a model using the DSKP-approach for maximizing the expected revenue with a given deadline and a multi-dimensional resource when the distribution functions of the demands and of the offered revenue are known and define conditions of a threshold policy that are necessary for any optimal policy.

1 Introduction

Revenue Management was first used to assign problems in airline industries (e. g. selling of seats). It has become more and more important in account of further research results, its practical successes and its wide fields of applications.

It addresses problems that occur when selling items that will become worthless at a known point in time (e. g. seats of aircrafts, hotel rooms). Some of the most important issues are overbooking (selling more items than available if it is known that some customers will not use their tickets), fleet assignment, defining the size of ticket or price classes or pricing the items that are going to be sold. McGill/van Ryzin give a good overview of these topics in [3].

In this article we address the question of accepting or denying requests for a multi-dimensional resource. Important applications for multi-dimensional resources are e. g. freight (volume, weight, containers, etc.) or items that depend on each other (seats on an aircraft and hotel rooms at the destination at vacation, tickets and hotels rooms at special events and so on). Obviously bad choices may result in being out of stock when getting lucrative requests or in denying too many requests and keeping unsold resources at the deadline.

We are going to maximize the revenue when selling a multi-dimensional perishable resource. There are requests/arrivals (offered price and demanded amount of resource) at different points in time. We have to decide on each request if we accept it (take the money and give away the requested resource) or deny it (keep the resource).

If we would know all upcoming requests at the beginning the problem would be to pick those arrivals that maximize our revenue without exceeding the available items.

In our particular problem these arrivals are stochastic and not known in advance. At each time our information consists of the amount of available items and the remaining time to sell those. Furthermore we assume that we know the distributions of the upcoming arrivals of demands.

For one-dimensional resources (e. g. seats or hotel rooms) this problem is addressed by the articles of Kleywegt/Papastavrou ([1] and [2]) and van Slyke/Young [4]. Both model the problem as a dynamic stochastic knapsack problem (DSKP). Van Slyke and Young give a hint how to apply their approach for a multi-dimensional resource.

In this article we will define a model for a multi-dimensional resource. First we will define the assumptions and variables of the model. We will name the variables analog to [2]. Next we will show some properties of these variables and give conditions that are necessary for maximizing the expected revenue. Afterwards we give some results from a similar approach with discrete periods of time.

2 Model

We are going to maximize the revenue of selling a resource of $k \in \mathbb{N}$ dimensions and an initial amount of $N_1 \in \mathbb{N}_0^k$ of this resource. Let $T \in (0, \infty)$ be the deadline for accepting these items. The remaining resources at T cannot be sold anymore and will be worthless.

Let $\{A_i\}_{i=1}^{\infty}$ describe the random arrival times and $\mathcal{N} \equiv \{n_0, n_1, \ldots, n_K\} \in (\mathbb{N}_0^k)^{K+1}$ the set of possible amounts that may be asked for. Each request consists of such an amount and a price offered for these items. With the state space Ω the random variables $S_i : \Omega \to \mathcal{N}$ and $R_i : \Omega \to \mathbb{R}$ give the asked amount and the offered price of the ith arrival (revenue or reward of this arrival). We assume the marginal probability distributions of S_i as $F_{S_i}(x) \equiv \mathrm{P}\{S_i < x\}$ and the conditional distribution of R_i given S_i as

$F_{R|S}(x) \equiv \mathrm{P}\{R_i < x \mid S_i = S\}$ are known. In this article we will not use a discount rate for the revenues.

To maximize the revenue of the accepted arrivals we have to decide for each arrival whether to accept or deny it. Let the decision variables D_i be 1 if we accept the ith arrival. If we deny it it will be 0.

To ensure that we do not sell more resources than available we define $\Pi_\mathrm{C}(N,t)$ as the class of admissible policies with resource N at time t. The expected reward when using an admissible policy π is given by $V_\mathrm{C}^\pi(t)$ and the optimal expected revenue by $V_\mathrm{C}^*(N,t)$ when N resources are left. N_i is the amount of available resources before the decision about arrival i, that is $N_i \equiv N_1 - \sum_{j=1}^{i-1} D_j S_j$.

By those definitions we get the set of arrival times $\mathcal{A} \equiv \{\{A_i\}_{i=1}^{\infty} : 0 < A_1 < A_2 < \cdots < \infty\}$, the set of rewards $\mathcal{R} \equiv \{\{R_i\}_{i=1}^{\infty} : R_i : \Omega \to \mathbb{R}\}$ and the set of the demanded resources $\mathcal{S} \equiv \{\{S_i\}_{i=1}^{\infty} : S_i : \Omega \to \mathcal{N}\}$. The set of decisions is given by $\mathcal{D} \equiv \{\{D_i\}_{i=1}^{\infty} : D_i \in \{0,1\}\}$. Therefore we can define the admissible policies if there has not been an arrival in t by $\Pi_\mathrm{C}(N,t) \equiv \{\pi : \mathcal{A}\times\mathcal{R}\times\mathcal{S} \to \mathcal{D} : \sum_{\{i:t<A_i\le T\}} D_i^\pi S_i \le N\}$ with $\{D_i^\pi\} \equiv \pi(\{A_i\},\{R_i\},\{S_i\})$ as the decisions when using policy π.

The expected reward using policy π is then given by $V_\mathrm{C}^\pi(t) \equiv \mathrm{E}[\sum_{\{i:t<A_i\le T\}} R_i D_i^\pi]$ while the optimal reward is defined by $V_\mathrm{C}^*(N,t) \equiv \sup_{\pi\in\Pi_\mathrm{C}(N,t)} V_\mathrm{C}^\pi(t)$.

To handle arrivals in t we define $\overline{\Pi_\mathrm{C}}(N,t) \equiv \{\pi : \mathcal{A}\times\mathcal{R}\times\mathcal{S} \to \mathcal{D} : \sum_{\{i:t\le A_i\le T\}} D_i^\pi S_i \le N\}$, $\overline{V_\mathrm{C}^\pi}(t) \equiv \mathrm{E}[\sum_{\{i:t\le A_i\le T\}} R_i D_i^\pi]$ and $\overline{V_\mathrm{C}^*}(N,t) \equiv \sup_{\pi\in\overline{\Pi_\mathrm{C}}(N,t)} \overline{V_\mathrm{C}^\pi}(t)$. If there is an j with $A_j = t$ the decision variable D_j^π must be 1 to accept or 0 to deny this arrival.

3 Threshold Policy

The following properties guarantee a 'good' behavior of the threshold we will define later. We will say $N \le N'$ for some $N, N' \in \mathbb{N}_0^k$ if each component of N is less or equal than the corresponding component of N' and $N \not\le N'$ if there is at least one component of N greater than the corresponding component of N' ($\ge$ and $\not\ge$ analog).

First we show that there are always admissible policies and that increasing our amount of resources can lead to more admissible policies.

LEMMA 1
Let $N \in \mathbb{N}_0^k$ and $t \in [0,T)$.

a) Then there is $\Pi_\mathrm{C}(N,t) \ne \emptyset$.

b) For each $N' \in \mathbb{N}_0^k$ with $N' \ge N$ there is $\Pi_\mathrm{C}(N,t) \subseteq \Pi_\mathrm{C}(N',t)$.

PROOF: a) The policy π_0 of denying all requests with $D_i^{\pi_0} \equiv 0\,\forall i$ is admissible for all resources $N \in \mathbb{N}_0^k$ and $t \in [0,T)$ with $V_\mathrm{C}^{\pi_0}(N,t) = 0$, hence $\Pi_\mathrm{C}(N,t) \ne \emptyset$.

b) Suppose π is some policy $\pi \in \Pi_C(N,t)$ for an arbitrary $N \in \mathbb{N}_0^k$ and $t \in [0,T]$. Then for all $N' \in \mathbb{N}_0^k$ with $N' \geq N$ follows from the definition of $\Pi_C(\cdot,t)$ that $\sum_{\{i:t<A_i \leq T\}} D_i^\pi S_i \leq N \leq N'$ and $\pi \in \Pi_C(N',t)$. □

Next we show that the expected revenue of an arbitrary policy V^π and the optimal expected revenue V^* are always defined.

LEMMA 2
Let $N \in \mathbb{N}_0^k$, $t \in [0,T)$ and $\mathrm{E}[\sum_{\{i:A_i \leq T\}} |R_i|] < \infty$
a) If $\pi \in \Pi_C(N,t)$, then $V_C^\pi(t) = \mathrm{E}[\sum_{\{i:t<A_i \leq T\}} R_i D_i^\pi] \in \mathbb{R}$.
b) $V_C^*(N,t) = \sup_{\pi \in \Pi_S(N,t)} \mathrm{E}[\sum_{\{i:t<A_i \leq T\}} R_i D_i^\pi] \in \mathbb{R}_{\geq 0}$.

PROOF: a) Let π be some policy $\pi \in \Pi_C(N,t)$ with resource $N \in \mathbb{N}_0^k$ and $t \in [0,T)$. Then we know with the assumption $\infty > \mathrm{E}[\sum_{\{i:A_i \leq T\}} |R_i|] \geq \mathrm{E}[\sum_{\{i:t<A_i \leq T\}} |R_i|] \geq \mathrm{E}[\sum_{\{i:t<A_i \leq T\}} |R_i|] \geq \mathrm{E}[\sum_{\{i:t<A_i \leq T\}} D_i^\pi |R_i|] = \mathrm{E}[\sum_{\{i:t<A_i \leq T\}} | D_i^\pi R_i|] \geq \mathrm{E}[| \sum_{\{i:t<A_i \leq T\}} D_i^\pi R_i|] \geq |\mathrm{E}[\sum_{\{i:t<A_i \leq T\}} D_i^\pi R_i]| \geq 0$.

Thus $V_C^\pi(t) = \mathrm{E}[\sum_{\{i:t<A_i \leq T\}} D_i^\pi R_i] \in \mathbb{R}$.

b) From Lemma 1 we know that $V_C^*(N,t) \geq V_C^{\pi^o}(t) = 0 \, \forall N \geq 0$. Furthermore $V_C^*(N,t) = \sup_{\pi \in \Pi_C(N,t)} V_C^\pi(t) = \sup_{\pi \in \Pi_C(N,t)} \mathrm{E}[\sum_{\{i:t<A_i \leq T\}} R_i D_i^\pi] \leq \mathrm{E}[\sum_{\{i:t<A_i \leq T \,\wedge\, R_i \geq 0\}} R_i] \leq \mathrm{E}[\sum_{\{i:t<A_i \leq T\}} |R_i|] < \infty$. □

Furthermore the optimal expected revenue is non-decreasing in the amount of available resources.

LEMMA 3 For each $N, N' \in \mathbb{N}_0^k$ with $N \leq N'$ and $t \in [0,T]$ follows $V_C^*(N,t) \leq V_C^*(N',t)$, that is $V_C^*(N,t)$ is non-decreasing in N.

PROOF: May $N, N' \in \mathbb{N}_0^k$ and $t \in [0,T)$ be arbitrarily chosen with $N' \geq N$. We know from Lemma 1 that $\Pi_C(N,t) \subseteq \Pi_C(N',t)$.

Let $\Delta \equiv \Pi_C(N',t) \setminus \Pi_C(N,t)$. Then $V_C^*(N',t) = \sup_{\pi \in \Pi_C(N',t)} V_C^\pi(t) = \sup_{\pi \in \Pi_C(N,t) \cup \Delta} V_C^\pi(t) = \max \left\{ \sup_{\pi \in \Pi_C(N,t)} V_C^\pi(t), \sup_{\pi \in \Delta} V_C^\pi(t) \right\} \geq \sup_{\pi \in \Pi_C(N,t)} V_C^\pi(t) = V_C^*(N,t)$.□

Now we define a class of admissible policies by a threshold r^*. Each arrival with a greater revenue than r^* will be accepted, each arrival with a lower revenue will be denied by this policy. We will show that each optimal policy must be such a threshold policy.

DEFINITION 1
Let the threshold $r^* : \mathbb{N}_0^k \times \mathcal{N} \times [0,T) \to \mathbb{R}_{\geq 0} \cup \{\infty\}$ be defined by

$$r^*(N,s,t) \equiv \begin{cases} V_C^*(N,t) - V_C^*(N-s,t) & ; s \leq N \\ \infty & ; s \nleq N. \end{cases} \tag{1}$$

Then we call a policy π' a r^*-threshold policy with the remaining resources N at time A_i if for each arrival $\{A_i, R_i, S_i\}$ with $t \leq A_i \leq T$ (and a positive probability of appearance) the implications $R_i > r^*(N, S_i, A_i) \Rightarrow D_i^{\pi'} = 1$ and $R_i < r^*(N, S_i, A_i) \Rightarrow D_i^{\pi'} = 0$ are true.

The class of all these r^*-threshold-policies is $\overline{\Pi'_C}(N,t)$ (and $\Pi'_C(N,t)$ if all $A_i \neq t \forall i$).

REMARK a) The values of r^* are nonnegative (follows from Lemma 1 to 3).

b) The policies are not different except of realizations where there is at least one $R_i = r^*(N, S_i, A_i)$. These arrivals can be accepted or denied.

LEMMA 4
$\Pi'_C(N,t) \subseteq \Pi_C(N,t)$ with $N \geq 0$ and $t \in [0,T]$, that is all r^*-threshold policies are admissible.

PROOF: Let $n \in \mathbb{N}$ be the number of arrivals after t and $\pi' \in \Pi'_C(N,t)$ some r^*-threshold-policy. We will show by induction that after every decision $N_n \geq 0$, that is we do not sell more resources than N.

i) $n = 1$: Consider arrival (A_1, S_1, R_1) with $N_1 = N$. If $S_1 \nleq N_1$ then $r^*(A_1, S_1, N_1) = \infty > R_1$ and $D_1^{\pi'} = 0$. Thus $N_2 = N_1 = N \geq 0$. If $S_1 \leq N_1$ both $D_1^{\pi'} = 0$ and $D_1^{\pi'} = 0$ lead in every dimension to a nonnegative stock of resources as $N_2 \geq N_1 - S_1 \geq 0$.

ii) $n \to n+1$: π' leads to admissible decisions for the first n arrivals: $N_{n+1} \geq 0$. If there is an arrival $(A_{n+1}, S_{n+1}, R_{n+1})$ again there are two cases. $S_{n+1} \nleq N_{n+1} \Rightarrow r^*(A_{n+1}, S_{n+1}, N_{n+1}) = \infty > R_{n+1} \Rightarrow D_{n+1}^{\pi'} = 0 \Rightarrow N_{n+2} = N_{n+1} \geq 0$, the decision is admissible. If $S_{n+1} \leq N_{n+1}$ again both accepting and denying are admissible as $N_{n+2} \geq N_{n+1} - S_{n+1} \geq 0$.

Hence π' never sells more resources than available, the policy is admissible.
□

PROPOSITION 1
Let $\pi^* \in \Pi_C(N,t)$ for some N and t be an optimal policy, that is $V_C^{\pi^*}(t) = V_C^*(N,t)$. Then $\pi^* \in \Pi'_C(N,t)$.

PROOF: Let $\pi \in \Pi_C(N,t)$ be some optimal policy with $\pi \notin \Pi'_C(N,t)$. Thus there is at least one realization with positive possibility and an $i : t < A_i \leq T$ that $D_i^{\pi'} \neq D_i^{\pi}$, $R_i \neq r^*(N_i, S_i, A_i)$ and $S_i \leq N_i$ $\forall \pi' \in \Pi'_C(N,t)$ (otherwise π would be a r^*-threshold-policy). Consider the following two possible cases for the smallest of these i:

i) r^*-threshold-policy accepts and π denies: $D_i^{\pi'} = 1$ and $D_i^{\pi} = 0 \Rightarrow R_i > r^*(N_i, S_i, A_i)$ and $\overline{V_C^{\pi}}(t) = 0 + V_C^{\pi}(t) \leq V_C^*(N_i, t) = r^*(N_i, S_i, A_i) + V_C^*(N_i - S_i, t) < R_i + V_C^*(N_i - S_i, t) = \overline{V_C^*}(N_i, t)$ showing that the denial of π is not optimal.

ii) r^*-threshold-policy denies and π accepts: $D_i^{\pi'} = 0$ and $D_i^{\pi} = 1 \Rightarrow R_i < r^*(N_i, S_i, A_i)$ and $\overline{V_C^{\pi}}(t) = R_i + V_C^{\pi}(t) \leq R_i + V_C^*(N_i - S_i, t) < r^*(N_i, S_i, A_i) + V_C^*(N_i - S_i, t) < V_C^*(N_i, t) = \overline{V_C^*}(N_i, t)$, accepting is not optimal.

In both cases the decisions that do not follow the r^*-threshold-policy lead to an expected revenue that is below the optimal expected revenue, hence π cannot be optimal. □

REMARK Under further assumptions it can be shown that a policy is optimal if it is a r^*-threshold policy (e. g. if there are at most $n \in \mathbb{N}_0$ further arrivals until T, by induction). More general assumptions are under current research by the author.

4 Extensions and Conclusions

In the last section we defined conditions for a threshold-policy that are necessary for policies creating an optimal expected revenue in our model. Unfortunately we do not have a formula to calculate the optimal revenues V^* that define the optimal thresholds.

If we use discrete time periods in our model above and choose these periods small enough to assume there is at most one arrival within one period then we can define an algorithm for calculating the optimal revenues $V^*(N, t)$ and the corresponding thresholds by a dynamic programming approach. This discrete model and the following observations are under current research by the author.

Depending on the current amount and the demand for the different resources the threshold may depend on all or only some of the demanded resources. E. g. if an arrivals demands one critical and one non critical resource the threshold will depend mainly (or only) on the critical resource. But the non critical resource might become critical by a high demand. There are examples where the thresholds react very sensitively to changes of the critical resource.

Another interesting observation is behavior of these thresholds depending on the remaining resources. In most cases a raising of the available resources leads to decreasing prices/thresholds (or to constant prices if there are plenty of this resource available). But in some cases several 'jumps' (up and down) can be observed when the available resource are raised (especially when demands for huge amounts of some resources with high revenues are possible, e. g. group flights, etc.).

References

1. Kleywegt A., Papastavrou J. (1998) The Dynamic and Stochastic Knapsack Problem. Operations Research **46**, January/February, 17–35
2. Kleywegt A., Papastavrou J. (2001) The Dynamic and Stochastic Knapsack Problem with Random Sized Items. Operations Research **49**, January/February, 26–41
3. McGill J., van Ryzin, G. (1999) Revenue Management: Research Overview and Prospects. Management Science **33**, May, 233–256
4. van Sylke R., Young Y. (2000) Finite Horizon Stochastic Knapsacks with Applications to Yield Management. Operations Research **48**, January/February, 155–172

A Note on Quantitative Stability and Empirical Estimates in Stochastic Programming

Vlasta Kaňková[1] and Michal Houda[2]

[1] Institute of Information Theory and Automation
Academy of Sciences of the Czech Republic
Pod vodárenskou věží 4, 182 08 Praha 8, Czech Republic
kankova@utia.cas.cz

[2] Charles University, Faculty of Mathematics and Physics
Sokolovská 83, 186 75 Praha 8, Czech Republic
houda@csmat.karlin.mff.cuni.cz

Abstract. The paper deals with a stability of stochastic programming problems considered with respect to a probability measure space. In particular, the paper deals with the stability of the problems in which the operator of mathematical expectation appears in the objective function, constraints set is "deterministic" and the probability measure space is equipped with the Kolmogorov or the Wasserstein metric. The stability results are furthermore employed to statistical estimates in the stochastic programming problems. Some results on a consistence and a rate of convergence are presented.

1 Introduction

From the mathematical point of view many applications correspond to deterministic optimization problems depending on a probability measure. To introduce this type of the problems let $(\Omega, \mathcal{S}, P)$ be a probability space, $\xi (= \xi(\omega) = [\xi_1(\omega), \ldots, \xi_s(\omega)])$ be an s–dimensional random vector defined on $(\Omega, \mathcal{S}, P)$, $F (:= F(z), z \in R^s)$ be the distribution function of ξ, P_F denote the probability measure corresponding to F, $Z_F \subset R^s$ denote the support of P_F. Let, furthermore, $g(x, z)$ be a real–valued function defined on $R^n \times R^s$, $X \subset R^n$ be a nonempty set. The symbol R^n, $n \geq 1$ is reserved for the n–dimensional Euclidean space.

A rather general above mentioned problem can be introduced in the form. Find

$$\varphi(F) = \inf\{\mathsf{E}_F g(x, \xi) | x \in X)\}. \qquad (1)$$

(The symbol E_F denotes the operator of mathematical expectation corresponding to F.)

To solve the problem (1) the complete information on F is necessary. However in applications very often at least one of the following cases happen: F must be replaced by its statistical estimate; F must be (for numerical difficulties) replaced by some simpler one; the actual distribution function is a little modified F. Consequently, it is suitable (or even necessary) to investigate the stability w.r.t. the probability measures space and statistical estimates. In the literature, a great attention was paid to these both problems. We can recall e.g papers [1], [4], [7], [8], [11].

2 Problem Analysis

To investigate the stability of the problem (1) we can state the assumptions under which there exists functions m_1, m_2 defined on R^n such that

$$|\varphi(F) - \varphi(G)| \leq m_1(d_s(P_F, P_G)), \quad \|x(F) - x(G)\| \leq m_2(d_s(P_F, P_G)), \tag{2}$$

where

$$x(F) = arg\min\{\mathsf{E}_F g(x, \xi) | x \in X\} \tag{3}$$

in the case when $x(F)$ is well-defined, G is s–dimensional distribution function "near" to F and d_s is a "suitable" metric; say Kolmogorov or Wasserstein. It is known that the Kolmogorov metric is defined by

$$d_K(F, G) := d_K(P_F, P_G) = \sup_{z \in R^s} |F(z) - G(z)|. \tag{4}$$

To define the Wasserstein metric $d_{W_1}(F, G)$, let $\mathcal{P}(R^s)$ denote the set of all (Borel) probability measures in R^s. If $\mathcal{M}_1(R^s) = \{\nu \in \mathcal{P}(R^s) : \int_{R^s} \|z\| \nu(dz) < \infty\}$ and $\mathcal{D}(\nu, \mu)$ denotes the set of those measures in $\mathcal{P}(R^s \times R^s)$ whose marginal measures are ν and μ.

$$d_{W_1}(\nu, \mu) = \inf\{ \int_{R^s \times R^s} \|z - \bar{z}\| \kappa(dz \times d\bar{z}) : \kappa \in \mathcal{D}(\nu, \mu)\}, \quad \nu, \mu \in \mathcal{M}_1(R^s).$$

The both metrics "measure" the distance between two probability measures. However, a "strange" situation can happen.

Example 1. Let $s = 1$. Let, moreover, $\varepsilon > 0$ and $a \in R^1$ be arbitrary and

$$\begin{array}{llll} F(z) = 0 & \text{for } z \in (-\infty, a), & G(z) = F(z - \varepsilon), & \text{for } z \in (-\infty, +\infty) \\ \phantom{F(z) = {}} 1 & \phantom{\text{for }} z \in \langle a, +\infty) & & \end{array}$$

then $d_{W_1}(F, G) = \varepsilon$ and $d_K(F, G) = 1$.

Example 2. Let $s = 1$. Let, moreover, $M < 0$, $\varepsilon \in (0, 1)$ be arbitrary and

$$\begin{array}{ll} F(z) = -\frac{\varepsilon}{z-(M+1)} & \text{for } z \in (-\infty, M) \\ \varepsilon & z = M \\ \text{arbitrary} & z \in (M, +\infty), \\ G(z) = 0 & z \in (-\infty, M), \\ F(z) & z \in \langle M, +\infty). \end{array}$$

Evidently, $d_{W_1}(F, G) = +\infty$, $d_{K_1}(F, G) = \sup_z |F(z) - G(z)| = \varepsilon$.

3 Stability Results

To recall the stability results we introduce the system of the assumption

A.1 a. $g(x, z)$ is a uniformly continuous function on $X \times R^s$,
b. for every $x \in X$, $g(x, z)$ is a Lipschitz function of $z \in R^s$ with the Lipschitz constant L not depending on $x \in X$,
c. X is a convex set and simultaneously $g(x, z)$ is a strongly convex function on X.

Proposition 1. *[6] Let $\delta > 0$ be arbitrary, $X \subset R^n$ be a nonempty, compact set. If A.1a, A.1b are fulfilled, then there exist $F^s_\delta(z)$, $F^\delta_s(z)$, $z \in E_s$ ($F^s_\delta(\cdot) \le F^\delta_s(\cdot)$) such that if G is s–dimensional distribution function fulfilling the relations*

$$G(z) = \langle F^s_\delta(z), F^\delta_s(z)\rangle, \; z \in R^s, \quad P_F, P_G \in \mathcal{M}_1(R^s),$$

then

$$|\varphi(F) - \varphi(G)| \le \delta L.$$

If, moreover, the assumption A.1c is fulfilled, then also

$$\|x(F) - x(G)\|^2 \le \frac{4}{\varrho}\, \delta L.$$

Employing the Kolmogorov metric it follows from Proposition 1.

Theorem 1. *[5] Let $X \subset R^s$ be a compact set, a finite $\mathsf{E}_F\xi$ exists. If*

1. *the assumptions A.1a and A,1b are fulfilled,*
2. *$Z_F = \prod_{j=1}^{s} \langle c_j, c_j'\rangle$, $c_j, c_j' > 0$, $c_j < c_j'$, $j = 1, 2, \ldots, s$ and, moreover, there exists a constant $\vartheta > 0$ such that $h(z) \ge \vartheta$ for every $z \in Z_F$,*
3. *$G(z)$ is an arbitrary s-dimensional distribution such that*

$$Z_G \subset Z_F(\delta') \quad \text{for}\, \delta' = 2\,(\frac{2d_K(F, G)}{\vartheta})^{\frac{1}{s}}, \quad 2(\frac{2d_K(F, G)}{\vartheta})^{\frac{1}{s}} \le \min_j(c_j'-c_j),$$

then

$$|\varphi(F) - \varphi(G)| \le 16\sqrt{s}\, L(\frac{2d_K(F, G)}{\vartheta})^{\frac{1}{s}}.$$

If, moreover, the assumption A.1c is fulfilled, then also

$$\|x(F) - x(G)\|^2 \le \frac{64}{\varrho}\, L\sqrt{s}(\frac{2d_K(F, G)}{\vartheta})^{\frac{1}{s}}.$$

($Z_F(\delta)$ denote the δ–neighbourhood of the set Z_F, the symbol h is reserved for the probability density corresponding to P_F.)

Theorem 2. *Let X be a nonempty compact set, $P_F \in \mathcal{M}_1(E_s)$. If the assumptions A.1a and A.1b are fulfilled, $P_G \in \mathcal{M}_1(R^s)$ is an arbitrary, then*

$$|\varphi(F) - \varphi(G)| \leq L d_{W_1}(F, G).$$

If, moreover, the assumption A.1c is fulfilled, then also

$$\|x(F) - x(G)\|^2 \leq \frac{16}{\varrho} L d_{W_1}(F, G).$$

Theorem 2 is only a little generalized the assertion of Römisch and Schulz proven originally for the complete linear recourse problems (see e.g [3]).

Example 3. Let $s = 1$. Let, moreover, $\delta \in (0, 1)$ and $a \in R^1$ be arbitrary. If

$$\begin{array}{llll} F(z) = 0 & \text{for} & z \in (-\infty, 0) \\ \qquad z & & z \in \langle 0, 1\rangle \\ \qquad 1 & & z \in \langle 1, +\infty) \\ G(z) = F(z - \delta) & & z \in (-\infty, +\infty), \end{array}$$

then $\vartheta = 1$ and simultaneously $d_{W_1}(F, G) = d_K(F, G) = \delta$ The upper bounds given by the Wasserstein and the Kolmogorov metrics are the same for every g fulfilling the system of assumptions A.1a and A.1b.

Example 4. Let $s = 1$. If N is an odd natural number $(N > 3)$, $F(z) := \bar{F}_N(z)$, $G(z) := \bar{G}_N(z)$, $N = 5, 7, \ldots$ successively,

$$\begin{array}{ll} \bar{F}_N(z) = 0, & z \in (-\infty, 0), \\ \qquad = \frac{z}{2(N-1)}, & z \in (0, \frac{N-1}{2}), \\ \qquad \frac{1}{4} + \frac{1}{4}[z - \frac{N-1}{2}], & z = \langle \frac{N-1}{2}, \frac{N+3}{2}\rangle, \\ \qquad \frac{3}{4} + \frac{1}{2(N-3)}[z - \frac{N+3}{2}], & z \in (\frac{N+3}{2}, N), \\ \qquad 1, & z \in \langle (N, +\infty), \\ \bar{G}_N(z) = \bar{F}_N(z), & z \in (-\infty, \frac{N-1}{2}) \cup (\frac{N+3}{2}, +\infty), \\ \qquad \frac{1}{4}, & z \in \langle \frac{N-1}{2}, \frac{N+3}{2}\rangle, \end{array}$$

then $d_{W_1}(F, G) = \frac{1}{2}$; $\sup_z |F(z) - G(z)| = d_K(F, G) = \frac{1}{2}$.

Consequently, the upper bound given by the Wasserstein metric is $\frac{1}{2}L$. However, since $\vartheta_N := \vartheta = \frac{1}{2(N-1)}$; the upper bound given by the Kolmogorov metric is $L(N-1)$.

4 Relationship between Stability and Statistical Estimates

If $\{\xi^k\}_{k=-\infty}^{\infty}$ is a sequence of s–dimensional random vectors with common distribution function F, F_N denotes a "good" statistical estimate of F determined by $\{\xi^k\}_{k=1}^{N}$ and F is replaced by F_N^{ζ}, then we obtain (under rather general assumptions) "good" estimates $\varphi(F_N)$ of $\varphi(F)$, $N = 1, 2, \ldots$

To present the corresponding assertion we introduce the assumptions

B.1 a. $\{\xi^k\}_{k=-\infty}^{\infty}$ is a sequence of s–dimensional random vectors with common distribution function F,

b. F_N is a statistical estimate of F determined by $\{\xi^k\}_{k=1}^{N}$.

Evidently, we can obtain.

Corollary 1. *Let $X \subset E_n$ be a compact set, $t > 0$, a finite $\mathsf{E}_F\xi$ exist. If the assumptions 1, 2 of Theorem 1 are fulfilled an if, moreover $F_N(\cdot)$ fulfills the system of the assumptions B.1 such that*

$$Z_{F_N} \subset Z(\delta^{'}),\ \delta^{'} = 2\,(\frac{2d_K(F,\, F_N)}{\vartheta})^{\frac{1}{s}}, \quad 2(\frac{2d_K(F,\, F_N)}{\vartheta})^{\frac{1}{s}} \le \min_j(c_j^{'} - c_j) \quad \text{a.s.},$$

then

$$P\{|\varphi(F) - \varphi(F_N)| > t\} \le P\{16\sqrt{s}\, L(\frac{2d_K(F,\, F_N)}{\vartheta})^{\frac{1}{s}} > t\}$$

Corollary 2. *Let X be a nonempty compact set, $P_F \in \mathcal{M}_1(E_s)$, $t > 0$. If the system of the assumptions A.1a and A.1b are fulfilled, and if, moreover, the system of the assumptions B.1 is fulfilled, $P_{F_N} \in \mathcal{M}_1(R^s)$a.s., then*

$$P\{|\varphi(F) - \varphi(F_N)| > t\} \le P\{L d_{W_1}(F,\, F_N) > t\}.$$

Employing the assertion of Corollary 1 and the known properties of the Kolmogorov metric we can obtained (for "empirical estimates") the upper bound independently on P_F (of course under some additional assumptions); see e.g. [2]. In special case we can even employ the limit Kolmogorov distribution function; see e.g. [9]. In the case of Wasserstein metric this upper bound depend on the distribution function P_F; see [9]. However, the following assertion can be proven.

Theorem 3. *[5] Let X be a compact set, a finite $\mathsf{E}_F\xi$ exist. If the assumptions 1, 2 of Theorem 1 and the assumptions B.1a and B.1b are fulfilled and, moreover,*

$$P\{N^{vs} \sup|F(z) - F_N(z)| > t\} \to_{(N\to+\infty)} 0 \quad \text{for a } v > 0 \text{ and every } t > 0,$$
$$P\{Z_{F_N} \subset Z_F(\tfrac{\sup|F(z)-F_N(z)|}{2})\} \to_{(N\to+\infty)} 1,$$

then

$$P\{N^{v}|\varphi(F) - \varphi(F_N)| > t\} \to_{(N\to+\infty)} 0. \tag{5}$$

Theorem 4. *Let X be a nonempty compact set, $P_F \in \mathcal{M}_1(R^s)$, $t > 0$. If*

1. *the system of the assumptions A. 1a, A. 1b are fulfilled,*
2. *$\{\xi^k\}_{k=1}^{\infty}$ is sequence of independent identically distributed s–dimensional random vectors with common distribution function F,*
3. *F_N is the empirical distribution function determined by $\{\xi^k\}_{k=1}^{\infty}$,*

then

$$P\{|\varphi(F) - \varphi(F_N)| \to_{(N \to +\infty)} 0\} = 1$$

Remark 1. Of course, the corresponding results can be proven (under the assumption A.1c) for the astimate $x(F_N)$ of $x(F)$.

Acknowledgement: This research was supported by the Grant Agency of the Czech Republic under Grants 402/02/1015, 402/01/0539 and and by Grant 436TSE113/40of the Deutsche Forschungsgemeinschaft.

References

1. Dupačová, J. and Wets, R. J.-B. (1984): Asymptotic behaviour of statistical estimates and optimal solutions of stochastic optimization problems, Ann. Statist., **16**, 1517–1549.
2. Dvoretzky, A., Kiefer, J. and Wolfowitz, J. (1956): Asymptotic minimax character of the sample distribution function and the classical multinomiial estimate. Ann. Math. Statist., **27**, (1956) 642–669.
3. Houda, M. (2001): Stability and Estimates in Stochastic Programming (Spesial cases) (in Czech). Diploma Work. Faculty of Mathematics and Physics, Charles University.
4. Kaňková, V. (1994): A note on estimates in stochastic programming, J. Comput. Math., **56**, 97–112.
5. Kaňková, V. (1994): A note on the relationship between distribution function estimation and estimation in stochastic programming. *Trans. 12th Prague Conf. 1978*, Academia, 122–125.
6. Kaňková (1994): On Distribution Sensitivity in Stochastic Programming. VZ ÚTIA AV ČR No. 1826.
7. Römisch, W. and Schulz, R. (1993): Stability of solutions for stochastic programs with complete recourse. Math. Oper. Res., **18**, 590–609.
8. Shapiro, A. (1994): Quantitative stability in stochastic programming, Math. Progr., **67**, 99–108.
9. Shorack, G. R. and Wellner, J. A. (1986): Empirical Processes with Applications to Statistica. Wiley, London.
10. Schulz, R. (2000): Some aspects of stability in stochastic programming, Ann. Oper. Res., **100**, 55–84.
11. Vogel, S. (1992): On stability in multiobjective programming – a stochastic approach, Math. Progr., **56** (1992) 91–119.

Splitting and Localization of the Epi-Topology Combined with Randomness *

Petr Lachout[1 2 3]

[1] Department of Probability and Statistics, Charles University of Prague, Sokolovská 83, 186 75 Praha 8, Czech Republic
[2] Institute of Information Theory and Automation, Czech Academy of Sciences, Pod vodárenskou věží 4, 182 08 Praha 8, Czech Republic.
[3] e-mail: lachout@karlin.mff.cuni.cz

Abstract. The paper introduces an extension of the epi-convergence, the lower semicontinuous approximation and the epi-upper semicontinuous approximation of random real functions in probability. The new notions could be helpful tools for sensitivity analyzes of stochastic optimization problems. The research is evoked by [4] and continuous the research started in [6], [7] and [5].

1 Introduction

The epi-convergence of real functions is a powerful tool for studying sensitivity of optimization functions; see e.g. [1], [2]. Provided nice topological space the epi-convergence is induced by the epi-topology, i.e. a restriction of Fell topology on sets, see [1], [2], [3] for details. The epi-convergence is originally defined by two different properties. The first one is induced by miss-part of Fell topology and the other by hit-part of Fell topology. In [4] it is suggested to employ these parts separately, "splitting", and consider them onto a given set, "localization". The splited and localized parts of the epi-convergence are called lower and epi-upper semicontinuous approximations. Unfortunately, these notions are not induced by any topology. The paper offers a general concept how to overcome the difficulty and presents an extension to random functions to receive desired consequences for sensitivity of stochastic optimization problems.

The epi-convergence itself is based on epi-topology. Hence, extension of it to random functions can be done in standard way, see [3]. Our case is not directly based on topology, because of restriction to a given set. We have to seek for more sophisticated description. A convenient description was introduced in [4], [6], [7] provided Euclidean space. The concept was generalized in [5].

* The research has been partially supported by Deutche Forschungsgemeinschaft grant No. 436TSE113/40, the project AV: K1019101, the project CEZ: MSM 113200008 and the Czech Republic Grant 201/02/0621.

2 Epi-convergence and lower and epi-upper semicontinuous approximations

This section deals with the epi-convergence of functions. We will work at a topological space $\mathcal{X}$.

Definition 1. Let us denote by $\mathsf{FCE}(\mathcal{X})$ the set of all functions $f : \mathcal{X} \to \overline{\mathbb{R}}$ and

$$\varphi(f; A) = \inf_{x \in A} f(x), \text{ where } A \subset \mathcal{X},$$
$$\Phi(f) = \{x \in \mathcal{X} : f(x) = \varphi(f; \mathcal{X})\},$$
$$\mathsf{epi}(f) = \left\{(x, \eta) \in \mathcal{X} \times \overline{\mathbb{R}} : f(x) \leq \eta\right\}$$

for any function $f \in \mathsf{FCE}(\mathcal{X})$.

Let us recall that Fell topology restricted to the set of all epi-graphs is called epi-topology. In the paper, we treat more finer objects.

Definition 2. Function $f : \mathcal{X} \to \overline{\mathbb{R}}$ having epi-graph closed in $\mathcal{X} \times \overline{\mathbb{R}}$ is called lower semicontinuous (l.s.c.). We denote the set of all l.s.c. functions on $\mathcal{X}$ by $\mathsf{LSC}(\mathcal{X})$.

Now, we introduce a lower and an epi-upper semicontinuous approximations in accordance with [4].

Definition 3. Let $< f_\lambda >_{\lambda \in \Lambda}$ be a net in $\mathsf{FCE}(\mathcal{X})$, $f \in \mathsf{FCE}(\mathcal{X})$ and $\emptyset \neq Y \subset \mathcal{X}$. We say:

(i) f_λ is a lower semicontinuous approximation to f at Y, notation is $f_\lambda \xrightarrow[\lambda \in \Lambda; Y]{l} f$, if

$$\sup_{x \in G, G \in \mathcal{G}(\mathcal{X})} \liminf_{\lambda \in \Lambda} \inf_{y \in G} f_\lambda(y) \geq f(x) \quad \forall x \in Y;$$

(ii) f_λ is an epi-upper semicontinuous approximation to f at Y, notation is $f_\lambda \xrightarrow[\lambda \in \Lambda; Y]{epi-u} f$, if

$$\sup_{x \in G, G \in \mathcal{G}(\mathcal{X})} \limsup_{\lambda \in \Lambda} \inf_{y \in G} f_\lambda(y) \leq f(x) \quad \forall x \in Y.$$

(iii) f_λ is epi-convergent to f at Y, notation is $f_\lambda \xrightarrow[\lambda \in \Lambda; Y]{epi} f$, if $f_\lambda \xrightarrow[\lambda \in \Lambda; Y]{l} f$ and $f_\lambda \xrightarrow[\lambda \in \Lambda; Y]{epi-u} f$.

The introduced semicontinuous approximations are closely connected to sensitivity analysis of optimization problems.

Lemma 1. *Let $\mathcal{X}$ be a compact, $< f_\lambda >_{\lambda\in\Lambda}$ be in* $\mathsf{FCE}(\mathcal{X})$ *and* $f \in \mathsf{LSC}(\mathcal{X})$.

(i) If $f_\lambda \xrightarrow[\lambda\in\Lambda;\mathcal{X}]{l} f$ *then* $\liminf_{\lambda\in\Lambda} \varphi(f_\lambda;H) \geq \varphi(f;H)$ *for all* $H \in \mathcal{F}(\mathcal{X})$.

(ii) If there is $z \in \Phi(f)$ *such that* $f_\lambda \xrightarrow[\lambda\in\Lambda;z]{epi-u} f$ *then* $\limsup_{\lambda\in\Lambda} \varphi(f_\lambda;\mathcal{X}) \leq \varphi(f;\mathcal{X})$.

(iii) If $f_\lambda \xrightarrow[\lambda\in\Lambda;\mathcal{X}]{l} f$ *and there is* $z \in \Phi(f)$ *such that* $f_\lambda \xrightarrow[\lambda\in\Lambda;z]{epi-u} f$ *then* $\lim_{\lambda\in\Lambda} \varphi(f_\lambda;\mathcal{X}) = \varphi(f;\mathcal{X})$ *and* $\operatorname{K-limsup}_{\lambda} \Phi(f_\lambda) \subset \Phi(f)$, *where* K-limsup *denotes Kuratowski limes superior of sets.*

A proof of Lemma can be done in similar way as the proof of Theorem 7.33, p.266 in [2]. The semicontinuous approximations possesses equivalent definitions employing hit- or miss- part of Fell topology.

Proposition 1. *Let $\mathcal{X}$ be a locally compact topological space,* $< f_\lambda >_{\lambda\in\Lambda}$ *be a net in* $\mathsf{FCE}(\mathcal{X})$, $f \in \mathsf{LSC}(\mathcal{X})$ *and* $\emptyset \neq Y \subset \mathcal{X}$.

Then $f_\lambda \xrightarrow[\lambda\in\Lambda;Y]{l} f$ *if and only if for each* $G \in \mathcal{G}\left(\mathcal{X}\times\overline{\mathbb{R}}\right)$, $Y\times\overline{\mathbb{R}} \subset G$ *and* $K \in \mathcal{K}\left(\mathcal{X}\times\overline{\mathbb{R}}\right)$ *such that* $\mathrm{epi}(f)\cap G\cap K = \emptyset$ *and* $(Y\times\overline{\mathbb{R}})\cap K \neq \emptyset$ *there is* $Q \in \mathcal{G}\left(\mathcal{X}\times\overline{\mathbb{R}}\right)$ *such that* $Y\times\overline{\mathbb{R}} \subset Q \subset G$ *and* $\mathrm{epi}(f_\lambda)\cap Q\cap K = \emptyset$ *eventually.*

Proposition 2. *Let $\mathcal{X}$ be a topological space,* $< f_\lambda >_{\lambda\in\Lambda}$ *be a net in* $\mathsf{FCE}(\mathcal{X})$, $f \in \mathsf{LSC}(\mathcal{X})$ *and* $\emptyset \neq Y \subset \mathcal{X}$.

Then $f_\lambda \xrightarrow[\lambda\in\Lambda;Y]{epi-u} f$ *if and only if for each* $H \in \mathcal{G}\left(\mathcal{X}\times\overline{\mathbb{R}}\right)$, $H\cap(Y\times\overline{\mathbb{R}}) \neq \emptyset$ *fulfilling* $\mathrm{epi}(f)\cap Q\cap H \neq \emptyset$ *for all* $Q \in \mathcal{G}\left(\mathcal{X}\times\overline{\mathbb{R}}\right)$, $Y\times\overline{\mathbb{R}} \subset Q$ *we have* $\mathrm{epi}(f_\lambda)\cap G\cap H \neq \emptyset$ *eventually for each* $G \in \mathcal{G}\left(\mathcal{X}\times\overline{\mathbb{R}}\right)$, $G \supset Y\times\overline{\mathbb{R}}$.

3 Combining with randomness

Now, we present a suggestion how to combine semicontinuous approximations with randomness. We consider net of random functions $F : \Omega \to \mathsf{FCE}(\mathcal{X})$. We will denote by $\mathsf{rFCE}(\mathcal{X})$ the space of all random functions and by $\mathsf{rLSC}(\mathcal{X})$ the set of all random l.s.c. functions.

Definition 4. Let $\mathcal{X}$ be a locally compact topological space, $< F_\lambda >_{\lambda\in\Lambda}$ be a net in $\mathsf{rFCE}(\mathcal{X})$, $F \in \mathsf{rFCE}(\mathcal{X})$ and $\emptyset \neq Y \subset \mathcal{X}$.

We define $F_\lambda \xrightarrow[\lambda\in\Lambda;Y]{l-prob} F$ if for each $\varepsilon > 0$, $G \in \mathcal{G}\left(\mathcal{X}\times\overline{\mathbb{R}}\right)$, $Y\times\overline{\mathbb{R}} \subset G$ and $K \in \mathcal{K}\left(\mathcal{X}\times\overline{\mathbb{R}}\right)$ such that $(Y\times\overline{\mathbb{R}})\cap K \neq \emptyset$ there is $Q \in \mathcal{G}\left(\mathcal{X}\times\overline{\mathbb{R}}\right)$ such that $Y\times\overline{\mathbb{R}} \subset Q \subset G$ and

$$\limsup_{\lambda\in\Lambda} \mathsf{P}^*(\mathrm{epi}(F_\lambda)\cap Q\cap K \neq \emptyset,\ \mathrm{epi}(F)\cap G\cap K = \emptyset) < \varepsilon.$$

Definition 5. Let $\mathcal{X}$ be a topological space, $< F_\lambda >_{\lambda\in\Lambda}$ be a net in $\mathsf{rFCE}(\mathcal{X})$, $F \in \mathsf{rFCE}(\mathcal{X})$ and $\emptyset \neq Y \subset \mathcal{X}$.

We define $F_\lambda \xrightarrow[\lambda\in\Lambda;Y]{epi-u-prob} F$ if and only if for each $\varepsilon > 0$, $H \in \mathcal{G}\left(\mathcal{X} \times \overline{\mathbb{R}}\right)$, $H \cap (Y \times \overline{\mathbb{R}}) \neq \emptyset$ there is $G \in \mathcal{G}\left(\mathcal{X} \times \overline{\mathbb{R}}\right)$, $G \supset Y \times \overline{\mathbb{R}}$ such that

$$\limsup_{\lambda\in\Lambda} \mathsf{P}^*(\mathsf{epi}(F_\lambda) \cap G \cap H = \emptyset,\ \mathsf{epi}(F) \cap G \cap H \neq \emptyset) < \varepsilon.$$

The proposed approximations yield desired convergences of optimal values and optimal solutions.

Theorem 1. *Let $\mathcal{X}$ be a compact, $< F_\lambda >_{\lambda\in\Lambda}$ be a net in $\mathsf{rFCE}(\mathcal{X})$ and $F \in \mathsf{rLSC}(\mathcal{X})$.*

(i) If $F_\lambda \xrightarrow[\lambda\in\Lambda;\mathcal{X}]{l-prob} F$ then for all $\Delta \in \mathbb{R}$

$$\lim_{\lambda\in\Lambda} \mathsf{P}^*(\varphi(F_\lambda; K) < \Delta < \varphi(F; K)) = 0$$

for all $K \in \mathcal{K}(\mathcal{X})$.

(ii) If there is $Z \subset \mathcal{X}$ such that $\mathsf{P}_*(Z \cap \Phi(F) \neq \emptyset) = 1$ and $F_\lambda \xrightarrow[\lambda\in\Lambda;Z]{epi-u-prob} F$ then for all $\Delta \in \mathbb{R}$

$$\lim_{\lambda\in\Lambda} \mathsf{P}^*(\varphi(F_\lambda; \mathcal{X}) > \Delta > \varphi(F; \mathcal{X})) = 0.$$

(iii) If $\varphi(F; \mathcal{X})$ is a random variable, $F_\lambda \xrightarrow[\lambda\in\Lambda;\mathcal{X}]{l-prob} F$ and there is $Z \subset \mathcal{X}$ such that $\mathsf{P}_*(Z \cap \Phi(F) \neq \emptyset) = 1$ and $F_\lambda \xrightarrow[\lambda\in\Lambda;Z]{epi-u-prob} F$ then

$$\varphi(F_\lambda; \mathcal{X}) \xrightarrow[\lambda\in\Lambda]{prob} \varphi(F; \mathcal{X}).$$

If moreover $\varphi(F; K)$ is a random variable for all $K \in \mathcal{K}(\mathcal{X})$ then

$$\limsup_{\lambda\in\Lambda} \mathsf{P}_*(\Phi(F_\lambda) \subset G) = 1$$

for all $G \in \mathcal{G}(\mathcal{X})$, $G \supset \Phi(F)$.

Proof. 1. In the case (i) the sets from the definition 4 must fulfill $G = Q = \mathcal{X} \times \overline{\mathbb{R}}$. Hence, $H = K \times [-\infty, \Delta] \in \mathcal{K}\left(\mathcal{X} \times \overline{\mathbb{R}}\right)$ and

$$\begin{aligned}
&\limsup_{\lambda\in\Lambda} \mathsf{P}^*(\varphi(F_\lambda; K) < \Delta < \varphi(F; K)) \leq \\
&\leq \limsup_{\lambda\in\Lambda} \mathsf{P}^*(\mathsf{epi}(F_\lambda) \cap (K \times [-\infty, \Delta]) \neq \emptyset, \mathsf{epi}(F) \cap (K \times [-\infty, \Delta]) = \emptyset) \\
&= \limsup_{\lambda\in\Lambda} \mathsf{P}^*(\mathsf{epi}(F_\lambda) \cap Q \cap H \neq \emptyset, \mathsf{epi}(F) \cap G \cap H = \emptyset) = 0,
\end{aligned}$$

since the choice is independent on $\varepsilon > 0$.

2. To show (ii) take $\varepsilon > 0$, $U \in \mathcal{G}(\mathcal{X})$, $U \supset Z$ and set $H = U \times [-\infty, \Delta)$. According to Definition 5 then there is a $V \in \mathcal{G}(\mathcal{X})$, i.e. $G = V \times \overline{\mathbb{R}}$, such that $U \supset V \supset Z$ and

$$\limsup_{\lambda\in\Lambda} \mathsf{P}^*(\varphi(F_\lambda;\mathcal{X}) > \Delta > \varphi(F;\mathcal{X})) \leq$$
$$\leq \limsup_{\lambda\in\Lambda} \mathsf{P}^*(\mathsf{epi}(F_\lambda)\cap(V\times[-\infty,\Delta))=\emptyset, \mathsf{epi}(F)\cap(V\times[-\infty,\Delta))\neq\emptyset)$$
$$= \limsup_{\lambda\in\Lambda} \mathsf{P}^*(\mathsf{epi}(F_\lambda)\cap G\cap H = \emptyset, \mathsf{epi}(F)\cap G\cap H \neq \emptyset) < \varepsilon.$$

3. Convergence of optimal values in probability in the case (iii) follows from

$$\limsup_{\lambda\in\Lambda}\ \mathsf{P}^*(\varphi(F;\mathcal{X}) + 2\Delta < \varphi(F_\lambda;\mathcal{X})) \leq$$
$$\leq \limsup_{\lambda\in\Lambda} \sum_{n=-N}^{N} \mathsf{P}^*(\varphi(F;\mathcal{X}) + 2\Delta < \varphi(F_\lambda;\mathcal{X}), n\Delta < \varphi(F;\mathcal{X}) \leq (n+1)\Delta)$$
$$+ \mathsf{P}(|\varphi(F;\mathcal{X})| \geq N\Delta) \leq$$
$$\leq \limsup_{\lambda\in\Lambda} \sum_{n=-N}^{N} \mathsf{P}^*(\varphi(F;\mathcal{X}) \leq (n+1)\Delta < (n+2)\Delta < \varphi(F_\lambda;\mathcal{X})) +$$
$$+ \mathsf{P}(|\varphi(F;\mathcal{X})| \geq N\Delta) = \mathsf{P}(|\varphi(F;\mathcal{X})| \geq N\Delta),$$

$$\limsup_{\lambda\in\Lambda} \mathsf{P}^*(\varphi(F_\lambda;\mathcal{X}) + 2\Delta < \varphi(F;\mathcal{X})) \leq$$
$$\leq \limsup_{\lambda\in\Lambda} \sum_{n=-N}^{N} \mathsf{P}^*(\varphi(F_\lambda;\mathcal{X}) + 2\Delta < \varphi(F;\mathcal{X}), n\Delta < \varphi(F;\mathcal{X}) \leq (n+1)\Delta)$$
$$+ \mathsf{P}(|\varphi(F;\mathcal{X})| \geq N\Delta) \leq$$
$$\leq \limsup_{\lambda\in\Lambda} \sum_{n=-N}^{N} \mathsf{P}^*(\varphi(F_\lambda;\mathcal{X}) \leq (n-1)\Delta < n\Delta < \varphi(F;\mathcal{X})) +$$
$$+ \mathsf{P}(|\varphi(F;\mathcal{X})| \geq N\Delta) = \mathsf{P}(|\varphi(F;\mathcal{X})| \geq N\Delta),$$

The property of optimal solutions can be shown directly

$$\limsup_{\lambda\in\Lambda} \mathsf{P}^*(\Phi(F_\lambda \not\subset G)) \leq$$
$$\leq \limsup_{\lambda\in\Lambda} \sum_{n=-N}^{N} \mathsf{P}^*(\Phi(F_\lambda \not\subset G), n\Delta < \varphi(F;\mathcal{X}) \leq (n+1)\Delta)$$
$$+ \mathsf{P}(|\varphi(F;\mathcal{X})| \geq N\Delta) \leq$$

$$\leq \limsup_{\lambda\in\Lambda} \sum_{n=-N}^{N} \mathsf{P}^*\left(\Phi\left(F_\lambda \not\subset G\right), n\Delta < \varphi(F;\mathcal{X}) \leq (n+1)\Delta,\right.$$

$$\left., \varphi(F_\lambda;\mathcal{X}\setminus G) \leq (n+1)\Delta\right) + \mathsf{P}(|\varphi(F;\mathcal{X})| \geq N\Delta) \leq$$

$$\leq \limsup_{\lambda\in\Lambda} \sum_{n=-N}^{N} \mathsf{P}(n\Delta < \varphi(F;\mathcal{X}) \leq (n+1)\Delta, \varphi(F;\mathcal{X}\setminus G) < (n+2)\Delta)$$

$$+ \mathsf{P}(|\varphi(F;\mathcal{X})| \geq N\Delta) \leq$$

$$\leq \mathsf{P}(\varphi(F;\mathcal{X}) > \varphi(F;\mathcal{X}\setminus G) - 2\Delta) + \mathsf{P}(|\varphi(F;\mathcal{X})| \geq N\Delta).$$

Hence,

$$\limsup_{\lambda\in\Lambda} \mathsf{P}^*(\Phi(F_\lambda \not\subset G)) \leq \mathsf{P}(\varphi(F;\mathcal{X}) \geq \varphi(F;\mathcal{X}\setminus G)) = 0,$$

since $\mathsf{P}(G \supset \Phi(F)) = 1$.

References

1. G. Beer (1993) *Topologies on Closed and Closed Convex Sets.* Kluwer Academic Publishers, Dordrecht.
2. T. Rockafellar, R. J-B Wets (1998) *Variational Analysis.* Springer-Verlag, Berlin.
3. G. Salinetti and R.J.-B. Wets (1981) *On the convergence of closed-valued measurable multifunctions*, Transactions of the American Mathematical Society, 266, pp. 275-289.
4. S. Vogel (1994) A stochastic approach to stability in stochastic programming. *J. Comp. Appl. Math.* **56**, 65-96.
5. P. Lachout (2000) Epi-convergence versus topology: splitting and localization. The research report of The Institute of Information Theory and Automation No.1987, Prague.
6. S. Vogel and P. Lachout (2000) On continuous convergence and epi-convergence of random functions I. The research report of The Institute of Information Theory and Automation No.1988, Prague.
7. S. Vogel and P. Lachout (2000) On continuous convergence and epi-convergence of random functions II. The research report of The Institute of Information Theory and Automation No.1989, Prague.

Standards für Modellierung und Simulation

Claus-Burkard Böhnlein

Lehrstuhl für BWL und Wirtschaftsinformatik, Universität Würzburg,
Neubaustraße 66, D-97070 Würzburg, boehnlein@wiinf.uni-wuerzburg.de

1 Entscheidungsunterstützung mittels Simulation

Simulation ist das systematische Durchspielen des Verhaltens von geplanten, sich in der Entwicklung befindlichen oder bereits existierenden Systemen. Dabei wird ein Simulationsmodell zugrundegelegt, das die für die Simulation relevanten Aspekte des betrachteten Systems nachbildet. Simulation ist somit das zielgerichtete Experimentieren an Modellen, die der Wirklichkeit nachgebildet sind (Oberweis et al. 1999; Bossel 1992).
Bei der Analyse betrieblicher Abläufe spielt die Simulation eine wichtige Rolle, da sie Aussagen über die Ablaufausführung unter verschiedenen Bedingungen und bezüglich verschiedener Aspekte, etwa Kosten, Durchlaufzeiten, Ressourcenverbrauch oder Maschinenauslastung, erlaubt. Die Simulation leistete daher in der Vergangenheit eine wichtige Unterstützung bei der Ablaufplanung, der Ablaufverbesserung sowie bei der Gestaltung von Informationssystemen zur Ablaufausführung.

1.1 Simulationszwecke

Je nach Zielsetzung nennt OBERWEIS unterschiedliche Simulationszwecke, die nachfolgend mit steigender Komplexität der Aufgabenstellung genannt werden (Oberweis et al. 1999):

- Das *Durchspielen des Verhaltens* real existierender Systeme wird eingesetzt bei Präsentationen und Schulungen. Hier steht die Animation zur Veranschaulichung des Systemverhaltens im Vordergrund.
- In großen Entwicklungsprojekten erlaubt die *Visualisierung* der Systemstruktur eine gezielte Kommunikation über geplante bzw. existierende Prozesse.
- Durch die *Validierung von Systementwürfen* kann am Simulationsmodell überprüft werden, ob sich das vorgeschlagene System wie geplant verhält.
- Durch *Antizipation von Systemverhalten* in Bereichen wie Baustatik, Crashverhalten oder Klimaentwicklung erlaubt die Simulation das Austesten von Systemgrenzen und das frühzeitige Erkennen von kritischen Systemzuständen. Dies ist dann bedeutsam, wenn Versuche am realen System zu teuer, zu risikoreich oder grundsätzlich nicht möglich sind.

□ Häufig wird die Simulation eingesetzt, um die *Reaktion eines Systems* auf externe Einflüsse oder Struktur- bzw. Regeländerungen im System zu untersuchen und zu verbessern.

1.2 Problemlösungsprozess

Ist ein simulationswürdiges Problem erkannt, dann wird es in einem mehrstufigen Prozess gelöst. Zuerst werden das Problem möglichst genau beschrieben und die erforderlichen Daten erhoben. Danach wird das Problem analysiert und in ein abstraktes Modell überführt. Das Modell wird mit Vergangenheitsdaten validiert und gegebenenfalls korrigiert. Verschiedene Modell- und Parametermodifikationen werden in iterativen Simulationsläufen überprüft, um gegenüber der Ausgangssituation verbesserte, gültige Lösungen zu ermitteln, und zur Umsetzung vorgeschlagen (vgl. Abb. 1).

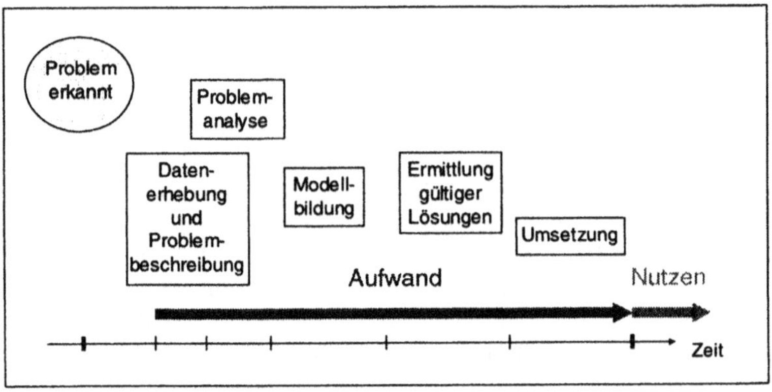

Abb. 1. Problemlösungsprozess

Bis zu diesem Zeitpunkt wurde nur Aufwand produziert, erst nach der Umsetzung kann die verbesserte Lösung auch nutzbringend eingesetzt werden. Dies setzt aber voraus, dass das Systemumfeld sich nicht verändert hat, andernfalls würde im ungünstigsten Fall die vorgeschlagene Lösung unbrauchbar. In dynamischen Systemumgebungen ist es deshalb entscheidend, dass ein Problem möglichst schnell erkannt und gelöst wird, damit die ermittelte Lösung unmittelbar nach der Umsetzung genutzt werden kann.

2 Märkte im Wandel

In der Vergangenheit vollzog sich ein Wandel vom Verkäufer- zum Käufermarkt, der durch eine Verschärfung des Wettbewerbs gekennzeichnet war. In zunehmend gesättigten Märkten sind Erzeugnisse schlechter absetzbar und die Abnehmer können zwischen verschiedenen Anbietern vergleichbarer Produkte auswählen. Die Hersteller versuchen in dieser Situation durch kundenindividuelle Produkte und kurze Produktlebenszyklen die Kundenbindung zu erhöhen und sich von ihren Mitbewerbern zu differenzieren. Da eine zunehmende Variantenvielfalt in der Regel mit sinkenden Losgrößen einher geht, steigt neben dem Koordinationsaufwand aber auch der Kostendruck in den Unternehmen. Diese beginnen deshalb sich auf ihre Kernkompetenzen zu konzentrieren, ihre Wertschöpfungstiefe zu reduzieren und beziehen zunehmend Leistungen von externen Anbietern. Kernkompetenzen sind in diesem Zusammenhang wesentliche technische, technologische und organisatorische Fähigkeiten, die durch eine hohe Wettbewerbswirksamkeit auf den Märkten und durch einen großen Wettbewerbsvorteil gegenüber der Konkurrenz gekennzeichnet sind.

3 Anforderungen an Unternehmen

Da mit wachsender Lieferantenanzahl auch der Koordinationsaufwand ansteigt, gehen die Hersteller Entwicklungs- und Produktionspartnerschaften mit sogenannten Systemlieferanten ein. Dadurch können aus Sicht der Hersteller einerseits die Entwicklungszeiten für neue Produkte gesenkt und andererseits die internen Beschaffungs- und Produktionsprozesse gestrafft werden. Diese Aufteilung der Wertschöpfung auf mehrere Unternehmen erfordert aber zusätzliche koordinierende Maßnahmen, denn die verteilte Leistungserbringung bedingt eine genaue Abstimmung der Termine, Kapazitäten und Mengen in den unterschiedlichen Bereichen aller beteiligten Unternehmen. Zur Lösung dieser Koordinationsaufgaben zeichnet sich für die kommenden Jahre ein weiteres Zusammenwachsen der integrierten Geschäftsabwicklung und der unternehmensübergreifenden Koordination der erforderlichen Material- und Informationsflüsse ab.

Konkret bedeutet dies für Unternehmen, sie müssen sich

- schnell und flexibel am Markt orientieren,
- Mitarbeiterpotentiale schöpfen und entwickeln,
- Kooperationen und Netzwerke nach innen und außen aufbauen und
- verstärkt integrierte Informationssysteme nutzen.

Für die Abstimmung und Koordination in derartigen unternehmensübergreifenden Netzwerken werden Funktionen zur Durchführung von What-If-Analysen und Simulationsmöglichkeiten benötigt, die nicht nur auf strategischer Ebene, sondern zunehmend auch auf taktischer und operativer Ebene eine Entscheidungsunterstützung bieten.

4 Status quo

Gemessen am Marktvolumen gibt es sehr viele verfügbare Simulationsumgebungen. Eine Möglichkeit zum Datenimport und -export über ASCII, DDE, ODBC oder SQL besitzen inzwischen alle Werkzeuge, ebenso eine leistungsfähige graphische Benutzeroberfläche (z. T. auch internetfähig). Vereinzelt werden auch integrierte Code-Generatoren angeboten, die direkt aus dem Modell ausführbaren

Programmcode erzeugen (z. B. PACE von IBE). Kommerzielle Simulationsumgebungen verfügen zum Teil auch über Schnittstellen zu gängigen ERP-Systemen. Einzelne Softwarehersteller bieten branchenspezifische Planungslösungen an, in die eine leistungsfähige Simulationsumgebung integriert werden kann (z. B. eM-Power und eM-Plant von Tecnomatix).

5 Simulation in integrierten IT-Umgebungen

GRAF und PUTZLOCHER unterscheiden den Produktentwicklungs-, Produktentstehungs-, Kundenauftrags- und Materialbeschaffungsprozess (vgl. Abb. 2). Der Erfolg einer Unternehmung hängt dabei entscheidend von einer ganzheitlichen Betrachtung und Koordination aller genannten Prozesse ab (Graf u. Putzlocher 2002). Schlagworte wie Digital Mock-Up und digitale Fabrik (Bracht U 2002) zeigen, dass in den Schlüsselbranchen wie der Automobilindustrie die Integration der Entwicklungssysteme vorangetrieben wird und neuere Planungssysteme sog. Advanced Planning and Scheduling Systems (APS) verfügen bereits über Simulationsfunktionen für What-If-Analysen.

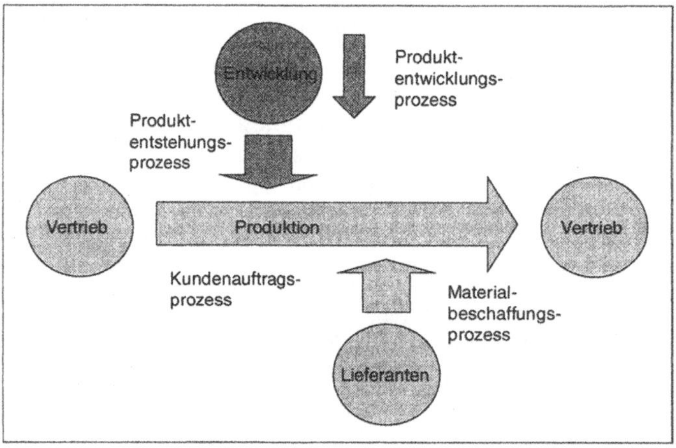

Abb. 2. Geschäftsprozesse (Graf u. Putzlocher 2002)

Als gemeinsame Datenquelle für die operativen Systeme wird zunehmend ein Produktdatenmanagement eingesetzt und mit Systemen für Business Intelligence werden Data Mining oder Text Mining Techniken genutzt, um aus Daten über Informationen Wissen abzuleiten (Bange C, Mertens H, Keller P 2001; Voß S, Gutenschwager K 2001).

Doch wo bleibt die dynamische Simulation, wenn operative Systeme zunehmend über Simulationsfunktionen verfügen und dadurch die zeitaufwändige Datenbeschaffung und Modellierung entfallen?

6 Anforderungen an Modellierung und Simulation

Leistungsfähige Simulationsumgebungen für dynamische Systemanalysen werden auch zukünftig ihre zentrale Bedeutung behalten, wenn die genannten Defizite behoben werden. Dazu müssen aus heutiger Sicht folgende Anforderungen umgesetzt werden:

1. Für den Datenaustausch mit operativen Systemen müssen einheitliche Schnittstellen geschaffen werden.
2. Durch die Entwicklung einer XML-basierten Sprache (Pardi W 1999) mit speziellen Dokumenttyp-Definitionen (DTD) können Simulationsanwendungen in einem betriebswirtschaftlichen Umfeld mit OLAP- oder Data-Mining-Werkzeugen sowie mit relationalen Datenbanken effektiv kommunizieren können.
3. Kurzfristige Aussagen und Analysen sind nur dann möglich, wenn der Zeitbedarf für Datenbeschaffung und Modellierung deutlich reduziert wird. Dazu müssen operative Systeme mit Simulationsumgebungen integriert werden.
4. Für eine systematische, effiziente Modellierung und Wiederverwendung von Teilmodellen muss nach dem Vorbild der Unified Modelling Language (UML) eine Unified Simulation Modelling Language (USML) entwickelt werden. Dies würde den Austausch von Modellen und die Interaktion zwischen verteilten Modellen unterstützen sowie die Entwicklung von Modellgeneratoren fördern.
5. Open Source Simulationsumgebungen müssen gefördert werden, damit nach dem Vorbild von LINUX bereits bei der Softwareentwicklung Standards etabliert werden und die Vielfalt der Simulationsumgebungen auf eine sinnvolle Größenordnung reduziert wird.

Aspekte der Standardisierung von Modellierungstechniken, Schnittstellen und Austauschformaten werden zukünftig sowohl für die Anwender, als auch für Anbieter von Simulationsumgebungen eine zentrale Bedeutung erlangen. Standards sind in integrierten IT-Umgebungen kooperierender Unternehmen eine notwendige Voraussetzung für eine Entscheidungsunterstützung im operativen Bereich.

Literatur

Bange C, Mertens H, Keller P (2001) OLAP und Business Intelligence. Oxygon, Feldkirchen

Bossel H (1992) Modellbildung und Simulation – Konzepte, Verfahren und Modelle zum Verhalten dynamischer Systeme. Vieweg, Braunschweig Wiesbaden

Bracht U (2002) Ansätze und Methoden der Digitalen Fabrik. Eröffnungsvortrag im Tagungsband "Simulation und Visualisierung 2002", 28.02 und 01.03 Uni Magdeburg

Graf H, Putzlocher S (2002) DaimlerChrysler: Integrierte Beschaffungsnetzwerke. In: Corsten D, Gabriel C (Hrsg.) Supply Chain Management erfolgreich umsetzen – Grundlagen, Realisierung und Fallstudien. Springer, Berlin Heidelberg New York

Oberweis A, Lenz K, Gentner C (1999) Simulation betrieblicher Abläufe. In: Hartmann-Wendels et al. (Hrsg.) wisu – das wirtschaftsstudium, 28 (1999) 02, Düsseldorf, S 216-223, 245

Pardi W (1999) XML in Action. Dynamische und datengestützte Webseiten mit der neusten Web-Technologie. Microsoft Press, Unterschleißheim

Voß S, Gutenschwager K (2001) Informationsmanagement. Springer, Berlin Heidelberg New York

System Dynamics (SD) – An Approach within Corporate Planning

Peter Bradl

Bavarian Research Center for Knowledge-based Systems – FORWISS, Information Systems Research Group, Äußerer Laufer Platz 13-15, 90403 Nuremberg, Germany

1 Corporate Planning

Increasing speed in business, engagement on international markets, as well as rising costs for mismanagement and wrong decisions require that effects of changes, both endogenous and exogenous, can be foreseen. At least, it is important to have the possibility to rewrite existing action plans according to the current situation almost instantly. The use of models that are built and validated according to the companies current situation (including market information etc.) not only allows playing with the status quo but is the basis for simulations that provide results on possible future outcomes.

1.1 Management Decisions and Planning

The difficult situation on the market and globalization are frequently mentioned by CEOs when they have to explain (e.g. to shareholders) why their company does or did not perform as expected. Enterprise Resource Planning (ERP) software packages already allow for static planning. Various statistical methods like exponential smoothing are known to correct planning figures; simulation tools are in use to some extent. Still, most of these tools are extremely useful predominately for operational planning. However, Corporate Planning which is generally accepted as core management functionality for a long time (Lyneis 1980, Perlitz 2000, Ross 1993) goes beyond this and yet is hardly supported by Information Systems.

1.2 Planning Approaches

In management literature several approaches to planning were introduced during the last decade (e.g. *Total Quality Management*, *Business Process Reengineering* or the *Balanced Scorecard* (see Maani a. Cavan 2000 pp. 4-5)). They mostly addressed a certain field or operational area. Besides the Balanced Scorecard there were only few individual techniques linking management aspects (Macharzina

1999 p. 166). Organizational theory itself hardly allows the usage of figures, however, figures carry and represent the type of information managers are used to. Since planning strategies change over time, (Macharzina 1999 p. 158), they require a method that takes this into account as well – SD.

2 System Dynamics

2.1 Systems Thinking

Underlying the method of SD and the tools to apply it there is the need to become sort of a systems thinker. This is e.g. to recognize that feedbacks are part of all systems (department, company, market) we belong to. It implies that our own actions will affect us later probably in a way we did not intend. An example:

> *In August 2000 the auction of UMTS licenses for Germany was held yielding about 100 billion Deutschmarks for the treasury. "Deutsche Telekom AG" alone had to pay more than 16 billion Deutschmarks which had to be financed. As a result of increasing debts Deutsche Telekom's capital structure changed. Consequently the rating agencies disrated the company, which in turn usually causes a risk premium for future debts – the feedback loop with external variables is closed. (Moody's 2002; Voba-Börse 2001)*

More often than not dependencies can not be followed as easy as in the case of the "Deutsche Telekom AG", however it is important to realize that those effects exist and that managers and everyone else involved into Corporate Planning have to consider this.

2.2 Modelling as Part of the Management Process

System thinking as the "starting point" for mental models has to be followed by some steps to finally derive Stock Flow Diagrams (SFD) representing SD model for simulations.

As intermediate stage towards SFDs so-called Causal Loop Diagrams (CLD) which allow identifying interactions and relationships between actions and e.g. their effect onto different departments, are commonly used. Richardson describes some serious problems that might occur when transforming CLDs into SFDs (Richardson 1986). But still there is a benefit when using CLDs, especially in Corporate Planning, since they may function as a type of communication tool. A main problem in decision-making processes (even on the corporate level) is participants having different understandings and terminologies. (E.g. sales may be interpreted as sales quantity or turnover. The use of CLDs by applying modelling techniques as described e.g. in (Gomez a. Probst 1999) or (Maani a. Cavan 2000) within a group enforces the members to use terms in their understanding within the design process team. Finally, when discussing dependencies, different interpretations become obvious. So modelling (especially CLDs) helps to gain a common understanding of relevant terms.[1]

2.3 Utilisability and Simulation

2.3.1 Hardware and Software

An important fact for the acceptance of SD in management – especially on corporate level – is that the models can be used anywhere and without detailed knowledge in the field of computer science (as it is desirable for most IT in management). Nowadays, several SD software packages (e.g. Vensim, Stella, or Powersim) run on standard PC or Macintosh computers, equipment that is available in almost every office – even on management level. Additionally, graphical user interfaces may be designed to support the manager or to provide help when entering figures e.g. when checking for consistency. Actually the underlying analytical model is usually hidden for the user and only sliders to change inputs and charts will be seen.

2.3.2 Validation and Simulation

Due to the wide-spread use of ERP-software in companies, data of almost every transaction may be stored in a database or data warehouse respectively. These figures are not only important for the profit and loss account or preparing the balance sheet but are extremely helpful for usage in models that support Corporate Planning. Modern SD-software packages allow to directly transfer data from the data warehouse into the modelling software, simulate and may even write the results back into the data warehouse. Hence, validation of models in financial planning might be much easier compared to topics like environmental issues. Simulation runs are the final stage in this process.[2]

3 Corporate Planning – Knowledge Management within a Software Company

At the Bavarian Research Center for Knowledge Based Systems (FORWISS) in cooperation with SAP AG we are developing reference models for financial planning that are based on relevant Key Performance Indicators and datasets transferred from the Business Information Warehouse (BW) of the SAP/SEM to support the whole Strategic Planning procedure. Since employees and knowledge represent important assets of a company we currently focus on these issues. An exemplary question within the planning process for human resource management is the following:

---

[1] General information about SD and details how to derive SFDs can be found in Coyle 1996, Ford 1999, Forrester 1968, Goodman 1974, Maani a. Cavan 2000; Sterman 2000.

[2] Details covering issues like simulation and scenarios can be found in Gausemeier et al. 1996, Mertens 1982.

What are the effects if management increases the salary of the employees by 10%?

Obviously enough, costs will increase. As an effect of assumed higher motivation, performance may improve. Additionally, the higher salary will lead to a positive reputation of the enterprise as potential employer and attract more and better-qualified employees while, in turn, fluctuation rates decrease resulting in overall declining recruitment costs. This way, the positive effects might go beyond the negative effects. The models will show whether this scenario is true and if it appears, to what extent and under which assumptions.

Presently, we focus on a specific topic within the field of Human Resource Management – Knowledge Management. Software firms strongly depend on manpower and knowledge – as a resource – linked to the employees. Although there are little investments to be made into machines within this industry – compared to e.g. printing companies – high costs (e.g. providing loans at attractive conditions to employees) have to be covered for being 'attractive' for the 'resource man'. So the strategic task and major point of interest is 'ensuring knowledge' by qualifying staff. The hiring process itself and the transition of newly hired employees to experienced workers have already been covered in several models (see Maani a. Cavan 2000; Sterman 2000).

3.1 Key Issues of Knowledge Management

In our current model we look at a department within a software company. Besides hiring and attrition we focus on the qualification process. A specific situation:

> *A department of about 100 software engineers which is currently responsible for a product that will be replaced. The new platform will be programmed in a different language. Management wants to know how to do the training.*

To answer questions that arise from this situation, we start with the description of a classification that distinguishes three different types of knowledge: *Programming, Business Administration, and Social Skills.*[3]

3.2 A Scale for 'Knowledge'

Referring to the availability of knowledge we developed a classification that uses five expressions: *Null, Poor, Average, Good, and Expert.* An employee holds a certain amount of 'knowledge units (KUs)' that lead to a position within the qualification groups. A change from one group into another can be achieved by various steps. It is possible to gain KUs as well as to lose them within a certain category. *Gaining KUs* is possible in two ways *Experience* and *Training*. The *loss of KUs* on the other hand is predominately an effect of time passing by. We distinguish the two alternatives *"Forgetting"* and *Obsolete Knowledge*.

[3] For details about assumptions and classifications for the case see Bradl 2002.

In order to be able to remove people from a group and add it to another we deplete them according to a distribution key that is initially set and adapted whenever peoples gain (or lose) qualification.

3.3 Costs for 'Knowledge'

As mentioned earlier Corporate Planning implies financial planning which is strongly related to human resource management. In our model we consider that laying off people is costly in several aspects. Compensation payment has to be calculated as well as additional costs for hiring. Additionally, so-called 'rookies' have to be integrated and must learn the rules of the company – which takes time – and are usually less efficient during induction period. On the other hand salaries are lower for new members of the staff (for non-specialists). Furthermore, there are fewer costs for training for this group since it is supposed that the new employees have the right qualification. Talking about training and costs leads to decisions about the type of training. On the job or fulltime; internal or external; team-by-team or only a certain percentage of the team at once? Revenue that might be lost due to reduced manpower while employees are in training is an important factor when talking about training costs.

3.4 Used Software and Data Support of the Model

At our institute we use Powersim as the environment for SD modelling. The expressions for knowledge are implemented with the array functionality in this software. We design personalized 'portals' to allow simulation for different members of the management and to ensure that the accessibility supports the use of the existing model. The parameters for dependencies within the model or historical data are put in via external data sources, like Excel or the BW. Besides the assurance that simulations run with real data, results will be written back into the BW for further calculations even within the planning area of the company's ERP or something else.

3.5 Scenario

Within one particular simulation run emphasis was put on the question which training approach should be preferred – full time (absence from daily work) or on the job. Assumptions were made about correlation between knowledge (as effect of training) and efficiency as well as the dependency of wages and efficiency. The scenarios showed that the chosen company will gain higher profits when training is performed on the job. It is clear that untrained employees perform not as well as others but they still cover workload and ensure that the company is able to complete orders in time.

4 Conclusions and Outlook

As mentioned above, dependencies in complex systems cannot intuitively be recognized. Hence, a method must be used to identify them. System Dynamics does this – and more. Even the simulation of only three types of scenarios – best, worst and most realistic case – for one variable will help to make plans in advance. Thus managers will be prepared what to do if the anticipated scenario arises in reality. The "low" Software and Hardware requirements nowadays allow the use of SD almost everywhere and should help to bring SD into management.

References

Bradl P (2002) Strategic Enterprise Planning with System Dynamics. In: Proceedings of the 2nd International Conference on Systems Thinking in Management. John Mason Printers, Skipton

Coyle RG (1996) System Dynamics Modelling – A Practical Approach, Chapman & Hall, London

Ford A (1999) Modelling the Environment, Island Press, Washington, D.C. et al.

Forrester JW (1968) Principles of System, Wright-Allen Press, Cambridge MA

Gausemeier J, Fink A, Schlake O (1996) Szenario-Management, Hanser, München

Gomez P, Probst G (1999) Die Praxis des ganzheitlichen Problemlösens, 3rd edn, Bern et al.

Goodman MR (1974) Study Notes in System Dynamics, Wright-Allen Press, Cambridge MA

Lyneis JM (1980) Corporate Planning and Policy Design, MIT Press, Cambridge, Massachusetts

Maani KE, Cavan RY (2000) Systems Thinking and Modelling – Understanding Change and Complexity, Pearson Education, New Zealand

Macharzina K (1999) "Unternehmensführung" Das internationale Managementwissen, 3rd edn, Gabler, Wiesbaden

Mertens P (1982) Simulation, Poeschel, Stuttgart

Moody's (2002) http://www.moodys.com (2002-08-29)

Perlitz M (2000) Internationales Management, 4th edn, Lucius u. Lucius, Stuttgart

Richardson GP (1986) Problems with Causal Loop Diagrams. In: System Dynamics Review 2, pp. 158-170

Ross SA, Westerfield RW, Jaffe JF (1993) Corporate Finance, 3rd edn, Irwin, Homewood, Illinois et al.

Sterman JD (2000) Business Dynamics – Systems Thinking and Modeling for a Complex World, McGraw-Hill, Boston et al.

Voba-Börse (2001) http://voba-boersedirekt.de/ticker00/20000821ticker.pdf (2001-12-20)

Optimal Decision Rules in a Monetary Union

Doris A. Behrens[1] and Reinhard Neck[2]

[1] Vienna University of Technology, Department of Operations Research and Systems Theory, Argentinierstrasse 8/119, A–1040 Vienna, Austria
[2] University of Klagenfurt, Department of Economics, Universitaetsstrasse 65–67, A–9020 Klagenfurt, Austria

Abstract. This paper develops a dynamic game model to study strategic interactions between the decision–makers in a monetary union. In such a union, governments of the participating countries pursue national goals when deciding on fiscal policies, whereas the common central bank's monetary policy aims at union–wide objective variables. Considering the example of a negative demand shock, we show how different solution concepts for the dynamic game between the common central bank and the national governments can be used as models of a conflict between national and supra-national institutions (noncooperative Nash equilibrium) and of coordinated policy–making (cooperative Pareto solutions).

1 Introduction

Policy–making for a national economy should be supported by a careful analysis of its objectives, its constraints, and the possibilities of achieving an outcome which is in some well–defined sense "better" than other available alternatives. Operations research and economics have produced many theoretical models in order to provide guidelines for arriving at "optimal" solutions for economic decision problems. In the context of fiscal and monetary policies in an international context, especially in a monetary union, where different objectives of policy–makers are nearly inevitable, game theory is an adequate tool to analyze and improve policy–making. Given the intertemporal nature of macroeconomic policy problems, the toolkit of dynamic game theory (see [1], for instance) recommends itself for obtaining insights and policy recommendations for decision–makers (governments of member countries and the common central bank) in a monetary union (see [2]).

Those mathematical models for such a macroeconomic system, which give a (largely) realistic picture for the real–world decision problem of concern, are rather soon reaching the limits of analytical tractability. Therefore, in this paper we will use the OPTGAME 2.0 algorithm [3] to analyze a simple policy problem in a two–country monetary union. This numerical algorithm is designed for determining solutions of dynamic difference games with a finite planning horizon. In particular, OPTGAME solves discrete–time LQ games, and approximates the solutions of nonlinear–quadratic difference games by iteration. At present, the algorithm calculates the open–loop and the feedback Nash equilibrium solution and the cooperative Pareto–optimal solutions

Operations Research Proceedings

for an arbitrary number of players; extensions to other solution concepts are being implemented. Here we will show that calculating different solution concepts for a dynamic game between the common central bank and national fiscal policy–makers can provide insights into possible conflicts and their solution in this context.

2 The Model

We consider a monetary union with two participating countries. *Monetary union* means that national currencies (national central banks) have been entirely replaced by a common currency (common central bank). Among others, this implies that the exchange rate has disappeared as an instrument of adjustment. In the following description of the model, capital letters indicate nominal values, while lower case letters correspond to real values. The two countries are assumed to be of equivalent size in terms of gross domestic product (GDP). The superscripts ^{d} and ^{s} denote demand and supply, respectively. The supply side is mostly exogenous.

The demand side goods market is modeled by a short–run income–expenditure equilibrium relation (IS curve), which is superimposed on an exogenous natural growth path. For $t = 1, ..., T$, real output in country i $(i = 1, 2)$ is given as the sum of the long–run equilibrium level of the real output, $\bar{y}_{it}$, and the short–term deviation therefrom, $\tilde{y}_{it}$, i.e.

$$y_{it} = \bar{y}_{it} + \tilde{y}_{it}, \tag{1}$$

where

$$\bar{y}_{it} = (1+\theta)\bar{y}_{i(t-1)}, \quad \bar{y}_{i0} \text{ given,} \tag{2}$$

$$\tilde{y}_{it} = \delta_i \left(\frac{P_{jt}}{P_{it}} - 1 \right) - \gamma_i (r_{it} - \theta) + \rho_i \tilde{y}_{jt} + \eta_i \tilde{f}_{it} + z_{it}, \tag{3}$$

for $i \neq j$ $(i, j = 1, 2)$. The variable $P_{it} (i = 1, 2)$ denotes country i's output price (its general price level), $r_{it} (i = 1, 2)$ represents country i's current real interest rate, and $\tilde{f}_{it} (i = 1, 2)$ denotes country i's short-term (deviation from a zero) real fiscal deficit. $\tilde{f}_{it} (i = 1, 2)$ in (3) is country i's fiscal policy instrument, i.e. its control variable. The natural real growth rate, θ, is assumed to be equal to the natural real rate of interest (assuming dynamic efficiency, in accordance with neoclassical growth theory). The parameters δ_i, γ_i, ρ_i, η_i $(i = 1, 2)$ in (3) are assumed to be positive. The variables z_{1t} and z_{2t} are not subject to control and represent exogenous shocks on the demand side goods market.

For $t = 1, ..., T$, the current real rate of interest for country i $(i = 1, 2)$ is given by

$$r_{it} = R_{Et} - X_{it}, \tag{4}$$

where R_{Et} denotes the common nominal rate of interest determined by the common central bank, and X_{it} $(i = 1, 2)$ represents country i's rate of inflation. Note that the equilibrium level of the natural long–run interest rate, $\bar{R}_{Et} = \bar{r}_{it} = \theta$ $(i = 1, 2)$, is "inflation–free", i.e. $\bar{X}_{it} = 0$ $(i = 1, 2)$. Output prices and inflation rates for $i = 1, 2$ and $t = 1, ..., T$ are determined according to a demand–pull relation:

$$P_{it} = (1 + X_{it})\, P_{i(t-1)}, \quad P_{i0} \text{ given}, \tag{5}$$

$$X_{it} = \xi_i \tilde{y}_{it}, \tag{6}$$

where ξ_1 and ξ_2 are positive parameters. We also define average variables as

$$y_{Et} = \omega y_{1t} + (1 - \omega)\, y_{2t}, \quad \omega \in [0, 1], \tag{7}$$

$$X_{Et} = \omega X_{1t} + (1 - \omega)\, X_{2t}, \quad \omega \in [0, 1]. \tag{8}$$

Money demand in country i $(i = 1, 2)$ is the sum of long–run and short–run money demand:

$$m_{it}^{\mathrm{d}} = \bar{m}_{it}^{\mathrm{d}} + \tilde{m}_{it}^{\mathrm{d}}. \tag{9}$$

Short–run money demand is determined by a Keynesian money demand function (LM curve):

$$\tilde{m}_{it}^{\mathrm{d}} = \kappa_i \tilde{y}_{it} - \lambda_i\, (R_{Et} - \theta). \tag{10}$$

Here κ_i, λ_i $(i = 1, 2)$ are positive parameters, θ is the natural rate of interest, and R_{Et} denotes the common nominal interest rate. Due to the long–run equilibrium relations, $\tilde{y}_{it} = 0$, $\bar{X}_{it} = 0$, and $\bar{r}_{it} = 0$ $(i = 1, 2)$, long–run equilibrium money demand is given by

$$\bar{m}_{it}^{\mathrm{d}} = \kappa_i \bar{y}_{it}. \tag{11}$$

Hence, there is no money illusion, and the Cambridge equation holds in the long run. This leaves us with the following equilibrium relationship for the long–run quantity of money (both demand and supply) in country i $(i = 1, 2)$:

$$\bar{M}_{it}^{\mathrm{s}} = \bar{M}_{it}^{\mathrm{d}} = P_{it} \bar{m}_{it}^{\mathrm{d}} = P_{it} \kappa_i\, (1 + \theta)\, \bar{y}_{i(t-1)}. \tag{12}$$

This means that in each country, the price level will stay constant in the long run if money supply $\bar{M}_{it}^{\mathrm{s}}$ grows at the natural rate θ. In a monetary union, the sum of the countries' money demands has to be equal to the monetary union's money supply.

In addition, we assume the money market always to clear in the short–run, too, and hence money supply to be equal to the sum of short–run money demands in countries 1 and 2,

$$M_{Et}^{\mathrm{s}} = M_{1t}^{\mathrm{d}} + M_{2t}^{\mathrm{d}}. \tag{13}$$

This leads to

$$M_{Et}^{\mathrm{s}} = \kappa_1 y_{1t} P_{1t} + \kappa_2 y_{2t} P_{2t} - (\lambda_1 P_{1t} + \lambda_2 P_{2t})\, (R_{Et} - \theta). \tag{14}$$

Note that this implies that the short–run real rates of interest in the two countries can considerably diverge both from each others and from the long–run (natural) real rate of interest.

The government budget constraint is given as an equation for government debt of country i $(i = 1, 2)$,

$$D_{it} = \left(1 + R_{E(t-1)}\right) D_{i(t-1)} + F_{it} - \beta_i \tilde{B}_{Et}, \quad D_{i0} \text{ given}, \tag{15}$$

where the nominal fiscal deficit of country i $(i = 1, 2)$ is determined by the identity

$$F_{it} = P_{it} f_{it} = P_{it} \tilde{f}_{it}. \tag{16}$$

$\tilde{B}_{Et}$ denotes the short–term deviations of high–powered money, B_{Et}, from its long–run equilibrium level, $\bar{B}_{Et}$. The equilibrium stock of high–powered money is assumed to grow geometrically at the natural rate θ. Hence,

$$B_{Et} = \bar{B}_{Et} + \tilde{B}_{Et} = (1 + \theta)\, \bar{B}_{E(t-1)} + \tilde{B}_{Et}, \quad \bar{B}_{E0} \text{ given}. \tag{17}$$

$\tilde{B}_{Et}$ represents the control variable of the common central bank. The change in high–powered money is distributed as seigniorage to the two countries according to given positive parameters $\beta_1 \in [0, 1]$ and $\beta_2 := 1 - \beta_1$. Assuming a constant money multiplier, ψ, the broad money supply of the monetary union is given by

$$M^s_{Et} = \psi B_{Et}. \tag{18}$$

Both national fiscal authorities are assumed to care about stabilization of inflation, output, debt, and fiscal deficits of their own countries. The common central bank is interested in the stabilization of inflation and output in the monetary union and in a low variability of its supply of high-powered money. Hence, the individual objective functions of the national governments and of the common central bank are given by

$$J_i = \frac{1}{2} \sum_{t=1}^{T} \left(\frac{1}{1+\theta}\right)^t \left(\alpha_{iX} X_{it}^2 + \alpha_{iy} \left(y_{it} - \bar{y}_{it}\right)^2 + \alpha_{iD} D_{it}^2 + \alpha_{if} \tilde{f}_{it}^2\right), \tag{19}$$

$$J_E = \frac{1}{2} \sum_{t=1}^{T} \left(\frac{1}{1+\theta}\right)^t \left(\alpha_{EX} X_{Et}^2 + \alpha_{Ey} \left(y_{Et} - \bar{y}_{Et}\right)^2 + \alpha_{EB} \tilde{B}_{Et}^2\right), \tag{20}$$

where all weights are positive numbers $\in [0, 1]$. The joint objective function for the calculation of the cooperative Pareto–optimal solutions is determined by $J = \mu_1 J_1 + \mu_2 J_2 + \mu_E J_E$ $(\mu_1, \mu_2, \mu_E \geq 0, \mu_1 + \mu_2 + \mu_E = 1)$.

The parameters of the model are specified numerically in the simplest possible way, leaving us with a symmetric monetary union (see Table 1). Lack of space precludes a detailed discussion of the parameter values chosen, the target values assumed for the objective variables of the players (which are basically the long–run equilibrium values of the respective variables), and the

Table 1. Parameter values for $i = 1, 2$

Parameter	T	θ	δ_i	γ_i	ρ_i	η_i	ξ_i	ω	κ_i	λ_i	β_i	ψ	α's	μ_i
Value	20	0.03	0.5	0.5	0.5	1.0	0.25	0.5	1.0	0.15	0.5	2.0	1.0	0.33

Table 2. Initial values for $i = 1, 2$

Variable	$\bar{y}_{i0}$	$\tilde{y}_{i0}$	P_{i0}	X_{i0}	D_{i0}	R_{E0}	$\bar{B}_{E0}$	$\tilde{f}_{i0}$	$\tilde{B}_{E0}$
Value	1	0	1	0	0	θ	1	0	0

initial values of the state variables (see Table 2) of the dynamic game model. Detailed information on this is available from the authors on request.

Equations (1)–(20) constitute a nonlinear dynamic game with a finite planning horizon, where the objective functions are quadratic in the deviations of state and control variables from their respective desired values.

3 Optimal Fiscal and Monetary Policies

Several experiments were performed with the model, using different assumptions about the paths of the exogenous non-controlled variables. For lack of space, we report only the results of one of them. This is a symmetric shock acting on both economies. In particular, we assume that autonomous real output (GDP) in both economies falls by 1.5% below the long-run equilibrium path for the first four periods and less for the next three periods: $z_{i0} = 0$, $z_{i1} = z_{i2} = z_{i3} = z_{i4} = -0.015$, $z_{i5} = -0.01$, $z_{i6} = -0.005$, $z_{i7} = -0.0025$, and $z_{it} = 0$ for $t \geq 8$ and $i = 1, 2$.

Without policy intervention, this demand side shock leads to lower output and inflation (compared to the long–run equilibrium path) during the first five periods, but higher output and inflation afterwards (see Fig. 3). That is, the uncontrolled dynamic system adjusts in dampened oscillations, getting close to the long–run path after twenty periods. The maximum deviation of output from its equilibrium path is approximately 0.33% (in the first period). That is, even without policy intervention there are sufficiently strong negative feedbacks in the system to reduce the impact of the shock on output to about one fifth of the original shock at most in the case of a temporary symmetrical shock. This is mainly due to a strong reaction of the rate of interest, which falls to values near zero in the first four periods (but rises above the long–run value afterwards). Due to the symmetry of the economies and of the shock, the reactions of all variables are identical in both economies.

When policy–makers are assumed to react on this shock according to their preferences as expressed in their objective functions, several outcomes are possible, depending on the assumptions made about the respective other policy–makers. Here we consider two noncooperative equilibrium solutions of the resulting dynamic game, the open–loop and the feedback Nash equilibrium solution, and one cooperative solution, the Pareto–optimal collusive solution (all players get the same weight $\mu_i = 1/3,\ i = 1, 2, E$). The feedback Nash equilibrium solution is more interesting than the open–loop one because the former is subgame perfect or Markov perfect, while the latter is only valid if it is assumed that all policy–makers commit themselves unilaterally and decide upon trajectories of their instrument variables once for all at $t = 0$.

The time paths of the control variables – real fiscal deficit (for either country) and additional high–powered money – under the three solution concepts considered are shown if Figs. 1 and 2, respectively, those of the state (and objective) variables – deviations from long–run equilibrium output and government debt – in Figs. 3 and 4, respectively. Inflation rates show the same qualitative pattern as outputs, price levels remain below the equilibrium value of one for all periods, and the common nominal rate of interest exhibits a behavior very similar to the uncontrolled case (falling to low values in periods one to four, rising up to about 3% later on). All country–specific variables show exactly the same time paths for both countries. More detailed results are available from the authors on request.

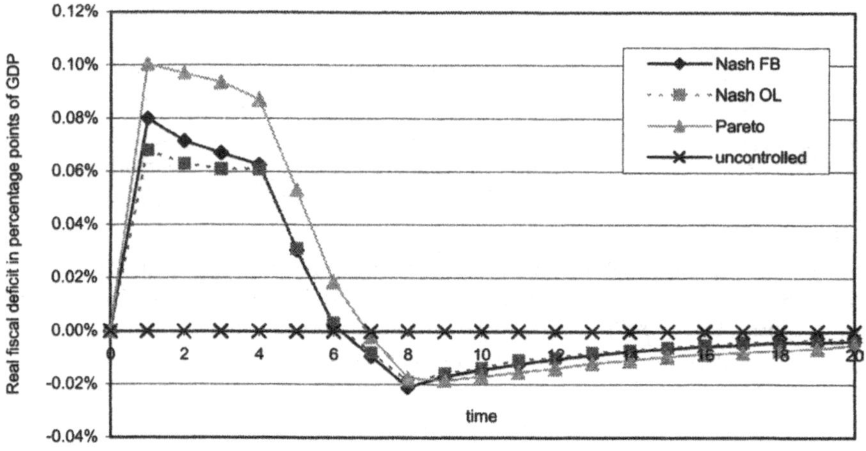

Fig. 1. Country i's real fiscal deficit

As can be seen from the graphs, both fiscal and monetary policies react on the negative demand shock in an expansionary and hence counter–cyclical way: both countries create fiscal deficits during the first five to six periods and surpluses afterwards, and the central bank raises its supply of high–powered money during the first six years and reduces it afterwards. This results in less output loss and lower deflation than in the uncontrolled solution. What

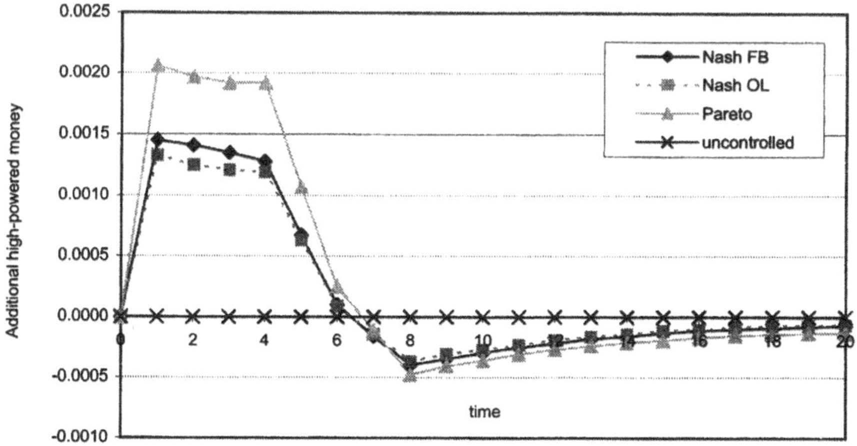

Fig. 2. Additional high–powered money

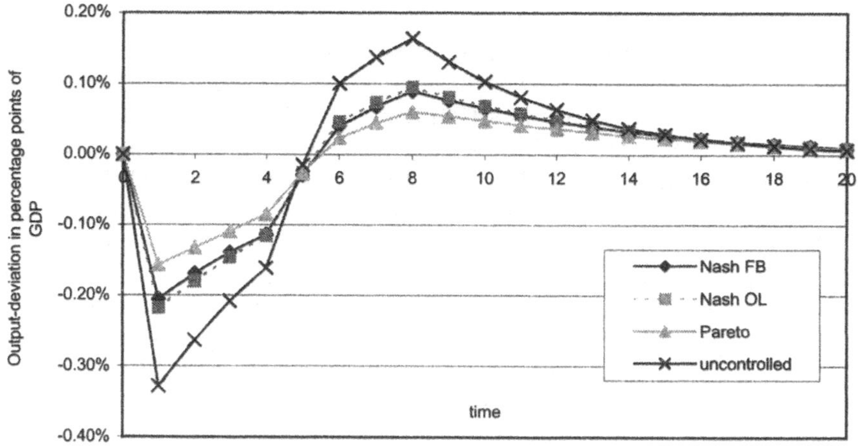

Fig. 3. Country i's output–deviation from its long–run equilibrium level

is remarkable is the small magnitude of the (absolute) values of the instruments involved: the highest value of the fiscal deficit created is one tenth of one percentage point of GDP (in period one), for example, which would be nearly invisible in terms of the Maastricht criteria if applied in the European Economic and Monetary Union. This is due to the strong self–stabilizing forces in the model used, acting especially through the interest rate channel, as noted already for the uncontrolled solution. As there is not much need for counter–cyclical action, it is not surprising that optimal (equilibrium) policies entail only cautious activities.

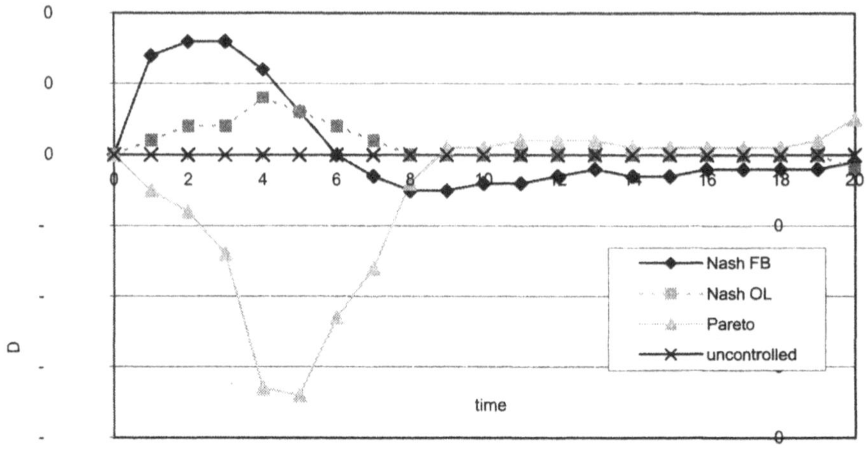

Fig. 4. Country i's debt

Comparing the noncooperative equilibrium solutions and the cooperative solution yields another interesting observation. All show qualitatively the same behavior, and the two noncooperative Nash equilibrium solutions are very close together in terms of all control and state variables. The collusive solution, although not too distant from the other two, exhibits more active policy–making (higher fiscal deficits and money creation in the first periods). This different policy–mix does not change the path of the rate of interest (higher deficits increase, higher money supply decreases the interest rate, ceteris paribus), but does so for the public debt trajectory: in the noncooperative solutions, government debt is increased, in the cooperative solution it is increased, reflecting the relatively higher increase of fiscal deficits as compared to monetary supply increases in the noncooperative solutions. It remains to be seen whether these policy patterns remain under alternative assumptions about the economic model or the shock.

4 Concluding Remarks

Applying dynamic game theory and the OPTGAME 2.0 algorithm to a simple macroeconomic model of fiscal and monetary policies in a monetary union, we obtained several insights into the design of economic policies facing a symmetric negative demand shock. In particular, optimal policies of both the governments and the common central bank are counter–cyclical but not very active, at least for the model under consideration. The outcomes of the different solution concepts of dynamic game theory are rather close to each other. In particular, a periodic update of information and related reduction of commitment (a change from an open–loop to a feedback Nash equilibrium solution) does not cause benefits or costs to either decision–maker. Cooperative economic policies (both fiscal and monetary ones) are more active or "aggressive" than noncooperative ones, resulting in a somewhat different policy–mix with higher stabilization effects. Further research will have to show how sensitive these results are with respect to the assumptions about the model and the shock.

Acknowledgement

This research was financially supported by the Austrian Science Foundation (project no. P12745–OEK). Any opinions, findings, conclusions or recommendations expressed in this paper are those of the authors and do not necessarily reflect the views of the Austrian Science Foundation.

References

1. Başar T., Olsder G. J. (1999) Dynamic Noncooperative Game Theory, 2nd edn. SIAM, Philadelphia
2. Petit M. L. (1990) Control Theory and Dynamic Games in Economic Policy Analysis. Cambridge University Press, Cambridge
3. Behrens D. A., Neck R. (2002) Approximating Equilibrium Solutions of Multi–Player Difference Games Using the OPTGAME Algorithm. Working paper, University of Klagenfurt

Impact of Feedback Loop on Group Decision Process when Applying System Dynamics Simulators

Andrej Škraba and Miroljub Kljajić

University of Maribor, Faculty of Organizational Sciences, Kidričeva cesta 55a, SI-4000 Kranj, Slovenia, e-mail: {andrej.skraba, miroljub.kljajic}@fov.uni-mb.si

Abstract. Influence of feedback information on group decision process supported by the application of system dynamics simulators is determined in present study. Experiment with decision groups was conducted under different experimental conditions: a_1) determination of business strategy without application of formal model, a_2) determination of the strategy with application of formal system dynamics model and a_3) determination of the strategy with additional application of the group feedback information. The hypothesis that model application and group feedback information positively influence the convergence of the decision process and yield higher values of criteria function was confirmed. Described group decision process was subsequently analyzed in order to determine its frequency and deviation.

1 Introduction and Methodology

There are several new approaches that expand System Dynamics (SD) methodology regarding exploration and development of SD models as for example Vennix's group model building [1] or application of AI [2]. The methods of scenario creation and selection applying multicriteria decision functions were developed in our previous research [3]. The present paper is the continuation of research conducted in the field of group exploration of SD models [4]. Experiments under different conditions were conducted in order to analyze the influence of information feedback and different methods of work in the decision process. Simulation system used at the experiments can be described with sets M, J and L, where M is the model, J is the set of criteria, and L is the set of limitations. The main task of efficient management, which participants should perform, is to determine optimal control based on the known M, J and L. Fig. 1 shows the model of production as the black box with input parameters u_1, u_2, u_3 and u_4 (where u_1 is Product Price, u_2 is Salary, u_3 are Marketing Costs and u_4 is Desired Inventory) and three different experimental conditions a_1, a_2 and a_3. Model can be described by the state equation:

$$x(k+1) = f(x(k), u(k), a_i). \tag{1}$$

where $x(k)$ represents the state of the system, $u(k)$ the control vector or alternative strategy and a_i the experimental condition.

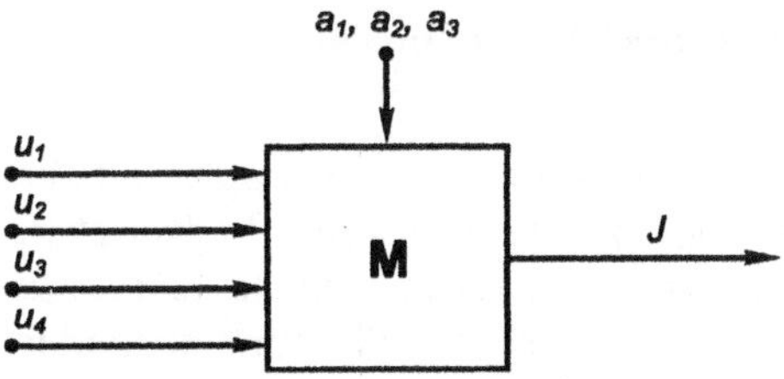

Fig. 1. Model with Input Parameters at Different Experimental Conditions

The model is controlled by a user-friendly interface in the form of a business simulator. Participants had a task to promote a product on the market, which lifecycle is one year. Criteria function was stated as:

$$\max_{u \in U} J = \frac{d_0 + \sum_{i=0}^{t_k} d(t_i)\, w_1}{c} + \frac{p_0 + \sum_{i=0}^{t_k} p(t_i)\, w_2}{o_0 + \sum_{i=0}^{t_k} o(t_i)} + \\ + \frac{s_0 + \sum_{i=0}^{t_k} s(t_i)\, w_3}{p_0 + \sum_{i=0}^{t_k} p(t_i)} + \frac{v_0 + \sum_{i=0}^{t_k} v(t_i)\, w_4}{p_0 + \sum_{i=0}^{t_k} p(t_i)} \quad (2)$$

where d_0 is the initial value of Income, $d(t_i)$ is the Income function where $d(t_i) = p(t_i) - o(t_i)$ where $p(t_i)$ is the Revenue function, $o(t_i)$ is the Expenses function, t_k is the final time of observation, c is the Capital, w_1, w_2, w_3 and w_4 are the weight values, p_0 is the initial value of Revenues, o_0 are the initial Expenses, s_0 are the initial workforce expenses, $s(t_i)$ is the workforce expense function, v_0 are the initial inventory costs, and $v(t_i)$ the inventory costs function. The sum of weights equals to $\sum_j^4 = 1$. Actual values of weights were determined as the constant factors. The criterion description and weight values are shown in Table 1.

Table 1. Criterions J_i and Weights w_i

Criterion	Description	Weight	Value of Weight
J_1	Capital Return Ratio	w_1	0.5
J_2	Overall Effectiveness Ratio	w_2	0.35
J_3	Workforce Effectiveness Ratio	w_3	0.10
J_4	$\frac{Inventory}{Income}$ Ratio	w_4	0.05

The goal for the participants was to maximize the stated criteria function in Eq. (2). The maximum value of the criteria function was determined by several optimization methods as $J = 1.5$.

1.1 Decision Task and Hypotheses

The decision task which was addressed by applying SD simulator was stated as follows: *Find the appropriate values for Product Price, Salary, Marketing Costs and Desired Inventory Level in order to achieve the maximum possible value of criteria function in Eq. (2).* The time of conducting the experiment with decision groups was $\frac{1}{2}$ hour for all three conditions, which can be described as follows:

Experimental condition a_1 assumes the individual assessment of the decisionmaker at the determination of the model parameters values u_1, u_2, u_3, u_4 through maximization of the criteria function in Eq. (2). The decision is made without a formal model.

Participants at the experimental condition a_2 search for the parameter values u_1, u_2, u_3, u_4 with the aid of the SD model through maximization of the criteria function. The interaction between subjects is not supported; therefore group feedback is not applied.

Experimental condition a_3 assumes application of the SD model by participants with group feedback information. Time of conducting the experiment under this condition was divided into four time intervals, $(8+8+8+6)\,min$. Each participant put the best-achieved set of parameter values u_1, u_2, u_3, u_4 to the network server where after each time interval the information about best parameter values is updated. The system provides feedback information to all participants regarding the individual best-achieved strategies of all other participants in the experiment. Participants get feedback on the defined strategies for all participants in group $S_i = \{u_{1i}, u_{2i}, u_{3i}, u_{4i}\}\,;\, i = 1, 2, \ldots, n$ as well as the aggregated values in the form of mean values of parameters. Mean values of parameters u are expressed as follows:

$$\bar{S} = \left\{ \frac{\sum_{i=1}^{n} u_{1i}}{n}, \frac{\sum_{i=1}^{n} u_{2i}}{n}, \frac{\sum_{i=1}^{n} u_{3i}}{n}, \frac{\sum_{i=1}^{n} u_{4i}}{n} \right\} \tag{3}$$

where $\bar{S}$ is the set of mean parameters' values and n is the number of participants. Such functional mapping could take many other forms, not only arithmetical mean [5].

The following statistical hypotheses about different experimental conditions were stated: a) *Hypothesis 1: There is significant difference in the results of the decision process conducted under experimental condition a_1 (individual assessment of the decision-maker at the business strategy determination with no formal model), and condition a_2 (individual application of SD model at the determination of the business strategy).* b) *Hypothesis 2: There is significant difference between the results of the decision process conducted under experimental conditions a_2 and a_3 (application of the SD model with group feedback information).*

1.2 Experimental Procedure

One hundred and forty seven senior graduate students (68 females and 79 males) from the University of Maribor participated in the research in order to satisfy the requirements of their regular study program. Subjects were introduced to the experimental problem of determining a business strategy according to the stated criteria function. Duration of introduction was 10 minutes. The structure of the considered system was presented and the main parameters in the model were explained. The evaluation criteria for the business strategies were also considered. The work with the simulator was explained at experimental conditions a_2 and a_3. After the introduction every participant formed a strategy according to the stated problem and passed these results to the network server.

2 Results

The results of the decision process conducted under experimental conditions a_1 ($N_1 = 52$), a_2 ($N_2 = 55$), and a_3 ($N_3 = 40$) are shown in Fig. 2 The values of criteria function J in Fig. 2 are ordered from highest to lowest (Y-axis). The X-axis shows the relative number of participating subjects. Results gathered from experimental condition a_1 are the lowest; their average is $M = 0.446$. Average results gathered under experimental condition a2 are higher than the results gathered under experimental condition a_1 ($M = 1.076$), while the standard deviation is smaller ($\sigma = 0.317$ for a_2 and $\sigma = 0.498$ for a_1). The highest values of criteria function were gathered under experimental condition a_3 ($M = 1.386$), while the deviation was the smallest ($\sigma = 0.073$).

Mann-Whitney nonparametric U-test was applied for the comparison of experimental conditions a_1, a_2 and a_3. Calculated values of U-test for three stated hypotheses were $U(N_1 = 52, N_2 = 55) = 410, p \leq 0.01$ for *H1* and $U(N_1 = 55, N_2 = 40) = 370, p \leq 0.01$ for *H2* therefore, both stated hypotheses are accepted. According to the tested hypotheses, it can be concluded that significant differences exist between the results of the decision process conducted under different experimental conditions. The main emphasis is on the difference between experimental conditions a_2 and a_3, where SD models are used without and with group informational feedback. These two groups have the same technical means to address the stated decision problem.

In order to analyze frequency and deviation of group decision process special network tool was developed which enabled us to sample the whole process with $[ms]$ accuracy. Every action of individual participant was recorded to the server database which also enabled the mediation of feedback information. The post experimental test was conducted with $N = 30$ subjects, which were also senior graduate University students. Left part of Fig. 3 represents the frequency of group decision process. There are two experimental conditions shown, a_2 and a_3. The frequency of process under condition a_3 is lower due to the fact that additional processing of feedback information was performed.

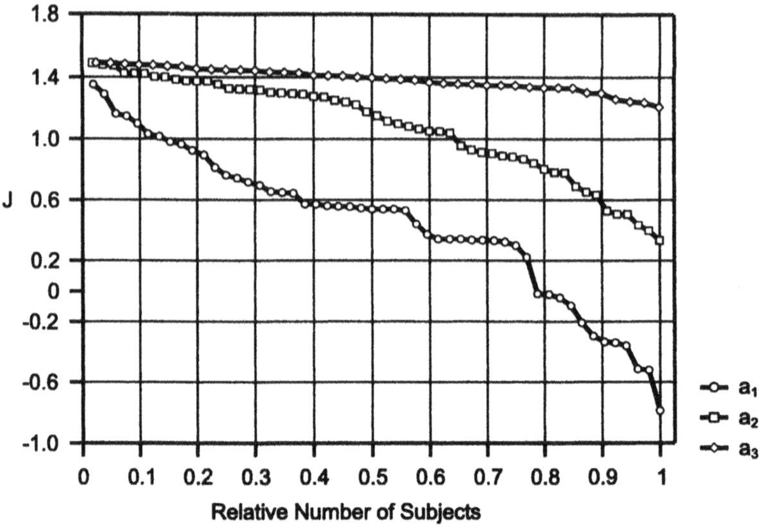

Fig. 2. Values of Criteria Function J for Different Experimental Conditions

This is evidently represented by oscillation and four peaks in performance which coincide with experimental phases of condition a_3. Therefore it could be concluded, that less processing of information was done i.e. tests on the simulator under condition a_3 but the feedback information contributed to the right orientation of decision group.

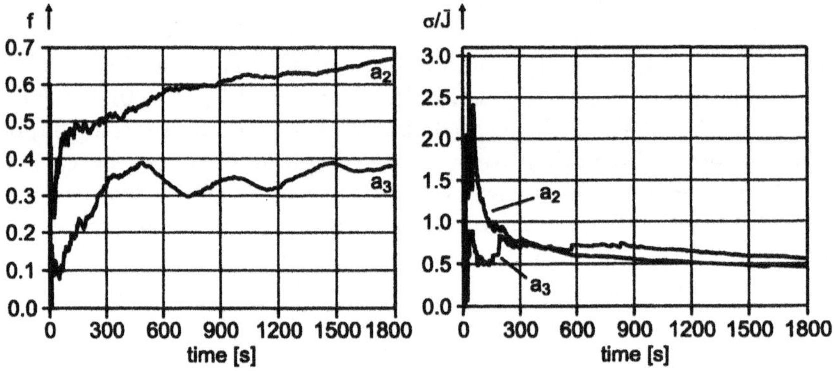

Fig. 3. Frequency *(left)* and deviation $\frac{\sigma}{J}$ *(right)* of strategy search process for conditions a_2 and a_3.

Right side of Fig. 3 shows the deviation of the process expressed as the ($\frac{\sigma}{J}$). This ratio shows that the deviation in initial phase of experiment is high in both cases a_2 and a_3 but later on, from time approx. 300 s stabilizes. The search space in both cases is thereon similar, in case a_3 slightly more perturbed on account of several stops at each experimental phase. In both cases the deviation gets smaller as the time progresses since the search space of optimal strategy is narrowly determined.

One could expect that cumulative frequency in the decision process under experimental condition a_3 will be lower on account of the subjective informational processing i.e. $\int_0^{t_k} f[a_3]dt < \int_0^{t_k} f[a_2]dt$. This determine the condition of information utility U which should be higher in the process where feedback is considered $U(I[a_3]) > U(I[a_2])$ if $J_2 = J_3$.

3 Discussion

Statistical analysis indicate that group feedback information significantly impacts group decision process supported by the SD models. The influence of information feedback results in higher convergence of the decision process. Feedback information is therefore the main component in the efficient decision support system based on the SD models. Results of experimental condition a_1 where no formal model was applied were an indicator of formal model importance.The effect of increasing system efficiency under three different conditions can be explained with the introduction of additional information into the system, which is the main component of learning and system's control. Information introduced by formal model as well as introduction of group feedback information contributed to the higher criteria function values and convergence. Conclusion that more feedback contributes to the better results would not be appropriate. Subjective limitation of information processing at designing feedback must certainly be considered. The question, which arose at this stage, is how to aggregate feedback information and mediate only the information, which is decisive for certain case. Mean values as the simplest example was used in our case and it was shown to be appropriate. Feedback information which is applied in similar decision environments should meet the conditions of utilization in order to positively enhance group decision process supported by the SD simulators. Feedback information mediated to the subjects was in our case the basis for faster solution of stated decision problem. It should be: *a.)* mediated to the group and *b.)* properly aggregated in order to prevent information overload of subjects and achieve goal seeking behavior.

Acknowledgement
This research was supported by the Ministry of Education, Science and Sport of the Republic of Slovenia (Program No. PP-0586-501 and Project No. Z5-3313).

References

1. Vennix, J. A. M. (1996) Group model building: facilitating team learning using system dynamics. Chichester Wiley
2. Hines, J., House, J. (2001) The source of poor policy. System Dynamics Review **17**, 3–32
3. Kljajić, M., Bernik, I., Škraba, A. (2000) Simulation Approach to Decision Assessment in Enterprises. Simulation **75**, 199–210
4. Škraba, A. (2000) Multicriteria Group Decision Making with System Dynamics Models. Doctoral Thesis. Kranj - SI, University of Maribor
5. Zeleny, M., (1982) Multiple Criteria Decision Making. New York Mc Graw-Hill

The Management Game SINTO-Market – Report on Some Recent Experiments

Otwin Becker[1], Tanja Feit[2], Vera Hofer[2], Ulrike Leopold–Wildburger[2], Susanne Lind–Braucher[2], Jörg Schütze[2], and Reinhard Selten[3]

[1] Alfred Weber Institute University of Heidelberg
[2] Department of Statistics and Operations Research, Karl–Franzens–University Graz, tanja.feit@uni-graz.at
[3] Laboratory for Experimental Economics, Friedrich–Wilhelms–University Bonn

Abstract. The management game SINTO-Market which was originally developed and programmed by Otwin Becker and Reinhard Selten was recently performed at the University of Kalmar/Sweden and University of Graz/Austria several times. The decisions were made by groups of students, each group representing one of the three firms. The electronic management game SINTO-Market puts the players in a competitive situation in the branded food product sector.

We will give a short description of the structure of the game. The game runs via internet and can be seen at the web-page of our department.

It is interesting to see that much can be learned about human behavior in complex economic decision situations from experimental research with management games. Surprisingly the results show some differences between genders.

1 Description of the Game SINTO-Market

Within a special food product sector, managers have to find the most successful strategies for production, prices and advertising expenses for up to 10 brands, as well as a suitable three-dimensional combination of brand components concerning their form, taste and quality.

SINTO is a synthetic protein product which does not exist in reality, although it may be developed at some time in the future. Experience gained in one sector cannot simply be transferred one to one to other sectors. Before beginning the game, the players receive a short game description.

The synthetic protein product SINTO was developed almost simultaneously by three companies. Each parent company has founded a subsidiary, further on referred to as the firm, which is given the task of producing and selling this new foodstuff. At the beginning of the game, each firm should bring only one brand onto the market. Then, as demand for the product grows over time it is worth offering two or more brands which differ in taste and packaging design. Each firm must decide how many brands to bring onto the market. Brand characteristics, production, price and advertising expenditure have to be fixed for each brand. The sales quantities are then calculated from this information. All brands are in competition with each other so that the sales of one brand are also dependent on the decisions made for all other brands.

The goal of each firm is the maximization of final owned capital.

At the beginning of the game all firms are in the same situation. Each firm has 10 registered brand names available. The choice of the brand name does not have any influence on the sales of the brand. Each firm must establish the following choices for each brand offered on the market:

1. The combination of the three brand characteristics
   a) taste (bitter - mild),
   b) form resp. the grainedness (fine - rough)
   c) quality according to the wrapping design (high - low).
2. The production quantity.
3. The amount of expenditure assigned for advertising.
4. The price.
5. The capital investment.

The choice of form and taste has no influence on the manufacturing costs, whereas the choice of the wrapping quality does. The wrapping design has a similar effect on sales as advertising.

The sales of a brand do not only depend on the decisions made concerning this brand but also on the decisions made for all the other brands. The demand for brands is not dependent on which firm produces the brand. The buyers only consider the brand names, not the manufacturer.

The demand for a single brand is greater, when there are less brands with similar taste characteristics on the market. A brand sells more, the higher the quality of its packaging design is in comparison to the other brands. Sales are larger, the lower the price is and the higher the prices of the other brands are. Sales of a brand can also be increased by spending more for advertising. The sales of a brand decrease if the advertising expenses for other brands increase. The competition between two brands is stronger, if their taste characteristics are similar. However, this is only the case for taste characteristics and not for price, packaging design and advertising. As regards these three variables, each brand competes with all the other brands, independent of their combination of taste characteristics.

The results of each period are given to each firm at the end of that period. This market overview is a print-out which gives information about the brands offered in that period together with their taste characteristics, and also about the following items:

the profit and loss account,
the turnover-based contribution cost analysis,
the production-based unit cost analysis,
the number of sales,
the advertising expenditures, and
the prices of all brands offered in a period.

Each firm should aim to attain the highest possible net assets by the end of the game. The amount of net assets equals the owner's capital (according to the latest balance sheet) plus the subsidiary's share of the activated "brand images" (that means net assets = owner's capital plus still effective advertisment). The latter is the value of money which represents the effect of the advertising if the game were to be continued. Due to this, it is still worth investing in advertising expenditure even in the last few periods of the game.

2 Results

The SINTO-Market was played once with 9 students at the University of Kalmar (Sweden). The next experiences with the electronic management game SINTO were made with students of Graz-University and documented in the following Table 1. We will show some interesting results.

Table 1. Composition of students at Graz-University

Game	female	male	Firm 1	Firm 2	Firm 3
818	3		1	1	1
222		3	1	1	1
111	7		3	2	2
444		7	2	2	3
777	9		3	3	3
999		9	3	3	3
666	10		6	3	1
333		11	3	1	7
Sum	29	30			

The success of each firm depends on the following strategies:

S1: A rapid increase of production capacity.
S2: An efficient protection of a good market position, combined with the relative standing of the others.
S3: The allocation of an appropriate amount of advertising expenditures.

Here we want to show in more detail some results. Obviously the price is responsible for the consumers behavior and the success of each firm. We figured out the following findings:

F0: Fundamental Finding:
Female participants differ significantly from male participants.
F1: Females set prices lower than males.

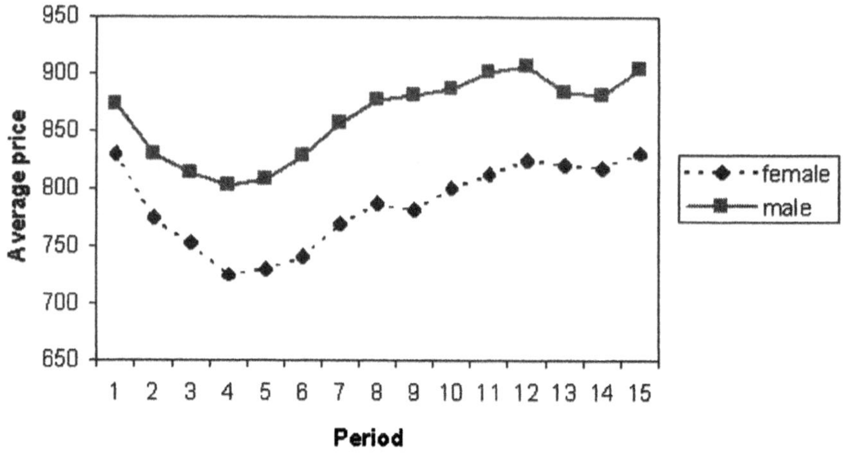

Fig. 1. Average price

The diagram of Figure 1 shows the average prices during the 15 periods. The high degree of statistical significance for the difference in the levels of male and female prices can be proved by several test. E.g. Friedman-Test on 0.01 level.

F2: The investment expenses of male subjects are higher than those of female groups.

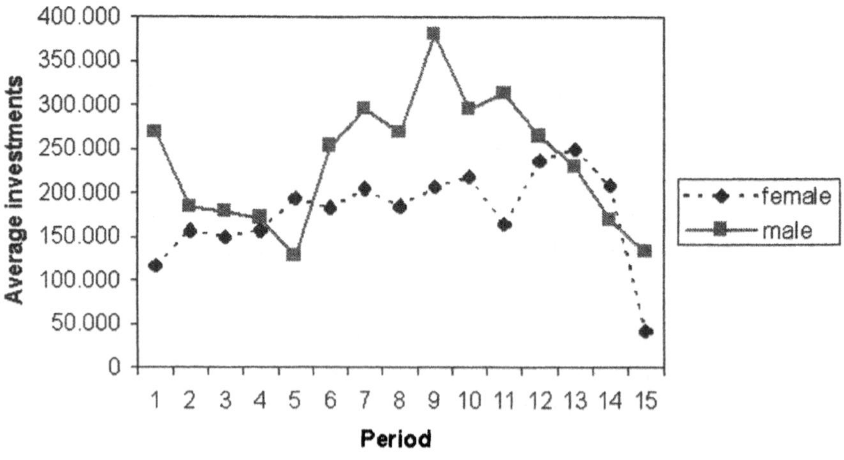

Fig. 2. Average investments

The diagram of Figure 2 shows the average investments during the 15 periods. Significant differences in the investment behaviour of male and female groups from period 5 up to period 12 can be shown on the level of 0.01 (eg. Friedman-Test). As in those periods the demand is growing, male groups profit from their higher investments.

F3: The net assets of male groups are higher than those of the female groups.

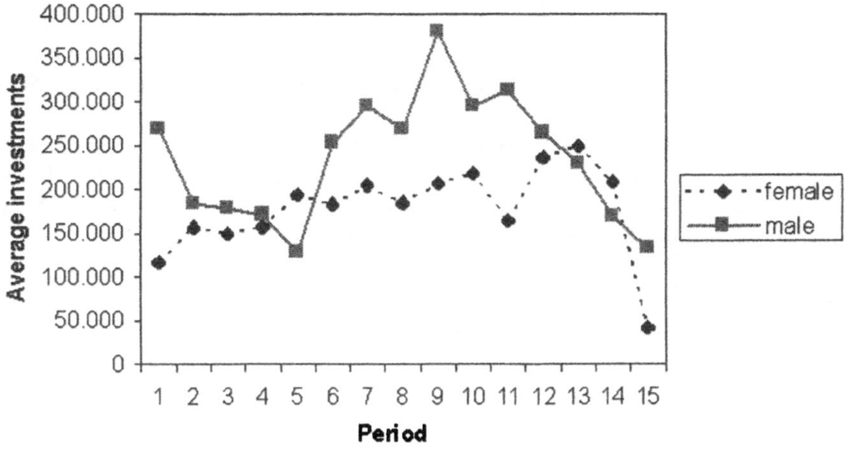

Fig. 3. Average net assets

The diagram of Figure 3 shows the average net assets during the 15 periods. Whereas in the first half of the game male and female groups have similar average net assets, average net assets of male and female groups differ on a significance level of 0.01 in the second half of the experiment.

Statistically it can be shown, that in the end of the game (after 15 periods), the male groups had a significantly higher average net asset than the female groups, (significance level 0.01), which means that male groups were significantly more successful than female groups.

Our final findings can be summarized in the following way: Female participants in the management game SINTO-Market are more cautious than their male competitors. Therefore their net assets and their profits are lower than those of male students.

References

1. Becker O., Selten R., (1970) Experiences with the Management Game SINTO–Market, 136–150. In: Sauermann, H. Contributions to Experimental Economics. Volume two. J. C. B. Mohr (Paul Siebeck) Tübingen
2. Becker O., Selten R., Leopold U., Lind S., (2001) The Management Game SINTO-Market revieved. Working Paper, Institute of Social Sciences and Management, University of Kalmar, Sweden
3. Blinder A. S., Canetti E. R. D., Lebow D. E., Rudd J. B., (1998) Asking About Prices. A New Approach to Understanding Price Stickiness. Russel Sage Foundation, New York
4. Selten, R. (1967) Investitionsverhalten im Oligopolexperiment, 60–102. In: Sauermann, H. Contributions to Experimental Economics. J. C. B. Mohr (Paul Siebeck) Tübingen
5. Selten, R. (1967) Ein Oligopolexperiment mit Preisvariation und Investition, 9–59. In: Sauermann, H. Contributions to Experimental Economics. J. C. B. Mohr (Paul Siebeck) Tübingen
6. Selten, R. (1967) Die Strategiemethode zur Erforschung des eingeschränkt rationalen Verhaltens im Rahmen eines Oligopolexperimentes, 136–168. In: Sauermann, H. Contributions to Experimental Economics. J. C. B. Mohr (Paul Siebeck) Tübingen

Bounds & Likelihood Procedure Revisited

Becker, O., Leitner, J., Leopold-Wildburger, U., Schuetze, J. H.,

BECKER, Otwin,
Alfred Weber Institut, Universität Heidelberg, Grabengasse, D-69117 Heidelberg.
o.becker.walldorf@t-online.de

LEITNER, Johannes; LEOPOLD-WILDBURGER, Ulrike;
SCHÜTZE, Jörg Hermann,
Institut für Statistik und OR, Universität Graz, Universitätsstrasse 15, A-8010 Graz, ulrike.leopold@uni-graz.at.

Abstract. In an experiment subjects are asked to predict the next value of an univariate time series on the basis of the past observations. The average forecasts of the subjects can be well described by a surprisingly simple rule, which is called the “Bounds and Likelihood procedure”(Becker/Leopold 1996). In a new series of the experiment, conducted 2002 in Graz with 72 participants, the behaviour and score of the subjects is further analysed with the Selective Attention Test (subset of D2, Brickenkamp 1962) and Intelligence Test for Reasoning (subset of PSB, Horn 1969).

Keywords: Experimental economics, forecasting, visual extrapolation, eye-balling, time series, concentration, intelligence, attention, gender.
JEL-classification: E27, D81, D7, C53, C92

1. Introduction

For a long time economists have shown increasing interest in the analysis of expectation formation. Following the approach of I. Fisher (1896, 1930) and F.A. Hayek (1948) a lot of macroeconomic theories have been developed which are at least partly based on variations in expectations.
There are two major approaches to forecasting: conjectural and structure-based procedures. Visual extrapolation ('eye-balling') belongs to conjectural procedures. The Bounds and Likelihood (B&L) procedure (Becker/Leopold 1996) is a heuristic to describe expectation formation in regard to average forecasts. In an experiment, subjects are asked to predict the next value of an univariate time series (see Figure 1). To make a forecast, the task of the respondents, depending solely on their visual processing of the information, then consists of extending the respective curve on the chart beyond the given endpoint. We focus on cases with limited information for the subjects. The only available information for the decisions are the past observations. No additional information about the meaning of the series and influential factors were given to the participants in the experiment. This reflects common cases from real economic life.

We analyse the individual behaviour in comparison to the expected group behaviour according to B&L. Therefore the B&L average forecast prediction is used to access and analyse individual forecasting behaviour.

We link the forecasting performance to certain measures of attention and fluid intelligence with the help of the Selective Attention Test (Test d2, Brickenkamp 1962) and Intelligence Test for Reasoning (subset of PSB, Horn 1969).

2. Description of experiments

The subject's task is the forecasting of a time series x_t over 42 periods. The first 6 periods serve as the phase of familiarisation. The time series is a realisation of the stochastic difference equation

$$x_t = x_{t-1} - \mathrm{INT}(0,5 \times x_{t-2}) + u_t$$

with the endogenous variable x_t and the white noise u_t. The variable u_t is uniformly distributed in the interval [1,6]. Consequently the random variable's realisations are integers. The B&L procedure is described in the appendix.

The subjects were given the following information: The time series can be interpreted arbitrarily, as share price, economic cycle or commodity price.

The gestalt of the time series can be seen more clearly from period to period. It is the only relevant information for the forecasting process. The subjects' forecasts do not have an influence on the progression of the time series.

The subjects had a strong financial incentive to forecast the time series correctly. They were paid 60 Eurocents for an exact prediction. A deviation of one (two) unit(s) was remunerated with 40 (20) Eurocents. The payments correspond to a function of prediction-errors based on absolute deviations, that is cut off to zero at the value of 3.

Figure 1 shows the subjects average forcasts, the B&L-values, the B&L errors in predicting the average opinion and the time series x_t.

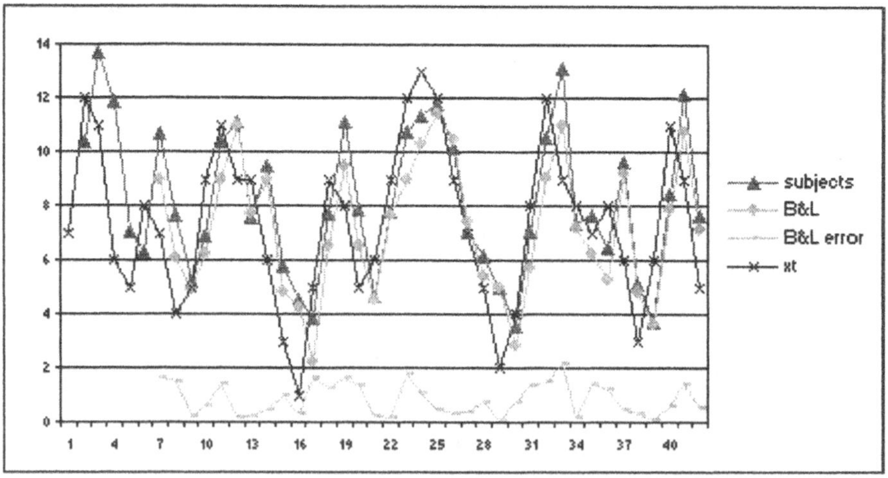

Figure 1: B&L and average opinion

Altogether 72 students of the faculty for social sciences and economics at Graz University took part in the experiment. 35 of the participants are male and 37 female.

3. Results

We do not take into account the first 6 periods since we regard these periods as a training time for the subjects. We use the remaining 36 periods to form four segments consisting of 9 periods. Figure 2 shows the small deviation error in regard to the B&L average forecast.

	Segments				
	1	2	3	4	All
Period	Per. 7-15	Per. 16-24	Per. 25-33	Per. 34-42	Per. 7-42
B&L sum	0,8	1,03	0,85	0,67	0,84

Figure 2: Average forecasts and deviation from B&L

We analyse the average error, distinguishing between gender. There are some interesting results:
As shown in the following boxblots of Figure 3a – 3d the forecasts of the female and male subjects show an interesting difference. The female participants show a higher deviation from B&L in the first segment while the male participants show a higher deviation in the last segment.

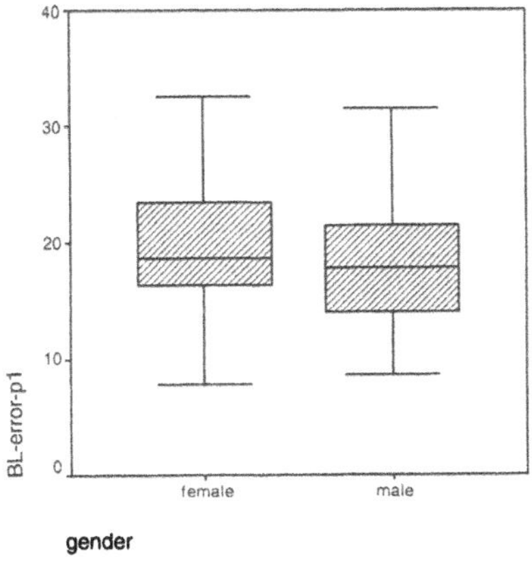

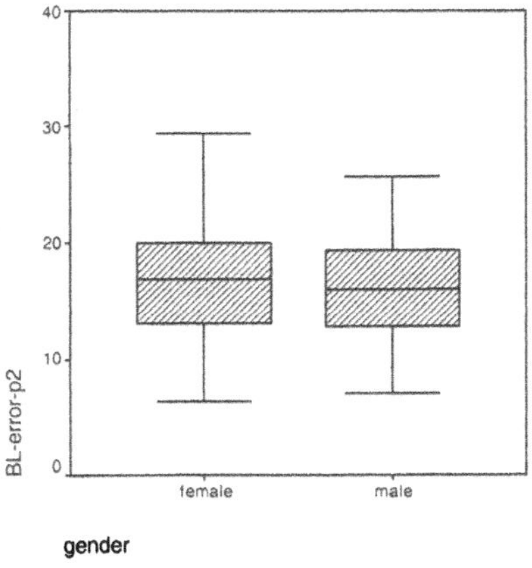

Figure 3a and 3b: Deviation from B&L in segment 1 and 2

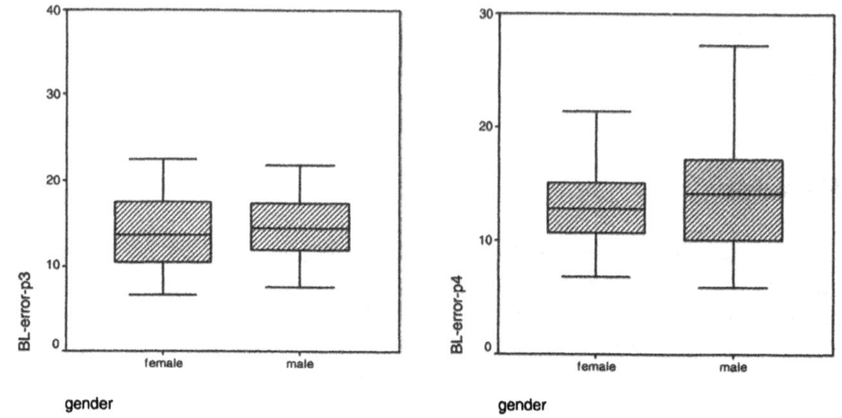

Figure 3c and 3b: Deviation from B&L in segment 3 and 4

We further analyse the results in regard to the intelligence tests: But we cannot reject the null hypothesis, that the applied tests are independent of the performance of the forecasts in regard to gender.

The turning points offer another point of investigation. These turning points are one of two main components of the B&L procedure: The extent of difference between the given period and the next period to be forecasted about the trend of the time series. The analysis concentrates on the turning points predicted by B&L and not on the turning points of the time series. There are 12 B&L turning points in the four segments covering the periods 7 to 42. The subjects score lies between 3 and 10 forecasted B&L turning points. Figure 4 illustrates how many turning points could be predicted correctly by the 72 subjects.

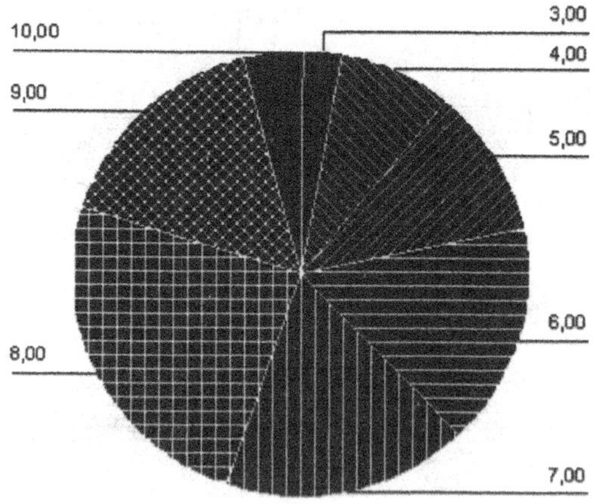

Figure 4: Correctly forecasted B&L turning points

According to Figure 5 female participants show better scores than their male counterparts. While females in average score about 8 turning points, males achieve at least one less.

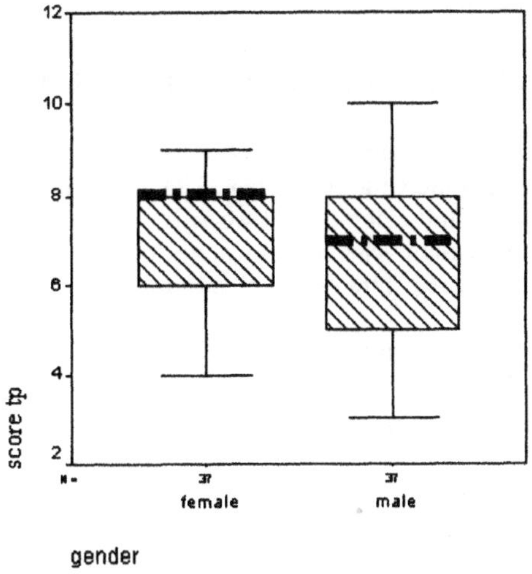

Figure 5: Correctly forecasted B&L turning points

Analysing the results with a U-test, we cannot reject the null hypothesis, that the score is independent of gender. The fourth segment, however, yields a significant difference between female and male participants (Z=-1,866, P=0,62, two tailed).

In regard to the turning points, the B&L procedure shows excellent results in predicting the behaviour of a subgroup. In average, 2/3 of the turning points are correctly anticipated. We could not find any link between the forecasting behaviour and the psychological tests applied.

Literature:

BECKER, O., LEOPOLD-WILDBURGER, U. (1996) The bounds & likelihood-procedure - A Simulation Study concerning the efficiency of visual forecasting techniques: *Central European Journal of Operations Research and Economics* 4, Iss. 2/3, 1996, 223-229.

BRICKENKAMP, R. (1962) Aufmerksamkeits-Belastungs-Test (Test d2). Göttingen, Hogrefe.

FISHER, I. (1896) Appreciation and Interest. New York.

FISHER, I. (1930) The Theory of Interest. New York.

HAYEK, F.A. (1948) Individualism and Economic Order. University of Chicago Press.

HORN, W. (1969) Prüfsystem für Schul- und Bildungsberatung PSB. Göttingen, Hogrefe.

SELTEN, R. (2001) What is Bounded Rationality, in Gigerenzer & Selten (Eds.), Bounded Rationality, MIT Press, 26-27.

Appendix

The bounds and likelihood procedure (Becker & Leopold, 1996):

With the B&L procedure we want to forecast the subject's prediction of a time-series x_t (t=1,2,...,42) for the next following period t+1. The value of the following period is x_{t+1}. The forecast value for the following period we will call f_{t+1}. Our forecasting technique is based on the time series' gestalts-characteristics. For period t+1 we start at the actual value x_t and consider the prior values of the time series as follows:

(1) $f_{t+1} = x_t + b_t \,.\, (1\text{-}2.l_t)\ \mathrm{sign}(x_t - x_{t-1})$

Obviously the estimated values b_t for the average variation within the time series are

$$(2)\ \mathbf{b}_t = \frac{1}{t-1}\sum_{j=2}^{t}\left|x_j - x_{j-1}\right|$$

and the turning point probabilities l_t

$$(3)\ \mathbf{l}_t = \begin{cases} \dfrac{1 + number of local \max ima \leq x_t}{2 + number of local \max ima} & for x_t > x_{t-1} \\ \dfrac{1 + number of local \min ima \geq x_t}{2 + number of local \min ima} & for x_t < x_{t-1} \end{cases}$$

If $x_t > x_{t-1}$ which means an upswing case, l_t is the propability that x_t is a upper turning point (peak). For description cases A-C:

A) all prior peaks above x_t
$l_t \rightarrow 0$ $f_{t+1} = x_t + b_t$
B) as many prior peaks above and below x_t
$l_t = 1/2$ $f_{t+1} = x_t$
C) all prior peaks below x_t
$l_t \rightarrow 1$ $f_{t+1} = x_t - b_t$

The coefficient b_t serves as a bound in both directions for the maximal predicted variation. b_t as explained in equation (2) is an estimator of the average variation of the time series. The relative frequency of the turning point forecasts is is given as l_t and defined as follows: If the time series increases from period t-1 to period t, then l_t is the relative frequency that a decrease in the base series is forecasted. If, on the other hand, the time series decreases from t-1 to t, then l_t is the relative frequency for forecasting an increase of the base series. Therefore, the coefficient c_t varies within the interval $[-b_t, b_t]$. In the extreme case there is a total uncertainty whether the time series will rise or fall and then, the naive prediction $f_{t+1} = x_t$ should be expected.

On the Allocation of Excesses of Resources in Linear Production Problems

F.R. Fernández[1], G. Fiestras[2], I. García-Jurado[3], and J. Puerto[1,4]

[1] Universidad de Sevilla, Facultad de Matemáticas, Spain. fernande@us.es
[2] Universidad de Vigo, Fac. Ciencias Económicas y Empresariales, Spain.
[3] Universidad de Santiago, Facultad de Matemáticas, Spain. ignacio@zmat.usc.es
[4] Corresponding author. puerto@us.es

Abstract. In this paper we consider non-centralized linear production situations. In one of those situations, each producer i of a set N has an optimal production plan x_0^i for a linear production problem given by $\max\{c^i x^i : A^i x^i \leq b^i, x^i \geq 0\}$. In order to maximize the benefits, the producers decide to take profit of their excess of resources $(b^1 - A^1 x_0^1, \ldots, b^n - A^n x_0^n)$. We study the games which describe this situation when players cooperate and side payments are possible, when players do not cooperate and when players cooperate and side payments are not possible.

Key words. Linear production models, linear programming, Nash equilibrium, core.

1 Introduction

Let us consider a set $N = \{1, \ldots, n\}$ of producers. For each agent i, $i = 1, \ldots, n$, let us denote by $A^i \in \mathbb{R}^{p \times q}$, $c^i \in \mathbb{R}^q$, and $b^i \in \mathbb{R}^p$ the technological matrix, the unit selling price vector and the resource bundle, respectively. We assume that there exist linear production technologies, the optimal individual production policy for agent i, $i = 1, \ldots, n$ is obtained solving the problem:

$$\begin{aligned} &\max c^i x^i \\ &s.t. : A^i x^i \leq b^i, \quad i = 1, \ldots, n, \\ &\qquad x^i \geq 0, \quad\quad i = 1, \ldots, n. \end{aligned} \qquad (P^i)$$

Let x_0^i be an optimal solution of (P^i) and b_0^i the vector of resources consumed to implement the optimal production plan x_0^i, i.e. $b_0^i = A^i x_0^i$. For any subset $S \subseteq N$, let us denote $b(S) = \sum_{i \in S} b^i$, and $b_0(S) = \sum_{i \in S} b_0^i$.

We assume that all the producers must maintain their production capabilities and then, they want to allocate their excesses of resources in order to improve benefits. We propose two ways of analyzing the allocation of the excesses. One of them assumes that producers act cooperatively. In this case, the agents within each coalition ensure their individual production plans and then they share their excesses in order to improve the summation of their benefits. The goal in this cooperative analysis is to identify some core allocations that can be realized by assignments of excesses. In the second way, producers behave non-cooperatively. This means that each agent independently requests a portion of the total excess of resources. The goal is to identify those claims leading to a rest point, and to establish the relationship between the payoff of some Nash equilibria of the non-cooperative case with the imputations of the cooperative situation.

2 The cooperative analysis

Given the production situation of n agents described above we assume that producers may cooperate. In this context cooperation means that agents within a coalition S can only distribute among themselves their total surplus in order to improve their joint benefits. In this way, we define the next game (N, v) in characteristic form, where $v(\emptyset) = 0$, and for any $S \subseteq N$

$$\begin{aligned} v(S) = \max\ & \textstyle\sum_{i \in S} c^i x^i \\ s.t.:\ & A^i x^i - \varepsilon^i \leq b_0^i,\ i \in S, \\ & \textstyle\sum_{i \in S} \varepsilon^i \leq b(S) - b_0(S), \\ & x^i \geq 0,\ \varepsilon^i \geq 0,\ i \in S. \end{aligned} \qquad (P^S)$$

We note in passing that even in the case where all the $c^i = c$, $A^i = A$, $i = 1, \ldots, n$, the game above differs from the classical linear production game (N, w) introduced by Owen (1975) [2]. (See [1] for an example that proves this claim.) In Owen's model, the production is centralized by one of the producers and the benefits are later allocated to the producers. Recall that in that game: $w(S) = \max\{cz : Az \leq b(S),\ z \geq 0\}$.

In general, if $c^i = c$, $A^i = A$, $i = 1, \ldots, n$, holds, it is clear that $v(S) \leq w(S)$, $\forall S \subseteq N$. This relationship does not hold in general when the technological matrices and price vectors are different.

Example 1. Let us consider the production system given by $N = \{1, 2\}$, $c^1 = (2, 6, 3)$, $b^1 = (7, 1)$, $c^2 = (2, 4, 6)$, $b^2 = (1, 5)$ and $A = \begin{pmatrix} 1\,2\,1 \\ 1\,1\,3 \end{pmatrix}$.

Next table below collects a pair of optimal solutions and their corresponding optimal values, individual consumptions and excesses of resources.

	P^1	P^2
optimal solution (x_0^i)	$(0, 1, 0)$	$(0, 0, 1)$
optimal value (cx_0^i)	6	6
consumption (b_0^i)	$(2, 1)$	$(1, 3)$
surplus $(b^i - b_0^i)$	$(5, 0)$	$(0, 2)$

In our approach, $v(1, 2)$ is given by the optimal value of the problem

$$\begin{aligned} \max\ & \sum_{i=1}^{2} c^i x^i \\ s.t.:\ & Ax^i - \varepsilon^i \leq b_0^i, i = 1, 2. \\ & \varepsilon^1 + \varepsilon^2 \leq (5, 2), \\ & x^i \geq 0, \varepsilon^i \geq 0, i = 1, 2 \end{aligned}$$

Its optimal value is 25.2, and an optimal solution is given by $(\tilde{x}^1, \tilde{x}^2, \tilde{\varepsilon}^1, \tilde{\varepsilon}^2) = ((0,3,0),(0,0.6,0.8),(4,2),(1,0))$. If we def. $w(1,2) = \max\{\bar{v}^1(1,2), \bar{v}^2(1,2)\}$, where $\bar{v}^1(1,2) = \max\{c^1x : Ax \leq (8,6),\ x \geq 0\}$, and $\bar{v}^2(1,2) = \max\{c^2x : Ax \leq (8,6),\ x \geq 0\}$, we get $\bar{v}^1(1,2) = 24$ and $\bar{v}^2(1,2) = 19.2$. Then, $w(1,2) = 24 < v(1,2)$.

In the game (N, v) the imputation set,

$$I(v) = \{r \in \mathbb{R}^n : \sum_{i \in N} r_i = v(N),\ r_i \geq c^i x_0^i,\ \forall i \in N\}.$$

In particular, given $(\bar{\varepsilon}^1, \dots, \bar{\varepsilon}^n)$ and $(\bar{x}^1, \dots, \bar{x}^n)$ an optimal solution of P^N, it is clear that $(c^1\bar{x}^1, \dots, c^n\bar{x}^n)$ is a n imputation of (N, v).

Let us denote by D^S the dual problem of P^S, for any coalition $S \subseteq N$. This is:

$$\begin{aligned} \min\ & \textstyle\sum_{i \in S} y^i b_0^i + \gamma(b(S) - b_0(S)) \\ s.t.:\ & y^i A^i \geq c^i,\ i \in S, \\ & -y^i + \gamma \geq 0, \\ & y^i \geq 0,\ i \in S,\ \gamma \geq 0. \end{aligned} \qquad D^S$$

Let $(y_{\bar{N}}, \bar{\gamma})$, with $\bar{y}_N = (\bar{y}_N^1, \dots, \bar{y}_N^n)$, be an optimal solution of Problem (D^N). This solution allows us to characterize a family of imputations of the game (N, v).

Proposition 1. *Let $(\bar{\varepsilon}^1, \dots, \bar{\varepsilon}^n)$ be an allocation of excesses satisfying $\sum_{i=1}^n \bar{\varepsilon}^i = b(N) - b_0(N)$, $\varepsilon^i \geq 0$, for any $i = 1, \dots, n$. Then,*

$$\bar{r} = (\bar{y}_N^1 b_0^1 + \bar{\gamma}_N \bar{\varepsilon}^n, \dots, \bar{y}_N^1 b_0^n + \bar{\gamma}_N \bar{\varepsilon}^n) \in I(v).$$

Proof. Since $(\bar{y}_N, \bar{\gamma})$ is an optimal solution of D^N then

$$v(N) = \sum_{i=1}^n \bar{y}_N^i b_0^i + \bar{\gamma}(b(N) - b_0(N)) = \sum_{i=1}^n \bar{y}_N^i b_0^i + \bar{\gamma} \sum_{i=1}^n \bar{\varepsilon}^i = \sum_{i=1}^n (\bar{y}_N^i b_0^i + \bar{\gamma}\bar{\varepsilon}^i).$$

Moreover, $(\bar{y}_N^i, \bar{\gamma}_N)$, is a feasible solution of D^i, the dual problem of P^i and thus, by the weak duality theorem $\bar{y}_N^i b_0^i + \bar{\gamma}_N \varepsilon^i \geq v(i)$. Since this argument does not depend on the particular choice of i, then $\bar{r}$ is an imputation of the game (N, v). □

In any case, we can characterize those imputations $\bar{r} = (\bar{r}_1, \dots, \bar{r}_n)$ that can be reached through allocations of excesses. Indeed, $\bar{r}$ can be reached if and only if Problem (P):

$$\begin{aligned} \max\ & \textstyle\sum_{i=1}^n c^i x^i \\ s.t.:\ & A^i x^i \leq b_0^i + \varepsilon^i,\ i \in N, \\ & \textstyle\sum_{i=1}^n \varepsilon^i \leq b(N) - b_0(N), \\ & c^i x^i \leq \bar{r}_i,\ i = 1, \dots, n, \\ & x^i \geq 0,\ \varepsilon^i \geq 0,\ i = 1, \dots, n, \end{aligned} \qquad (P)$$

has an optimal value of $v(N)$.

We note in passing that reachable imputations are rare because they must hold the condition above. A special case of allocation introduced in Proposition 1 gives us a core element in the game (N, v).

Proposition 2. *The imputation* $r = (y_N^1 b_0^1 + \gamma_N(b^1 - b_0^1), \ldots, y_N^1 b_0^n + \gamma_N(b^n - b_0^n)$ *where* (y_N, γ_N) *is an optimal solution of* (D^N) *belongs to the core of the game* (N, v).

Proof. It is sufficient to prove that for any $S \subseteq N$ then $\sum_{i \in S} y_N^i b_0^i + \gamma_N(b^i - b_0^i) \geq v(S)$. Indeed, this inequality follows because (y_N, γ_N) is a feasible solution of (D^S), the dual problem of (P^S). □

Notice that the game (N, v) is balanced. Since the class of balanced games coincides with the class of linear production games, there exists a linear production game that induces the game (N, v). In this case the construction is straightforward in terms of the elements of the production system (technologies, prices ad resources.)

3 The non-cooperative analysis

In this section, we analyze our model from a non-cooperative point of view. Assuming that the excesses of the different producers are available in a common bundle and that the agents behave non-cooperatively, their interaction can be modeled as the following n-person non-cooperative game in strategic form. For each agent i, $i = 1, \ldots, n$, its set of pure strategies is given by $S^i = \{\varepsilon^i \in \mathbb{R}^p : 0 \leq \varepsilon^i \leq b(N) - b_0(N)\}$, and given a profile of pure strategies $(\varepsilon^1, \ldots, \varepsilon^n)$ the payoff function is:

$$K^i(\varepsilon^1, \ldots, \varepsilon^n) = \begin{cases} \max\{c^i x^i : A^i x^i \leq b_0^i + \varepsilon^i,\ x^i \geq 0\} & \\ & \text{if } \sum_{i \in N} \varepsilon^i \leq b(N) - b_0(N), \\ c^i x_0^i & \text{otherwise.} \end{cases}$$

In order to obtain rest points of this conflict situation we look for Nash equilibria of this game.

Theorem 1. *Any profile* $(\varepsilon^1, \ldots, \varepsilon^n)$ *such that* $\sum_{i \in N} \varepsilon^i = b(N) - b_0(N)$, $\varepsilon^i \in S^i$, $i = 1, \ldots, n$, *is a Nash equilibrium.*

Proof. Let us take $(\varepsilon^1, \ldots, \varepsilon^n)$, $\sum_{i \in N} \varepsilon^i = b(N) - b_0(N)$. If we assume that agent j deviates and chooses $\bar{\varepsilon}^j \in S^j$, $\bar{\varepsilon}^j \neq \varepsilon^j$, two situations may occur: 1) $\sum_{i \neq j} \varepsilon^i + \bar{\varepsilon}^j \leq b(N) - b_0(N)$ and not equal; or 2) there exists a component, say k, such that $\sum_{i \neq j} \varepsilon_k^i + \bar{\varepsilon}_k^j > b(N)_k - b_0(N)_k$. In the latter case, $K^j(\varepsilon^{-j}, \bar{\varepsilon}^j) = c^j x_0^j \leq K^j(\varepsilon^1, \ldots, \varepsilon^n)$. In the former case, since any feasible x^j in the definition of $K^j(\varepsilon^{-j}, \bar{\varepsilon}^j)$ is feasible in $K^j(\varepsilon^1, \ldots, \varepsilon^n)$ then $K^j(\varepsilon^{-j}, \bar{\varepsilon}^j) \leq K^j(\varepsilon^1, \ldots, \varepsilon^n)$. Therefore, $(\varepsilon^1, \ldots, \varepsilon^n)$ is a Nash equilibrium.
□

Remark 1. In general null requests of all the producers is not a Nash equilibrium. On the other hand, any profile in which some players ask for too much is a Nash equilibrium. (Nevertheless, these Nash equilibria are not really interesting.)

Another interesting property of this model is that the set of undominated payoff equilibria is non-empty. Moreover, our next result characterizes the Nash equilibrium profiles in pure strategies whose corresponding payoff vector are undominated.

Theorem 2. *For any payoff undominated equilibrium $(\bar{\varepsilon^1}, \ldots, \bar{\varepsilon^n})$ there exists $(\bar{x^1}, \ldots, \bar{x^n})$ such that $(\bar{\varepsilon^1}, \ldots, \bar{\varepsilon^n}, \bar{x^1}, \ldots, \bar{x^n})$ is a Pareto-solution of*

$$\begin{aligned} &\max\ (c^1x^1, \ldots, c^nx^x) \\ &s.t.:\ A^ix^i \leq b_0^i + \varepsilon^i,\ i \in N \\ &\qquad \textstyle\sum_{i\in N} \varepsilon^i \leq b(N) - b_0(N), \\ &\qquad x^i \geq 0,\ \varepsilon^i \in S^i,\ \forall\, i \in N. \end{aligned} \qquad (UNP)$$

Conversely, for any Pareto-solution $(\bar{\varepsilon^1}, \ldots, \bar{\varepsilon^n}, \bar{x^1}, \ldots, \bar{x^n})$ of Problem (UNP) then $(\bar{\varepsilon^1}, \ldots, \bar{\varepsilon^n})$ is a undominated payoff Nash equilibrium.

The proof is straightforward from the definition of undominated payoff Nash equilibrium and Pareto-optimal solution of a vector-maximum problem.

The links between the non-cooperative and cooperative analysis is stated below.

Proposition 3. *If $(\bar{x}, \bar{\varepsilon})$ is an optimal solution of (P^N) then $\bar{\varepsilon}$ is an undominated payoff Nash equilibrium of the non-cooperative game and $K^i(\bar{\varepsilon}) = c^i\bar{x}^i$, $i = 1, \ldots, n$.*

Proof. The proof follows because $(\bar{x}, \bar{\varepsilon})$ is an optimal solution of the scalarization of (UNP) with the scalar weights $\lambda = (1, \ldots, 1)$. □

4 The non-transferable utility cooperative analysis

In this section we consider the production situation from a cooperative point of view, like in section 2, but assuming that side payments are not possible. Thus, an NTU game should be used to model this situation, more precisely, an NTU market game (see, for instance, Owen (1995) [3] for details on market games). We again assume that each producer has an optimal production plan (x_0^i being the optimal production plan of agent i). So, the original excess of resources of every player $i \in N$ is $b^i - b_0^i$. If a coalition S forms, players can redistribute their endowments in any way desired, i.e., they can obtain any tuple $(\varepsilon^i)_{i\in S}$ such that: 1) $\sum_{i\in S} \varepsilon^i = b(S) - b_0(S)$, and; 2) $\varepsilon^i \geq 0$, for all $i \in S$. Any such tuple is called a feasible allocation for S.

Then, for every S, $V(S)$ is the set of all $z \in \mathbb{R}^n$ for which there exists $(\varepsilon^i)_{i\in S}$, a feasible allocation for S, such that $z_i \leq u_i(\varepsilon^i)$, where

$$u_i(\varepsilon^i) = \max\{c^i x^i : A^i x^i \leq b_0^i + \varepsilon^i,\ x^i \geq 0\}, \quad \text{for all } i \in S.$$

It is immediate to check that, defined in this way, the functions u_i are concave (just note that $u_i(\varepsilon^i) = \min\{(b_0^i + \varepsilon^i)y^i : y^i A^i \geq c^i, : y^i \geq 0\}$), continuous and non-decreasing, so the core of (N, V) is non-empty, a competitive equilibrium exists and it provides a core allocation.

An interesting subset of $core(N, V)$ is the following set, that will be called $subcore(N, V)$:

$$\{z \in PB(V(N)) \ :\ (z_i)_{i\in S} \geq (y_i)_{i\in S}, \quad \text{for all } y \in V(S) \text{ and all } S \subset N\},$$

where $PB(V(N))$ denotes the Pareto boundary of $V(N)$. Next we provide a characterization of the non-emptiness of $subcore(N, V)$. For any $\hat{z} \in V(N)$, we define the family of TU-games $\{(N, v_{\hat{z}}^j),\ j \in N\}$ such that $v_{\hat{z}}^j(N) = \hat{z}_j$ and, for any non-empty $S \subset N$,

$$v_{\hat{z}}^j(S) = \begin{cases} u_j(b(S) - b_0(S)) & \text{if } j \in S, \\ 0 & \text{if } j \notin S. \end{cases}$$

Theorem 3. *It holds that $subcore(N, V) \neq \emptyset$ if and only if there exists $\hat{z} \in V(N)$ such that all the TU-games $(N, v_{\hat{z}}^j)$, $j \in N$, have a non-empty core.*

Proof. Assume that the games $(N, v_{\hat{z}}^j)$, $j \in N$, have a non-empty core. Then, for every $j \in N$, there exists $z^j = (z_1^j, \ldots, z_n^j)$ a core element of $(N, v_{\hat{z}}^j)$. Hence, for any $S \subset N$ and any $j \in S$,

$$\hat{z}_j = \sum_{i\in N} z_i^j \geq \sum_{i\in S} z_i^j \geq v_{\hat{z}}^j(S) \geq y_j,$$

for all $y \in V(S)$. Thus, there must exist $z \in subcore(N, V)$ with $z \geq \hat{z}$. Conversely, take $\hat{z} \in subcore(N, V)$ and define $\bar{z}^j = (0, \ldots, \overset{j}{\hat{z}_j}, \ldots, 0)$. Clearly $\bar{z}^j \in core(N, v_{\hat{z}}^j)$. □

References

1. Fernández F.R., Fiestras G., Garcia-Jurado I., Puerto J. (2002) Competition and Cooperation in non-centralized linear production games. Prepublicaciones de la Facultad de Matemáticas, July-01. Universidad de Sevilla
2. Owen G (1975) The core of linear production games. Mathematical Programming **9**, 358-370.
3. Owen G (1995) Game Theory. Academic Press.

Simulation eines CO_2-Zertifikatenhandels und algorithmische Optimierung von Investitionen

Silja Meyer-Nieberg and Stefan Pickl

Department of Mathematics, University of Cologne, Center for Applied Computer Science, Weyertal 80, 50931 Cologne, Germany

Abstract. Mithilfe der am Zentrum für Angewandte Informatik entwickelten Software TEMPI (Technology Emissions Means Process Identification) steht eine Modellierungsumgebung zur Verfügung, die es ermöglicht, finanzielle Investitionen und ihren Einfluss auf CO_2- Minderungsmassnahmen mithilfe eines zeitdiskreten Modells vergleichend gegenüberzustellen und zu optimieren.

1 Simulation eines internationalen CO_2-Zertifikatenhandels und Optimierung von Investitionen

TEMPI basiert auf den Grundgleichungen des TEM-Modells, die diese Wechselwirkung mathematisch beschreiben. Wesentlich ist, dass das TEM-Modell nur auf Grössen beruht, die empirisch erfassbar sind. Für eine ausführliche Beschreibung sei auf PICKL (1999) und PICKL (2002) verwiesen.

2 Umweltlizenzen - Zertifikatenhandel

Seit mehr als 30 Jahren werden innerhalb der Umweltökonomie Umweltlizenzen als Alternative zu existierenden ordnungsrechtlichen Verfahren angesehen. In dem Kyoto Protokoll werden handelbare Umweltrechte sogar explizit als marktwirtschaftliche Instrumente empfohlen. Damit kann über die ökonomischen Steuerungsanreize ein effizienter Beitrag zum Umweltschutz geleistet werden. Für die Entwicklung und Etablierung eines solchen Umweltlizenzsystems werden in Pickl (2002) die folgenden vier Phasen als notwendig angesehen:

1. Festlegung einer zulässigen Gesamtemissionsmenge
2. Verteilung der Lizenzen, z.B. durch ein *grandfathering*-Verfahren
3. Vorkehrungen für einen funktionierenden Handel
4. Entwicklung von Kontrollmechanismen

Desweiteren kann die Reduktion von CO_2-Emissionen als exemplarisch für die Etablierung eines solchen Umweltlizenzsystems angesehen werden. In diesem Zusammenhang kommt der qualitativen und quantitativen *Bewertung* von verschiedenen Technologiepfaden eine zentrale Bedeutung zu.

2.1 Leakage-Effekte und Portfolio-Optimierung

Da weltweit bisher nur geringe Projekterfahrung mit diesem geplanten Zertifikatenhandel vorliegt, ist man daher besonders auf eine Simulation von Handlungsszenarien angewiesen. Nach Oberthür (2000), *müsste darüber hinaus geklärt werden, wie sich die Emissionen ohne das Projekt entwickelt hätten* (Berücksichtigung sogenannter Leakage-Effekte), um ein erfolgreiches Gelingen zu garantieren.

Die Simulationseinheit TEMPI bietet nun die Möglichkeit, verschiedene Optionen mithilfe eines zeitdiskreten Modells vergleichend gegenüberzustellen. Dadurch kann es den Ausgestaltungsprozess für einen weltweiten CO_2-Zertifikatenhandel bei den folgenden Punkten vorbereiten und begleiten:

- Simulation von verschiedenen Anfangskontingenten
- Variation von Handelsrestriktionen
- Analyse von Diskontierungsmassnahmen

Desweiteren wird dadurch eine spätere rationale Entscheidungsfindung der einzelnen Akteure in der Portfolio-Optimierung beim zukünftigen Kraftwerksbau unterstützt. Da die aktuellen CO_2-Daten eine grosse Varianz aufweisen, werden in einer Pilotphase nur die im Clearing-House registrierten Daten berücksichtigt und diese in normalisierter Form verwandt. In einer zweiten Phase sollen die dann zur Verfügung stehenden realen Daten analysiert und in das Simulationsmodell aufgenommen werden.

2.2 Devisenspekulationsprobleme

Die Modellierung orientiert sich dabei an bestehenden Modellen zu Devisenspekulationsproblemen (DSP) Diese werden häufig als Maximalflussprobleme mit Flussmultiplikatoren im Rahmen der kombinatorischen Optimierung simuliert. Spezielle Netzwerk-Simplex Algorithmen können zu ihrer Lösung eingesetzt werden. Geht es bei DSPs jedoch ausschliesslich um eine Maximierung des betrachteten Geldwerts, so muss bei einem internationalen Zertifikatenhandel bei der optimalen Investitionsplanung das Vorhandensein von zulässigen Zertifikaten zusätzlich betrachtet werden. Der Einsatz von Investitionsmitteln bzw. der begleitende An- und Verkauf von Zertifikaten wird durch rationale Entscheidungen beeinflusst, die mit spieltheoretischen Methoden entsprechend berücksichtigt werden sollen. Insbesondere wird der Core als mögliche Allokation herangezogen, um ein gemeinsames kooperatives Vorgehen zu beschreiben.

2.3 Algorithmische Bestimmung von Core-Elementen

Im folgenden soll ein naheliegendes Verfahren beschrieben werden, das es ermöglicht, Core Elemente über eine geeignete Dualisierung zu bestimmen. Der Algorithmus bietet sich an, im Rahmen einer *experimental economics* umgesetzt zu werden. Hierzu beginnen wir wie folgt:

- Ordne die Teilmengen der Spielermenge $\mathcal{N}$, die mindestens 2 Elemente besitzen, auf die folgende Art und Weise

$$\text{Bilde die Sequenz} \quad K_1, \ldots, K_{2^n-n-1} \quad \text{d.d.}$$

$$|K_i| \leq |K_{i+1}| \quad \text{für} \quad i = 1, \ldots, 2^n - n - 2\,\text{gilt.}$$

  Dies impliziert $K_{2^n-n-1} = \mathcal{N}$
- Definiere nun die $(2^n - n - 1) \times n$-Matrix $A = (a_{ik})_{i=1,\ldots,2^n-n-1,k=1,\ldots,n}$ durch

$$a_{ik} = \begin{cases} 0, & \text{falls} \quad k \notin K_i \\ 1, & \text{falls} \quad k \in K_i \end{cases}$$

- und einen $(2^n - n - 1)$- Vektor $b = (b_i)_{i=1,\ldots,2^n-n-1}$ durch

$$b_i = v(K_i) \quad \text{für} \quad i = 1, \ldots, 2^n - n - 1$$

Es ist nun bekannt, dass ein Vektor $x \in R^n$ genau dann Element des Core ist, wenn die folgenden Bedingungen erfüllt sind:

- P1 $\sum\limits_{k=1}^{n} a_{ik}x_k \geq b_i \quad \text{für} \quad i = 1, \ldots, 2^n - n - 2$
- P2 $\sum\limits_{k=1}^{n} x_k = b_{2^n-n-1} = v(\mathcal{N})$
- P3 $x_k \geq 0 \quad \text{für} \quad k = 1, \ldots, n$

Ersetzt man nun P1 durch $\sum\limits_{k=1}^{n} a_{ik}x_k + x_{n+1} \geq b_i \quad \text{für} \quad i = 1, \ldots, 2^n - n - 2$ und P3 durch $x_k \geq 0 \quad \text{für} \quad k = 1, \ldots, n+1$ dann erhält man:

- Maximiere $\sum\limits_{i=1}^{2^n-n-1} b_i y_i$
- D1 $\sum\limits_{i=1}^{2^n-n-1} a_{ik} y_i \leq 0 \quad \text{für} \quad k = 1, \ldots, n$
- D2 $\sum\limits_{i=1}^{2^n-n-2} y_i \leq 1$ Duales Problem
- D3 $y_i \geq 0 \quad \text{für} \quad i = 1, \ldots, 2^{n-n-2}$

References

1. OBERTHÜHR, S.; OTT, H.E. (2000): *Das Kyoto Protokoll - International Climate Policy for the 21st Century (International and European Environmental Policy Series)*, Springer Verlag Berlin-Heidelberg.
2. PICKL, S. (1999) *Der τ-value als Kontrollparameter. Modellierung und Analyse eines Joint-Implementation Programmes mithilfe der kooperativen dynamischen Spieltheorie und der diskreten Optimierung*, Shaker Verlag Aachen.
3. PICKL, S. (2002) *Process Identification and Technical Investments with TEMPII - Bubbles, Quelros and Environmental Management. Operations Research Proceedings 2001 (Selected Papers)*, Springer Verlag Berlin, Heidelberg.

Indirect Expenditure Functions and Shephard's Lemma

S. Fuchs-Seliger[1]

Universität Karlsruhe (TH), Kaiserstr. 12, D-76128 Karlsruhe, Deutschland

Abstract. Based on preferences on the normalized price space, an indirect expenditure function will be defined. We will study the properties of the inverse demand function and of the indirect expenditure function following from hypotheses on normalized prices. It will also be shown that Shephard's lemma holds without assuming transitivity and completeness of the underlying preference relation or differentiability of the indirect expenditure function.

Journal of Economic Literature Classification: D11, D69

1 Introduction

In this article we will investigate consumer behavior when her or his preferences on the price space $\mathbb{R}^n_{++}$ are known. Every $p \in \mathbb{R}^n_{++}$ will be interpreted as an income-normalized price vector. It is appropriate to assume that, if normalized prices increase such that we have $p > p'$, then the well-being of the individuals in price situation p is worse than in price situation p'.

Imposing weak conditions on the preference relation on the price space, an indirect expenditure function and an inverse compensated demand function will be deduced and their properties will be studied. An inverse compensated demand function assigns, to every commodity bundle x, a price vector which minimizes the expenditure for x and makes the individual as well-off as in a reference price situation p^0.

Finally, we will be concerned with Shephard's Lemma which is an important tool in consumer theory as well as in producer theory. It will be shown that Shephard's lemma holds without imposing hypotheses on the differentiability of the indirect expenditure function.

2 Properties of the Price Space and Preliminary Results

In the following we will consider an individual having certain preferences with respect to the prices of goods in an economy. If a good becomes cheaper, then the individual's well-being rises. We will base our analysis on the following weak hypotheses (for a suitable interpretation assume that the individual has one unit of income):

(P1) R is a relation on the strictly positive n-dimensional real space $\mathbb{R}^n_{++}$. Every $p \in \mathbb{R}^n_{++}$ will be interpreted as an income-normalized price vector. Therefore, $p = \frac{P}{M}$, where P are the market prices and M the income of the individual.
(P2) R is reflexive.
(P3) R is lower semicontinuous, i.e. $\{p \mid p^0 \, R p\}$ is closed in $\mathbb{R}^n_{++}$ for every $p^0 \in \mathbb{R}^n_{++}$
(P4) R is decreasing, i.e. for all $p^1, p^2 \in \mathbb{R}^n_{++}$: $p^1 < p^2 \Rightarrow p^1 P p^2$ [1], where $p^1 P p^2$ means, $p^1 R p^2 \wedge \neg(p^2 R p^1)$.

We can also interpret $p^1 R p^2$ as p^1 is at least as good as p^2 in the eyes of the individual. The above hypotheses are appropriate for a preference relation on the price space. Apparently, the individual will prefer lower prices to higher prices when income does not change but is a fixed amount of money.

We are going to define a function
$C : \mathbb{R}^n_{++} \times \mathbb{R}^n_{++} \to \mathbb{R}_+$ by $C(x,p) = \min\limits_{p' \in \mathbb{R}^n_{++}} \{p'x \mid p R p'\}$ for $x \in \mathbb{R}^n_{++}$ and $p \in \mathbb{R}^n_{++}$ under the supposition that the minimum exists.
$C(.,.)$ will be considered as an indirect expenditure function.
Preliminarily, one can show

Lemma 1. *Under* (P1), (P2) *and* (P4)

$$C^0(x,p) = \inf_{p' \in \mathbb{R}^n_{++}} \{p'x \mid p R p'\}$$

is well-defined, for all $x \in \mathbb{R}^n_{+}$, *and* $C^0(x,p) > 0$ *for all* $(x,p) \in \mathbb{R}^{2n}_{++}$.

In order to guarantee the existence of $\min\limits_{p' \in \mathbb{R}^n_{++}} \{p'x \mid p R p'\}$ one can assume the further condition:

(P5) For every sequence $<p^k>$, $p^k \in \mathbb{R}^n_{++}$, such that $\lim\limits_{k \to \infty} p^k = p^0 \not> 0$ it follows:
For every $p \in \mathbb{R}^n_{++}$ there exists a positive integer N such that $p^k P p$ for all $k > N$.
Note that the indirect utility function $v(p) = \frac{1}{p_1 \cdot p_2}$ which belongs to the utility function $u(x) = x_1 \cdot x_2$ satisfies (P5).

It should be stressed that the following analysis can be also done for $\inf\{p'x \mid p R p'\}$ instead of $\min\{p'x \mid p R p'\}$ if we assume (P1) to (P4) only. Assuming hypothesis (P5) instead of (P4) the preference field of R may posses "thick" indifference surfaces. This is excluded if we require R to be decreasing.

Based on the hypotheses (P1), (P2), (P3) and at least on one of the hypotheses (P4) or (P5) we will develop a model of consumer behavior when a preference relation on income-normalized prices are given.

[1] $p^1 < p^2$ means $p^1_i < p^2_i$ for all $i \leqq n$

Lemma 2. *Under* (P1), (P2), (P3) *and* (P5) *the indirect expenditure function*

$$C(x,p) = \min_{p' \in \mathbb{R}^n_{++}} \{p'x \mid p R p'\} \; for \; (x,p) \in \mathbb{R}^{2n}_{++}$$

is well-defined.

We will define a correspondence δ by

$$\delta(x,p) = \arg \min_{p' \in \mathbb{R}^n_{++}} \{p'x \mid p R p'\} \quad , \quad \forall \, (x,p) \in \mathbb{R}^{2n}_{++} \quad .$$

Note that $\delta(x,p)$ can be many-valued.

δ will be called a "compensated price correspondence" or an " inverse compensated demand correspondence". By interpretation, $\delta(x,p)$ assigns, to every commodity bundle x, all those price situations p' which minimize the expenditure for commodity bundle x and makes the individual feel not better than he or she felt in price situation p. Denote by I the symmetric part of R, i.e. $xIy :\Leftrightarrow xRy \wedge yRx$, then I means indifference.

Lemma 3. *If (P1) to (P5) is assumed, and if additionally R is complete* [2] *and upper semicontinuous* [3], *then*

$$\delta(x,p) = arg \min_{p' \in \mathbb{R}^n_{++}} \{p'x \,|\, p'Ip\}.$$

Proof: By contradiction, suppose $p P p^0$ for some $p^0 \in \delta(x,p)$ then, in view of the upper semicontinuity, the completeness and the decreasingness of R, there would exist a $\tilde{p}$ in an ε-neighborhood of p^0 such that $\tilde{p} < p^0$ and $p \, P \, \tilde{p}$. Since then $\tilde{p}x < p^0x$ we obtain a contradiction to the definition of $\delta(x,p)$.

The following theorems base heavily on the concavity of $C(\cdot, p)$[4] which can be easily demonstrated.

Lemma 4. *Under the conditions (P1) to (P3) and (P5)*

a) $C(\cdot, p)$ *is concave on* $\mathbb{R}^n_{++}$,

b) $C(\cdot, p)$ *is continuous on* $\mathbb{R}^n_{++}$.

Remark 1. Concavity of $C(\cdot,p)$ plays an important role for the differentiability of $C(\cdot,p)$. Since $C(\cdot,p)$ is concave, the function $(-)C(\cdot,p)$ is convex. Therefore, we can apply the results on convex functions. Especially, we obtain that the subdifferential (or more precisely superdifferential) $\partial C(x,p)$ of $C(x,p)$ for every $x \in \mathbb{R}^n_{++}$ is non-empty (see Rockafellar [10], Theorem 23.4). We also

[2] $R \subseteq X \times X$ is complete if for all x, $y \in X$, $xRy \vee yRx$ holds.

[3] $R \subseteq X \times X$ is upper semicontinuous, if $\{z \,|\, zRz^0\}$ is closed in X for all $z^0 \in X$.

[4] a function $f : \mathbb{R}^n \to \mathbb{R}$ is called concave, if for all x, $x' \in \mathbb{R}^n$, $x \neq x'$:
$f(\lambda x + (1-\lambda)x') \geqq \lambda f(x) + (1-\lambda) f(x') \quad , \quad \forall \, \lambda \in [0,1]$,

have that if $C(\cdot,p)$ has a unique subgradient (supergradient) at x, then $C(\cdot,p)$ is differentiable at x. Moreover, then $\nabla C(x,p) = \left(\frac{\partial C(x,p)}{\partial x_1},\ldots,\frac{\partial C(x,p)}{\partial x_n}\right)$ is the unique subgradient, and thus $C(z,p) \leqq C(x,p) + \nabla C(x,p)\cdot(z-x)$, $\forall z \in \mathbb{R}^n_{++}$. (see Rockafellar [10], Theorem 25.1. p. 242).

3 On Shephard's Lemma

It is well-known that Shephard's lemma is an important tool in both consumer theory and production theory. In our context Shephard's lemma means, that the partial differentiation of the indirect expenditure function $C(x,p^0)$ with respect to the i-th good x_i for every $i \leqq n$, yields a price vector $\delta(x,p^0)$ which minimizes the expenditure for x and makes the individual as well-off as in the price situation p^0. We will deduce the differentiability of $C(\cdot,p)$ directly without using the concept of subgradients.

Theorem 1. *Assume* (P1), (P2), (P3) *and* (P5), *and additionally let* $\delta(x,p)$ *be single-valued. Then*

a) $\delta(\,\cdot\,,p^0)$ *is continuous on* $\mathbb{R}^n_{++}$ *for every* $p^0 \in \mathbb{R}^n_{++}$,

b) Shephard's Lemma: $\dfrac{\partial C(x,p)}{\partial x_i} = \delta_i(x,p)$, $\qquad \forall x \in \mathbb{R}^n_{++}$.

Proof for a):
Let us consider a price vector $p^0 \in \mathbb{R}^n_{++}$ and a sequence $<x^k>\subseteq \mathbb{R}^n_{++}$ such that $\lim_{k\to\infty} x^k = x^0 \in \mathbb{R}^n_{++}$. The definition of δ implies $0 \leqq x^k\delta(x^k,p^0) \leqq x^k p^0$. Hence, there exists a bounded subsequence of $<\delta(x^k,p^0)>$. Without loss of generality let this be $<\delta(x^k,p^0)>$ itself. Hence, there exists $p^* \in \mathbb{R}^n_{++}$ such that $\lim_{k\to\infty} \delta(x^k,p^0) = p^*$. Since $p^0 R\,\delta(x^k,p^0)$, for all k, there does not exist a positive integer N such that for all $i \geqq N$, $\delta(x^i,p^0)\,P\,p^0$. This together with (P5) implies $p^* > 0$. Using the lower semicontinuity of R, we obtain $p^0 R p^*$. From the continuity of $C(\cdot,p^0)$ it follows, $p^*x^0 = \lim_{k\to\infty}\delta(x^k,p^0)\cdot\lim_{k\to\infty}x^k = \lim_{k\to\infty}(x^k\cdot\delta(x^k,p^0)) = \lim_{k\to\infty}C(x^k,p^0) = C(x^0,p^0)$. Since by assumption $\delta(x,p^0)$ is single-valued, we must have $p^* = \delta(x^0,p^0)$ and thus $\delta(x^0,p^0) = \lim_{k\to\infty}\delta(x^k,p^0)$.

Proof for b):
Let us consider $p^0 \in \mathbb{R}^n_{++}$ and the commodity vectors x and $x + \Delta x$, and write $\Delta C(x,p^0) = C(x+\Delta x,p^0) - C(x,p^0)$. Hence,

$$\begin{aligned}\Delta C(x,p^0) &= (x+\Delta x)\cdot\delta(x+\Delta x,p^0) - x\cdot\delta(x,p^0) \\ &= x\cdot\delta(x+\Delta x,p^0) - x\cdot\delta(x,p^0) + \Delta x\cdot\delta(x+\Delta x,p^0)\quad .\end{aligned} \tag{1}$$

The definition of δ yields $p^0\; R\;\delta(x+\Delta x,p^0)$ and $p^0\; R\;\delta(x,p^0)$, and thus we have $x\cdot\delta(x+\Delta x,p^0) - x\cdot\delta(x,p^0) \geqq 0$. In view of (1) we therefore obtain

$\Delta C(x,p^0) \geqq \Delta x \cdot \delta(x+\Delta x, p^0)$. For any $j \leqq n$ consider $\Delta x = (0,\dots,0,\Delta x_j, 0,\dots,0)$. Then we obtain $\Delta C(x,p^0) \geqq \Delta x_j \cdot \delta_j(x+\Delta x,p^0)$. If $\Delta x_j > 0$, then we have $\dfrac{\Delta C(x,p^0)}{\Delta x_j} \geqq \delta_j(x+\Delta x,p^0)$. If $\Delta x_j < 0$, then we obtain the converse. Since the concavity of $C(\,\cdot\,,p^0)$ implies that $C(\,\cdot\,,p^0)$ is right- and left-hand differentiable (see Fenchel [4], pp. 71, 79) and since $\delta(\,\cdot\,,p^0)$ is continuous on $\mathbb{R}^n_{++}$

$$\lim_{\substack{\Delta x_j>0\\ \Delta x_j\to 0}} \frac{\Delta C(x,p^0)}{\Delta x_j} \geqq \lim_{\Delta x_j\to 0} \delta_j(x+\Delta x,p^0) = \delta_j(x,p^0) \geqq \lim_{\substack{\Delta x_j<0\\ \Delta x_j\to 0}} \frac{\Delta C(x,p^0)}{\Delta x_j} . \tag{2}$$

Moreover, in view of the concavity of $C(\,\cdot\,,p^0)$ we have

$$\lim_{\substack{\Delta x_j<0\\ \Delta x_j\to 0}} \frac{\Delta C(x,p^0)}{\Delta x_j} \geqq \lim_{\substack{\Delta x_j>0\\ \Delta x_j\to 0}} \frac{\Delta C(x,p^0)}{\Delta x_j} .$$

From this together with (2) the assertion follows. □

Recalling the remark of the second section we immediately obtain, that $C(\cdot,p)$ is differentiable at x for all $x \in \mathbb{R}^n_{++}$ and

$$C(z,p) \geqq C(x,p) + (\delta_1(x,p), \ \dots, \ \delta_n(x,p))\,(z-x), \ \forall\, z \in \mathbb{R}^n_{++}.$$

Without requiring single-valuedness of $\delta(x,p)$ one can also show that the conditions (P1), (P2), (P3) and (P5) imply that $\delta(\cdot,p^0)$ is a compact-valued and upper hemicontinuous correspondence.

In order to deduce Shephard's Lemma we assumed that $\delta(\cdot,\,p^0)$ is single-valued. We will now establish the single-valuedness of $\delta(\cdot,\,,p^0)$ requiring, additionally, that R is complete, transitive, continuous and strictly concave. By the latter we mean the following:

R is strictly concave on $\mathbb{R}^n_{++}$ iff for all $p^1, p^2 \in \mathbb{R}^n_{++}$:

$$p^2 R p^1 \ \wedge \ p^1 \neq p^2 \quad \Rightarrow \quad p^2 P\,(\lambda p^1 + (1-\lambda)p^2), \ \forall\, \lambda \in]0,1[\,.$$

Theorem 2. *Assume (P1) to (P5) and additionally let R be complete, transitive, continuous and strictly concave.*
Then $\delta(x,p^0)$ is single-valued for every $(x,p^0) \in \mathbb{R}^{2n}_{++}$.

Proof:

Consider $p^0 \in \mathbb{R}^n_{++}$. Then under the above conditions $C(x,p^0)$ is well-defined for every $x \in \mathbb{R}^n_{++}$. Suppose that there exist at least two different price vectors $p^1, p^2 \in \delta(x,p^0)$. Hence $p^0 R p^1$ and $p^0 R p^2$. Since R is complete, without loss of generality let us suppose that $p^2 R p^1$. Then strict concavity of R implies $p^2 P\,(\lambda p^1 + (1-\lambda)p^2)$, $\forall\, \lambda \in]0,1[$. In view of transitivity and completeness of R, $p^0 P\,(\lambda p^1 + (1-\lambda)p^2)$, $\forall\, \lambda \in]0,1[$. Consider $\lambda = \frac{1}{2}$. Then

$p^0 P \frac{p^1+p^2}{2}$. Set $\tilde{p} = \dfrac{p^1+p^2}{2}$. Since R is continuous and decreasing, there exists $\hat{p} \in \mathbb{R}^n_{++}$ such that $\hat{p} < \tilde{p}$ and $p^0 P \hat{p}$. From this follows $\hat{p}x < \tilde{p}x$. This together with $p^0 P \hat{p}$ contradicts the definition of $C(x, p^0)$. □

Remark 2. Comprising the results of this section, it was shown that under the hypotheses of Theorem 2 there exists an inverse compensated demand function $\delta(x, p^0)$, continuous with respect to x, and the partial derivatives of the indirect expenditure function $C(x, p^0)$ with respect to x_i are equal to the components of the inverse compensated demand function. Therefore, $\left(\dfrac{\partial C(x,p^0)}{\partial x_1}, \ldots, \dfrac{\partial C(x,p^0)}{\partial x_n}\right)$ equals the price system which makes the individual as well-off as in the price situation p^0 and minimizes the expenditure for commodity bundle x.

References

1. Blackorby, Ch.; Primont, D. and Russel, R.R.: Duality, Separability, and Functional Structure: Theory and Economic Applications. New-York : North-Holland, 1978.
2. Diewert, W.E.: Duality Approaches to Microeconomic Theory in K.J. Arrow and M.D. Intriligator: "Handbook of Mathematical Economics", Amsterdam : North-Holland, 1982.
3. Färe, R.; Primont, D.: Multi-Output Production and Duality: Theory and Applications. Boston : Kluwer Academic Publishers, 1995.
4. Fenchel, W.: Convex Cones, Sets and Functions. Mimeographed lecture notes, Princeton University, 1951
5. Fuchs-Seliger, S.: Compensated and Direct Demand without Transitive and Complete Preferences. Annals of OR 23 (1990), p.p. 199-310.
6. Fuchs-Seliger, S.: A further Remark on Shephard's Lemma. Economics Letters 56 (1997), p.p. 359-365.
7. Fuchs-Seliger, S.: A Note on Duality in Consumer Theory. Economic Theory 13 (1999), p.p. 239-246.
8. Hanoch, G.: Symmetric Duality and Polar Production Functions, in: M. Fuss and D. McFadden (eds.), "Production Economics: A Dual Approach to Theory and Applications", vol. 1, Amsterdam : North-Holland, p.p. 111-131.
9. Martinez-Legaz, J. E., Santos, M. S.: On Expenditure Functions, Journal of Mathematical Economics 25 (1996), p.p. 143-163.
10. Rockafellar, R. T.: Convex Analysis. Princeton, New Jersey, Princeton University Press, 1970.
11. Shephard, R.W.: Cost and Production Functions. Repr. of the 1. ed., Berlin : Springer, 1981.
12. Shephard, R.W.: Indirect Production Functions. Meisenheim am Glan : Anton Hain Verlag, 1974.

Bayesian Estimation of the Heston Stochastic Volatility Model

Sylvia Frühwirth-Schnatter[1] and Leopold Sögner[2]

[1] Email: Sylvia.Fruehwirth-Schnatter@wu-wien.ac.at, Vienna University of Economics and B.A., Augasse 2-6, A-1090 Vienna, Austria

[2] Email: soegner@ibab.tuwien.ac.at, Vienna University of Technology, Theresianumgasse 27, A-1040 Vienna, Austria

Abstract. The goal of this article is an exact Bayesian analysis of the Heston (1993) stochastic volatility model, where different parameterizations of the latent volatility process and the parameters of the volatility process will be used to improve convergence and the mixing behavior of the sampler. We apply the sampler to simulated data and to DM/US$ exchange,rate data.

1 The Heston model

In this paper we investigate the following model, based on [2]:

$$dx^\star(t) = (\mu + \beta\sigma(t)^2)dt + \sigma(t)dW_0(t)\ , \tag{1}$$

where actual volatility is given by a superposition of k independent square root processes:

$$\begin{aligned} \sigma^2(t) &= \sum_{i=1}^{k} \sigma_i^2(t) \\ d\sigma_i^2(t) &= \lambda_i(\alpha_i - \sigma_i^2(t))dt + \tau_i\sigma_i(t)dW_i(t)\ , \end{aligned} \tag{2}$$

where $i = 1, \ldots, k$; λ_i is often interpreted as the speed of mean reversion, α_i is the mean of the instantaneous volatility process and τ_i is the volatility parameter; $W_j(t), j = 0, 1, \ldots, k$, are independent Brownian motions.

For the underlying stochastic volatility model, the parameters μ, β, α, λ, and τ, where $\alpha = (\alpha_1, \ldots, \alpha_k)$, etc. are unknown. These parameters – abbreviated by θ – have to be inferred from the data available. For financial time series these are (equidistant) observations of asset yields (y_n), where the step-width will be denoted by Δ in the further analysis. The index n is used to abbreviate the points of time $\ldots, (n-1)\Delta, n\Delta, \ldots$, i.e. if N yields are observed, then $Y = (y_n)_{n=1}^N$ within the time span $[\Delta, T]$, where $T = N\Delta$. In this paper we abbreviate increases in Brownian motion by $\omega_j(s,t) = W_j(t) - W_j(s)$, $j = 0, 1, \ldots, k$. If a constant step width $\Delta = t - s$ is used, then simply $\omega_{j,n}(\Delta) = W_j(n\Delta) - W_j((n-1)\Delta)$ will be used. Note that $\omega_{j,n}(\Delta)$ has the same distribution as $\sqrt{\Delta}\omega_{j,n}(1)$.

The goal of the following paragraphs is the specification of the conditional distributions, such that an exact Bayesian analysis can be performed. Let us define integrated volatility:

$$\sigma^{\star,2}(t) := \sum_{i=1}^{k} \sigma_i^{\star,2}(t); \quad \sigma_i^{\star,2}(t) := \int_0^t \sigma_i^2(u)du \; . \tag{3}$$

Conditionally on the increases in integrated volatility h_n from $t = (n-1)\Delta$ to $t = n\Delta$, the asset returns are normally distributed, i.e.

$$y_n \sim N(\mu\Delta + \beta h_n, h_n) \; . \tag{4}$$

The conditional distribution of the increases in integrated volatility fulfills $\pi(h_n|h_{n-1}, \sigma_{n-1}^2; \theta) = \Pi_{i=1}^k \pi(h_{i,n}|\sigma_{i,n-1}^2; \theta_i)$. Although the conditional density $\pi(h_{i,n}|\sigma_{i,n-1}^2; \theta_i)$ cannot be derived analytically, the Fourier transform is available as described in [4]. The conditional distribution of the instantaneous volatility process $\pi(\sigma_n^2|\sigma_{n-1}^2; \theta)$ is a non-central χ^2 distribution. Last but not least a stationary law of $(\sigma_i^2(t))$ exists if the parameters satisfy: $\lambda_i > 0$ and $2\lambda_i\alpha_i/\tau_i^2 \geq 1$. The stationary law is a Gamma distribution. For $b_i := 2\lambda_i/\tau_i^2$ and $a_i := b_i\alpha_i$, the marginal densities of the instantaneous volatilities are given by:

$$\pi(\sigma_i^2(t)) = \frac{b_i}{\Gamma(a_i)}(\sigma_i^2(t))^{a_i-1} \exp(-b_i\sigma_i(t)^2) \; . \tag{5}$$

Remark 1. Despite the fact that only the conditional distributions are required to apply MCMC, the restriction $2\lambda_i\alpha_i/\tau_i^2 \geq 1$ – implying stationary instantaneous volatility processes with marginal gamma law – is plausible from an economic point of view. If $2\lambda_i\alpha_i/\tau_i^2 \geq 1$ is met, the asset yields will remain stochastic since $\sigma_i^2(t) > 0$ with probability one.

2 Bayesian estimation of the Heston model

Parameter estimation for stochastic volatility models is known to be a difficult problem (see e.g. [8]). The problem stems from the fact that the conditional distribution $f(y_n|h_n, \mu, \beta)$ of the observed returns y_n depends on the unobservable integrated volatility $\sigma^{2\star}(t)$. To apply the method of maximum likelihood, where

$$L(\theta) = g(y_1, \ldots, y_N|\theta) = \prod_{n=1}^{N} f(y_n|y_1, \ldots, y_{n-1}, \theta) \; , \tag{6}$$

the closed forms of the conditionals $g(y_n|.)$ are required. To derive $h_n|\sigma_{n-1}^2$ we have to invert the characteristic function. To derive $g(y_n|.)$, the integrated volatilities have to be integrated out in (4). Since this integration is practically impossible, the application of maximum likelihood becomes infeasible.

2.1 Estimation of the parameters of the volatility processes

To sample from the joint posterior distribution

$$\pi(X, \theta|Y) \propto f(Y|X, \theta)\pi(X|\theta)\pi(\theta) \ , \tag{7}$$

the "complete data" likelihood $f(Y|X,\theta)$ as well as the "prior density" $\pi(X|\theta)$ of the joint distribution of the latent variables X under θ has to be known explicitly. $\pi(\theta)$ is the prior of θ. The "complete data" likelihood $f(Y|X,\theta)$ is easily obtained from (4) as the product of N densities from a normal distribution,

$$f(Y|X, \theta) = \prod_{n=1}^{N} f(y_n|\mu\Delta + \beta h_n, h_n) \ . \tag{8}$$

W.l.g. we restrict to $k = 1$ for notational simplicity in this subsection. The natural candidates to describe the latent process are the integrated volatilities $X_1^C := (h_{1,C}, \ldots, h_{n,C}, \ldots, h_{N,C})$ and the instantaneous volatilities $X_2^C = (\sigma^2_{0,C}, \sigma^2_{1,C}, \ldots, \sigma^2_{n,C}, \ldots, \sigma^2_{N,C})$. This version to express X_1 and X_2 will be called to centered version, where $h_n = id(h_{n,C})$, etc. Following [5] and [6], another version – called the non-centered version X_1^{NC}, X_2^{NC} – uses a starting value σ_0^2 and increases in Brownian motion $\omega_n(1)$, to calculate (h_n) and (σ_n^2) by means of the model parameters θ. When simulating (σ_n^2) by means of the Euler scheme, the map $\omega_n(1) \mapsto \sigma^2_{n-1}\exp(-\lambda\Delta) + (1 - \exp(-\lambda\Delta))\alpha + \tau\sigma_{n-1}(\Delta)^{0.5}\omega_n(1) =: \sigma^2_{n,NC}$, where $\omega_n(1) \sim \mathcal{N}(0,1)$, provides us with an (approximation of the) instantaneous volatility process. Since this map is one-to-one it is indeed possible to derive the process $(\omega_n(1))$ from the instantaneous volatilities (σ_n^2). An integration of σ_n^2 over the time interval Δ provides us with $h_{n,NC} = \alpha\Delta + (\sigma^2_{n-1} - \alpha)\frac{1-\exp(-\lambda\Delta)}{\lambda} + \tau\sigma_{n-1}(\Delta)^{1.5}\tilde{\omega}_n(1)$. The reader should note that we use $\omega_n(1)$ and $\tilde{\omega}_n(1)$ in the above expression. If Δ approaches 0, then the equality $\omega_n(1) = \tilde{\omega}_n(1)$ has to hold. However, for observations on a discrete grid, h_n remains a random variable given σ^2_{n-1}. To account for this fact we used $\omega_n(1)$ and $\tilde{\omega}_n(1)$. In the non-centered version the yields and (h_n) depend on the parameters θ. This is not the case when the centered version is applied. Following [7], there exists the opportunity to construct a partially non-centered parameterization. This can be obtained by a convex combination of centered and non-centered terms, i.e. $X_1^{PC} = (1-\nu)X_1^C + \nu X_1^{NC}$ and $X_2^{PC} = (1-\nu)X_2^C + \nu X_2^{NC}$, $h_n = (1-\nu)h_{n,C} + \nu h_{n,NC}$, etc.

After we have discussed how volatilities can be represented in our analysis, let us investigate the "prior" $\pi(X|\theta)$, where $X = (X_1^j, X_2^j)$ with $j \in \{C, NC, PC\}$. From the above definitions $\pi(X|\theta)$ is nothing more than the product of N conditional densities $\pi(h_n, \sigma_n^2|h_{n-1}, \sigma^2_{n-1}; \theta)$, where h_n and σ_n^2 are conditionally independent, i.e.

$$\pi(h_n, \sigma_n^2|h_{n-1}, \sigma^2_{n-1}; \theta) = \pi(h_n|\sigma^2_{n-1}; \theta)\pi(\sigma_n^2|\sigma^2_{n-1}; \theta) \ . \tag{9}$$

From section 1 we already known that the first density in (9) can be reconstructed by means of Fourier inversion while the second density is given by a non-central χ^2 law.

2.2 Priors

In the ongoing analysis we put the following priors on the model parameters: $\mu \sim \mathcal{N}(m_0, M_0)$, $\beta \sim \mathcal{N}(b_0, B_0)$ and $\sum_{i=1}^{k} \alpha_i \sim \mathcal{N}(a_0, A_0)$. For λ_i and τ_i we use a gamma prior $\mathcal{G}(l_0, L_0)$ and $\mathcal{G}(d_0, D_0)$. $\mathcal{N}$ stand for the normal law, $\mathcal{G}$ for the gamma distribution, and $\mathcal{U}$ for the uniform distribution. The parameters μ and β will not estimated and set to zero, i.e. $m_0 = b_0 = 0$ and $M_0 = B_0 = 0$. We use $l_{0,i} = 1$, $L_{0,i} = 1/50$, $d_{0,i} = 0.01$ and $D_{0,i} = 1/5$. For $h_{i,n,C}$, $\sigma_{i,n,C}$, $h_{i,n,NC}$ and $\sigma_{i,n,NC}$ we construct the priors by geometric averaging. I.e. $\pi(h_n) = \pi_u(h_{i,n,C})^{1-\nu_i} \pi_u(h_{i,n,NC})^{\nu_i}$, $\pi(\sigma_n^2) = \pi_u(\sigma_{i,n,C}^2)^{1-\nu_i} \pi_u(\sigma_{i,n,NC}^2)^{\nu_i}$. For $\pi_u(.)$ we use a uniform distribution on $[0, S]$, S is sufficiently large.

For $\sum_{i=1}^{k} \alpha_i$ we use an informative prior. The importance of using an informative prior with the mean volatility parameters can be motivated by the fact that (i) $\sum \alpha_i$ equals to the expected sum of instantaneous volatilities and (ii) also the properties of a maximum likelihood estimation (with a fixed process (σ_n^2) and $k = 1$) improve if α is a-priori fixed at the sample mean. To incorporate this 'prior' knowledge into a Bayesian model, we use strong priors on $\sum_i \alpha_i$, more precisely we set $a_0 = y'y/T/\Delta$ and $A_0 = 0.1\ a_0$.

2.3 MCMC estimation

The Subsections 2.1 and 2.2 have already described the conditional distributions and the priors of the underlying stochastic volatility model. In this paper we restrict ourselves to the estimation of $(h_{i,n}, \sigma_{i,n}^2)$, and the parameters α, λ and τ. β and μ are fixed at 0. We derive samples of the parameters X, θ as follows:

Step 1: h_n from $\pi(h_n | Y, h_{j,j \neq n}, \sigma_0^2, \ldots, \sigma_T^2, \theta)$
Step 2: σ_n^2 from $\pi(\sigma_n^2 | Y, X_1, \sigma_{j,j \neq n}^2, \theta)$
Step 3: α from $\pi(\alpha | Y, X_1, X_2, \lambda, \tau, \nu)$
Step 4: λ from $\pi(\lambda | Y, X_1, X_2, \lambda, \tau, \nu)$
Step 5: τ from $\pi(\tau | Y, X_1, X_2, \lambda, \tau, \nu)$

3 Results

First we tested our sampler in simulated data. For the one factor setting ($k = 1$), we generated paths with $N = 400$, $\alpha = 0.0012$, $\lambda = 50$, $\tau = 0.5$, $\beta = 0$ and $\mu = 0$. Throughout this paper, we work with a step-width $\Delta = 1/252$. For the two factor setting ($k = 2$), we set $\alpha = (0.006\ 0.007)'$, $\lambda = (50\ 150)'$,

$\tau = (0.5\ 1)'$, $\beta = 0$ and $\mu = 0$. The parameters are updated by means of the Metropolis-Hastings (MH) algorithm. In the MH algorithm we use smoother densities $\pi(h_n, \sigma_n^2 | h_{i,i\neq n}, \sigma_{i,i\neq n}^2, Y, \theta)$ as described in [3] to update the parameters of the latent process $X = (X_1^j, X_2^j)$ with $j \in \{C, NC, PC\}$. In all update steps random-walk proposals are used.

The most important point of this analysis is the *parameterization of the latent process* and *blocking.* For a centered parameterization we observe that λ and τ are positively correlated, their posterior modes are much too large. This can neither be improved by blocking steps 3 and 4 nor by blocking steps 3-5. For the non-centered parameterizations the acceptance rates in the updates of θ are very small and the posterior distributions of λ, τ have modes much smaller than the true parameter values. This motivated us to construct the partially centered parameterization. With a mean of the hyper-parameter $\nu_i = 0.05$ the partially centered version results in good acceptance rates, and the modes or means of the estimated posterior of the parameters are close to the true parameter values. This can be observed for both the one and the two factor setting.

After we have tuned our sampler with simulated data, we applied this algorithm to daily DM/US$ exchange rate data from January 1998 to December 1999, where $N = 504$. Note that the squared variation $y'y/T$ was equal to $3.4224e-5$, such that a posterior mean for $\sum_i \alpha_i$ of $y'y/T/\Delta = 0.0086$ can be expected. For the empirical data a small ν_i, $\nu_i = 0.025$, was necessary to get considerable acceptance rates. We observe good mixing for the one factor model, while the mixing for the first factor (λ_1 small) in the two factor setup has to be improved by further tuning. We claim that this can be done by tuning ν_i. Table 1 presents the mean and the standard deviation of the posterior for the corresponding model parameters for the one factor and the two factor setting. Comparing our results to the estimates of [1] we observe that our one factor estimates are very close to the estimates presented in [1] while for the two factor model, the both persistence parameter λ_i are larger than in [1].

Table 1. DM/US$ exchange rates form January 1998 to December 1999, $N = 504$. Posterior means, standard deviations and parentheses. 10000 simulation steps, 1000 burn-in steps.

$k=1$			$k=2$					
α	λ	τ	α_1	λ_1	τ_1	α_2	λ_2	τ_2
0.0089	61.2167	11.5113	0.0044	43.7663	0.3094	0.0040	203.1584	1.0441
(5e-4)	(11.511)	(0.413)	(3e-4)	(4.475)	(0.011)	(3e-4)	(23.928)	(0.052)

4 Conclusions

This article has implemented an exact Bayesian analysis of the Heston (1993) stochastic volatility model. We observe that different parameterizations of the latent volatility process and the parameters of the volatility process result in very different convergence behavior of the MCMC sampler; the best performance has been observed with a partially centered version of the latent process. We apply the sampler to simulated data and to DM/US$ exchange rate data.

References

1. Bollerslev, T., H. Zhou (2002) Estimating stochastic volatility diffusion using conditional moments of integrated volatility. *Journal of Econometrics*, **109**, 33–65.
2. Heston, S.L. (1993) A closed-form solution for options with stochastic volatility with applications to bond and currency options. *Review of Financial Studies*, **6**, 327–343.
3. Jong, P., N. Sheppard (1995) The simulation smoother for time series models. *Biometrika*, **82**, 339-350.
4. Lamberton, D., B. Lapeyre (1996) *Introduction to Stochastic Calculus – Applied to Finance*. Chapman & Hall, London, 1st edition.
5. Papaspiliopoulos, O., P. Dellaportas, G. Roberts (2001). Bayesian inference for non-gaussian ornstein-uhlenbeck stochastic volatility processes. *mimio, Lancaster.*
6. Roberts, G., and O. Stramer (2001). On inference for partially observed nonlinear diffusion models using the Metropolis-Hastings algorithm. *Biometrika*, **88**, 603–621.
7. Papaspiliopoulos, O., G. Roberts, and M. Sköld (2002). Non-centered parameterisations for hierarchical models and data augmentation. *mimeo, Lancaster.*
8. Shephard, N. (1996) Statistical aspects of arch and stochastic volatility. *In Cox,D., D. Hinkley and O. Barndorff-Nielsen (Eds.): Time Series Models in econometrics, finance and other fields. Chapman & Hall, London.*

Forecasting with Leading Economic Indicators – A Neural Network Approach

Timotej Jagric

Department for quantitative economic analysis, Faculty of Economics and Business, University of Maribor, Razlagova 14, 2000 Maribor, Slovenia, Tel.: +386 2 22 90 343 Fax.: +386 2 25 16 141, E-mail: timotej.jagric@uni-mb.si

1 Introduction

There is variety of important issues associated with the problem of business cycle forecasting, especially regarding forecast methodology and forecast evaluation. Overall, we can say that macroeconomic forecasting has a fairly poor reputation (Granger 1996). Still, even with the recognition that forecasting business cycles is a very difficult task, we find some hopeful signs for future progress. Our research on forecasting has focused on development of new approach in forecasting with classical NBER leading indicators by applying neural networks.

The decision to focus on the neural networks arises directly from the features of these models, as described by Bishop (1995). First, neural networks are data-driven and can "learn" from, and adapt to, underlying relationships. This is useful in contexts where one does not have any a priori beliefs about functional forms. Second, when properly specified they are universal functional approximates. Finally, neural networks are non-linear, which seems to be a case for many macroeconomic time-series.

The paper is organized as follows. In section two the structure of the database and the selection of reference series are explained. The selected input variables of the model are presented in section three. The suggested model is explained in section four. In section five of the paper we present the results.

2 Data

Important step in the process of construction the model is development of a broad database, which should cover all crucial fields of economic activity. The database, which was used in the model, includes 365 time series. To ensure sufficient transparency, the time series are classified in categories. The database can be divided in two major groups of time series: time series representing Slovene economic activity, time series representing foreign economic activity. Since Slovenia got independence in October 1991, the time series start with January 1992. The database covers the period 1992:01 – 2001:08. In the final version of the database, all time series were transformed into growth rates.

The aim of the model is to forecast a selected reference variable, which is the benchmark that indicates fluctuations in the economic activity. We selected monthly index of total industrial production. Extensive analysis (Jagric 2002) of such reference variable also gave support to our decision, since it was discovered, that industrial production has same cyclical characteristics as GDP in Slovenia.

3 Scoring System for Business Cycle Indicators

To construct a forecasting model, a selection of input variables is needed. In our study we extended the use of criteria employed by NBER. The scoring of each series reflects our desire not only to make as explicit as possible the criteria for selecting indicators but also to increase the amount of information available to the user in order to aid in evaluating their current behavior. The scoring plan includes five major elements: economic significance, statistical adequacy, promptness of publication, smoothness, conformity and timing. When the subheads under these elements are counted, eight different properties of series are rated in all.

We decided to score all time-series. The total score of time-series, theoretical lead-time, and the results of graphical analysis, were than used to form the group of leading indicators, which includes 58 time series from database. The average lead-time is determined by cross-spectral analysis (phase and coherency) and Granger test of causality, where two criteria were used: the value of adjusted determination coefficient, and Akaike information criteria. The average lead-time is only an estimate of actual lead-time for selected time-series. The scoring system, we have used, ensured that the selected indicators posses the best characteristics among all time-series in database.

4 Neural Network Model

The scoring system, which was used, determined the selection of input variables in our model. The target variable is the monthly index of industrial production. The input and target variables are not seasonally adjusted. As Stock and Watson (1998) noted, seasonal adjustment procedure applied to the data may be cleansing them of any underlying non-linearities. In our case many selected variables show an underlying trend. As there is no universal trend function, which could be applied to all variables, we decided to use one-year and monthly growth rates (in decimal). These transformations also ensured that the values of variables do not differ significantly from each other and are between −1 and 1.

Another important reason for pre-processing is phenomenon known as the 'course of dimensionality'. If we are forced to work with a limited quantity of data, as we are in practice, then the dimensionality of the input space can rapidly lead to the point where the data is very sparse, in which case it provides a very poor representation of the mapping. Therefore our goal is, to map vectors $\mathbf{x}^n$ in a d-dimensional space $(x_1,...,x_d)$ onto vectors $\mathbf{z}^n$ in an M-dimensional space $(z_1,...,z_M)$, where $M < d$. To achieve this goal, we use unsupervised linear transformation technique known as principal component analysis (PCA), where a set of data are summarized as a linear combination of an orthonormal set of vectors.

More challenging was to determine the design of neural network. Changes to the architecture can fundamentally alter the forecasts produced by the network, even when no changes are made to the inputs, outputs or sample size. It is important to distinguish between two distinct aspects of the architecture selection problem. First, we need a systematic procedure for exploring some space of possible architectures, and this forms the subject of this selection. Second, we need some way of deciding which of architectures considered should be selected. This is determined by the requirement of achieving the best possible generalization.

In our case we first formed some basic requirements for the network. First, we will use feed-forward back-propagation network, since this type of network is mostly used in forecasting applications. Due to the type of input and output data, we will only use two types of transfer function: pure linear and tan-sigmoid activation function. Next, we decided that only the output-layer will have a bias. The output layer will consist of one neuron, since we try to predict the future value of one reference series. Last, the network will not have more than three layers of neurons, since such network is a universal approximator.

The above requirements have greatly reduced the space of possible architectures. Therefore we could apply a simplified tiling algorithm (Mezard and Nadal 1989). We build the network in successive layers with each layer having fewer units than the previous layer. When a new layer was constructed, a single unit, called the master unit is added. Then step-by-step additional units are added. Every step the network is trained and the forecasting performance is estimated. The whole process is repeated until a larger network does not sufficiently contribute to the forecasting performance.

To train our network, we use quasi-Newton approach, since it has a significant advantage over the conjugate gradient method, which is normally used for training. One of the problems that occur during neural network training is called overfitting - the network has memorized the training examples, but it has not learned to generalize to new situations. One method for improving network generalization is to use a network that is just large enough to provide an adequate fit (Hagan et al. 1996). Since we do not know how large a network should be in our application, we selected regularization. It has been found empirically that by using a regularizer one can significantly improve the network generalization (Hinton 1987).

5 Results

In the process of network architecture selection, we tested every form for 6, 9, 12, and 15 months forecast (all calculations were performed with Matlab v.6.0 R12). We developed special program, which automatically supervised the testing and recorded the performance. The performance was measured with different criteria: total mean square error for in- and out-of-sample data, determination coefficient for regression between target variable and estimated variable, and comparison of a spectrum for target and estimated variable. Testing was performed on two computers. Major testing was performed on IBM Server X220 with Pentium X-III 866 MHz. The stability of results was tested on the second computer with Pentium III 500 MHz. Testing required about 600 computing hours on the server since every estimation of a possible architecture was repeated max. 5000 times. To avoid idle processing time, the developed program controlled this process. After each cycle the program reported final results, which were then tested manually on the second computer.

It has frequently been argued that statistical error measures do not measure the right thing. Therefore many authors developed additional evaluation methods (Stekler 1991, Leitch and Tanner 1991). In the experimental phase of testing, we discovered that even if a network scores high, the results may not be stable or are not well suited for forecasting. Therefore we developed a new testing routine for networks with two or more layers.

After extensive testing of possible forms of network architectures, we selected a neural network, which can be represented with following equations:

$$\mathbf{y} = \mathbf{f}^{3,1}(\mathbf{w}_{3,2}\mathbf{f}^{2,2}(\mathbf{w}_{2,6}\mathbf{f}^{1,3}(\mathbf{w}_{1,12}\mathbf{x})) + \mathbf{b}_{3,1}) \tag{1}$$

$$f^{i} = \text{purelin}(n) = n \qquad i = 3,1 \tag{2}$$

$$f^{j} = \text{tansig}(n) = (2/1 + e^{-2n}) - 1 \qquad j = 2 \tag{3}$$

where $\mathbf{y}$ is output vector, $\mathbf{f}^{i,z}$ is matrix of z activation functions of neurons in layer i, and $\mathbf{w}_{i,q}$ is matrix of q weights in layer i. Only output layer of neurons has a bias vector, which are represented by $\mathbf{b}_{i,z}$. As it can be seen, we selected a three-layer network. The input layer has three neurons (each with pure linear activation function), hidden layer has two neurons (each with non-linear tansig activation function), and the output layer has one neuron (with pure linear activation function). The selected model performed best for 9, 12 and 15 months forecast. Forecasts for longer periods did not produce good results – as we expected, since we did not have long time series.

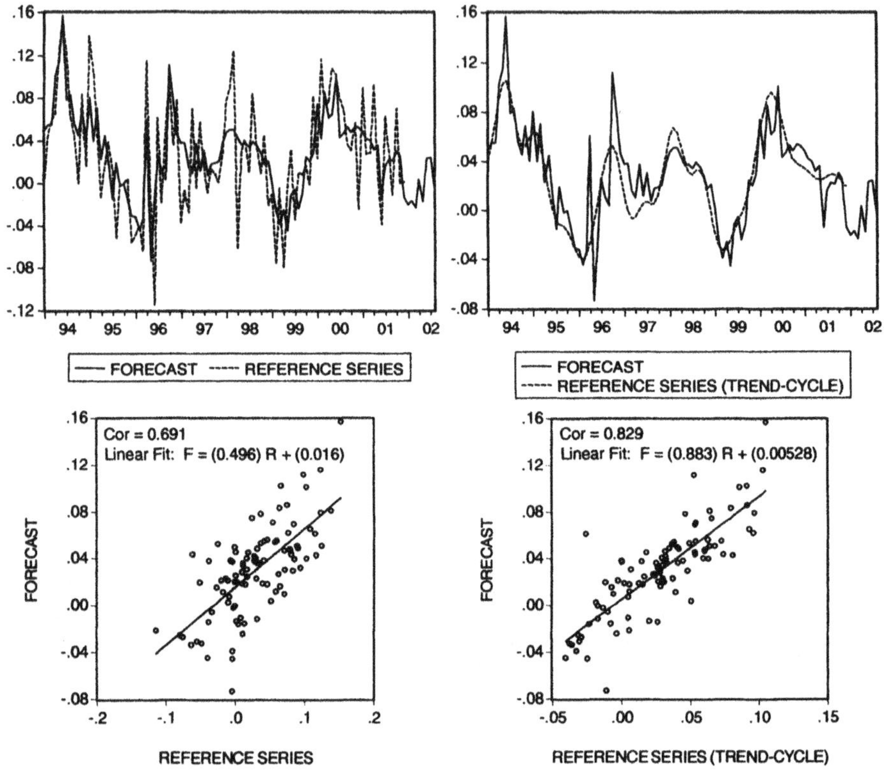

Fig. 1. Performance of neural network model

By using PCA we were able to reduce the input space significantly. Instead of 58 input variables, we used only seven principal components, since we eliminated those principal components that contribute less than 10% to the total variation in the data set of input variables.

The forecast performance of the new model of leading indicators for Slovenia is presented in Figure 1. The parameters of the model were estimated on the data from 1993:01 to 1997:08. As the upper two graphs in Figure 1 suggest, the model capture the dynamics of the reference series well. All turning points were detected and the forecasted value follows the dynamics of the reference series.

The statistical properties of the estimated model are presented in lower two graphs. We performed post-regression analysis, were we compared original (target) data with forecasted (estimated) data for in- and out-of-sample data. As it can be seen, the model captures best the business cyclical frequencies of the reference series. The high frequencies are not detected. This was expected, since we use principal components – non-measurable indicators, which do not have all the information of original leading indicators. In addition, we only searched for linkage between reference and leading indicators in business cycle frequencies.

In future work, we hope to further refine the best neural network models in this research (by considering additional types of networks, different training methods, etc.) for use as forecasting tools to exploit readily available data in order to gauge future economic activity. Forecast comparisons with other models, such as a vector error-correction models may also be undertaken. As well, future projects may involve the construction of neural net models to forecast other important macroeconomic variables.

References

Bhishop CM (1995) Neural Networks for Pattern Recognition. Oxford University Press: New York

Charemza W,Deadman DF (1992) New directions in econometric practice: General to specific modelling, cointegration and vector autoregression. Edward Elgar, Aldeshot

Granger CWJ (1996) Can We Improve the Predictive Quality of Economic Forecasts? Journal of Applied Econometrics 11: 455-473

Hagan MT, Demuth HB, Beale MH (1996) Neural Network Design. PWS Publishing, Boston

Hinton GE (1987) Learning translation invariant recognition in massively parallel networks. In: Bakker JW, Nijman AJ, Treleaven PC (eds) Proceedings PARLE Conference on Parallel Architectures and Languages Europe. Springer Verlag, Berlin

Jagric T (2002) Measuring Business Cycles. Eastern European Economics 40: 63-87

Leitch G, Tanner JE (1991) Economic Forecast Evaluation: Profits Versus The Conventional Error Measures. American Economic Review 81: 581-590

Mezard M, Nadal JP (1989) Learning in feedforward layered networks: The tiling algorithm. Journal of Physics A 22: 2191-2203

Stekler HO (1991) Macroeconomic Forecast evaluation techniques. International Journal of Forecasting, 7: 375-384

Stock JH, Watson MW (1989) New Indexes of Coincident and Leading Economic Indicators. In: Blanchard O, Fischer S (eds) NBER Macroeconomics Annual. MIT Press, Cambridge

Estimating Multivariate Conditional Distributions - An Application to the Truck Sales Forecast

Eric A. Stützle and Tomas Hrycej

Information Mining
DaimlerChrysler AG, Research & Technology Ulm,
P.O. Box 2360, 89081 Ulm, Germany
E-mail: Eric.Stuetzle@DaimlerChrysler.com

Abstract. A concept for forecasting the conditional multivariate distribution has been developed. It allows the forecast of the joint distribution of target variables in dependence on explaining variables. The concept can be applied to general distribution families such as stable or hyperbolic distributions. The conditional distribution parameters are estimated by a global optimization method, using neural networks for functional approximation. The information about a complete distribution of forecasts can be used to quantify the reliability of the forecast. A comparison with conventional forecasting concepts is done and the additional benefit of forecasting conditional distribution in general, and of hyperbolic distribution in particular is shown. The concept is illustrated on a case study concerning the future truck demand. In this application, the distribution parameters are conditional on properties of the product and information about existing orders.

1 Introduction

It is a widespread practice to understand a forecast for a variable of interest as a single number. For example, an inflation rate forecast may amount to "three per cent". The most common meaning of this single number is the estimated mean value.

In contrast to this simplification, the statistical view of forecast implies that the forecast variable (or, generally, forecast vector) y is random with a certain distribution conditional on explaining variables x [6]:

$$f(y|x) \tag{1}$$

For decision making about future actions, every action can be assigned a value resulting from this action assuming a certain state of the world (described by y). With a forecast distribution of the state of the world, a decision optimal in some sense can be taken. For example, an optimality criterion may be the minimum expected loss. Let us define the loss for action a and state of the world y, conditioned on explaining variables x, as $L(a, y, x)$. Then, the optimum decision is

$$\arg\min \int_y L(a, y, x) f(y|x) dx,$$

which is obviously not calculable without knowing the conditional probability density function $f(y|x)$.

Unfortunately, it is not easy to figure out a general law expressing the dependence of the distribution on the given attribute vector x. If we had a set of cases for each attribute vector with different outcomes, we would be able to compute the conditional distribution over this set in a straightforward way. But unfortunately this comfortable setting occurs scarcely in practice. This is why we have to find out the distribution in a less direct way: as a function of the attribute vector x.

The need for identification of a complete and general conditional distribution (1) motivated the development of the forecasting model presented here. We are roughly following the concepts developed by [8]. Nonlinear dependencies are represented by a multi-layer perceptron [4], and estimator parameters are identified numerically with help of the maximum likelihood principle and a global optimization method [2]. The developed general concept can be used under many distribution assumptions, such as the multivariate Gaussian, Student, stable [3], hyperbolic distributions [1], or to nonparametric mixture distribution approximators ([7]).

In Section 2 we develop a concept of the underlying general probabilistic model. The method to estimate the distribution parameters with the help of a global optimization method, neural networks for functional approximation and a developed transformation is specified in Section 3 and 3.2. A short overview of applicable probability distributions is given in Section 5. The application of the truck demand forecast is presented in Section 6.

2 Probabilistic Model

2.1 A general model of probability distributions

To generate distributions from a certain class, it is frequently possible to find a "canonical form" as a representative of this class, and to derive all distributions from this class by a linear transform of the random vector. It is important that this canonical density has only few parameters, or even is completely parameter-free. This is the way we are going to follow in this subsection.

Suppose there is a input vector $x \in I\!R^m$ and a n-dimensional random vector Z with some simple conditional probability density function (PDF)

$$g(Z = z). \tag{2}$$

This canonical density function may be, for example, a standard multivariate Gaussian. Also other density families (stable, hyperbolic) possess such simple canonical special cases with few parameters (basic distribution parameters), based on which the whole distribution class can be generated with the aid of a linear transformation.

The nonsingular conditional matrix $A(x)$ and the conditional position parameter vector $\mu(x)$ form a linear transformation $y = \mu(x) + A^{-1}(x)z$, which generates the dependent random variables $Y_i = y_i, i = 1, \ldots, n$ with the following relationship:

$$z = A(x)(y - \mu(x)). \tag{3}$$

Applying the transformation theorem , we get the conditional PDF f of the transformed random variable Y as

$$f(Y = y|x) = |A(x)|g(Z = z), \tag{4}$$

with $|A(x)| \neq 0$ the determinant of $A(x)$. The conditional PDF f in (4) represents the whole probability distribution class generated by the canonical density $g(z)$ with the aid of (3).

According to (1) and (3), $\mu_{\omega_1}(x)$ and $A_{\omega_2}(x)$ are to be viewed as parametric functions of the input vector x. By existence of basic distribution parameters, we obtain a further parametric function $v_{\omega_3}(x)$. These mappings can be modelled by an appropriate approximator.

3 Functional Approximation

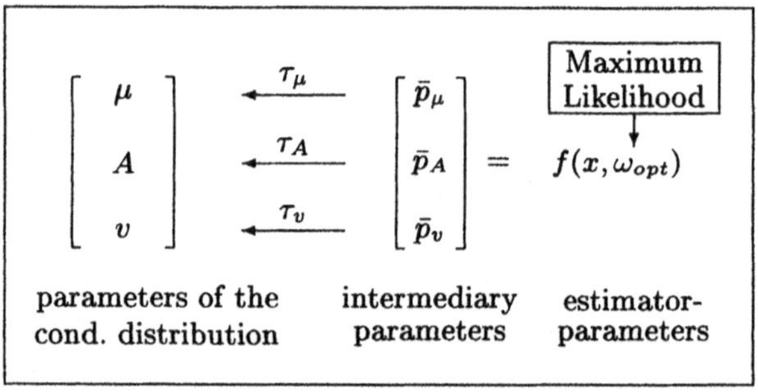

Fig. 1. The estimation model for the conditional distribution

Due to functional approximators are real-valued mappings, but the distribution parameters are subject to certain restrictions, a transformation τ of the outputs is required. τ is described in Section 3.2.

For this reason the concept shown in Figure 1 is used. The intermediary parameters $\bar{p} = (\bar{p}_\mu, \bar{p}_A, \bar{p}_v)$ are the outputs of the mapping approximator.

3.1 Functional approximation with neural networks

The real-valued mapping f_ω maps from the input variable space $I\!R^m$ to the intermediary parameter space $I\!R^{1/2n(n+3)+u}$, with u the amount of basic distribution parameters.

This mapping can be represented by a multi-layer perceptron, whose outputs are the intermediary parameters $\bar{p}$. Suppose the use of a one-layer perceptron with l hidden neurons. Then the number of perceptron parameters to be determined by an estimation procedure amount to $ml + l1/2n(n+3) + u$.

3.2 Mapping real-valued vectors to distribution parameters

Following [8], suppose $A \in I\!R^{n \times n}$ is a upper triangular with strict positiv diagonal elements. This is not restrictive under the distribution classes treated here.

Further matrix A is factorized into the following form $A = UD$ where U is upper triangular with $u_{ii} = 1$ and D is diagonal. It can be shown , that the strictly positive parameters d_{ii} play the role of scaling parameters. This representation has lead to a significant improvement of the fit to the data in comparison to the method of [8].

The components of the conditional position parameter vector $\mu(x)$ are arbitrary real values and so the corresponding transformation is, trivially, an identity transformation. To enforce the strict positiveness of the diagonal elements of D and thus of A, d_{ii} are received with help of an exponential transformation. Therefore the transformation $\tau = (\tau_{\mu_i}, \tau_{\alpha_{ii}}, \tau_{\alpha_{ij}})^T$, with $\mu_i = \tau_{\mu_i}(\bar{p}_{\mu_i}) = \bar{p}_{\mu_i}, i = 1 \dots n$, $\alpha_{ii} = d_{ii} = \tau_{\alpha_{ii}}(\bar{p}_{d_{ii}}) = e^{\bar{p}_{d_{ii}}}, i = 1 \dots n$ and $\alpha_{ij} = d_{ii}u_{ij} = \tau_{\alpha_{ij}}(\bar{p}_{\alpha_{ij}}) = e^{\bar{p}_{\alpha_{ii}}}\bar{p}_{\alpha_{ij}}, j > i$, maps the components of the intermediary parameters to the position parameter vector $\mu(x)$ and to the components of the matrix $A = UD$. Due to the dependence to the chosen distribution class, the transformation to the components of the basis distribution parameter vector v is not treated in detail.

4 Parameter Estimation

4.1 Maximum likelihood principle

The fitting of the approximator is done via the maximum likelihood principle by minimizing the cross entropy. With help of a data set containing K pairs x_k, y_k, which can be viewed as columns of matrices X and Y, i.e. $X \in I\!R^{m \times K}$ and $Y \in I\!R^{n \times K}$.

Independent data samples lead to the negative log-likelihood as

$$L(X,Y) = -\sum_{k=1}^{K} log(f(y_k|x_k)). \tag{5}$$

Substituting the conditional PDF of (4) into (5) we obtain the parameterized negative log-likelihood as

$$L(X,Y,\omega) = -\sum_{k=1}^{K} log(|A_{\omega_2}(x_k)|) - \sum_{k=1}^{K} log(g_{\omega_3}(A_{\omega_2}(x_k)(y - \mu_{\omega_1}(x_k)))) \tag{6}$$

4.2 Estimation procedure

In Equation (6), the log-likelihood function is expressed with help of the parameter functions $\mu_{\omega_1}(x)$, $A_{\omega_2}(x)$ and $v_{\omega_3}(x)$. The task of functional approximation is solved by numerical optimization methods.

Minimizing (6) over the parameter set $(\omega_1, \omega_2, \omega_3) = \omega \in \Omega$ maximizes the likelihood. As estimation procedure, the modified global optimization method "Multi-Level Single-Linkage" has been used. A local method is part of the global procedure (e.g. a conjugate gradient method).

The gradient formula used in the local search follows the chain rule

$$\frac{\partial L(X,Y,\omega)}{\partial \omega} = \sum_{k}^{K} \frac{\partial L(x_k, y_k, \omega)}{\partial \bar{p}(x_k, \omega)} \frac{\partial \bar{p}(x_k, \omega)}{\partial \omega}.$$

The first term in the last summation is computed by taking the corresponding derivatives of formulas of Section 2. The second term is received by usual backpropagation formulas [5].

5 Probability Distribution Classes

The developed concept can be applied to a big variety of probabilistic distributions. The well known Gaussian distribution is one special case of the canonical PDFs of the general concept above.

Likewise the multivariate Student, stable or the hyperbolic distribution can be applied to this concept. To generate the family of generalized hyperbolic probability distributions, we need to use the spheric form of the PDF as

$$g(Z = z) = a \frac{K_{\lambda - n/2}(\alpha \sqrt{z^T z})}{(\alpha^{-1} \sqrt{z^T z})^{n/2 - \lambda}} e^{\beta^T z},$$

with $a = \sqrt{\alpha^2 - \beta^T \beta}^{\lambda} / (2\pi)^{n/2} K_\lambda(\sqrt{\alpha^2 - \beta^T \beta})$ and K_ν denotes the modified Bessel function of the third kind with index ν. The parameters have the following domain of variation: $\lambda, \alpha \in I\!R$, $\beta \in I\!R^n$ with $\beta^T \beta > \alpha^2$.

The parameters α and β and *gamma* are basic distribution parameters.

6 Application: Truck Sales Forecast

The distribution forecast concept has been applied to the demand forecast of trucks. The forecast model for the sales of the next month has been identified. The illustrated results were received under the assumption of different underlying classes of distributions and different information levels. Explaining variables x have been for example the demand over the past six months (pd), seasonality (s) and the existing orders (eo) of the upcoming three months. Table 1 shows the results for a homoscedastic model (with only conditional mean) as well as heteroscedastic (i.e., with conditional scaling parameter) models with several distribution types and various information levels. The forecasting error is improved by 45 percent by using available information. A further improvement of 21 percent through using conditional variance, of 76 percent through using stable distribution instead of Gaussian, and of 84 percent through using hyperbolic distribution.

Table 1. Forecast improvement due to information, conditionality and non-Gaussian distributions

Distribution	Information	error
homoscedastic Gaussian	pd	0.295
homoscedastic Gaussian	pd + s	0.284
homoscedastic Gaussian	pd + s + eo	0.161
heteroscedastic Gaussian	pd + s + eo	0.127
heteroscedastic stable distribution	pd + s + eo	0.039
heteroscedastic hyperbolic distribution	pd + s + eo	0.025

References

1. Ernst Eberlein and Karsten Prause. The generalized hyperbolic model: financial derivatives and risk measures. *Mathematical Finance*, 2000.
2. A.H.G. Rinnooy Kan and G.T. Timmer. Stochastic global optimization methods, part i: Clustering methods, part ii: Multi-level methods. *Mathematical Programming*, 39(1):26–78, 1987.
3. Svetlozar Rachev and Stefan Mittnik. *Stable Paretian Models in Finance.* Wiley & Sons, Inc., 2000.
4. D.E. Rumelhart, G.E. Hinton, and R.J. Williams. Learning internal representations by error propagation. In D.E. Rumelhart and J.L. McClelland, editors, *Parallel Distributed Processing*, volume 1, chapter 8. MIT Press, Cambridge MA, 1986.
5. D.E. Rumelhart and J.L. McClelland, editors. *Parallel Distributed Processing*, volume 1. MIT Press, Cambridge, 1986.
6. Alan Stuart, Keith Ord, and Steven Arnold. *Kendall's Advanced Theory of Statistics, Volume II A: Classical Inference and the Linear Model.* Arnold, London, 1999.
7. D. M. Titterington, A. F. M. Smith, and U. E. Markov. *Statistical Analysis of Finite Mixture Distributions.* Chichester: John Wiley, 1985.
8. Peter M. Williams. Using neuronal networks to model conditional multivariate densities. *Neural Computation*, 8:843–854, 1996.

Application of Techniques of Functional Data Analysis to Spectroscopic Data

Vera Hofer

Karl-Franzens University, Graz, Austria

Abstract. A classification and identification problem is solved in the case of functional data. Functional data are data records considered as functions rather than vectors of observations at different values of the argument. This research deals with the spectral lines of rocks, being treated with light in the visible and mid-infrared regions. In a first step Principal Component Analysis in the functional data case will be applied to spectral lines of rock samples of various rock types to find out, whether the spectral lines can be separated.

1 Introduction

Functional Data Analysis stands for statistical techniques, being applied when the features observed have a characteristic structure [1,2]. We do not meassure a scalar but a function. In this study ten rock samples of six rock types each were selected and treated with light of various wavelengths in visible and mid-infrared region. The reflectivity of this samples was meassured for these wavelengths in some proper unit. Overlapping all curves does

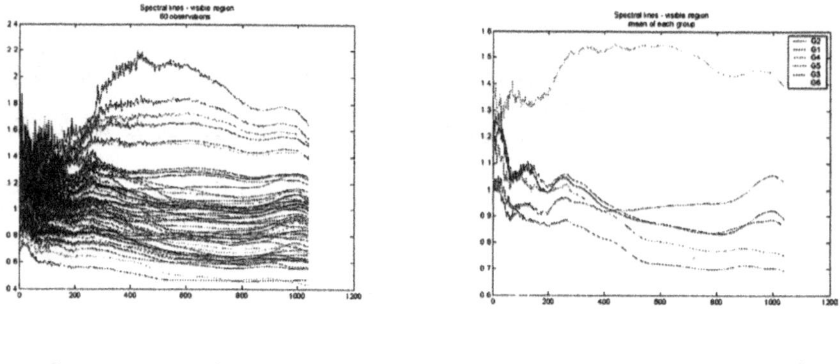

Fig. 1. Spectral lines of ten samples of six rock types each – visible region – (**left**) and the mean functions for the six groups (**right**).

not suggest separation's being possible. We cannot find any structure except the lines of group six that have a completely different shape than the others. Looking at the mean functions imposes that separation of groups could probably be successful. Now the task is to identify each class by its typical spectral lines so that by means of spectral lines one can say, which class a sample of unknown class membership belongs to. This problem is of great interest because mineral aggregates are the most commonly used construction materials worldwide. Improving quality control would reduce not only production costs but also costs occuring as a result of improper exploitation of the material or of environmental impacts of the production process.

Why is Functional Data Analysis applied in this study? One might suggest, using techniques of Multivariate Analysis. Then each wavelength corresponds to a treatment. But you have to bear in mind that there are much more treatments (predictors) than observations. Furthermore the variables are highly correlated so that the matrices, occuring in Multivariate Analysis, are badly conditioned [5].

2 First Analysis

If we look at our functions, we see that four steps have to be carried out first: bundling smoothing, registration, Principal Component Analysis. We have observed

$$y(t) = g(t) + \varepsilon. \tag{1}$$

t stands for the wavelength. Our functions $g(t)$, i.e. the spectral lines, are arbitrary, that means that their shape is determined mainly by the class membership. But there are also other factors affecting the shape of the lines such as elements that can be included or not (e.g.: Fe, Si, Hydroxyls), depending on the growth conditions in the mines the samples were taken from. This fact calls for a careful choice of mines to garantee representative samples.

Variability is also caused by the meassurement applied. This error ε (roughness of the curves) is called white noise. As this white noise affects the correlation to a high extend, we should try to eliminate it by smooting. Besides we are only interested in the characteristic shape of the curves $g(t)$ [4]. But before proceeding in this way, we have to eliminate a parallel effect: The lines are parallel which can be shown by a scatter plot (Fig. 2). After bundling this effect has disappeared. Bundling means that the average distance of each curve to either the class mean function or the overall mean function is calulated. This is done in a regionwhere the lines are parallel.

The next step is the elimination of white noise. There are different ways of smoothing. Moving average has two disadvantages: Firstly, we loose observation points and secondly, we get a discretized function so that we have to interpolate afterwards to be able to evaluate the function at any point or compute derivatives. This is why smoothing is carried out by regression analysis. Polynomial regression is strongly affected by extreme observations. So we use spline interpolation that means that the regression function is piecewise a cubic polynomial. If we don't restrict the regression function, we get the interpolation function for all points observed and the squared error is

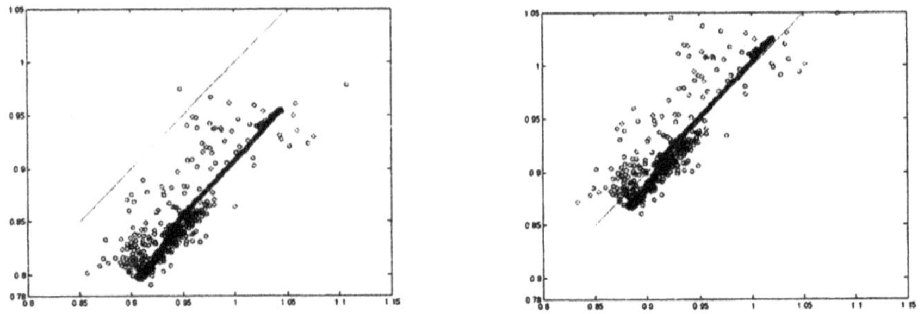

Fig. 2. Scatter plot of two observations of the first group (**left**). The red line indicates the first median. After bundling to the group mean this effect has disappeared (**right**) – visible region.

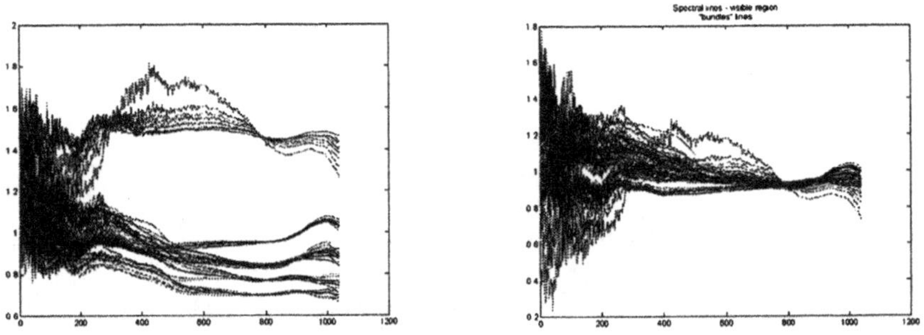

Fig. 3. Bundling all curves to their group mean functions (**left**) and to the common mean function (**right**) – visible region.

zero. Now we can calculate the first two derivatives but the white noise has not disappeared. That means that we could not extract the characteristics of the curves observed. Therefore we add a penalty to the sum of squares to penalize the curvature [3].

$$\sum_{i=1}^{n}\{Y_i - g(t_i)\}^2 + \alpha \int_a^b \{g''(x)\}^2\,dx \qquad a \le t_1 < t_2 < \cdots < t_n \le b \quad (2)$$

To avoid huge systems of linear equations and to be able to use different penalties as well as to simplify further analysis such as Principal Component Analysis, we want the regression function to be a basis expansion. The basis functions $\beta_j(t)$ are splines with compact support.

$$g(t) = \sum_{j=1}^{q} \delta_j \beta_j(t) \qquad (3)$$

Five knots within the interval observed are used for creation of each basis functions. So we have to estimate only q Paramters. The number of basis functions will depend on the oscillation of the curves. The choice of a too small number of basis functions for highly oscillating curves results in a smooth curve that has less characteristics (maxima, minima) than the original curve.

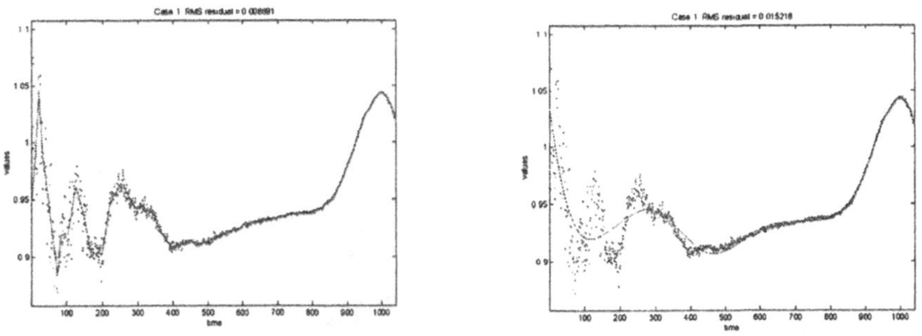

Fig. 4. Basis expansion of the spectral line of the first sample of the first group with 100 basis functions (**left**) and with 10 basis functions (**right**) – visible region.

After smoothing the curves could be registered. Registration is a transformation of the abscissa to aline the characteristic features of the curves (e.g. maxima, minima). In this study the variability in phase is as important as the variability in amplitude. So registration was not carried out in the first approach.

3 Principle Component Analysis

Having estimated our curves for the various samples, we apply the Principle Component Analysis. Similar to the multivariate case we look for the direction of the greatest variability.

$$\max \frac{1}{N} \sum_{i=1}^{N} \quad \text{subject to} \quad ||\beta||^2 = 1 \tag{4}$$

In the functional case [2] the scores are

$$f_i = \langle \beta, x_i \rangle = \int \beta(s) x_i(s) ds. \tag{5}$$

For the mid-infrared meassurements we don't get clusters by potting the scores of the first principle compontent against the scores of the second one. After bundling as discribed before, the first, second, third and sixth group are clustered whereas we cannot separate the fourth and fifth group although they can be distinguished from the others. The use of more basis functions doesn't improve separation. Principal Component Analysis in the visible re-

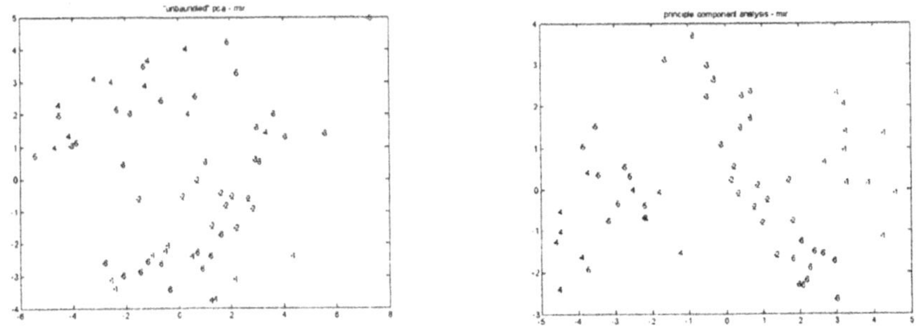

Fig. 5. Principle Component Analysis in the mid-infrared region without bundling (ten basis functions) (**left**) and after bundling (**right**).

gion shows that there is no difference in separation whether we bundle around the common mean function or use the data after smoothing but without further manipulation. By applying Principal Component Analysis to the group-bundled data it is also possible to separate group four and group five.

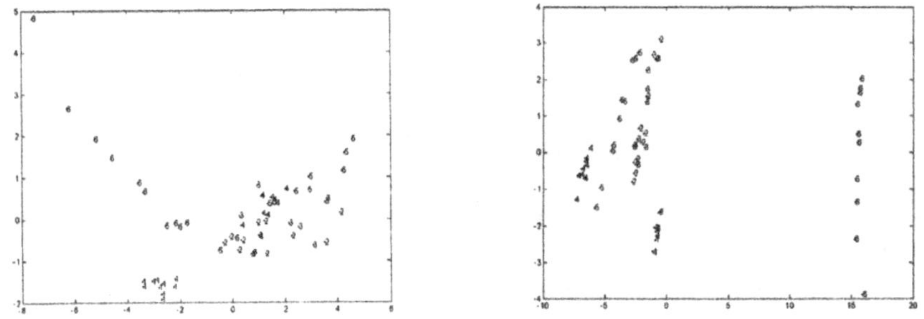

Fig. 6. Principal Component Analysis of the spectral lines in the visible region without bundling (**left**) and after bundling around the group mean functions (**right**). In both cases 100 basis functions were used.

According to the Principal Component Analysis before we see the proportion of variability which is explained by the first principal component. This is 93%, together with the second principal component we have explained 98 % of variability. In Functional Data Analysis the principal components are functions. The first principal component is low at the beginning and in the end

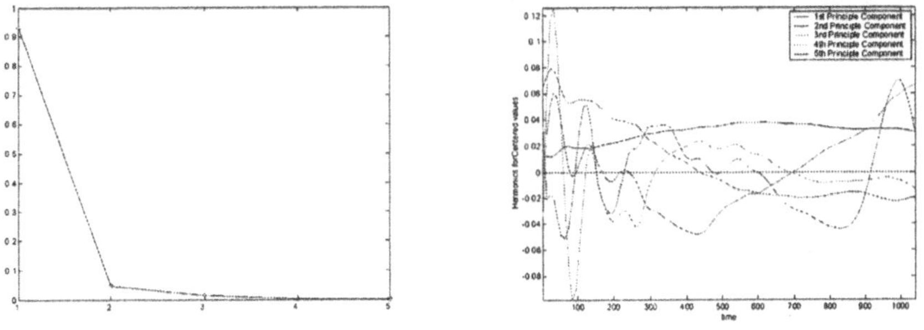

Fig. 7. Proportion of variance explained by each eigenfunction (**left**), the five principal components (**right**).

and higher in the middle. So, samples with high reflectivity at the beginning and in the middle of the interval will have high first scores. That means that the first principle component discribes variability according to the common mean function whereas the second principal component indicates the variability of the function values throughout the interval. Those lines will have high second scores that form a broad image [2].

4 Conclusion

Separation was possible by using bundling and smoothing. Further steps are to optimize the interval where the bundling technique is carried out and to implement a cross validation. When assigning a curve of unknown class membership to the group, it belongs to, one could proceed stepwise: If the curve cannot be alined to group 1 or group 6, the bundling technique has to be applied. Besides the combination of various information of the samples could sucessfully be done by a Canonical Correlation Analysis. Anyway, a Discriminat Analysis should make a quick separation be possible.

References

1. Ramsay J. O., Silverman B. W. (2002) Applied Functional Data Analysis. Methods and Case Studies. Springer, Berlin Heidelberg
2. Ramsay J. O., Silverman B. W. (1997) Functional Data Analysis. Springer, Berlin Heidelberg
3. Green P. J., Silverman B. W. (1994) Nonparametric Regression and Generalized Linear Models. Monographs on Statistics and Applied Probability **58**. Chapman and Hall
4. James G. M., Hastie T. J. (2001) Functional Linear Discriminat Analysis for Irregular Sampled Curves. Internet
5. Hastie T. J., Buja A., Tibshirani R. (1995) Penalized Discriminant Analysis. Annals of Statistics, **23**, 73 — 102

Integrating Exchange Rate Theory in Data Mining

Bernd Brandl

Department of Government, University of Vienna, Brünnerstraße 72, A-1210 Vienna (bernd.brandl@univie.ac.at)

Abstract. This paper focuses on an integration of exchange rate theory in a data mining process for the purpose of forecasting. The applied approach is centred by a Genetic Algorithm (GA) and Neural Networks (ANN), which allows the identification of relationships that are not describable by economic theory. As experience showed, relationships derived by data mining are often not convincing as regards their correctness and effectiveness. Most data mining approaches do not contribute much to persuade otherwise. In this work, it is tried to remove parts of this limitation by combining economic theory with data mining. However, usually the role of economic theory in exchange rate forecasting is to identify a list of relevant variables to be included in the analysis, with possibly and plausible signs of their coefficients. Previous research documents the failure of this way of analysis and thus for exchange rate theory in forecasting exchange rates at frequencies up to one month. Consistent with these results, in this paper the role of economic theory is extended as it is implemented in data mining as a framework in which and among which the possibilities of data mining are exploited. Other findings include: (i) During the years 2000 and 2001 countries relative economic growth is most significant in forecasting exchange rates one period ahead, (ii) the financial market is of major concern when explaining exchange rate movements as it may be used to proxy real economic activity as well as it maps massive capital flows between countries, which in turn affect exchange rates. (iii) fundamental forecasting is more effective on lower frequencies. The approach is illustrated in some detail for five exchange rates on a daily, weekly and monthly frequency.

Data Mining

The definition says that data mining searches for unknown and interesting relationships. This is true and it is the standard way data mining is viewed in literature. See for example Hand et al (2001). However, for the purpose of this work, data mining is defined as the semi-automatic search for relationships among many time series which are possibly important to explain future exchange rate fluctuations. For a similar definition of data mining see for example Chereb (1998). Yet, as the last years showed, in economics and finance, the possibilities of data mining are seen quite skeptical. Also, because relatively often the results of data mining are not startling. Furthermore, economic theory offers many (profound) relationships and, of course, if relationships are already known it makes no sense to search for them again. While, as the definition says, data mining may find undetected relationships, economic literature often tends to interpret them as temporal or too case specific. In general, the relationships uncovered by data mining are seen as less apparent and they are expected to be more case specific and valid only temporarily. The problem in context with exchange rate forecasts, however, is that forecast models which are based on economic theory, in general, perform weakly to predict exchange rate movements out-of-sample on frequencies (at least) up to one month. Even though relationships described by economic theory are often found to hold in the long run and thus can be interpreted as a kind of long-run gravitation center. This pessimistic view of the possibilities of exchange rate theory is the mainstream notion in economic literature since the seminal work of Meese and Rogoff (1983).

Dangers of data mining

Even though the critique on the relationships found by data mining has to be taken seriously, other work showed that the utilization of different data mining techniques and data mining approaches, indeed, can be very effective for many fields of applications. Own experiences in context with building forecast models on basis of fundamentals showed that data mining, indeed, is also very effective in providing forecast models with a remarkable high goodness of fit in out-of-sample evaluations. Yet, it has to be stressed that sometimes a high goodness of fit was valid only temporarily (over specific evaluation periods) and may also be the result of *data snooping*, *overfitting* and *spurious relationships*. However, the latter problems are inherent questions of statistical forecasting and are almost impossible to solve, see for example White (2000), and therefore not only valid in context with the data mining forecast models. So, nevertheless, data mining is a very useful tool of analysis. However, one limitation of data mining is that it uses only information that is covered in the data set. From a theoretical perspective this limitation on the data set available is unproblematic. Whereas, in practice data sets that are usually available to exploit unknown relationships are often problematic since they are too small or not sufficiently clean. This limitation on the data set available is a crucial point for the combination of data mining with economic theory as the latter is able to add information that lies outside of too small data sets.

Using exchange rate theory

In this work the forecast equations of the *Purchasing Power Parity* (PPP), the *Uncovered Interest Parity* (UIP), the general *Monetary Model*, the *Dornbusch* (1976) *Overshooting Model* and the *Frankel* (1979) *Real Interest Differential Model* have been used to provide a framework in which data mining not only searches among various time series to express the formulated aggregates of explanatory variables best, but also among which data mining searches for further relationships to explain one-period-ahead exchange rate movements. These five theoretical approaches have been selected as literature promises them to be most effective, or as they are central in the discipline of international economics.[1] However, applying economic theory constrains the forecast models to behave according to described laws. For example, PPP says that when the price level in country A is increasing, with other things held equal, the currency of country A is expected to depreciate. This translates into forcing the model to yield a negative coefficient. The data set, however, and data mining results based solely on statistical criteria (measured on basis of limited evaluation periods) will permit a coefficient different to that from economic theory, depending on the mix of series used in the found forecast equation. Within the applied GA the correct coefficient was taken into consideration making sure that economic theory is used adequately. Coefficients other than those of economic theory would violate economic laws and therefore would be more vulnerable to lead to ridiculous results. As mentioned above, the forecast equations provided by economic theory are extended by adding time series selected by data mining. Because of the dangers of data mining the applied data mining approach is based on various steps of analysis which all try to filter out unimportant and spurious relationships. Therefore, the applied approach consists of a combination of traditional data mining tools such as cluster analysis, (rolling) correlation analysis and "newer" tools such as the core of the data mining approach, the constrained GA. By applying different methods and techniques of analysis it was possible to find time series out of the vast universe of possibly relevant time series and thus to reduce the search space in which combinations of variables are checked on their effectiveness to explain future exchange rate movements. However, after having reduced the universe of possibly interesting time series to a pool of time series which have a minimum correlation to the respective exchange rate, the following (constrained) GA was used.

Designing the GA

The applied GA was designed to build exchange rate forecast models that proof to be successful in out-of-sample tests, which means that they have to provide a forecasting performance better than the naïve forecast. To do so, the GA was constraint in such way that the forecast models constructed meet the demands of the forecast equations provided by economic theory, but allowed to include other sorts of influence, which means additional series than those suggested by economic theory. The idea behind to build forecast models which, first, are based on profound economic relationships, provided by economic theory, that behave relative stable over a long time but may temporarily be covered by other sources of influence and second, to consider temporal and/or case specific sorts of influences which, indeed, can become important in some sub-periods. Therefore in the GA fitness function a penalty term is included which ensures that the performance of the models does not differ significantly over different sub-out-of-sample evaluation periods. For reasons of space a more detailed discussion of the fitness function is forgone. For the exploitation of temporal and/or case specific influences

[1] For recent literature on the applied theoretical approaches see for example Dutt and Ghosh (2000), Groen (2000) and Rogoff (2002). For a description of the variables used in the forecast equations see p. 5.

and relationships automated data mining appears to be perfect since it allows (but also demands) a continuous check on the validity and correctness of relationships. However, the combination of theoretical relationships with additional relationships has the advantage that the goodness of fit of the theoretical models can be raised without endangering that profound relationships offered by economic theory get lost. To search among combinations between those two sorts of influences, in this work a GA was applied. However, to constrain the GA, the search space is divided into several clusters or groups from each of which the GA has to select variables to build forecast models. Usually one series from each cluster, while the clusters are assembled according to theoretical issues. This means for example that if economic theory demands a consideration of interest rates in a forecast equation, a cluster consisting of a variety of bonds with varying maturity and other series to express countries' interest rate, such as for example prime lending rates or interbank rates, is used from which the GA tries to find the most appropriate series. A cluster representing a country's real economic activity usually contains series of industrial production, leading indicators such as housing starts, car sales and registrations, composite indicators and other series that express economic growth. The advantage of using a GA for this optimization task is that many combinations between series from different clusters can be evaluated without loosing the structure of the theoretical models. The number of clusters, depends on the theory used, and the number of series in the cluster depends on how abstract the theory is formulated, as well as on issues such as availability of data on different frequencies. Especially on higher frequencies many theories were only applicable by using proxies. However, as mentioned, in addition to considering the series in accordance with economic theory, other sources of influence are evaluated. Usually other exchange rates and financial market series such as stock market indices have been taken into account, because they not only are said to have the capacity to proxy (or even anticipate) real economic activity but also are mapping capital movement between countries, which in turn affect exchange rates (S). This idea of constraining the GA can be expressed analytically. The general representation of the forecast equation to be optimized is of the following form:

$$S_{t+1} = \sum_{j=0}^{k_1} \sum_{i=1}^{n_1} \beta_{ij}^1 F_{i,t-j}^1 + \sum_{j=0}^{k_2} \sum_{i=1}^{n_2} \beta_{ij}^2 F_{i,t-j}^2 + \cdots + \sum_{j=0}^{k_m} \sum_{i=1}^{n_m} \beta_{ij}^m F_{i,t-j}^m + \sum_{j=0}^{k_1} \sum_{i=1}^{n_1} \alpha_{ij} X_{i,t-j} \, .$$

With F indicating series from groups according to economic theory and X indicating all other series used. Accordingly β denotes the coefficients of the theoretical variables and α that of all others. m stands for the number of theoretical clusters, n for the number of variables in one cluster and k for the number of lagged series considered. The GA is used for the model selection, whereas is constrained to select exactly one variable from every theoretical cluster, which means $\forall m \, \exists! i,j : \beta_{ij}^m \neq 0$. The number of series that can be selected from cluster X is unconstrained. As heard the number clusters depends on which theory used. This means for PPP that $F^1 = \Delta P$ and $F^2 = \Delta P^*$. Whereas P denotes time series which express changes in the domestic price level and P^* in the foreign price level. For UIP $F^1 = r_{dif}$. Whereas r_{dif} expresses time series which consider the interest rate differential between the two countries. For the Monetary Model $F^1 = M$, $F^2 = M^*$, $F^3 = Y$, $F^4 = Y^*$, $F^5 = r$ and $F^6 = r^*$. M denotes time series for series for the domestic money supply (M^* for the foreign money supply), Y are time series which express domestic real economic activity (Y^* for foreign real domestic activity) and r, respectively r^* denotes all time series which express the rate of interest in the home country, respectively foreign country. For the Dornbusch Model $F^1 = P_{ratio}$, $F^2 = r^*$ and $F^3 = r$. Series for P_{ratio} all express the ratio between domestic and foreign price level. Last, for the Frankel Model $F^1 = M$, $F^2 = M^*$, $F^3 = Y$, $F^4 = Y^*$, $F^5 = r_{long}$, $F^6 = r^*_{long}$, $F^7 = r_{real}$, $F^8 = r^*_{real}$, $F^9 = r_{short}$ and $F^{10} = r^*_{short}$. Whereas in the Frankel approach the interest rate factor is divided into three clusters to consider the real rate of interest (subscript *real*), short term interest rates (subscript *short*) as well as long term interest rates (subscript *long*) separately.

Summary of the Results

Applying the GA within the theoretical constraints not only resulted in considerably well performing forecast models, but also revealed some interesting features. The performance was evaluated using linear methods (OLS regression) and non-linear methods (ANNs). Firstly, the out-of-sample performance achieved on a monthly frequency was significantly superior compared to that on the daily and weekly frequency. The reasons for this are manifold and can be found in the richer availability of data on lower frequencies (on a daily and weekly frequency often only proxies were available), but also on the fact that, except for the UIP, the theories integrated are not focusing on a daily and weekly perspective, but on much longer perspectives. This lead to the result that in general only for the exchange rates on a monthly frequency the naive forecast could be outperformed significantly. However, regarding the applicability of the five theoretical approaches the general Monetary Model in combination with other sources of influence achieved the best performance. The forecast equation of the Monetary Model only in one case, namely of the Euro/US dollar on a monthly frequency, was outperformed by the PPP approach as theoretical core. Secondly, when looking at the results, it is worth noting that during the evaluation periods (the years 2000 and 2001) variables expressing relative economic growth between the two countries in the bilateral exchange rate relationship have been selected very often by the GA and provide much explanatory power. This result is especially interesting as it was achieved by machinery learning but also reflects the general view in economic literature as the main source of influence on exchange rate movements (during that evaluation period). Thirdly, it was discovered that aside from the inclusion of variables expressing real economic activity, data mining also resulted often in the inclusion of stock market indices. Interesting in context of the importance of variables expressing economic growth is that the presence of stock market indices can be explained by the fact that the stock market is said to exhibit predictive power for real economic activity. However, the application of the constrained GA increased the out-of-sample performance of the theoretical approaches of exchange rate determination substantially. In conclusion, the combined approach, in general, resulted in a better out-of-sample performance than both approaches applied separately. By combining the two approaches (data mining and economic theory) it was possible to construct forecast models on a monthly frequency which behave stable over time and have a high goodness of fit.

References

Chereb, D., 1998, *Does data mining improve business forecasting?*, Presentation at the 18th international symposium on forecasting, Edinburgh, Scotland

Dornbusch, R., 1976, *Expectations and exchange rate dynamic*, Journal of Political Economy 84:6, 1161-1176

Dutt, S.D., Ghosh, D., 2000, *An empirical note on the monetary exchange rate model*, Applied Economics Letters 7, 669-671

Frankel, J.A., 1979, *On the mark: A theory of floating exchange rates based on real interest differentials*, American Economic Review 69:4, 610-622

Groen, J.J.J., 2000, *The monetary exchange rate model as a long-run phenomenon*, Journal of International Economics 52, 299-319

Hand, D., Mannila, H., Smyth, P., 2001, *Principles of Data Mining*, MIT Press, Cambridge, Massachusetts

Meese, E., Rogoff, K., 1983, *Empirical exchange rate models of the seventies: do they fit out of sample?*, Journal of International Economics 14, 3-24

Rogoff, K., 2002, *Dornbusch's overshooting model after twenty-five years*, IMF Working Paper, WP/02/39

White, H., 2000, *A reality check for data snooping*, Econometrica 68, 1097-1126

The Ability of Artificial Neural Networks to Exploit Non-Linearities by Data Mining Models Compared to Statistical Methods

Lutz Beinsen and Bernd Brandl

Department of Economics, University of Graz, Universitätsstraße 15, A-8010 Graz, Austria, lutz.beinsen@uni-graz.at / Department of Government, University of Vienna, Brünnerstraße 72, A-1210 Vienna, Austria, bernd.brandl@univie.ac.at

Abstract. This paper discusses the question whether *Artificial Neural Networks* (ANNs) have the capacity to exploit (additional) non-linear information on the basis of exchange rate forecast models selected on linear criteria. This includes a comparison of exchange rate forecasts between ANNs and linear statistical methods and it is asked whether linear relationships serve as an approximation. The focused forecast models are selected by a data mining approach which combines fundamentals from economic theory, respectively building blocks from economic theory, with "fundamentals" derived solely by statistical criteria. This combination of theoretical and statistical relationships in data mining makes sure that both long run determinants on exchange rate behaviour as well as current influences are integrated in the forecast models. The results are evaluated for five exchange rates on a monthly frequency. The results favour the use of ANNs as they slightly improve the out-of-sample performance of the forecast models. What is more, linear as well as non-linear methods can be applied and the advantages from both methods can be used, which means that statistical methods allow a more detailed analysis of the results whereas ANNs offer a slightly better forecast performance.

Introduction

For almost twenty years now research on ANNs in context of exchange rates can be found in economic literature. At the same time, forecasting exchange rates has become a challenge among economists and the use of new methods, such as ANNs, has emerged as one of the most exciting areas of the discipline. However, during the last years one could also witness an interesting reorientation in research and thinking about the nature of exchange rate behaviour. Throughout the last two decades, numerous papers have been produced on comparisons between traditional linear methods and non-linear methods such as ANNs. See for example Rehkugler and Poddig (1990), Franses and van Griensven (1998) and Zhang et al (2001). However, such research has shed new light on the dynamics and notions between fundamentals and exchange rates. Nevertheless, non-linear analysis is not rarely accompanied by a variety of problems and disadvantages, especially in the process of model selection, which traditional (linear) methods and procedures do not show. "Traditional methods" and "new methods" both reveal advantages as well as disadvantages, so the question which method is *strictly* to prefer in most cases remains open.

The forecast models

The question of the future behaviour of exchange rates is interesting not only for market practitioners but also for theorists. Because of the substantial macroeconomic effects of exchange rate movements, it is not surprising that several theories, accompanied by a variety of statistical methods have been applied. It has to be stressed that, though, since the work of Meese and Rogoff (1983) economic literature, in general, sees most exchange rate forecasting attempts as relatively unsuccessful compared to the effectiveness of the naïve forecast. This is still the mainstream notion in literature, even though recent literature rather frequently claims to beat the random walk hypothesis. However, within this article a discussion on the variety of existing approaches of forecasting exchange rates is foregone. As mentioned within this paper fundamental forecast models are used which are derived by a combination of data mining with economic theory. To be more precise, two popular approaches to explain exchange rate movements, that is in their expression as forecast equation, namely the *Purchasing Power Parity* (PPP) and the so called *monetary approach* are used as a core among which a *Genetic Algorithm* (GA) is applied to extend these theories by additional explanatory variables.

PPP and the monetary approach

The monetary approach states that since the exchange rate is the relative price of two currencies, it has to reflect the willingness to hold these monies. The determinants therefore should be the supply and demand for these monies. Economic theory tells us that demand for money depends on the main economic variables such as income, prices and interest rates. Hence these variables should indicate changes in nominal exchange rates and therefore are constituting the forecast equation. The theory of PPP states that exchange rates between currencies are in equilibrium when their purchasing power, or the amount of goods and services that one unit of money can buy, is the same in each of the two countries. This means that the exchange rate between two countries should equal the ratio of the two countries' price level of a fixed basket of goods and services. When a country's domestic price level is increasing (i.e. a country is experiencing inflation), that country's exchange rate must depreciate in order to return to PPP. However, in absence of transportation and other transaction costs, competitive markets will equalise prices of identical goods and services in two countries when the prices are expressed in the same currency. So, according to the relative simple PPP approach the parsimonious forecast equation consists only of the two countries' price levels. Hence, as in real world there are transportation and transaction costs which may disturb or weaken such relationships and may provoke non-linearity. See for example Iannizzotto (2001) for evidences regarding non-linearity in PPP adjustments. However, in this sense it appears to be appealing to search among such relationships for non-linearity, which possibly may improve results.

The data mining extension

The fact that there is correlation, dependency and interaction between variables, other than those cited in economic theory and integrated in the forecast equations, is obvious and newer literature showed that such variables can have an important, and statistically significant, bearing on the determination of exchange rates in out-of-sample tests. However, the role of such alternative variables provokes heavy discussions in exchange rate forecasting literature. Nevertheless, as the sources of influence are manifold and the relationships on the Foreign Exchange market often are abstruse, complex and temporary, data mining can be very powerful to detect *current* relationships in data. To build the forecast models, in this work, a GA is applied, which has the capacity to find such (unknown) relationships and patterns between independent variables and dependent variables, but which is constrained to search for those variables consistent with the forecast equations from economic theory. Additional to the use of a GA typical data mining methods such as cluster analysis and (rolling) correlation analysis have been applied in order to reduce the GA search space on a computable size. However, a principle idea in the process of finding exchange rate forecast models was to combine economic theory with data mining. The reason for this combination is the attempt to find and consider different sources of influence on future exchange rate behaviour. This means that the variables in the forecast equations are selected by theoretical criteria and/or statistical criteria to form a broad basis of influences. As experience showed, the combined approach performs better than both others applied separately.

Input selection on basis of linear criteria

In economic literature usually linear models have been used to model relationships between economic variables. This way of doing can be justified by the following facts: Linear models represent in many cases a reasonable approximation of non-linear models, whose functional form in general is not known a priori. Due to the lack of computing techniques it is very costly to specify and estimate complex non-linear models. Nevertheless, in order to make exchange rate forecast models to beat the naïve forecast, a linear description of the dependencies between variables may not be sufficient. However, as already mentioned, by the use of cluster analysis and (rolling) correlation analysis the GA search space was minimised to a computable size. Nevertheless, as ANNs demand an exponentially higher degree of computational capacity than traditional statistical methods, the GA fitness function was set on basis of linear regression. Apart from computational reasons for the use of linear criteria instead of non-linear criteria another advantage of regression analysis in the process of model selection is that it can use statistical tests to show the importance of each variable. In addition, linear regression analysis is a very flexible method with well-known properties. Moreover, it is easy to understand and it explains its results. These are principle advantages of regression analysis, but with the disadvantage that non-linear relationships are forgone. Whereas the question is whether there is non-linearity in data or not, and if so, how can these relationships be exploited without superproportional increasing computational capacity. Therefore the applied approach of selecting forecast models on linear criteria and then applying ANNs promises to be more effective. Especially ANNs, not demanding strong assumptions about the nature of the data as traditional statistical techniques do, appear to be adequate as they may use the linear relationship as approximation to exploit additional non-linearity in data. Moreover, as recent literature stressed, there are empirical as well as theoretical evidences for their existence. See for example Ma and Kanas (2000) for a recent investigation on non-linearity in the context of exchange rates. However, beside the existence of transportation costs and transactions costs, as recently stressed in literature, the use of non-linear methods can be especially fruitful in case of the existence of phenomena such as *rational bubbles*, *stop-loss trading strategies of traders*, *interventions by central banks* and *noise traders*, as all four are sources of non-linearity in data.[1] So, a comparison of the out-of-sample forecast performance of ANNs in comparison to statistical methods should show whether non-linearity can be found and in case of finding it if the former can improve forecasts.

Are ANNs able to find additional non-linearity?

ANNs are highly sophisticated computer applications that borrow features from the physiology of the human brain which enables them to recognise patterns from large amounts of data. They are made up of simple processing elements called *neurons* named after the most basic cells in the human brain. By adding connections between these artificial neurons an ANN is created. ANNs have some advantages that suggest them to have the capacity to find *additional* non-linearity. First, using an ANN does not require making strong assumptions about the nature of the underlying time series as traditional statistical techniques do. In this sense they are more flexible. Second, and maybe more important in this context, an ANN does not require the user to choose a priori what variables are important, what variables provide non-linear information and how the functional form has to be. The network itself *should* indicate which variables comprehend non-linear information and how this non-linearity can be incorporated in the whole set of explanatory variables in a model. So, ANNs seem to be ideal to advance models if linear relationships found are only an approximation. Therefore a comparison between a linear regression analysis and an ANN analysis should show whether it is possible to exploit additional non-linear information behind existing linear relationships.

[1] For recent literature on empirical as well as theoretical evidences regarding the existence of non-linearity in connection with exchange rate modeling see also Boero and Marrocu (2002), Guerra (2001), Mahajan and Wagner (1999), Qi and Wu (2001), Sarantis (1999), and Taylor and Peel (2000).

Empirical results

The empirical results of the out-of-sample one step forecasts have been evaluated for five exchange rates on a monthly frequency.[2] The statistical measures are expressed in **Table 1** and **2.**[3]

Table 1. Econometric forecast results

	ME	MSE	CORR(F,R)	R^2	HR
Euro/US dollar	-0.0026	0.0009	-0.0934	0.0377	53%
Pound/US dollar	-0.0035	0.0004	-0.1619	0.0823	60%
Yen/US dollar	-0.0030	0.0015	-0.5938	0.0952	50%
Euro/Pound	-0.0036	0.0005	-0.2193	0.0831	53%
Euro/Yen	-0.0069	0.0015	-0.5341	0.3134	53%

Table 2. ANN forecast results

	ME	MSE	CORR(F,R)	R^2	HR
Euro/US dollar	-0.0026	0.0009	-0.0521	0.0510	73%
Pound/US dollar	-0.0023	0.0003	-0.1759	0.1986	63%
Yen/US dollar	0.0008	0.0009	-0.0663	0.2088	60%
Euro/Pound	-0.0036	0.0004	-0.0457	0.1965	70%
Euro/Yen	-0.0005	0.0015	-0.0477	0.0630	60%

To assess the difference between the non-linear and the linear approach a variety of statistical measures are compared. Although the predicted values track fairly closely along each other the hitrate of the direction of change is substantially higher with the ANN while the other measures not rarely are better in the linear approach, especially for the Euro/Yen exchange rate. However, the empirical out-of-sample one-step ahead forecast results of this work compared the ability of ANNs to exploit non-linearity of exchange rate forecast models on a monthly frequency selected on linear criteria. The direct comparison between ANNs and statistical methods reveals that ANNs are able to outperform the statistical procedures in terms of commonly applied performance measures. We conclude that ANNs offer an alternative and often superior predictive capacity over traditional statistical methods but are stressing the effectiveness and importance of statistical methods for selecting, interpreting and analysing. As the improved performance of the non-linear ANN approach showed, indeed, linear models and linear forms of analysis can be used as they complement each other. Statistical models and methods allege good approximations on which non-linear methods such as ANNs can build up even better performances.

---

[2] All data is provided by Reuters Terminal and the length of series used on a monthly frequency was 1991-02-28 – 2001-09-28. The out-of-sample evaluation periods were the last 30 periods back from 2001-09-28.

[3] ME (*Mean Error*), MSE (*Mean Square Error*), CORR(F,R) (*Correlation between Forecast and Residuals*), R^2 (*R-square*), HR (*Hitrate*)

References

Boero, G., Marrocu, E., 2002, *The performance of non-linear exchange rate models: a forecasting comparison*, Journal of Forecasting, forthcoming

Franses, P.H., van Griensven, K., 1998, *Forecasting Exchange Rates Using Neural Networks for Technical Trading Rules*, Studies in Nonlinear Dynamics and Econometrics, 2:4, 109-114

Guerra, R., 2001, *Fundamentals and Exchange Rates: How About Nonlinear Adjustment?*, Working Papers 01.04, Department of Economics, University of Geneva

Iannizzotto, M., 2001, *Exchange rate misalignment and nonlinear convergence to purchasing power parity in the European exchange rate mechanism*, Applied Financial Economics 11:5, 511-526

Mahajan, A., Wagner, A., 1999, *Nonlinear dynamics in foreign exchange rates*, Global Finance Journal 10:1, 1-23

Ma, Y., Kanas, A., 2000, *Testing for a nonlinear relationship among fundamentals and exchange rates in the ERM*, Journal of International Money and Finance 19, 135-152

Meese, E., Rogoff, K., 1983, *Empirical exchange rate models of the seventies: do they fit out of sample?*, Journal of International Economics 14, 3-24

Qi, M., Wu, Y., 2001, *Nonlinear Prediction of Exchange Rates with Monetary Fundamentals*, Working Paper

Rehkugler, H., Poddig, T., 1990, *Statistische Methoden versus Künstliche Neuronale Netzwerke zur Aktienkursprognose: Eine vergleichende Studie*, Bamberger Betriebswissenschaftliche Beiträge 73, Universität Bamberg 1990

Sarantis, N., 1999, *Modelling non-linearities in real effective exchange rates*, Journal of International Money and Finance 18, 27-45

Taylor, M.P., Peel, D.A., 2000, *Nonlinear adjustment, long-run equilibrium and exchange rate fundamentals*, Journal of International Money and Finance 19, 33-53

Zhang, G.P., Patuwo, B.E., Hu, M.Y, 2001, *A simulation study of artificial neural networks for nonlinear time-series forecasting*, Computers & Operations Research 28, 381-396

Preference Measurement with Conjoint Analysis and AHP: An Empirical Comparison

Roland Helm, Laura Manthey, Armin Scholl, Michael Steiner

Friedrich-Schiller-University Jena,
{roland.helm, l.manthey, a.scholl, m.steiner}@wiwi.uni-jena.de

Summary: Conjoint analysis (CA) and analytic hierarchy process (AHP) are common methods for measuring preferences with CA dominating marketing research and practice and AHP becoming more and more relevant as a tool of decision analysis.
Both methods mainly differ with respect to their basic conception: AHP is a compositional method whereas CA is designed in a decompositional manner. Our study aims at comparing the methods as instruments of preference measurement on a fair basis therefore being designed as similar as possible. As decision problem we use the question which university to prefer or how to design preferable universities, respectively. The results show a high degree of predictive and convergent validity of both methods. However, inspecting the results in detail reveals considerable differences which indicate that AHP performs slightly better.

1 Introduction

Measuring the subjective preference of individuals and groups of individuals is an important task in several scientific disciplines like psychology, behavioural research, and economics. From the economic point of view preference measurement is a central issue both in decision analysis and marketing.

Table 1. Preference measurement in decision analysis and marketing

	decision analysis	marketing
problem	selection of an alternative	design of products/services
objective	maximal subjective utility	maximal profit
core problem	modelling and measuring preferences	
selection of methods	scoring methods multiattribute utility theory analytic hierarchy process (AHP)	self explicated methods multidimensional scaling variants of conjoint analysis (CA)

Table 1 contrasts both fields concerning the role of preference measurement. Within *decision analysis* the problem consists of selecting a utility maximizing alternative among a set of feasible ones, whereas *marketing* aims at designing products and services which promise a maximal profit within a market or market segment. In both cases the most critical task is to model and measure the preferences of decision makers and/or customers (DM for short).

Each of both research areas has developed its own set of methods for measuring preferences based on different *attributes* or *criteria* being important for judging alternatives. Despite the obvious similarity mentioned above only a few researchers have recognized and discussed this correspondence (Wind and Saaty 1980; Tscheulin 1991; Mulye 1998). Table 1 shows some typical methods in both areas.

Considering the similarity of the problems to be solved, their importance and the almost complete separation of the research in both fields it seems to be very interesting to examine methods from both areas in a common problem context. Therefore, we conducted an empirical study comparing AHP and CA in order to find out whether both methods reach similar (converging) results or which method gives the better ones.

As decision problem we consider the task of choosing the individual best university (from the decision analytic point of view) and designing the most attractive university (from the marketing point of view), respectively. We call this problem the *university selection problem* (USP).

In Sect. 2, descriptions of AHP and CA are given. Sect. 3 presents the design of the study and a selection of results. Concluding remarks are given in Sect. 4.

2 Selection and Description of the Methods

For our analysis, we choose AHP and CA, because both are popular and successful methods in decision analysis and marketing, respectively. AHP has wide applications in different areas of decision making (Zahedi 1986; Vargas 1990). CA is the most used research method for measuring customer preferences in marketing with many real-world applications (Wittink et al. 1994; Green et al. 2001).

In order to perform a fair comparison and to find out fundamental methodical differences between AHP and CA we choose a basic version of each method, which has proven to be successful, such that their similarity is as large as possible.

2.1 Basic Version of AHP

From the marketing point of view, AHP is a *compositional* method transforming and combining elementary preference judgments to total utilities of alternatives, which may solve the problem. In the following, we give an outline of the essential steps of AHP (building a hierarchy of attributes, computing relative utilities of attribute levels and relative weights of attributes, and deriving total utilities). For more detailed descriptions see Saaty (1980) and Helm et al. (2002).

At first, the attributes being relevant for judging alternatives have to be determined and organized within a *hierarchy*. At the top level, the *main objective* of the decision problem has to be specified and decomposed into several second level attributes. Each attribute may also be subdivided generating a tree of attributes with an arbitrary number of levels. The subdivision stops as soon as *elementary attributes* are found which may directly be used for evaluating alternatives.

At the lowest level of a *complete hierarchy* all alternatives are arranged each being connected to every elementary attribute. In an *incomplete* hierarchy only the different levels of the attributes are integrated in the hierarchy and connected to the respective attribute. Fig. 1 shows an example of an incomplete hierarchy with three levels thereby concretizing our design of the USP used within the study. The attributes h=1,2,3,4,6 have M_h=3 and the attribute 5 has M_5=2 levels.

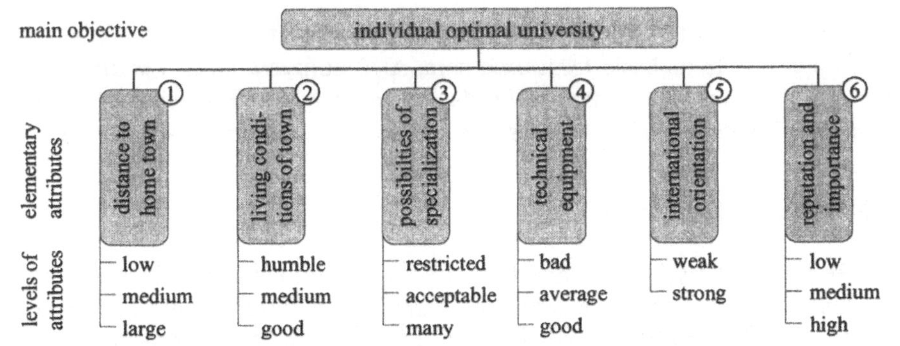

Fig. 1. Hierarchy of Attributes for University Selection Problem

In order to compute utilities and weights, the hierarchy is evaluated in a *bottom up* manner by *comparing all pairs* of elements being related to the same element at the next level.

Table 2. Saaty's scale

verbal judgment	v_{jk}	v_{kj}
j and k are equal	1	1
j is weakly better	3	1/3
j is strongly better	5	1/5
j is very strongly better	7	1/7
j is absolutely better	9	1/9

At first, the decision maker (DM) compares every *attribute level* $j=1,\ldots,M_h$ with each other for all elementary attributes $h=1,\ldots,H$. For measuring the relative preference the DM has for an attribute level j versus a level k, the 9-point ratio scale of Saaty (1980, ch. 1-4) is used. It transforms verbal judgments by the DM into *priority ratios* v_{jk} (Table 2). Since v_{jk} is defined as a ratio, the reciprocal value expresses that k is preferred over j, i.e. $v_{kj} = 1/v_{jk}$ for all pairs (j,k).

By means of these ratios we can compute *relative utility values* u_{hj} (with sum 1) of the attribute levels j with respect to an attribute h by the relation $u_{hj}/u_{hk} = v_{jk}$ for all pairs (j,k) of levels of h (note that for each h another matrix $V=(v_{jk})$ is built).

However, this computation requires the *consistency condition* $v_{jk} = v_{ji} \cdot v_{ik}$ to be fulfilled for all triplets of levels (j,i,k). To measure the usual inconsistency of human judgments, Saaty (1980, ch. 3) proposed the computation of the largest *eigenvalue* λ_{max} for the comparison matrix V. The degree of inconsistency can be measured by the *consistency ratio CR* which relates the difference $\lambda_{max}-M_h$ to the average difference of random matrices with dimension M_h. If $CR \leq 0.1$, the matrix is considered to be sufficiently consistent, otherwise the DM should rework it.

A vector of reasonable relative utilities is obtained from normalizing the *eigenvector* connected to the largest eigenvalue (Saaty 1980, ch. 7).

In an analogous manner, *relative weights* w_h of the attributes h can be computed level-by-level starting with elementary attributes. For this purpose, the scale of Table 2 has to be modified such that it verbally expresses *importance relations* transformed into importance ratios v_{hk}.

Finally, the utilities and weights computed while traversing the hierarchy can be used for computing total utilities of alternatives. Within our study, we collected preference data based on an incomplete hierarchy. Therefore we are interested in *partial utilities (part-worths)* of all attribute levels in order to be able to evaluate real universities (alternatives) with pre-specified levels of the different elementary attributes. The part-worth of level j=1,...,M_h of attribute h=1,...,H is given by $U_{hj} = w_h \cdot u_{hj}$. In its basic form used here, AHP utilizes a *linear additive utility function*. This means that the *total utility* of an alternative is computed by summing up the part-worths U_{hj} of the active levels j of the elementary attributes h.

2.2 Basic Version of CA

CA is a *decompositional* method that asks for general judgments on alternatives (*stimuli*, products), which are decomposed into part-worths for single attribute levels. It is a general concept being applicable in many variations (Green and Srinivasan 1990, Mulye 1998, Hensel-Börner 2000).

We only describe a basic version of CA which is well suited to performing a fair comparison to AHP (for details see Helm et al. 2002). To rebuild the basic operation within AHP, we choose a *full profile approach* with *paired comparisons* of given stimuli as evaluation tasks within CA (Hausruckinger and Herker 1992).

In the first step of CA, the attributes and their levels have to be defined. This step is crucial for the success of a study. Therefore, we performed a preliminary study based on several methods to determine the six most important attributes (Helm et al. 2002). The selected attributes and their different levels are also used for AHP (cf. Fig. 1).

The total number of stimuli which may be designed by combining the different levels of all attributes is $3^5 \cdot 2^1 = 486$. This is far too much to be evaluated by any DM, because there exist $(486 \cdot 485)/2$=117,855 pairs of different alternatives. In order to obtain a manageable set of stimuli and pairs, respectively, we apply standard reduction techniques based on principles defined by Addelman (1962). In this study a 3^7 Addelman basis plan is applicable which contains 18 stimuli.

The pairs of stimuli to be compared are defined by a *difference design* with 24 comparisons. The design has been constructed using a randomised procedure based on the construction principles given in Hausruckinger and Herker (1992). Within this design, the DMs have to evaluate utility differences u_{ik} between stimuli i and k on a *bipolar rating scale*. In order to get the best possible similarity with AHP, a nine-point scale is used (Table 3).

Table 3. Bipolar 9-point scale

verbal judgment	u_{ik}	u_{ki}
i and k are equal	0	0
i is weakly better	1	-1
i is strongly better	2	-2
i is very strongly better	3	-3
i is absolutely better	4	-4

In the last step, the *part-worths* of all attribute levels are estimated. As indicated by the scales of the attributes we employ an ordinary least square (OLS) *regression* which is based on an *additive linear utility function*.

Using the part-worths, the best possible stimulus (e.g. an ideal product) may be derived for any individual by setting each attribute to its level which has the largest part-worth.

3 Design and Results of the Empirical Study

We consider the *decision problem* of selecting a university for studying economics (*USP*). The study was organized as a paper-and-pencil interview with more than 300 students of economics from the university of Jena (Germany) participating. After eliminating incomplete or incorrect questionnaires, a sample of 232 DMs remained which served as basis for our analysis. In order to examine and equalize sequence effects, the methods were presented in different orders, i.e., one half of the DMs started with AHP, the other one with CA.

To evaluate the results of AHP and CA on an objective basis, a *reference method* (RM) is needed which gives realistic preference information. For this purpose, we used a *direct rating* method: While AHP and CA were based on fictitious universities only being characterized by certain attribute levels, RM asked DMs to express preferences on more realistic alternatives by distributing a total of 100 points among six *real German universities*. Because of the different levels of knowledge students have about the real properties of the universities considered, they were asked to assign attribute levels to each university in order to reveal their basis of judgment. Due to the increase in realism and the higher level of information, it can be expected that RM measures the real preferences with sufficient reliability despite of the methodological weaknesses of direct rating approaches.

The main goal of preference elicitation methods is to get a valid and reliable model of the DM's preference structure. In the following, we consider different measures for judging and comparing the methods with respect to several types of *validity* (Green and Srinivasan 1990) and discuss the results.

Table 4. Hit rates

HR [%]	CA	AHP
HR1	54.3	57.3
HR12	31.0	35.3
HR123	23.7	28.0

Predictive Validity: Comparing AHP and CA to RM may give insight into their ability to model the real preferences of a DM and to predict his/her behaviour (*predictive validity*). Taking the ranking of universities given by RM as a surrogate of the reality, the predictive validity may be judged on the basis of several *hit rates*. HR1 measures the frequency with which AHP or CA generates the same first-ranked university as RM. Both methods show high hit rates indicating a good predictive performance with AHP being slightly better. This impression is intensified considering the requirement of reproducing the first two ranks (HR12) or first three ranks (HR123) of RM. This finding is supported by further measures including Spearman's *coefficient of rank correlation* (Helm et al. 2002).

A quite different way of evaluating a method's predictive validity is to examine its ability to fulfil obvious preference statements of the DM. A very evident assumption is that a rational DM always prefers an alternative that dominates all other ones. For performing such a *dominance test*, we construct sets of realistic alternatives one dominating each other based on elementary and reliable statements asked for within AHP and CA, respectively. Via the part-worths β_{hj} and U_{hj} provided by CA and AHP the total utilities of all alternatives within the test can be computed. Whenever the dominant alternative gets a lower value than any dominated alternative, the method examined has failed the test.

- *AHP-based dominance test:* AHP directly asks a DM for comparing the 2 or 3 levels of every attribute with each other. From those statements it is possible to deduce a reliable ranking of the levels for each attribute separately. Now we construct a realistic alternative *X* which shows the second-ranked levels of each attribute except for h=5 where it gets the first-ranked level (due to M_5=2). Six dominated alternatives Y_h are derived from *X* by setting attribute h=1,...,6 to the worst level, respectively. Whenever a method prefers at least one Y_h to *X*, it is not able to reproduce the most elementary preference statements of the DM and fails the test. *PA* denotes the frequency of failing these tests.
- *CA-based dominance test:* Within CA, 24 paired comparisons of stimuli (*i*,*k*) are performed where DMs are directly asked for their relative preference. Due to this explicit statement, any pair constitutes a reliable dominance test with *i* dominating *k*, *k* dominating *i* or both judged as being equal. For both tested methods, we check whether or not the pre-defined relationship is reproduced by the computed total utilities. Let *PC* be the frequency of failing these tests.

Table 5. Dominance tests

	CA	AHP
PA [%]	73.3	0.0
PC [%]	17.5	28.5

Table 5 shows that CA prefers a dominated alternative for more than 73% of all DMs. By construction, AHP passes all such tests, because the partial utilities U_{hj} are monotonically decreasing within the ranking constructed. The bad result of CA is due to the compositional approach where complex evaluation tasks are given to the DMs. The majority of them do not succeed in judging trade-offs such that the elementary relationships on attribute levels are correctly reflected. Considering *PC* we recognize that CA performs better than AHP but is not able to fulfil all preference orderings that serve as input for the estimation of part-worths. Possible reasons are inconsistent statements of DMs, the complexity within the decompositional approach and the linear utility model used. Therefore, PC can be interpreted as a measure for internal validity in case of CA.

Internal validity: Generally speaking, the *internal validity* aims at evaluating the internal information processing within a preference elicitation method. One aspect is the *consistency* of preference statements achieved. As described in Sect. 2, a consistency ratio *CR* is computed for each comparison matrix of AHP with $CR > 0.1$ indicating inconsistent judgments. In an interactive application of AHP such a matrix is given back to the DM for modification until it fulfils the consistency condition. In our paper-and-pencil study, this was not possible. Therefore, we had to accept all preference statements during the interview and use *CR* only as a means for an ex post evaluation of the consistency achieved. In order to evaluate the degree of consistency of the entire hierarchy of a DM, we use the arithmetic mean of the consistency ratios *CR* of all comparison matrices. An alternative way not applied here is proposed by Saaty (1980, ch. 4.5).

Both measures and their thresholds are difficult to interpret. In order to give an alternative, which is directly derived from the consistency conditions of AHP, we define a *geometric consistency index GI.* For each subset of three attributes or attribute levels $\{j,i,k\}$ the consistency condition $v_{jk} = v_{ji} \cdot v_{ik}$ is considered defining a ratio $g_{jik} = \dfrac{\min\{v_{jk}, v_{ji} \cdot v_{ik}\}}{\max\{v_{jk}, v_{ji} \cdot v_{ik}\}} \leq 1$ with perfect consistency indicated by $g_{jik}=1$.

The geometric average of all g_{jik} within any comparison matrix V leads to an index $GI(V)$. To judge the overall consistency of a DM's preference statements the geometric average on all $GI(V)$ is built to compute the index GI. In order to exclude only DMs who did not cope with the evaluation task or were not willing to give useful answers we set $GI \geq 0.35$ as threshold for acceptable consistency.

In case of CA, the standard measure for internal validity is the *degree of determination* R^2, which is computed within the regression anyway. A value close to 1 shows that it was possible to find part-worths that match the input relationships well. As a threshold for acceptable consistency we set a value of 0.8.

Table 6 shows averages of the measures on the entire sample and the quotas of DMs considered as sufficiently consistent. While R^2 and GI indicate that most of the DMs were able to cope with the evaluation tasks sufficiently, CR proposes to eliminate the majority of DMs from the study. However, within a marketing research context AHP should be able to get reasonable results even for DMs that do not fulfil the hard requirements set by CR. Indeed, further investigations reveal that both methods are robust in the sense that there is no fundamental difference between results obtained for a small subsample of very consistent DMs and the entire sample. For respective results see Helm et al. (2002).

Table 6. Consistency indices

CA	R^2	arith. Ø	0.83
		$R^2 \geq 0.8$	74.1%
AHP	CR	arith. Ø	0.13
		$CR \leq 0.1$	40.1%
	GI	geom. Ø	0.41
		$GI \geq 0.35$	68.5%

Convergent validity: We directly compare the results obtained by CA and AHP when evaluating the set of six real universities: For 68.9% of all DMs the first-ranked universities are identical. Completely the same ranking of alternatives is obtained in 15.9%. For the remaining DMs a high Spearman's coefficient of rank correlation of 0.77 is obtained on average. These results strongly indicate that there is a high degree of *convergent validity* between both methods.

Since both methods use completely different approaches, we can support the previous observation that AHP and CA provide valid models of the DM's preferences and thus have good predictive capabilities.

Applicability of the methods: Additionally, it is important to examine factors that determine the applicability of the methods, because valid measurements are only possible if DMs are able and willing to apply the method in a serious manner. Hence, we consider factors in the four categories motivation, information content, cognitive burden and clearness. We find that AHP is at least slightly preferred for all indicators except for "degree of realism". Mainly, the important motivation related indicators show a clear advantage for AHP (cf. Helm et al. 2002).

A further indicator strongly favouring AHP is the time necessary for giving the preference statements. While AHP required 5.9 minutes on average, the DMs have to spend 9.2 minutes for CA. That means that AHP is able to gather more information in a given time span. However, it has to be recognized that the information content provided by CA per paired comparison is greater. Since the previous findings on the validity support the assumption that AHP is at least on a par with CA, the latter seem not to derive advantage from this greater information content.

4 Conclusion

From our study, we derive that AHP is a good alternative to CA for measuring preferences in case of a relatively complex problem with more than the usual number of attributes and levels within standard CA studies. With respect to almost every measure AHP performs slightly better. In particular, AHP seem to have a better applicability with respect to motivation and time.

Notwithstanding the similarity of total utilities computed by CA and AHP for complete alternatives, significant differences occur when considering partial utilities of single attributes. CA produces considerable contradictions with the consequence that dominated alternatives are preferred to dominating ones.

References

Addelman S (1962) Orthogonal main-effect plans for asymmetrical factorial experiments. Technometrics 4:21-46

Green PE, Srinivasan V (1990) Conjoint analysis in marketing: New developments with implications for research and practice. Journal of Marketing 54 (October):3-19

Green PE, Krieger AM, Wind YL (2001) Thirty years of conjoint analysis: Reflections and prospects. Interfaces 31/3:S56-S73

Hausruckinger G, Herker A (1992) Die Konstruktion von Schätzdesigns für Conjoint-Analysen auf der Basis von Paarvergleichen. Marketing ZfP 14/2:99-110

Helm R, Manthey L, Scholl A, Steiner M (2002) Empirical evaluation of preference elicitation techniques from marketing and decision analysis. Working Paper, FSU Jena

Hensel-Börner S (2000) Validität computergestützter hybrider Conjoint-Analysen. Gabler, Wiesbaden

Mulye R (1998) An empirical comparison of three variants of the AHP and two variants of conjoint analysis. Journal of Behavioral Decision Making 11:263-280

Saaty TL (1980) The analytical hierarchy process. McGraw-Hill, New York

Tscheulin K (1991) Ein empirischer Vergleich der Eignung von Conjoint-Analyse und "Analytic Hierarchy Process" (AHP) zur Neuproduktplanung. Zeitschrift für Betriebswirtschaft 61:1267-1280

Vargas LG (1990) An overview of the analytic hierarchy process and its applications. European Journal of Operational Research 48:2-8

Wind Y, Saaty TL (1980) Marketing applications of the analytic hierarchy process. Management Science 26:641-658

Wittink DR, Vriens M, Burhenne W (1994) Commercial use of conjoint analysis in Europe: Results and critical reflections. International Journal of Research in Marketing 11:41-52

Zahedi F (1986) The analytic hierarchy process - A survey of the method and its application. Interfaces 16:96-108

Further Development of MADM-Approaches in China and in Germany

Jutta Geldermann, Kejing Zhang, Otto Rentz

University of Karlsruhe,
{jutta.geldermann, kejing.zhang, otto.rentz}@wiwi.uni-karlsruhe.de

Abstract: This paper seeks to give an overview and compare the main stream of thought in Multi Attribute Decision Making (MADM) theory and practice in China and Germany. MADM approaches are suitable for evaluation of a set of discrete alternatives. Widely applied approaches, including Multi Attribute Utility Theory (MAUT), Analytic Hierarchy Process (AHP) and the outranking methods ELECTRE, PROMETHEE, TOPSIS, find applications in the strategic production planning, site selection or technique assessment under economic, ecological and technical attributes. In China, MAUT and AHP are predominantly applied, while in Germany outranking methods are more popular. As further development, the integration of Fuzzy theory and artificial intelligence in MADM are discussed, thus imprecise information can be considered.

1 Introduction

MADM methods are suited to the problem of evaluating a set of well-defined, discrete alternatives. With the developments of MADM, two different philosophies have been distinguished in the last few decades [9]:

- The American school assumes, that the DMs have precise ideas about the utility of performance and weights. The approaches are called in general "classical MADM approaches". MAUT and AHP belong to this group. These approaches are appropriate to incorporate a hierarchical structure of attributes.
- The European philosophy supposes, that the DMs are not aware of the preferences, so that they need decision support to structure the decision situation. The significant contribution of outranking methods is the introduction of incomparability and weak preference. ELECTRE and PROMETHEE are two important outranking methods.

Starting point of MADM is a decision matrix, which is composed of alternatives and decision relevant attributes, by which the alternatives are compared.

Along with the decision procedure, such fundamental issues of MADM as

1. Problem structuring: definition of alternatives and structuring of attributes
2. Determination of weights
3. Consideration of imprecision
4. Preference aggregation, recommendation
5. MADM in the context of group decisions
6. Sensitivity analysis

are discussed in the following sections, relating to the research work.

2 Commonly Applied MADM Methods and Applications in China

MAUT and AHP are two popular MADM techniques in China. Especially, AHP is most frequently analysed and applied. So far many suggestions for improvement have been discussed.

In the literature, problem structuring received little attention. A lot of efforts have been made to derive preferences in the normalization. Liu and Qiu [15] studied the theories on attributes. Six types of attributes, such as profit, cost, fixation, interval, deviation, and deviating interval attribute, with the generalised preference functions, have been presented.[1]

2.1 The Treatment of Imprecision in MADM

The imprecision of preference information is one fundamental problem of MADM. The imprecision can be represented in terms of interval numbers or fuzzy numbers or verbal inputs, which seem more natural and may better represent the reality. Consequently, Fuzzy theory is integrated into MADM.

Fan and co-workers [8, 33] conducted much research on MADM problems with interval numbers. A linear programming method and a ranking approach with possibilities have been proposed to solve MADM problems with uncertain intervals.

A lot of research work has been done concerning AHP, especially regarding its consistency ratio and fuzzy extension. A new Extent Analysis Method for handling fuzzy AHP has been introduced, by applying the principle of the comparison of fuzzy triangular numbers [34]. Leung and Cao [14] introduced Sinarchy to address the earlier criticisms on AHP. It is identified that in AHP, tradeoffs between attributes vary amongst individual alternatives and are dependent on the alternatives' proportion of contribution to each attribute. Sinarchy should be used for problems, where tradeoffs between attributes are in terms of their relative measurements. Sinarchy also can be used to prevent rank reversal.

In Saaty's opinion, a consistency ratio (C.R.) of no more than 0.1 is acceptable. Leung and Cao [13] defined fuzzy consistency, based on tolerance deviation.[2] Xu and Wei [31] proposed a method, by which the C.R. of the modified matrix is less than that of the original one.

Besides the above mentioned approaches, Fuzzy integrated evaluation methods are widely applied in China. The notion of ideal-point, introduced by TOPSIS, has been used in Set Pair Analysis to deal with stability analysis under uncertainty.

---

[1] Furthermore, for each type of attribute, a new method of qualitative-quantitative analysis in indices is proposed, based on different characters of the indices [24].

[2] Fuzzy ratios of relative importance, allowing certain tolerance deviation, are formulated as constraints on the membership values of the local priorities. The fuzzy local and global weights are determined via the extension principle. The alternatives are ranked on the basis of the global weights by applying a max-min set ranking method.

2.2 Incorporation of MADM in Group Decision Making

MADM has been extended into group decision making situations. A great deal of research work has been done in this area.

Lai et al. [12] applied the AHP technique to support the selection of a multimedia authorizing system in a group decision. The post-study survey suggested that AHP is more preferable than the Delphi technique. AHP is found to be more beneficial to consensus building in group decision settings. Xu [30] developed a theoretical basis for the application of the Weighted Geometric Mean Method in group decision making, by proving that the weighted geometric mean complex judgement matrix is of acceptable consistency. Wei et al. [25] proposed a min-max principle based procedure of preference adjustments to find the compromise weight in decision groups.

2.3 Weighting Methods

Attribute weights play an important role in measuring the overall preference values of the alternatives in MADM models. Many distinct procedures for deriving weights have been proposed, such as trade-off, swing, direct-ratio and Saaty's eigenvector approach. In China, there is a widely accepted weighting method, by which objective weight-set is derived through linear programming [8]. Ma et al. [17] proposed a subjective and objective integrated weighting method, by forming a two-objective optimisation model.

Wei et al. [26] analysed the structure of the weight-set with parameter β while keeping the preference orders on alternatives, where β is a differential amount of value that makes one alternative preferable to the other. It is proved that weight-set can be written in the standard form of linear programming problems, and determined by a convex combination of the extreme points according to the corresponding intervals of β.

2.4 New MADM Models

One of the newly proposed methods is Superiority and Inferiority Ranking (SIR) [28], which is proved to be an extension of PROMETHEE. SIR has more choices at the aggregation step. When using Simple Additive Weighting (SAW) as the aggregation procedure, the model SIR·SAW coincides with PROMETHEE, while TOPSIS can be used to build a new model SIR·TOPSIS. Further, the relationships between these methods have been explored. SIR appears to be not a single method, but a general MADM approach.

Xu et al. [29] proposed a MADM procedure based on distances between partial preorders. Firstly, the DM ranks alternatives with a preorder for each attribute and provides complete/partial linear information about weights. Secondly, a distance procedure is used to aggregate the above individual rankings into a global ranking.

Widely accepted, weights have to be determined in usual MADM. However, Du and Yu [4] proposed an approach, in which weight information is formulated in the form of facts and rules, without converting them into weights. Thus, ranking preference can be obtained by the reasoning of the knowledge.

The following table gives a brief overview of the applications of MADM in China.

Table 1. Applications of MADM in China

Type of problem	Method	Reference
Blank selection in DFM	Integrated Fuzzy approach	[21]
Evaluation of CIMS investment strategies	Fuzzy approach	[27]
Software selection	Group AHP	[12]
Oil-field evaluation	Fuzzy AHP and Extend Analysis Method	[34]
Evaluation of institutes	Linear Programming	[8]
Evaluation of environment, resources and sustainable development	Building of hierarchical attributes	[32]

3 Commonly Applied MADM Methods and Applications in Germany

Although classical methods, such as MAUT and AHP, are also applied, the outranking methods (ELECTRE, PROMETHEE) are much more popular in Germany than in China. MADM researchers in Germany have laid stress on:

1. Application to real world problems, especially in environmental assessment.
2. Comparative studies of different MADM methods.

3.1 Theoretical Research

In MADM, DMs are often assumed to be faced with a situation in which the well-defined alternatives already exist. However, in cases that the set of available operation possibilities are not known, DMs need to define appropriate alternatives. Dyckhoff and Ahn [5] proposed the principle of successive revelation of decision relevant alternatives, which is characterised by integration of generation and evaluation phases.

Preference relations are defined as strict preference, weak preference, indifference and incomparability. Two preference relations may contradict each other. Esser [7] developed a framework to systemize these connections between preference relations, by defining and analysing when a preference relation is more complete, consistent and compatible in respect of another one. This conceptual framework can be applied to compare the preference relations.

The wide variety of available techniques confuses the users. Distinct MADM techniques usually yield different results when applied to the same problem. A certain MADM method may appear to be appropriate for a particular case.

Pudenz et al. [20] applied MADM methods to evaluate sustainability of management strategies. MAUT, PROMETHEE, and AHP are presented and compared with other mathematical methods.[3]

[3] Whereas the mathematical approach (Hasse diagram technique) is directed to the scientifically given data matrix and therefore yields an objective and transparent evaluation, MADM approaches have a higher participation degree by DMs in the decision process.

Eickemeier [6] studied different weighting methods in MADM and compared the nine-point-scale of Saaty's AHP, Barzilai/Lootsma's multiplicative AHP, and the exchange ratio in form of a membership value by Kaprcyk/Nurmi.

3.2 MADM in Group Decision Making

Chwolka and Raith [3] extended different group aggregation procedures applied in AHP to multiple-issue decision problems,[4] and developed a utilitarian weighted arithmetic mean method for preference aggregation.

Geldermann et al. [9] studied incorporation of MADM into group decision, in which common procedures are proposed to aggregate group preference for MAUT, AHP and PROMETHEE.

3.3 Fuzzy MADM

German researchers have done much research work in the field of Fuzzy MADM. Rommelfanger and Eickemeier [23] extended fuzzy decision theory and integrated Fuzzy theory in MADM. Buckley et al. [2] proposed a method to find the fuzzy weights in fuzzy AHP. Geldermann and co-workers [11, 10] developed Fuzzy PROMETHEE, dealing with imprecise information formulated in trapezoidal fuzzy numbers. This approach is applied to environmental assessment of production techniques and determination of Best Available Techniques.

3.4 New Research Area in MADM

MADM has been used in environmental decision situations, which offer a new challenge for MADM in ethical aspects. Rauschmayer [22] argues that decision analysis has to reflect on its normative foundation. As a prerequisite for a normative argument of the DMs, the attribute has to reflect not only the interests but possibly also all values stemming from normative arguments of the DMs, which is especially important for environmental decision. The integration of values will result in changes of the decision process and will not be possible without analytical capabilities of the decision analyst in ethics.

[4] It is demonstrated how existing procedures will generally fail to generate Pareto optimal agreements when applied to multiple users. The approach provides a theoretical basis for designing the AHP to implement social choice functions in practice.

The following table shows some selected applications of MADM in Germany.

Table 2 Applications of MADM in Germany

Type of problem	Method	Reference
Software evaluation	AHP	[19]
Synergy allocation within partner in a merger	AHP[5]	[18]
Environmental assessment of production techniques in iron and steel industry, Determination of Best Available Techniques	Fuzzy PROMETHEE	[10], [11]
Evaluation of sustainability of water management strategies	MAUT, AHP, PROMETHEE	[20]
Two-issue marketing negotiation	Group AHP	[3]

4 Sensitivity Analysis

Sensitivity analysis plays an essential role in MADM, it can provide the DMs insight in the decision. Till now, much research is dedicated to this subject.

Wei et al. [26] proposed a method to determine the weight set, while keeping the preference orders on alternatives. The results are important for sensitivity analysis. Liu and Qiu [16] presented a sensitivity analysis method for TOPSIS and Double Base Points Ordering Method.

Geldermann and co-workers [11, 9] developed a graphical sensitivity analysis instrument, delivering insensitivity intervals, and a systematic sensitivity analysis method, by which a sensitivity degree of each attribute can be derived.

5 Conclusions

The literature on theory and applications of MADM in both countries has been reviewed. This study reflects the recent contributions made by MADM researchers in both countries. Fundamental issues such as MADM models, weighting methods, attribute generation, preference elicitation and integration of fuzzy theory are presented. Each country has its own emphases on the MADM research.

Taking into account of the recent developments of problem structuring, future interdisciplinary research should be dedicated to cultural attitudes towards the interpersonal process of decision making– especially in the light of globalisation. [1]

[5] Ossadnik and Lange [19] applied AHP to evaluate AHP Software products -AutoMan, ECPro and HIPRE 3+. The results delivered from the Software were employed to generate pairwise comparisons in AHP model. The relevant attributes are derived from ISO/IEC 9126.

6 References

[1] Belton V, Stewart TJ (2001) Multiple criteria decision analysis: An integrated approach. Kluwer Academic Publishers, London

[2] Buckley JJ, Feuring T, Hayashi Y (2001) Fuzzy hierarchical analysis revisited. European Journal of Operational Research 129:48-64

[3] Chwolka A, Raith MG (2001) Group preference aggregation with the AHP - Implications for multiple-issue agendas. European Journal of Operational Research 132:176-186

[4] Du XM, Yu YL (1999) Intelligent multi-attribute decision making. Journal of Armament 20:90-93

[5] Dyckhoff H, Ahn H (1998) Integrierte Alternativen-Generierung und -Bewertung. Die Betriebswirtschaft 58:49-63

[6] Eickemeier S (2001) Bestimmung der Gewichte bei Mehrzielentscheidungen. Eine vergleichende Analyse ausgewählter Verfahren. In: Chamoni P, Leisten R, Martin A, Minnemann J, Stadtler H (Hrsg) Operations Research Proceedings, Springer-Verlag, Berlin, 389-396

[7] Esser J (2001) Vollständigkeit, Konsistenz und Kompatibilität von Präferenzrelationen. OR Spectrum 23:183-201

[8] Fan ZP, Zhang Q (1998) A linear programming method for uncertain multiple attribute decision making. Journal of North-Eastern University 19:419-421

[9] Geldermann J, Zhang KJ, Rentz O (2002) Entwicklung eines integrierten multikriteriellen Gruppenentscheidungsunterstützungssystems (MGDSS). In: Fichtner W, Geldermann J (Hrsg) Einsatz von OR-Verfahren zur Techno-ökonomischen Analyse von Produktionssystemen, Verlag Peter Lang, Frankfurt, 169-186

[10] Geldermann J, Spengler T, Rentz O (2000) Fuzzy outranking for environmental assessment case study: Iron and steel making industry. Fuzzy Sets and Systems 115:45-65

[11] Geldermann J, Rentz O (2001) Integrierte Technikbewertung bei unvollständigen Informationen als Unterstützung für die Bestimmung von Besten Verfügbaren Techniken (BVT). OR Spectrum 23:137-157

[12] Lai VS, Wong BK, Cheung W (2002) Group decision making in a multiple attribute environment: A case using the AHP in software selection. European Journal of Operational Research 137:134-144

[13] Leung LC, Cao D (2000) On consistency and ranking of alternatives in fuzzy AHP. European Journal of Operational Research 142:102-133

[14] Leung LC, Cao D (2001) On the efficacy of modelling multi-attribute decision problems using AHP and Sinarchy. European Journal of Operational Research 132:39-49

[15] Liu SL, Qiu YH (1998) Studies on the basic theories for MCDM. Theory and Practice of System Engineering 1:38-43

[16] Liu SL, Qiu YH (1998) Generalization for the double base points ordering method for MADM. Theory and Practice of System Engineering 2:23-25

[17] Ma J, Fan ZP, Huang LH (1998) A subjective and objective integrated approach to determine attribute weights. European Journal of Operational Research 112:397-404

[18] Ossadnik W (1996) AHP-based synergy allocation to the partners in a merger. European Journal of Operational Research 88:42-49

[19] Ossadnik W, Lange O (1999) AHP-based evaluation of AHP-software. European Journal Of Operational Research 118:578-588

[20] Pudenz S, Brüggemann R, Voigt K, Welzl G (2002) Nachhaltige Entwicklung von Managementstrategien - Multikriterielle Bewertungs- und Entscheidungshilfe-Instrumente. Umweltwissenschaften und Schadstoff-Forschung 14(1) 52-57

[21] Qiao LH, Ma T, Wang SC (1999) An approach to multiattribute evaluation in manufacturability analysis for blank selection. Chinese Journal of Mechanical Engineering 35(4):42-46

[22] Rauschmayer F (2001) Reflections on ethics and MCA in environmental decisions. Journal of Multi-Criteria Decision Analysis 10(2):65-74

[23] Rommelfanger HJ, Eickemeier SH (2002) Entscheidungstheorie – Klassische Konzepte und Fuzzy Erweiterungen. Springer-Verlag, Berlin

[24] Wang P, Wang LL, Tian H, Luo BY (1999) Study of qualitative quantitative normalization of targets in multi-objective evaluation. Journal of Wuhan Automotive Polytechnic University 21(6): 37-40

[25] Wei QL, Han H, Ma J, Fan ZP (2000) A compromise weight for multi-attribute group decision making with individual preference. Journal of the Operational Research Society 51:625-634

[26] Wei QL, Ma J, Fan ZP (2000) A parameter analysis method for the weight-set to satisfy preference orders of alternatives in additive multi-attribute value models. Journal of Multi-Criteria Decision Analysis 9(5):181-190

[27] Wu YH, Fu YJ, Zhou JR (1998) Research on Fuzzy multiobjective method for the evaluation of CIMS. Theory and Practice of System Engineering 9:55-60

[28] Xu XZ (2001) The SIR method: A superiority and inferiority ranking method for multiple attribute decision making. European Journal of Operational Research 131:587-602

[29] Xu XZ, Martel JM, Lamond BF (2001) A multiple criteria ranking procedure based on distance between partial preorders. European Journal of Operational Research 133:69-80

[30] Xu Z (2000) On consistency of the weighted geometric mean complex judgement matrix in AHP. European Journal of Operational Research 126:83-687

[31] Xu ZS, Wei CP (1999) A consistency improving method in the analytic hierarchy process. European Journal of Operational Research 116:443-449

[32] Yue CY, Li H (1998) Multi-attribute evaluation for environment, resource and sustainable development. Monitoring and Assessment 12:28-30

[33] Zhang F, Fan ZP, Pan DH (1999) A ranking approach with possibilities for multi attribute decision making problems with intervals. Control and Decision 14(6):703-706

[34] Zhu KJ, Jing Y, Chang DY(1999) A discussion on Extent Analysis Method and applications of fuzzy AHP. European Journal of Operational Research 116:450-456

Wissensrevision in einer MaxEnt/ MinREnt-Umgebung

Elmar Reucher und Wilhelm Rödder

FernUniversität in Hagen,
Lehrstuhl für Betriebswirtschaftslehre, insb. Operations Research
elmar.reucher@fernuni-hagen.de, wilhelm.roedder@fernuni-hagen.de

Zusammenfassung Die Abbildung menschlichen Wissens und seine maschinelle Verarbeitung sind ein wichtiger Forschungsgegenstand auf dem Gebiet der Künstlichen Intelligenz (KI). Im allgemeinen unterliegt das Wissen zeitlichen Veränderungen, so dass zu einem späteren Zeitpunkt nach Erfahren neuer Sachverhalte bisher bekannte nicht mehr gültig sind und das "alte" Wissen revidiert werden muss. Zur Formulierung von Sachverhalten in einer Domäne werden in diesem Aufsatz probabilistische Konditionale verwendet, mit denen eine Sprache zur Kommunikation zwischen Mensch und Maschine zur Verfügung steht, die eng der menschlichen Denkweise angelegt ist. Das auf diese Weise formulierte Wissen wird dabei informationstreu durch das Entropieprinzip verarbeitet. In dem vorliegenden Aufsatz wird gezeigt, wie der Wissensrevisionsprozess in einer entropieoptimalen Umgebung durchgeführt wird. Anhand eines ökonomischen Beispiels wird der Wissensrevisionsprozess näher illustriert; sämtliche hierfür erforderlichen Berechnungen werden in der Expertensystem-Shell SPIRIT vollzogen.

1 Einführung

Die Grundidee des *Wissenserwerbs* in dem vorliegenden Beitrag liegt im Aufbau einer Verteilung maximaler Entropie (MaxEnt-Verteilung) auf dem Wertebereich einer endlichen Menge endlichwertiger Variabler, bei Respektierung vorgegebener probabilistischer Konditionale. Nun ist Wissenserwerb keine einmalige Angelegenheit, sondern unterliegt im Laufe der Zeit durch das Bekanntwerden immer neueren Wissens einer dynamischen Veränderung. Gelegentlich ist neueres Wissen mit älterem kompatibel, gelegentlich nicht. In der Literatur findet man dazu verschiedene Adaptionsformen, die unter dem Begriff *Wissensrevision* zusammengefasst sind [3]. In dem vorliegenden Beitrag liegt der Fokus auf der Abbildung einer speziellen Form der Wissensrevision: Neu erworbene Sachverhalte sollen unter dem aktuellen Wissensstand gültig sein und die alten Sachverhalte dabei "so wenig wie möglich" revidiert werden.

2 Grundlagen

2.1 Probabilistische Konditionale

Auf dem Ereignisfeld über einer Variablenmenge $\mathcal{V} = \{V_1, ..., V_n\}$, wobei jede Variable V_i eine endliche Anzahl diskreter Ausprägungen v_i besitzt, sei eine

Verteilung P definiert. p bezeichne dabei die Wahrscheinlichkeitsfunktion mit $P(\mathcal{V} = \mathbf{v}) = p(\mathbf{v})$ mit Vollkonjunktion $\mathbf{v} = v_1, ..., v_n$. Fasst man die Variablenrealisationen v_i als Literale auf, so lässt sich eine Sprache $\mathcal{L}$ bestehend aus syntaktischen Ausdrücken formulieren, indem die Literale durch Disjunktion $\vee$, Negation $\neg$ und Konjunktion $\wedge$ miteinander verknüpft werden. Solche aussagelogischen Formeln werden als Fakten bezeichnet und die Elemente aus $\mathcal{L}$ werden im folgenden durch Großbuchstaben $A, B, C, ...$ symbolisiert. Probabilistische Sachverhalte sind nun Ausdrücke der Form $B|A\,[x]$, wobei das Konditional $B|A$ entsprechend der dreiwertigen Logik von CALABRESE wohldefiniert ist. Der Ausdruck *wenn A dann B*, $B|A(v)$ ist wahr, falls das Faktum BA unter v wahr ist, falsch für $\overline{B}A$ unter v wahr und nicht definiert für $\overline{B}$ unter v [1]. x steht für die Wahrscheinlichkeit, dass in einer Domäne das Konditional gilt.

2.2 Das Entropieprinzip

Wissen in Form einer Menge von Sachverhalten $\mathcal{R} = \{B_i|A_i\;[x_i],\; i = 1, ..., I\}$ wird in diesem Beitrag entropieoptimal durch Lösen der Aufgabe

$$P^* = \arg\max H(Q) = -\sum_{\mathbf{v}} q(\mathbf{v}) \cdot \log q(\mathbf{v}) \qquad (1)$$

$$\text{u. d. N.:} \qquad Q \,\text{erfüllt}\, \mathcal{R} \quad (Q \models \mathcal{R})$$

verarbeitet [6]. Das Prinzip *Maximaler Entropie* haben verschiedene Autoren unabhängig voneinander axiomatisch fundiert; man vergleiche hierzu [2], [8]. Die auf diese Weise erzeugte Verteilung P^* ist gerade jene unter allen Verteilungen Q, in der alle in $\mathcal{R}$ explizit formulierten – und keine weiteren – Abhängigkeiten repräsentiert sind. Die in (1) formulierte Optimierungsaufgabe garantiert somit eine informationstreue Wissensverarbeitung. Liegen nun situativ gültige Sachverhalte $\mathcal{R}_E$ vor, so wird das bekannte Wissen aus P^* durch Lösen der Aufgabe

$$P^{**} = \arg\min R(Q, P^*) = \sum_{\mathbf{v}} q(\mathbf{v}) \cdot \log(\frac{q(\mathbf{v})}{p^*(\mathbf{v})}) \qquad (2)$$

$$\text{u. d. N.:} \qquad Q \models \mathcal{R}_E$$

informationstreu an das neue Wissen aus $\mathcal{R}_E$ adaptiert. Man vergleiche hierzu auch [6].

3 Wissensrevision

3.1 Formen der Wissensrevision

Gegeben sei eine Wahrscheinlichkeitsverteilung P auf dem Grundraum $\mathbf{V}$, in der bekannte Sachverhalte $\mathcal{R}$ gültig sind. Werden neue Sachverhalte $\mathcal{R}'$

bekannt, so erfährt das Wissen aus P i. a. eine Veränderung. In der Literatur unterscheidet man hier zwischen verschiedenen Formen der Wissensrevision [3].

- *Wissensfokussierung* (Focusing)
  Hier erfolgt die Adaption des schon bekannten Wissens P an neue situative Sachverhalte $\mathcal{R}_E$. Diese Form der Wissensrevision ist direkt in einer entropieoptimalen Umgebung durch Lösen der Aufgabe (2) möglich.

- *Wissenserweiterung* (Expansion)
  Neue Sachverhalte $\mathcal{R}'$ ergänzen schon bekannte in $\mathcal{R}$; alle Sachverhalte in $\mathcal{R} := \mathcal{R} \cup \mathcal{R}'$ sind unter dem neuen Wissen gültig. Die Wissenserweiterung erfolgt entropieoptimal durch Lösen der Aufgabe (1). Man vergleiche hierzu [4], [5] (Seite 100ff.).

- *Wissensabgleich* (Updating)
  Angenommen, zwei Experten besitzen unterschiedliches Wissen über die gleiche Domäne, so dass die formulierten Sachverhalte $\mathcal{R}$ des einen Experten und die Sachverhalte $\mathcal{R}'$ des anderen nicht konsistent zueinander sind. Dann existiert keine Verteilung P, die $\mathcal{R} \cup \mathcal{R}'$ erfüllt. Eine Möglichkeit zur Repräsentation beider Wissensteile besteht nun darin, das Expertenwissen so aneinander anzupassen, dass leicht modifizierte Sachverhalte $\tilde{\mathcal{R}}$ und $\tilde{\mathcal{R}}'$, also auch $\tilde{\mathcal{R}} \cup \tilde{\mathcal{R}}'$ gelten. In ihrer ursprünglichen Form sind sie dann i.a. nicht mehr erfüllt. Rödder und Xu geben in [7] verschiedene Methoden an, wie diese Form der Wissensrevision in einer entropieoptimalen Umgebung durchgeführt werden kann.

- *Wissensrevision im engeren Sinne* (Revision)
  Die alten Sachverhalte $\mathcal{R}$ sind mit neuen Sachverhalten $\mathcal{R}'$ ebenfalls nicht kompatibel. Aber anders als beim Wissensabgleich wird bei dieser Form der Wissensrevision gefordert, dass die neuen Sachverhalte $\mathcal{R}'$ unter dem neuen Wissen gültig sind. Dies entspricht dem beim menschlichen Lernprozess häufigen Fall, dass man neue Sachverhalte erfährt und frühere Überzeugungen teilweise aufgeben muss; vgl. hierzu das in Abschnitt 4 vorgestellte Beispiel.

Wie die Wissensrevision im engeren Sinne in einer entropieoptimalen Umgebung durchgeführt werden kann, wird im folgenden Abschnitt beschrieben.

3.2 Wissensrevision in einer MaxEnt-Umgebung

Wir betrachten einen Wissensrevisionsprozess, der sich zu aufeinander folgenden Zeitpunkten $t = 1, ..., T$ vollzieht, wobei zu jedem Zeitpunkt t neue Sachverhalte $\mathcal{R}_t = \{B_i^t | A_i^t [x_i^t], i = 1, ..., I_t\}$ bekannt werden. Ziel ist es, die einzelnen Stufen der Wissensrevision in einer entropieoptimalen Umgebung

so abzubilden, dass zum aktuellen Zeitpunkt t die Sachverhalte $\mathcal{R}_t$ gültig sind – in einem noch zu spezifizierenden Sinn –, wobei die zu früheren Zeitpunkten gültigen Sachverhalte $\mathcal{R}_{t-1}, ..., \mathcal{R}_1$ so wenig wie möglich revidiert werden. Es bezeichnen W und W_t, $t = 1, ..., T-1$ binäre (Welt-)variable und $\mathcal{R}_t | W_t := \{B_i^t | A_i^t W_t \, [x_i^t] \, i = 1, ..., I_t\}$ sind gleichsam unter solchen Weltvariablen konditionierte Sachverhalte. Dann lässt sich ein T−stufiger Wissensrevisionsprozess wie folgt durchführen:
In $t = 1$ werden die Sachverhalte $\mathcal{R}_1$ durch Lösen der Aufgabe eingelernt:

$$\begin{aligned} P^1 &= \arg\max H(Q) \\ \text{u. d. N. } &: Q \models \mathcal{R}_1. \end{aligned} \tag{3}$$

Die Lösung (3) entspricht genau der in (1), wird jedoch hier der Vollständigkeit halber wiederholt.
In $t = 2$ wird die Aufgabe

$$\begin{aligned} P^2 &= \arg\max H(Q) \\ \text{u. d. N. } &: Q \models \mathcal{R}_2 \cup \mathcal{R}_1 | W_1 \cup \{W_1 | W \, [x_1^{max}]\} \cup \{W \, [1.0]\}. \end{aligned} \tag{4}$$

gelöst. Mit der Weltvariablen W wird über $\mathcal{R}_1 | W_1$ mittels des Konditionals $W_1 | W \, [x_1^{max}]$ erreicht, dass die Sachverhalte aus $\mathcal{R}_1$ revidiert werden und die Sachverhalte $\mathcal{R}_2$ unter dem Wissen P^2 gültig sind. Wählt man dabei das Maximum x_1^{max} für die Wahrscheinlichkeit, mit der die Welt W_1 wahr sein kann, so wird die Philosophie einer gleichmäßig über alle Sachverhalte aus $\mathcal{R}_1$ geringstmöglichen Wissensrevision verfolgt. In dem Beispiel in Abschnitt 4 werden die Zusammenhänge noch näher verdeutlicht.

Einschub: Der Wahrscheinlichkeit x_1^{max} kommt dabei folgende Bedeutung zu.

- $x_1^{max} = 0$. Die Sachverhalte $\mathcal{R}_1$ werden vollständig aufgegeben.
- $0 < x_1^{max} < 1$. Es existiert ein echtes Maximum für die Wahrscheinlichkeit, mit der die Welt W_1 unter W und somit die Sachverhalte aus $\mathcal{R}_1$ wahr sind. Dabei gilt, je größer x_1^{max}, desto weniger werden die Sachverhalte aus $\mathcal{R}_1$ revidiert.
- $x_1^{max} = 1$. Einige Sachverhalte aus $\mathcal{R}_1$ sind unter dem neuen Wissen P^2 substantiell nicht mehr enthalten; es liegt dann der Fall bedingter oder schwacher Inkonsistenz vor. Dieser Fall wird ausführlich in [7] behandelt.

Berechne nun die veränderten Wahrscheinlichkeiten, mit denen die Sachverhalte $\mathcal{R}_1$ unter dem Wissen P^2 gelten durch $P^2(B_i^1 | A_i^1)$ für $i = 1, ..., I_1$. Dann bezeichnet $\mathcal{R}_1 := \{B_i^1 | A_i^1 \, [P^2(B_i^1 | A_i^1)] \, i = 1, ..., I_1\}$ die Sachverhalte in $\mathcal{R}_1$, jedoch mit gemäß (4) modifizierten Wahrscheinlichkeiten.
In $t = 3$ löst man schließlich

$$\begin{aligned} P^3 &= \arg\max H(Q) \\ \text{u. d. N. } &: Q \models \mathcal{R}_3 \cup (\mathcal{R}_2 \cup \mathcal{R}_1) | W_2 \cup \{W_2 | W \, [x_2^{max}]\} \cup \{W \, [1.0]\}. \end{aligned} \tag{5}$$

Die Revision erfolgt nun iterativ analog zu (5) bis zum Zeithorizont T.

4 Ein Beispiel

In einer (fiktiven) statistischen Erhebung für die Automobilbranche wurde untersucht, inwieweit Fahrzeugmerkmale wie *Farbe*, *Comfort*, *Extras* und der *Preis* den Kauf für ein *Fahrzeug* beeinflussen. Die Ergebnisse dieser Unter-

Index	P act	
0	0,90000	Farbe | Fahrzeug
1	1,00000	Verbrauch | Fahrzeug
2	0,50000	Comfort | Fahrzeug
3	0,10000	Extras | Fahrzeug
4	1,00000	Preis | Fahrzeug

Abbildung1. Sachverhalte $\mathcal{R}_1$

suchung zeigt Abbildung 1. So hängt die Kaufentscheidung für alle Fahrzeuge (100%) vom *Preis* und dem *Verbrauch* ab, bei fast allen Fahrzeugen (90%) ist die *Farbe*, bei der Hälfte der Fahrzeuge der *Comfort*, und bei 10% aller Fahrzeuge sind die Extras für die Kaufentscheidung relevant. Da die

Index	P act		Index	P act	
5	0,99000	Farbe | Luxusklasse	21	0,05000	Preis | Luxusklasse
6	0,01000	Farbe | Nutzfahrzeuge	22	0,98000	Preis | Nutzfahrzeuge
7	1,00000	Farbe | Sportwagen	23	0,05000	Preis | Sportwagen
8	0,70000	Farbe | Kleinwagen	24	1,00000	Preis | Kleinwagen
9	0,05000	Verbrauch | Luxusklasse	25	1,00000	Fahrzeug | Luxusklasse
10	0,70000	Verbrauch | Nutzfahrzeuge	26	1,00000	Fahrzeug | Nutzfahrzeuge
11	0,01000	Verbrauch | Sportwagen	27	1,00000	Fahrzeug | Sportwagen
12	1,00000	Verbrauch | Kleinwagen	28	1,00000	Fahrzeug | Kleinwagen
13	0,99000	Comfort | Luxusklasse	29	1,00000	Fahrzeug | LKW
14	0,10000	Comfort | Nutzfahrzeuge	30	1,00000	Luxusklasse | Ferrari
15	0,05000	Comfort | Sportwagen	31	1,00000	Farbe | Ferrari
16	0,10000	Comfort | Kleinwagen	32	0,00000	Farbe | LKW
17	0,99000	Extras | Luxusklasse	33	1,00000	Verbrauch | LKW
18	0,50000	Extras | Nutzfahrzeuge	34	0,90000	Comfort | LKW
19	0,80000	Extras | Sportwagen	35	0,05000	Extras | LKW
20	0,01000	Extras | Kleinwagen	36	0,90000	Preis | LKW

Abbildung2. Sachverhalte $\mathcal{R}_2$

hieraus gewonnenen Informationen noch sehr allgemein sind, erhofft man durch nähere Spezifizierung der Fahrzeugklassen weitere Erkenntnisse über das Käuferverhalten zu gewinnen und gibt dazu eine zweite Erhebung in Auftrag. Die Ergebnisse sind in Abbildung 2 zur Regelmenge $\mathcal{R}_2$ zusammengefasst. Der Wissensrevisionsprozess erfolgt nun durch Lösung der Aufgabe (4) mit $x_1^{max} = 0.99999$, was zugleich der maximal zulässige Wert ist. Unter dem neuen Wissensstand P^2 gelten jetzt die Sachverhalte $\mathcal{R}_2$, wohingegen die in $\mathcal{R}_1$ revidiert werden müssen, man vergleiche Tabelle 1. Schien der Preis nach der ersten Erhebung noch bei *allen* Fahrzeugen für die Kaufentscheidung relevant, so ist diese Aussage jetzt nur noch für *fast alle* Fahrzeuge richtig (98%),

Tabelle1. Sachverhalte und revidierte Sachverhalte

		P^1	P^2
1	*Farbe|Fahrzeug*	0.90	0.87
2	*Verbrauch|Fahrzeug*	1.00	0.98
3	*Comfort|Fahrzeug*	0.50	0.49
4	*Extras|Fahrzeug*	0.10	0.12
5	*Preis|Fahrzeug*	1.00	0.98

womit die frühere sichere Überzeugung aufgegeben wurde. Entsprechend lesen sich die übrigen Ergebnisse in Tabelle 1. Nach Revision nehmen übrigens nicht stets die Wahrscheinlichkeiten für die Gültigkeit der Sachverhalte ab, man vergleiche hierzu Regel 4. Sämtliche Berechnungen wurden dabei mit der Expertensystem-Shell SPIRIT durchgeführt [9].

5 Fazit

In diesem Artikel wurde gezeigt, wie Wissen bei Erfahren neuer Sachverhalte, die im Widerspruch zu schon bekannten stehen, revidiert werden kann. Das Ergebnis ist eine die aktuellen Sachverhalte repräsentierende Wahrscheinlichkeitsverteilung, wobei das früher gültige Wissen so wenig wie möglich revidiert wird. Insbesondere lassen sich so auch sichere Überzeugungen aufgeben, was die Qualität des hier vorgestellten Wissensrevisionsprozesses in einer entropieoptimalen Umgebung eindrucksvoll unterstreicht.

Literatur

1. Calabrese, P. M. (1991) Deduction and Inference Using Conditional Logic and Probability, Conditional Logic in Expert Systems.I. R. Goodman, M. M. Gupta, H. T. Nguyen, G. S. Rogers (editors), Elsevier Science Publishers B. V.
2. Kern-Isberner, G. (1998) Characterizing the principle of minimum cross-entropy within a conditional-logical framework. Artificial Intelligence, Vol. 98, 169-208.
3. Kern-Isberner, G. (2001) Conditionals in Nonmonotonic Reasoning and Belief Revision. Springer Berlin Heidelberg
4. Reucher, E., Rödder, W. (2001) Relevanz von Information in konditionalen Entscheidungsmodellen. OR Proceedings, Springer Berlin Heidelberg 379-386
5. Reucher, E. (2002) Modellbildung bei Unsicherheit und Ungewissheit in konditionalen Strukturen. Dissertation am Fachbereich Wirtschaftswissenschaft der FernUniversität in Hagen
6. Rödder, W. (2000) Conditional Logic and the Principle of Entropie. Artificial Intelligence, 117, 83-106
7. Rödder, W., Xu, L. (2000) Behebung von Inkonsistenzen in der probabilistischen Expertensystem-Shell SPIRIT OR Proceedings, Springer Berlin Heidelberg 260-265
8. Shore, J.-E., Johnson R.-W. (1980) Axiomatic Derivation of the Principle of Maximum Entropy and the Principle of Minimum Cross-Entropy. IEEE Transact Inf Theory IT-26(1): 26-37
9. SPIRIT, http://www.fernuni-hagen.de/BWLOR/spirithome.html

Innovation, Operations Research & Decision Support in the Military

Heiner Micko

National Defense Academy
A - 1070 Vienna

Operations Research as a scientific discipline initially started some decades ago within the Military, focussing that time on logistics and questions about effective distribution of resources.

Nowadays Operations Research has won a strong role within nearly all fields of innovative economy, and has also gained a certain acceptance at different departments of the governments.

Military Applications of Operations Research recently include as well all topics of governmental Decision Support in peacetime as some more operational tasks of Decision Support in wartime and War gaming Applications, the latter often being called – and mixed up with – Cyber war.

Main purpose of Operations Research in the Military can be seen as a knowledge base for decisions, innovation and change management, promoting and helping decisions about organizational change.

This paper tries to explain the theoretical role of decision support within an innovation context, and to give some possible practical options for operations research in the Military.

1. The Science of National Economy and Early Operations Research

During 19th and early 20th century many if not all Armed Forces tried to get scientific methods into the military planning process[1], thus defining tasks and interrelationship of politics, strategy, and what was called the operational art.

The emergence of capitalistic and Marxist theories about global and national economies as well as a lot of questions arising from the world economic crisis in 1920/1930 gave strong input to general explanation efforts and approaches[2].

Logistic calculus became commonly used in the daily life of business management[3], in order to make optimized use of limited resources. In cases of more complex decisions mathematicians and engineers were called upon together. Operational research was done explicitly for the first time in the military[4].

Further on the concept of total war and the concentration of all national resources to the main goal of winning second world war brought closely together all questions about strategic disposition from the military and national economy.

---

[1] see General Theories "On War" by CLAUSEWITZ 1832 and his perception till nowadays

[2] see Economic Theories by TAYLOR 1911 and KEYNES 1936

[3] see Fordism, Industry and Technical Intelligence by OTTILIENFELD 1926

[4] see Equations of Combat by LANCHESTER 1916, and see Comparative Studies of World War Casualties by GILCHRIST 1928

Not only the that time global players Germany and America but also the European defense analysis community as a whole were engaged with mathematical models in economic and military logistics on strategic and operational levels.

After second world war military applications of mathematical models were extended extensively to all fields of defense, including primarily all Cold War intelligence results and "what if scenarios". Such models sought to compare existing weaponry on both sides of the iron curtain, and led to the development of strategic arms control treaties and concluding verification efforts.

Due to scientific progress, esp. in the computer science, the Military during the 1960ies went into euphorical expectations concerning artificial intelligence, war gaming and Operations Reseach. Such overestimation came down to a more realistic attitude soon, but led to frequent use of expert systems. Especially the realization of cybernetic measuring and controls into early autonomous agents or robots showed the complexity of the man-nature-system. On the one hand the rise of Systems Thinking[5], on the other hand fuzzy logic and fractals to describe chaotic conditions gave mathematical attempts at hand to deal with the recognition of limited predictability as well as limited resources[6] worldwide.

2. Knowledge Based Society, Innovation and Change Management

Economic Cycles[7] caused "basic innovations" in a 30 to 60 years sequence, such as steam machine and railroads, electricity, chemical synthesis of organic materials like aniline and rubber, combustion engine and motor car, transistor and computer, nuclear fission and atomic bomb or lately genetic engineering.

The development of „incremental innovations“out of the basic innovations can be recognized at early stages when published or patents pending (invention), or later on when they are marketable (innovation). The novelty of any invention is stated legally at the patent application procedure, but de facto economy is following different strategies to dominate the market – therefore competition on a specific field of technology often causes the secrecy of significant details without a patent even for a long time after successful introduction to the market.

There is a difference between objective and subjective innovation. Who innovates objectively is using a product or process uniquely for the first time world wide. To innovate subjectively is the case, when individuals, enterprises or regional entities such nation states pick up an idea as a possibility shown earlier by an objective innovation: The subjects in that sense are using objective, already existing and somewhere realized technical knowledge and abilities.

Research about innovations looks closely at the diffusion process of each innovation, also known as transfer of technology. The diffusion of an innovation over space and time follows usually the typical, well known growth-patterns.
Some preconditions are needed to implement an innovation on a specific market:

---

[5] see General Systems Theory by BERTALANFFY 1968, and see The Transfer of Systems Thinking from the Pentagton to the Great Society by JARDINI, in: HUGHES 2000

[6] see Limits to Growth by MEADOWS, MEADOWS et al. 1972

[7] see Theory of Economic Cycles by KONDRATIEFF 1926, SCHUMPETER 1935

Table 1. Innovation Preconditions (MICKO 2002)

Precondition:	functions concerning the innovation process:
feasibility	technology push
affordability	market pull, economic demand
entrepreneurship	individual initiative of an innovator
resources	- physical + political infrastructure
	- material + capital
	- personnel + human capital
capability	know-how transfer

Whenever the preconditions are given the innovation diffusion will be as follows:

Table 2. Innovation Diffusion (see BRASCHE 1989)[8]

User segments, varying along time in the innovation diffusion process	
a -	first user, innovator
b -	early users
c -	early majority
d -	„take off" of this innovation at this market
e -	late majority
f -	stragglers
g -	market saturation
h -	non-user segment, rejecters of an innovation

The non-user segment is important in cases of rejected or in a specific market forbidden technologies, e.g. nuclear fission, genetic engineered food, or some international contracted obligations to non-proliferation of defense technologies.

Fig. 1. Innovation between Research, Education, Politics and Economy (MICKO 2002)

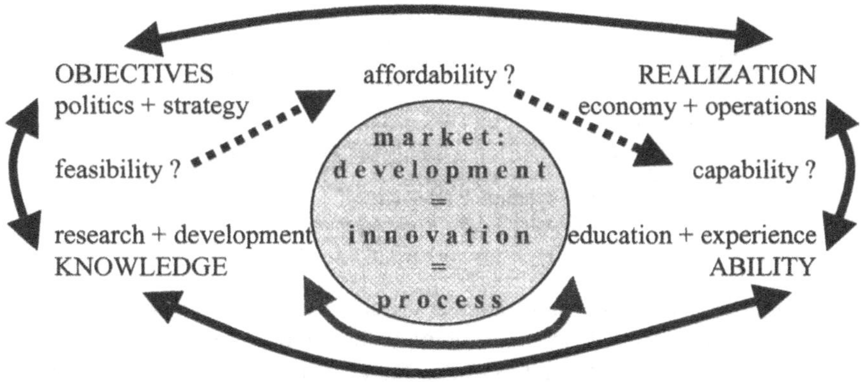

Realization of technical feasibility is an early phase of the innovation process, where the initiative of an innovator meets new knowledge from research, and with political consent a strategic donation of resources (means for development) is allocated to support the innovation and realize first prototypes.

[8] see Qualification, Innovation Processes and Diffusion by BRASCHE 1989

Realization of market chances and affordability is possible only after realization of the feasibility, and depends on first results of development. The influence of politico-strategic objectives at this stage is getting less in favor of the economical and operational adoption process within the innovator's enterprise.

Implementation of the production capability is possible only after realization of the market chances and the affordability, and thus depends on further results of development. Critical for success is the development of organization, and mainly of the human capital („organizational learning"), within the innovator's enterprise.

The last step is deemed as the utmost challenge for every today's enterprise, to meet the „**knowledge based society**" and enable a good „**change management**". Notwithstanding that civil service in general tends not to be innovative or adaptive and not to follow best practice from industry, it's obvious to state the same requirements to a change management for the public, and also for armed forces: Organizational learning always combines proven with the newly recognized.

Whenever an organization is going to be changed, there is a need to have a look at the state of the art, at science and technology, at the practice, or at least at flops. Using terms of the Military, an estimate of situation needs not only a task, own and enemy troops, and the environment, but especially to include a sound estimate of the options given to act or to react: High Commander's decisions need support.

Scientific Support certainly will not be given to low or intermediate echelons, but to high or top levels only. Thus we state, that all resources given to scientific research and development in an enterprise are taken by the strategic leadership. Sure the result depends mainly on the efficiency of operational implementation, but the responsibility for science in any economic process always is on boss level.

Fig. 2. Development of Organization as Center of Innovation Processes (MICKO 2002)

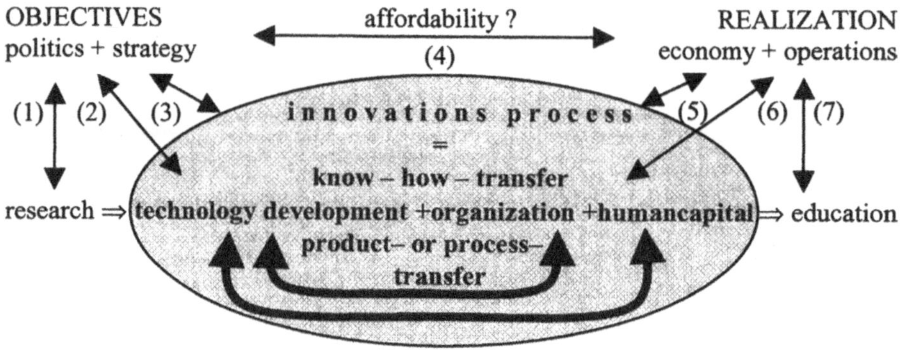

The initiative for the transition from research (1) to technical development (2) during initial innovative phases („feasibility, affordability, capability") always is **strategic responsibility** of the top management. Further the **strategic influence** to innovation processes consists mainly of strategic controlling (3) and periodical re-assessment of the central question (4): The market chances („affordability").

Operational fields of the enterprise have to deal with the innovation process in the first run when contributing to economical assessment of planned initiatives (4). The **operationalization of technical developments** during early phases of the innovation process (5) has to be done mainly through development of the organization (6). To implement new **structural and/or workflow organization** and to build an efficient and finally effective innovation, the decisive point is more than anything else development of human capital (7), starting as early as possible in close **conjunction with late phases of research** and technical development.

The aspects of **know-how-transfer** from technical development to development of the human capital, and the **product-transfer** from procurement to development of new structural organization are the broadly well aware components of organizational change. Less superficial, but nevertheless important is the development of processes, the development of new workflow organization, the **transfer of procedural knowledge**.

An additional important aspect, which represents an optional, suitable starting point for **Operations Research in the innovation process**, is the necessary meshing together of procedural organization with the experience present in the enterprise. Difficult to measure and thus difficult to promote, too, is the heavily discussed **„tacit knowledge"**, which is the non-codified technological know-how in all enterprises. It is very much essential and crucial to all innovation processes. Quality management, performance evaluation, and other instruments of operative controlling also undergo this main purpose: To actively search for, identify and shape these **„weak" and scarcely quantifiable criteria for success or failure**.

3. Operations Research and Decision Support in the Military

The Military estimate of situation needs not only a task, own and enemy troops, terrain and environment, but especially needs to include a sound estimate of the options given to act or to react: A high Commander's decision needs support.

As described before, the use of scientific resources for a specific innovation project as an operational decision support task always is a strategic investment decision (capital and human capital, budget and time budget) within an enterprise. Under certain circumstances it can be crucial for competition, if a specific innovation process with all steps was invested: crucial for victory or defeat in war. On the other hand for sure it's possible that some of the efforts may prove useless.

That's why Decision Support in wartime, and esp. war gaming applications already in peace times, are given a high priority whenever Armed Forces are able to allocate some additional new resources, and to re-develop their organization.

To give some examples, recently all Defense Organizations are re-orienteering themselves under the evolving changes of threats towards constabulization, operations other than war, anti-terrorism, anti-organized-crime and peace support.

Specifically the anti-terrorism task for the Military needs not only to deal with conventional or strategic defense questions as in the years of the Cold War, but to engage in the matters of interagency operations. This means to include all national resources, not only the military ones, in a coordinated effort of fighting the threat.

The Austrian Defense in the European context is still maintaining neutrality in case of war – but in the much more likely cases of non-war, non-declared war and subconventional attack or threat we are to contribute to the overall efforts for construction of a comprehensive security umbrella to the Austrian population.

The role of Operations Research in this regard is to give estimates about options to act or to react, using the full spectrum of quantitative and qualitative methodology which was developed by this scientific discipline in the past.

Main purpose of Operations Research in the Military in general is to be a knowledge base for decisions, innovation and change management, promoting and helping decisions about organizational change as described in detail above.

Literature:

BERTALANFFY, L: General Systems Theory, New York 1968

BRASCHE, U: Qualifikation – Engpass im Innovationsprozess ? Die Diffusion von Mikroelektronik und die Veränderung der Qualifikationsanforderungen, Berlin 1989

CLAUSEWITZ, On War, Berlin 1832

GILCHRIST, H.L: A Comparative Study of World War Casualties from Gas and Other Weapons, Chemical Warfare School 1928

JARDINI, D: Out of the Blue Yonder: The Transfer of Systems Thinking from the Pentagton to the Great Society, in: HUGHES, T., HUGHES, A. (eds.): Systems, Experts, and Computers, MIT 2000

KEYNES, J.M: The General Theory of Employment, Interest, and Money, New York 1936

KONDRATIEFF, N.D: Die langen Wellen der Konjuktur, in: Archiv für Sozialwissenschaft und Sozialpolitik, Tübingen 1926

LANCHESTER, F.W: Equations of Combat, in: Aircraft in Warfare, London 1916

MEADOWS, D.H., MEADOWS, D.L. et al.: Limits to Growth, New York 1972

OTTILIENFELD, F.G: Fordism, Industry and Technical Intelligence, Jena 1926

SCHUMPETER, J.A: The Analysis of Economic Change, in: Review of Economic Statistics, MIT 1935

TAYLOR, F.W: Scientific Management, New York 1911

Von der Prädikatenlogik zur unternehmerischen Entscheidungsunterstützung

Friedhelm Kulmann und Wilhelm Rödder

Lehrstuhl für Betriebswirtschaftslehre, insb. Operations Research
FernUniversität in Hagen

friedhelm.kulmann@fernuni-hagen.de

Zusammenfassung Mit der Expertensystemshell SPIRIT steht ein mächtiges Instrument zur unternehmerischen Entscheidungsunterstützung auf der Basis probabilistischen konditionalen Wissens zur Verfügung. In mehreren Arbeiten - wie zum Beispiel [5], [3] - wurde über ihren professionellen Praxiseinsatz berichtet. Wegen der hohen Komplexität der Entscheidungsmechanismen bedarf es zumeist intensiver Gespräche bis zur Akzeptanz der zugrunde liegenden Theorien; letzte Zweifel werden jedoch häufig nicht ganz ausgeräumt. Mit diesem Beitrag wird nun für die Verwendung von SPIRIT ein neues Einsatzgebiet gewählt und damit ein noch nicht da gewesener Zugang geschaffen. Konditionalmodelle sind auch zur Lösung prädikatenlogischer Aufgaben geeignet, wie sie beispielsweise in sogenannten Logikrätseln gestellt werden. Damit wird gleichsam für derartige Aufgaben eindrucksvoll nachgewiesen, dass die Fähigkeit des Menschen zur Deduktion von Wissen aus Wissen auf dem Computer formalisierbar ist. Nach der verbalen Formulierung einer beispielhaft ausgewählten prädikatenlogischen Aufgabe wird diese in der Shell SPIRIT kausal modelliert und für verschiedenartige Fragestellungen eindeutig gelöst. Im letzten Teil des Beitrags wird durch Aufzeigen von Analogien deutlich, dass der Weg von »Logeleien« zu ernsthaften ökonomischen Anwendungen nicht weit ist. Mit dem Verständnis des Mechanismus zur Modellierung und Lösung prädikatenlogischer Aufgaben kann so vielleicht der Widerstand gegen die Verwendung probabilistischer, mathematisch fundierter Konzepte abgebaut werden.

1 Einleitung

Zentraler Bestandteil einer wissensbasierten Modellierung unternehmerischer Entscheidungsprobleme ist die Erfassung der Zusammenhänge durch Bausteine, die als prädikatenlogische Formeln interpretiert werden können. Die Prädikatenlogik ist in idealer Weise geeignet, die interne Struktur von Aussagen zu berücksichtigen und Beziehungen zwischen Eigenschaften von Objekten zum Ausdruck zu bringen (vgl. [1]). Letztlich ist es auch möglich, mit ihr Fragen zu formulieren und Antworten aus dem bisher »Gelernten« abzuleiten. Sie kann damit gleichsam als Vorstufe zur Umsetzung des Entscheidungsproblems in die probabilistische Konditionallogik genutzt werden.

Leider muss bei Unternehmen, die potentielle Nutzer leistungsstarker Werkzeuge zur Entscheidungsunterstützung sind, einerseits eine generelle Skepsis hinsichtlich der Anwendbarkeit und andererseits Misstrauen bezüglich vorgeschlagener Entscheidungen festgestellt werden. Wir haben deshalb in diesem Beitrag nicht den klassischen Zugang über die ausführliche Darstellung der verwendbaren reichhaltigen Sprache und des eingesetzten Inferenzmechanismus gewählt, sondern nutzen das jedem vertraute weite Feld der Logikrätsel, um das System SPIRIT als Instrument der Entscheidungsunterstützung vorzustellen (vgl. [6]).

Im Kapitel 2 erfolgt nach der Präsentation eines Rätsels dessen Formalisierung in der Notation der Prädikatenlogik. Die Umsetzung als Wissensbasis des probabilistischen Systems SPIRIT wird im Kapitel 3 vorgenommen; außerdem wird der Inferenzmechanismus erläutert, und beispielhaft werden einige Anfragen gestellt. Kapitel 4 zeigt - wie eingangs angedeutet - logisch gleich strukturierte, unternehmerische Fragestellungen auf, die analog beantwortet werden können.

2 Vom Wortmodell zu logischen Strukturen

Zur Vorbereitung der Nutzung von Computersystemen bei der Lösung von Aufgaben aus dem unternehmerischen Umfeld müssen die relevanten Größen und Einflussfaktoren zunächst verbal formuliert werden. Das daraus resultierende, noch nicht formalisierte sogenannte Wortmodell ist daraufhin in seiner kausalen Struktur zu analysieren und in eine Sprache zu übersetzen, die mit ihrem Formalismus diese Abhängigkeiten unmittelbar zum Ausdruck bringt.

Bei Aufgaben aus dem Bereich der Logik ist das Wortmodell zumeist auf ein Minimum reduziert, und man ist zur Lösung gestellter Fragen aufgefordert, die oben genannte Analyse und Umsetzung - zumindest gedanklich - vorzunehmen. Besonders schwierig sind diese Punkte dann, wenn das Lösen der Aufgabe durch fehlende Assoziationsmuster erschwert wird. Als Beispiel diene das nun folgende, von Janko, A. und Janko, O. im Internet veröffentlichte Rätsel, das mit »Pprills, Squirde und Glopps« überschrieben ist (s. [2]).

»Alle gebildeten Leute wissen nun, dass Pprills, Squirde und Glopps einfach nur Formen von Nahfen sind. Es ist außerdem bewiesen, dass Squirde sowohl Glopps als auch Nahfen sind.
Allerdings gibt es eine Komplikation: Neuerdings fand man heraus, dass es Glopps gibt, die weder Squirde, Pprills noch Gdynxe sind. Zusätzlich gibt es noch Squirde, die weder Gdynxe noch Pprills sind.
Zugegebenermaßen, manche Pprills und manche Gdynxe sind Glopps, und manche Squirde auch. Aber jetzt wissen wir mehr über Gdynxe: Manche sind Squirde, manche Glopps und manche komischerweise sowohl Pprills als auch Squirde.

a) Gibt es unter den Gdynxen, die keine Nahfen sind, Glopps?
b) Wenn ein Pprill ein Squird ist, ist es dann auf jeden Fall ein Glopp? [...]

c) Gibt es in diesem Universum tatsächlich eine Kreatur, die in der Situation ist, von sich behaupten zu müssen, sowohl ein Pprill und ein Nahf als auch ein Squird und ein Glopp und obendrein auch noch ein Gdynx zu sein?« (vgl. [2])

Der erste Schritt der Formalisierung besteht in der Beschreibung der einstelligen Prädikate **Nahf**, **Glopp**, **Squird**, **Pprill**, **Gdynx**, die zur Charakterisierung von Objekten verwendet werden. Das vollständige Ergebnis der Umsetzung sind die folgenden Formeln 1 bis 10.

1. $\forall x$: $\texttt{Glopp}(x) \longrightarrow \texttt{Nahf}(x)$
2. $\forall x$: $\texttt{Squird}(x) \longrightarrow \texttt{Nahf}(x)$
3. $\forall x$: $\texttt{Pprill}(x) \longrightarrow \texttt{Nahf}(x)$
4. $\forall x$: $\texttt{Squird}(x) \longrightarrow (\texttt{Glopp}(x) \wedge \texttt{Nahf}(x))$
5. $\exists x$: $\texttt{Glopp}(x) \longrightarrow (\neg\texttt{Squird}(x) \wedge \neg\texttt{Pprill}(x) \wedge \neg\texttt{Gdynx}(x))$
6. $\exists x$: $\texttt{Squird}(x) \longrightarrow (\neg\texttt{Gdynx}(x) \wedge \neg\texttt{Pprill}(x))$
7. $\exists x$: $\texttt{Pprill}(x) \longrightarrow \texttt{Glopp}(x)$
8. $\exists x$: $\texttt{Gdynx}(x) \longrightarrow \texttt{Glopp}(x)$
9. $\exists x$: $\texttt{Gdynx}(x) \longrightarrow \texttt{Squird}(x)$
10. $\exists x$: $\texttt{Gdynx}(x) \longrightarrow (\texttt{Pprill}(x) \wedge \texttt{Squird}(x))$

Es handelt sich ausnahmslos um mit dem Allquantor gebundene Implikationen (1 bis 4) oder um Existenzaussagen (5 bis 10). Im nun folgenden Kapitel 3 wird mit der Einbettung in ein System zur probabilistischen Wissensverarbeitung die Möglichkeit der Beantwortung komplexer Fragestellungen in diesem Universum skuriler Kreaturen geboten.

3 Die logische Struktur als Wissensbasis

Zur Umsetzung in eine probabilistische Wissensbasis werden zunächst die einstelligen Prädikate als boolsche Variable interpretiert. **Nahf**, **Glopp**, **Squird**, **Pprill**, **Gdynx** bilden im obigen Beispiel die vollständige Variablenmenge, deren positive und negative Ausprägungen somit den Ereignisraum Ω repräsentieren. Formeln, die durch die Verknüpfung mittels Konjunktion, Disjunktion und durch Negation entstehen, sind in natürlicher Weise mit Ereignissen aus Ω identifizierbar. Zur Repräsentation der in prädikatenlogischen Formeln zum Ausdruck gebrachten Implikation wird das Konditional verwendet. Der Allquantor bringt die Allgemeingültigkeit einer Formel zum Ausdruck; bei der Einbettung in die probabilistische Wissensbasis gilt das entsprechende Konditional mit Sicherheit. Durch den Existenzquantor wird eine schwächere Forderung für das Konditional aufgestellt; die Wahrscheinlichkeit muss echt positiv sein. Die damit auftretenden Modellierungsfragen werden im späteren Beispiel diskutiert.

Aus den Formeln 1 bis 10 ergeben sich für SPIRIT unmittelbar die in der Abbildung 3 gezeigten Konditionale. Da im Beispiel nur qualitative Angaben gemacht werden, wird bei der Umsetzung der Existenzaussagen in die

SPIRIT-Syntax eine positive (kleine) Wahrscheinlichkeit für die Gültigkeit der konditionierten Aussage angenommen. Da der numerische Wert das Ergebnis nicht beeinflusst, wurde hier 0, 01 gewählt.

P prescr	P act	Rule text
1	1,000	Nahf | Glopp
1	1,000	Nahf | Squird
1	1,000	Nahf | Pprill
1	1,000	(Glopp ∧ Nahf) | Squird
0,01	0,010	(¬Squird ∧ ¬Pprill ∧ ¬Gdynx) | Glopp
0,01	0,010	(¬Gdynx ∧ ¬Pprill) | Squird
0,01	0,010	Glopp | Pprill
0,01	0,010	Glopp | Gdynx
0,01	0,010	Squird | Gdynx
0,01	0,010	(Pprill ∧ Squird) | (Gdynx)

Abbildung 1. Wissensbasis zum Logikrätsel »Pprills, Squirde und Glopps«

SPIRIT ermöglicht auf der Basis gesicherter informationstheoretischer Erkenntnisse die Generierung einer Wissensbasis, die alle geforderten Kausalitäten mit den angegebenen Wahrscheinlichkeiten repräsentiert und darüber hinaus keine nicht intendierten Abhängigkeiten zwischen Variablen anlegt. Mathematisch geschieht dieser informationstreue Wissensaufbau durch die Wahrung des Prinzips maximaler Entropie (siehe [6]). Abbildung 3 zeigt den im System angezeigten sogenannten (ungerichteten) Abhängigkeitsgraphen mit den Variablen als Knoten, die als zusätzliche Angabe die jeweilige Randverteilung ausweisen (zum theoretischen Hintergrund siehe auch [4]).

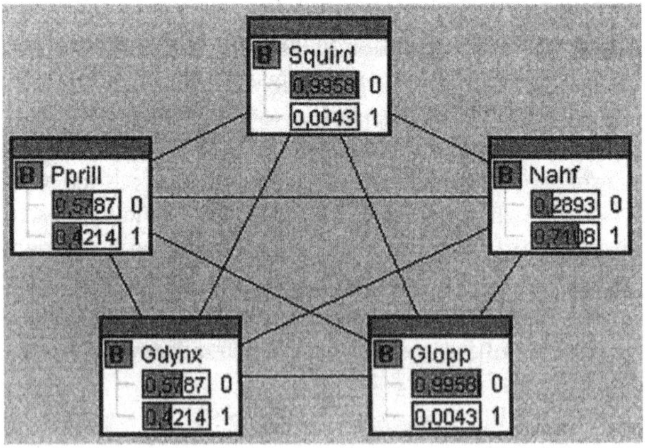

Abbildung 2. Abhängigkeitsgraph zum Logikrätsel

Bisher unbeantwortet geblieben sind die in Kapitel 2 gestellten Fragen. Wahrscheinlichkeitstheoretisch formuliert lautet beispielsweise Teil a): »Wie groß ist die Wahrscheinlichkeit für ein Glopp, unter der Bedingung, dass es Gdynx und kein Nahf ist?« Als Anfragekonditional für die Wissensbasis ist die formale Schreibweise: [?] `Glopp`| (`Gdynx` ∧ ¬`Nahf`) (siehe auch Abbildung 3). Unter Verwendung des in SPIRIT implementierten Inferenzmechanismus können die im Abschnitt 2 unter a), b), c) gestellten Fragen nun beantwortet werden (vgl. hierzu [6], [7]).

<table>
<tr><th>P act</th><th>Rule text</th></tr>
<tr><td>0,000</td><td>Glopp | (Gdynx ∧ ¬Nahf)</td></tr>
<tr><td>1,000</td><td>(Glopp | Pprill) | (Squird | Pprill)</td></tr>
<tr><td>0,004</td><td>Pprill ∧ Nahf ∧ Squird ∧ Glopp ∧ Gdynx</td></tr>
</table>

Abbildung 3. Anfragen zum Logikrätsel

a) Es gibt unter den Gdynxen, die keine Nahfen sind, **keinen** Glopp.
b) Wenn ein Pprill ein Squird ist, dann ist es **auf jeden Fall** ein Glopp.
c) Es ist zwar unwahrscheinlich, aber es gibt in diesem Universum **tatsächlich** eine Kreatur, die in der Situation ist, von sich behaupten zu müssen, sowohl ein Pprill und ein Nahf als auch ein Squird und ein Glopp und obendrein auch noch ein Gdynx zu sein.

Nach Behandlung der logisch abstrakt gestellten Aufgabe bleibt noch die Frage zu beantworten, welcher Realitätsbezug dabei existieren könnte.

4 Das unternehmerische Entscheidungsproblem

Das Management eines Unternehmens steht - so ist jedenfalls zu hoffen - nur in den seltensten Fällen vor einem Rätsel; dennoch sind die logischen Problemstrukturen oft ähnlich. Deshalb soll das obige Beispiel in die Terminologie des Marketings übertragen und als Analyse des Kundenstruktur einer Supermarktkette zwecks Durchführung zielgerichteter Werbemaßnahmen interpretiert werden. Die logische Äquivalenz der Aussagen lässt sich beim direkten Vergleich mit den ursprünglichen Formulierungen im Kapitel 2 unmittelbar nachvollziehen.

Laut Umfrage gibt es u.a. drei typische Zielgruppen von Kundinnen, nämlich Mütter, Halbtagsbeschäftigte und Hausfrauen. Die halbtags Beschäftigten unter den Kundinnen pflegen aufgrund ihrer besonderen Situation auch den Haushalt zu führen. Andererseits gibt es durchaus Hausfrauen, die weder halbtags beschäftigt sind, noch ein Kind haben, noch alleinerziehend sind. Weiterhin sind gewisse halbtags Beschäftigte weder alleinerziehend noch haben sie ein Kind. Mütter und Alleinerziehende können ebenso Hausfrauen sein wie manche Halbtagsbeschäftigte auch. Ist eine Kundin Alleinerziehende, so kann sie durchaus halbtags beschäftigt oder Hausfrau oder sogar halbtagsbeschäftigte Mutter sein,

Diese etwas unstrukturierten Gruppenabhängigkeiten sind durch offene Interviews ermittelt worden und können als gesichert gelten. Erstaunlicherweise lassen sich aus den unvollständigen Angaben wichtige Folgerungen über die Gesamtstruktur der Kundinnen im Hinblick auf Werbemaßnahmen ziehen. Die Aussagen ergeben sich unmittelbar aus der Übertragung der zum Ende des letzten Kapitels gegebenen Antworten.

a) Die Neukundenwerbung für die Zielgruppe »Alleinerziehende« erreicht keinesfalls auch »Hausfrauen«.
b) Mütter, die halbtags beschäftigt sind, werden automatisch auch über die Werbebotschaft für Hausfrauen angesprochen.
c) Es gibt durchaus eine Gruppe von Kundinnen - wenn auch eventuell eine kleine -, die auf jede zielgruppengerichtete Werbekampagne reagiert.

5 Fazit und Ausblick

Mit dem Einstieg in die komplexe Thematik der probabilistischen Wissensverarbeitung und der Modellierung konditionaler Strukturen über eine als Logikrätsel gestellte Aufgabe wurde eine neuartige Sichtweise auch für die Lösung unternehmerischer Fragestellungen eröffnet. Bereits das vorgestellte einfache Beispiel hat eindrucksvoll die Notwendigkeit der Erfassung komplexer logischer Abhängigkeiten deutlich gemacht. Außerdem wurde gezeigt, dass wissensbasierte Systeme nicht nur in der Lage sondern auch dringend erforderlich sind, um geeignete Strukturierungshilfe bei Entscheidungsproblemen zu bieten.

Literatur

1. Genesereth, M.R.; Nilsson, N.J. (1989): Logische Grundlagen der Künstlichen Intelligenz, *Vieweg, Braunschweig, 1989.*
2. Janko, A.; Janko, O. (2002): Rätsel und Denksport. *http://www.janko.at/Raetsel/Logik/003.a.htm (Stand August 2002).*
3. Kulmann, F. (2002): Wissen und Information in konditionalen Modellen - zur Entscheidungsvorbereitung im Anfrage- und Auftragsmanagement, *DUV Gabler, Wiesbaden, 2002.*
4. Meyer, C.-H. (1998): Korrektes Schließen bei unvollständiger Information, *Europäische Hochschulschriften, Peter Lang, Frankfurt a.M., 1998.*
5. Kulmann, F.; Reucher, E. (2000): Computergestützte Bonitätsprüfung bei Banken und Handel, *DBW - Die Betriebswirtschaft, 60 (2000) 113-122.*
6. Rödder, W. (2000): Conditional Logic and the Principle of Entropy, *Artificial Intelligence, 117 (2000) 83-106.*
7. Rödder, W. (2001): Knowledge Processing under Information Fidelity, *Proc. IJCAI 2001 - Seventeenth International Joint Conference on Artificial Intelligence, Seattle, Washington (2001) 749-754.*

GPSR Compliance

The European Union's (EU) General Product Safety Regulation (GPSR) is a set of rules that requires consumer products to be safe and our obligations to ensure this.

If you have any concerns about our products, you can contact us on ProductSafety@springernature.com

In case Publisher is established outside the EU, the EU authorized representative is:

Springer Nature Customer Service Center GmbH
Europaplatz 3
69115 Heidelberg, Germany

Batch number: 10761599

Printed by Printforce, the Netherlands